T0202970

Lecture Notes in Physics

Founding Editors

Wolf Beiglböck

Jürgen Ehlers

Klaus Hepp

Hans-Arwed Weidenmüller

Volume 995

The series Lecture Notes in Physics (LNP), founded in 1969, reports new developments in physics research and teaching - quickly and informally, but with a high quality and the explicit aim to summarize and communicate current knowledge in an accessible way. Books published in this series are conceived as bridging material between advanced graduate textbooks and the forefront of research and to serve three purposes:

- to be a compact and modern up-to-date source of reference on a well-defined topic;
- to serve as an accessible introduction to the field to postgraduate students and non-specialist researchers from related areas;
- to be a source of advanced teaching material for specialized seminars, courses and schools.

Both monographs and multi-author volumes will be considered for publication. Edited volumes should however consist of a very limited number of contributions only. Proceedings will not be considered for LNP.

Volumes published in LNP are disseminated both in print and in electronic formats, the electronic archive being available at springerlink.com. The series content is indexed, abstracted and referenced by many abstracting and information services, bibliographic networks, subscription agencies, library networks, and consortia.

Proposals should be sent to a member of the Editorial Board, or directly to the responsible editor at Springer:

Dr Lisa Scalone
Springer Nature
Physics
Tiergartenstrasse 17
69121 Heidelberg, Germany
lisa.scalone@springernature.com

More information about this series at https://link.springer.com/bookseries/5304

Tatsuo Kobayashi · Hiroshi Ohki ·
Hiroshi Okada · Yusuke Shimizu ·
Morimitsu Tanimoto

An Introduction to Non-Abelian Discrete Symmetries for Particle Physicists

Second Edition

 Springer

Tatsuo Kobayashi
Department of Physics
Hokkaido University
Sapporo, Japan

Hiroshi Okada
Asia Pacific Center for Theoretical Physics
Pohang, Korea (Republic of)

Morimitsu Tanimoto
Department of Physics
Niigata University
Niigata, Japan

Hiroshi Ohki
Department of Mathematical and Physical
Sciences
Nara Women's University
Nara, Japan

Yusuke Shimizu
Graduate School of Advanced Science
and Engineering
Hiroshima University
Hiroshima, Japan

ISSN 0075-8450   ISSN 1616-6361   (electronic)
Lecture Notes in Physics
ISBN 978-3-662-64678-6   ISBN 978-3-662-64679-3   (eBook)
https://doi.org/10.1007/978-3-662-64679-3

Preface to the Second Edition

Non-Abelian discrete groups are quite useful tools in particle physics. The purpose of the first edition was to introduce non-Abelian discrete symmetries for particle physicists, and to present important application in particle physics. After the first edition, lots of new applications of non-Abelian discrete groups have been studied in particle physics, e.g. generalized CP and modular flavor symmetries. To meet these developments, we have added Chap. 16 modular symmetry, Chap. 17 automorphism, and Chap. 20 generalized CP transformation. Also we have rewritten Chap. 19 to explain new models and developments on modular flavor models. Furthermore, we have added Chap. 15 on finite subgroups of continuous groups.

We hope that the second edition with these additional chapters could be very useful for researchers in these subjects.

It is pleasure to acknowledge fruitful discussions with Hiroyuki Abe, Takeshi Araki, Gustavo C. Branco, Mu-Chun Chen, Kang-Sin Choi, Kenji Hashimoto, Yasuhiro Daikoku, Gui-jun Ding, Ferruccio Feruglio, Walter Grimus, Martin K. Hirsch, Gino Isidori, Hajime Ishimori, Shota Kikuchi, Stephen F. King, Jiske Kubo, Manfred Lindner, Ernest Ma, Hans P. Nilles, Takaaki Nomura, Pavel Novichkov, Yuta Orikasa, Hajime Otsuka, João T. Penedo, Serguey T. Petcov, Felix Ploger, Stuart A. Raby, Saul Ramos-Sanchez, Michael Ratz, M. Nesbitt Rebelo, Werner Rodejohann, Takashi Shimomura, Takuya H. Tatsuishi, Kenta Takagi, Arsenii V. Titov, Andreas Trautner, Hikaru Uchida, Shohei Uemura, Jose W. F. Valle, Patrick K. S. Vaudrevange, Takahiro Yoshida, and Shun Zhou.

Sapporo, Japan   Tatsuo Kobayashi
Nara, Japan   Hiroshi Ohki
Pohang, Korea (Republic of)   Hiroshi Okada
Hiroshima, Japan   Yusuke Shimizu
Niigata, Japan   Morimitsu Tanimoto
September 2021

Preface to the First Edition

The purpose of the lecture notes is to introduce the basic framework of the non-Abelian Discrete Symmetries, and to present some important application in Particle Physics. Non-Abelian discrete groups have been playing an important role in particle physics. However, those may not be familiar to all of particle physicists compared with non-Abelian continuous symmetries. Therefore, the lecture notes are written for particle physicists. In this respect, it is different from the standard books of the group theory. However, preliminary knowledge of the group theory is not required in advance to understand the non-Abelian Discrete Symmetries.

We hope that our lecture notes could play a crucial role as a handbook in serious learners, and also as a helpful reference textbook in experts in addition to providing triggers for subject of researches.

It is pleasure to acknowledge fruitful discussions with H. Abe, T. Araki, K. S. Choi, Y. Daikoku, K. Hashimoto, J. Kubo, H. P. Nilles, F. Ploger, S. Raby, S. Ramos-Sanchez, M. Ratz, and P. K. S. Vaudrevange.

September 2012

Hajime Ishimori
Tatsuo Kobayashi
Hiroshi Ohki
Hiroshi Okada
Yusuke Shimizu
Morimitsu Tanimoto

Contents

Introduction

These lecture notes aim to provide a pedagogical review of non-Abelian discrete groups and show some applications for physical issues. Symmetry constituents a very important principle in physics. In particular, it has played an essential role in constructing the framework of the particle physics. For example, continuous (and local) symmetries such as Lorentz, Poincaré and gauge symmetries are crucial to understand several phenomena, such as the strong, weak and electromagnetic interactions among particles. On the other hand, discrete symmetries such as C, P and T are also vital concepts in particle physics. Abelian discrete symmetries, Z_N, are also often imposed in order to control allowed couplings for particle physics, in particular model-building beyond the standard model. In addition to Abelian discrete symmetries, non-Abelian discrete symmetries have also been applied for model-building in particle physics, in particular to understand the three-generation flavor structure.

There are many free parameters in the standard model including its extension with neutrino mass terms. Most of them are Yukawa couplings of quarks and leptons to the Higgs boson. The quark and lepton sector is called as the flavor sector. Flavor physics is challenging aspect of the construction of the theory beyond the standard model. If a symmetry is imposed on the flavor sector, one can control Yukawa couplings in the three generations, although the origin of the generations remains unknown. Therefore, the quark masses and mixing angles have been studied from the standpoint of flavor symmetries.

In addition, the discovery of neutrino oscillation [1] has stimulated the work of the flavor symmetries since the information of neutrino mass squared differences and neutrino mixing [2,3] have been obtained. Now, experiments of the neutrino oscillation go into a new phase of precise determination of mixing angles and mass squared differences [4]. Global analyses of neutrino data at first suggested the specific neutrino mixing pattern, which is called tri-bimaximal mixing for three flavors in the lepton sector [5–8]. These large mixing angles can be understood in the framework

T. Kobayashi et al., *An Introduction to Non-Abelian Discrete Symmetries for Particle Physicists*, Lecture Notes in Physics 995, https://doi.org/10.1007/978-3-662-64679-3_1

of the flavor symmetry. Therefore, it is very important to find a natural model that leads to observed large mixing angles.

Non-Abelian discrete symmetries are considered to be the most attractive choice for the flavor sector. Model builders have tried to derive experimental values of quark/lepton masses and flavor mixing angles by assuming non-Abelian discrete flavor symmetries of quarks and leptons. In particular, lepton mixing has been intensively discussed in the context of non-Abelian discrete flavor symmetries, as seen, e.g. in the reviews [9–13].

Particle physicists may be interested in the origin of the non-Abelian discrete flavor symmetry for flavors. One of the most interesting one is the higher dimensional space-time symmetry. After it has been broken down to the 4D Poincaré symmetry through compactification, e.g. via orbifolding, a remnant symmetry appears in the flavor sector. This remnant symmetry is often the non-Abelian symmetry. Actually, it has been shown how the flavor symmetry A_4 (or S_4) can arise if the three fermion generations are taken to live on the fixed points of a specific 2D orbifold [14]. Further non-Abelian discrete symmetries can arise in a similar setup [15] (see also [16]).

Superstring theory is a promising candidate for unified theory including gravity. Certain string modes correspond to gauge bosons, quarks, leptons, Higgs bosons and gravitons as well as their superpartners. Superstring theory predicts six extra dimensions. Certain classes of discrete symmetries can be derived from superstring theories. A combination among geometrical symmetries of a compact space and stringy selection rules for couplings enhances discrete flavor symmetries. For example, D_4 and $\Delta(54)$ flavor symmetries can be obtained in heterotic orbifold models [17–19]. In addition to these flavor symmetries, the $\Delta(27)$ flavor symmetrtry can be derived from magnetized/intersecting D-brane models [20–22].

In addition, certain compact spaces such as tori and orbifolds have the modular symmetry. Its quotients Γ_N includes S_3, A_4, S_4, A_5, $\Delta(96)$, and $\Delta(384)$, and also their covering groups can be realized. Thus, the modular symmetries can be origins of flavor symmetries. Indeed, modular flavor symmetries were studied in heterotic orbifold models [23–26] and magnetized D-brane models [27–36].

Thus, a non-Abelian discrete symmetry can arise from the underlying theory, e.g. the string theory or compactification via orbifolding. In addition, the non-Abelian discrete-symmetries are interesting tools for controlling the flavor structure in model building using the bottom-up approach. Hence, the non-Abelian flavor symmetries could become a bridge between the low-energy physics and the underlying theory. Therefore, it is quite important to understand the properties of non-Abelian groups for the particle physics.

Continuous non-Abelian groups are well-known and of course there are several good reviews and books. On the other hand, discrete non-Abelian symmetries may not be so familiar to particle physicists as continuous non-Abelian symmetries. However, discrete non-Abelian symmetries have become important tools for model building, as discussed above, in particular in the context of the flavor physics. The purpose of this lectures notes is therefore to provide a pedagogical review of non-Abelian discrete groups with particle phenomenology in mind, and to exhibit the group-theoretical aspects of many concrete groups explicitly, including, for example, representations and their tensor products [10,37–43]. We present these aspects in detail for the groups S_N [44–142] A_N [142–252], T' [42,253–272], D_N [273–294], Q_N [295–309], QD_{2N}, $\Sigma(2N^2)$ [310], $\Delta(3N^2)$ [311–324], T_N [311–313,321,325–334], $\Sigma(3N^3)$ [327,335], and $\Delta(6N^2)$ [311–313,321,336–343]. We explain pedagogically how to derive conjugacy classes, characters, representations and tensor products for these groups (with a finite number) when algebraic relations are given. Thus, it will be straightforward for readers to apply this to other groups. For other groups we do not mention in these lecture notes, see [38,40,43,145]. Also see [344,345] for $PSL_2(7)$, [346,347] for the double cover of S_4 group and [348–350] for the double cover of A_5 group.

In applications for particle physics, the breaking patterns of discrete groups and decompositions of multiplets are often required to understand the low energy phenomena. Such aspects are given in these lectures notes.

There is another possibility, namely that non-Abelian discrete groups may originate from the breaking of continuous (gauge) flavor symmetries [351–353]. For example, discrete subgroups of $SU(3)$ would be also interesting from the viewpoint of phenomenological applications for the flavor physics. Continuous flavor symmetries such as the $SU(3)$ group have long been considered for quarks and leptons. We study non-Abelian discrete groups and their representations as finite subgroups of continuous groups. In the framework of the quantum field theory this could occur due to the spontaneous symmetry breaking of a non-zero vacuum expectation value (VEV) of a scalar field. This leads to a reduction of the original group to a subgroup of unbroken symmetry in the low-energy physics.

The group theoretical investigation of the CP transformation in the flavor group G leads to the study of the automorphism of a group of G, which is a *bijective* homomorphism: $G \rightarrow G$. Indeed, the generalized CP transformation [354–358] has been discussed in the context with the automorphism of a flavor group. We summarize the automorphism for non-Abelian groups explicitly.

Symmetries at the tree level are not always symmetries in quantum theory. If symmetries are anomalous, breaking terms are induced by quantum effects. Such anomalies are important in applications for particle physics. Here, we study such anomalies for discrete symmetries [42,359–372] and show anomaly-free conditions explicitly for the above concrete groups. If flavor symmetries are stringy symmetries, these anomalies may also be controlled by string dynamics, i.e. anomaly cancellation.

We also present flavor models with discrete non-Abelian symmetry as typical examples. One can see how to use the discrete non-Abelian symmetry for flavors. A lot of references are available to understand the model building here. We also add the recent discussion of the generalized CP transformation in the framework of the modular symmetry of flavors.

This article is organized as follows. In Chap. 2, we summarize basic group-theoretical aspects used in the subsequent of chapters, and also present some examples to provide a more concrete understanding. Readers familiar with group theory can skip Chap. 2. In Chaps. 3–13, we present non-Abelian discrete groups, S_N, A_N, T', D_N, Q_N, QD_{2N}, $\Sigma(2N^2)$, $\Delta(3N^2)$, T_N, $\Sigma(3N^3)$, and $\Delta(6N^2)$, respectively. In each chapter, groups with specific values of N are also discussed for typical examples. Chap. 14 discusses the breaking patterns of the non-Abelian discrete groups. Chap. 15, presents finite subgroups of continuous groups of $SO(3)$, $SU(2)$ and $SU(3)$. Chapter 16 introduces the modular symmetry. Chapter 17 presents the automorphism of relevant groups. In Chap. 18, we review the anomaly of non-Abelian flavor symmetries, which is an important topic in the particle physics, and exhibit the anomaly-free conditions explicitly for the above concrete groups. Chapter 19 presents typical flavor models with the non-Abelian discrete symmetries. Chapter 20 discusses the generalized CP transformation and related flavor models. Appendix A gives some useful theorems on finite group theory, while Appendices B, C, D, E and F provide the representation bases of S_4, A_4, A_5, T' and $\Delta(96)$, respectively, which are different from those in Chaps. 3–5 and 13. Appendix G presents other smaller groups are presented in details. Appendix H discusses generators of the modular group. Appendix I gives a brief review on modular forms.

References

1. Fukuda, Y. et al.: [Super-Kamiokande], Phys. Rev. Lett. **81**, 1562–1567 (1998). https://doi.org/10.1103/PhysRevLett.81.1562, arXiv:hep-ex/9807003 [hep-ex]
2. Pontecorvo, B.: Sov. Phys. JETP **6**, 429 (1957). [Zh. Eksp. Teor. Fiz. **33** (1957) 549]
3. Maki, Z., Nakagawa, M., Sakata, S.: Prog. Theor. Phys. **28**, 870 (1962)
4. Esteban, I., Gonzalez-Garcia, M.C., Maltoni, M., Schwetz, T., Zhou, A.: JHEP **09**, 178 (2020). [arXiv:2007.14792 [hep-ph]]
5. Harrison, P.F., Perkins, D.H., Scott, W.G.: Phys. Lett. B **530**, 167 (2002). [arXiv:hep-ph/0202074]
6. Harrison, P.F., Scott, W.G.: Phys. Lett. B **535**, 163 (2002). [arXiv:hep-ph/0203209]
7. Harrison, P.F., Scott, W.G.: Phys. Lett. B **557**, 76 (2003). [arXiv:hep-ph/0302025]
8. Harrison, P.F., Scott, W.G.: arXiv:hep-ph/0402006
9. Altarelli, G., Feruglio, F.: Rev. Mod. Phys. **82**, 2701–2729 (2010). [arXiv:1002.0211 [hep-ph]]
10. Ishimori, H., Kobayashi, T., Ohki, H., Shimizu, Y., Okada, H., Tanimoto, M.: Prog. Theor. Phys. Suppl. **183**, 1–163 (2010). [arXiv:1003.3552 [hep-th]]
11. King, S.F., Luhn, C.: Rept. Prog. Phys. **76**, 056201 (2013). [arXiv:1301.1340 [hep-ph]]
12. King, S.F., Merle, A., Morisi, S., Shimizu, Y., Tanimoto, M.: New J. Phys. **16**, 045018 (2014). [arXiv:1402.4271 [hep-ph]]
13. Petcov, S.T.: Eur. Phys. J. C **78**(9), 709 (2018). [arXiv:1711.10806 [hep-ph]]
14. Altarelli, G., Feruglio, F., Lin, Y.: Nucl. Phys. B **775**, 31 (2007). [arXiv:hep-ph/0610165]
15. Adulpravitchai, A., Blum, A., Lindner, M.: JHEP **0907**, 053 (2009). [arXiv:0906.0468 [hep-ph]]
16. Abe, H., Choi, K.-S., Kobayashi, T., Ohki, H., Sakai, M.: Int. J. Mod. Phys. A **26**, 4067–4082 (2011). [arXiv:1009.5284 [hep-th]]
17. Kobayashi, T., Raby, S., Zhang, R.J.: Nucl. Phys. B **704**, 3 (2005). arXiv:hep-ph/0409098
18. Kobayashi, T., Nilles, H.P., Ploger, F., Raby, S., Ratz, M.: Nucl. Phys. B **768**, 135 (2007). arXiv:hep-ph/0611020
19. Ko, P., Kobayashi, T., Park, J. h., Raby, S.: Phys. Rev. D **76**, 035005 (2007); [Erratum-ibid. D **76**, 059901 (2007)] [arXiv:0704.2807 [hep-ph]]
20. Abe, H., Choi, K.S., Kobayashi, T., Ohki, H.: Nucl. Phys. B **820**, 317 (2009). [arXiv:0904.2631 [hep-ph]]
21. Abe, H., Choi, K.S., Kobayashi, T., Ohki, H.: Phys. Rev. D **80**, 126006 (2009). [arXiv:0907.5274 [hep-th]]
22. Abe, H., Choi, K.S., Kobayashi, T., Ohki, H.: arXiv:1001.1788 [hep-th]
23. Lauer, J., Mas, J., Nilles, H.P.: Phys. Lett. B **226**, 251 (1989)

24. Lauer, J., Mas, J., Nilles, H.P.: Nucl. Phys. B **351**, 353 (1991)
25. Lerche, W., Lust, D., Warner, N.P.: Phys. Lett. B **231**, 417 (1989)
26. Ferrara, S., Lust, D., Theisen, S.: Phys. Lett. B **233**, 147 (1989)
27. Kobayashi, T., Nagamoto, S.: Phys. Rev. D **96**(9), 096011 (2017). [arXiv:1709.09784 [hep-th]]
28. Kobayashi, T., Nagamoto, S., Takada, S., Tamba, S., Tatsuishi, T.H.: Phys. Rev. D **97**(11), 116002 (2018). [arXiv:1804.06644 [hep-th]]
29. Kobayashi, T., Tamba, S.: Phys. Rev. D **99**(4), 046001 (2019). [arXiv:1811.11384 [hep-th]]
30. Kariyazono, Y., Kobayashi, T., Takada, S., Tamba, S., Uchida, H.: Phys. Rev. D **100**(4), 045014 (2019). [arXiv:1904.07546 [hep-th]]
31. Ohki, H., Uemura, S., Watanabe, R.: Phys. Rev. D **102**(8), 085008 (2020). [arXiv:2003.04174 [hep-th]]
32. Kikuchi, S., Kobayashi, T., Takada, S., Tatsuishi, T.H., Uchida, H.: Phys. Rev. D **102**(10), 105010 (2020). [arXiv:2005.12642 [hep-th]]
33. Kikuchi, S., Kobayashi, T., Otsuka, H., Takada, S., Uchida, H.: JHEP **11**, 101 (2020). [arXiv:2007.06188 [hep-th]]
34. Kikuchi, S., Kobayashi, T., Uchida, H.: [arXiv:2101.00826 [hep-th]]
35. Almumin, Y., Chen, M.C., Knapp-Pérez, V., Ramos-Sánchez, S., Ratz, M., Shukla, S.: JHEP **05**, 078 (2021). [arXiv:2102.11286 [hep-th]]
36. Tatsuta, Y.: JHEP **10**, 054 (2021). [arXiv:2104.03855 [hep-th]]
37. Ramond, P.: Group Theory: A Physicist's Survey. Cambridge University Press (2010)
38. Miller, G.A., Dickson, H.F., Blichfeldt, L.E.: Theory and Applications of Finite Groups. Wiley, New York (1916)
39. Hamermesh, M.: Group Theory and its Application to Physical Problems. Mass, Addison-Wesley, Reading (1962)
40. Fairbairn, W.M., Fulton, T., Klink, W.H.: Finite and disconnected subgroups of $SU(3)$ and their application to the elementary-particle spectrum. J. Math. Phys. **5**, 1038 (1964)
41. Georgi, H.: Lie algebras in particle physics. From isospin to unified theories. Front. Phys. **54**, 1 (1982)
42. Frampton, P.H., Kephart, T.W.: Int. J. Mod. Phys. A **10**, 4689 (1995). [arXiv:hep-ph/9409330]
43. Ludl, P.O.: arXiv:0907.5587 [hep-ph]
44. Pakvasa, S., Sugawara, H.: Phys. Lett. B **73**, 61 (1978)
45. Bhupal Dev, P.S., Mohapatra, R.N., Severson, M.: [arXiv:1107.2378 [hep-ph]]
46. Tanimoto, M.: Phys. Rev. D **41**, 1586 (1990)
47. Lee, C.E., Lin, C.L., Yang, Y.W.: The minimal extension of the standard model with $S(3)$ symmetry. Prepared for 2nd International Spring School on Medium and High-Energy Nuclear Physics, Taipei, Taiwan, 8–12 May 1990
48. Kang, K., Kim, J.E., Ko, P.: Z. Phys. C **72** (1996) 671. [arXiv:hep-ph/9503436]
49. Kang, K., Kang, S.K., Kim, J.E., Ko, P.W.: Mod. Phys. Lett. A **12** 1175 (1997). [arXiv:hep-ph/9611369]
50. Fukugita, M., Tanimoto, M., Yanagida, T.: Phys. Rev. D **57**, 4429 (1998). [arXiv:hep-ph/9709388]
51. Tanimoto, M., Watari, T., Yanagida, T.: Phys. Lett. B **461**, 345 (1999). [arXiv:hep-ph/9904338]
52. Tanimoto, M.: Phys. Lett. B **483**, 417 (2000). [arXiv:hep-ph/0001306]
53. Tanimoto, M.: Phys. Rev. D **59**, 017304 (1999). [arXiv:hep-ph/9807283]
54. Xing, Z.Z.: Phys. Lett. B **533**, 85–93 (2002). [arXiv:hep-ph/0204049 [hep-ph]]
55. Tanimoto, M., Yanagida, T.: Phys. Lett. B **633**, 567 (2006). [arXiv:hep-ph/0511336]
56. Honda, M., Tanimoto, M.: Phys. Rev. D **75**, 096005 (2007). [arXiv:hep-ph/0701083]
57. Dong, P.V., Long, H.N., Nam, C.H., Vien, V.V.: Phys. Rev. D **85**, 053001 (2012). [arXiv:1111.6360 [hep-ph]]
58. Xing, Z.Z., Yang, D., Zhou, S.: Phys. Lett. B **690**, 304–310 (2010). [arXiv:1004.4234 [hep-ph]]
59. Dicus, D.A., Ge, S.-F., Repko, W.W.: Phys. Rev. D **82**, 033005 (2010). [arXiv:1004.3266 [hep-ph]]
60. Meloni, D., Morisi, S., Peinado, E.: J. Phys. G **38**, 015003 (2011). [arXiv:1005.3482 [hep-ph]]

61. Jora, R., Schechter, J., Shahid, M.N.: Phys. Rev. D **82**, 053006 (2010). [arXiv:1006.3307 [hep-ph]]
62. Bhattacharyya, G., Leser, P., Pas, H.: Phys. Rev. D **83**, 011701 (2011). [arXiv:1006.5597 [hep-ph]]
63. Watanabe, A., Yoshioka, K.: Prog. Theor. Phys. **125**, 129–148 (2011). [arXiv:1007.1527 [hep-ph]]
64. Kaneko, T., Sugawara, H.: Phys. Lett. B **697**, 329–332 (2011). [arXiv:1011.5748 [hep-ph]]
65. Adulpravitchai, A., Batell, B., Pradler, J.: Phys. Lett. B **700**, 207–216 (2011). [arXiv:1103.3053 [hep-ph]]
66. Teshima, T., Okumura, Y.: Phys. Rev. D **84**, 016003 (2011). [arXiv:1103.6127 [hep-ph]]
67. Dev, S., Gupta, S., Gautam, R.R.: Phys. Lett. B **702**, 28–33 (2011). [arXiv:1106.3873 [hep-ph]]
68. Zhou, S.: Phys. Lett. B **704**, 291–295 (2011). [arXiv:1106.4808 [hep-ph]]
69. Koide, Y.: Phys. Rev. D **60**, 077301 (1999). [arXiv:hep-ph/9905416]
70. Chalut, K., Cheng, H., Frampton, P.H., Stowe, K., Yoshikawa, T.: Mod. Phys. Lett. A **17**, 1513 (2002). [arXiv:hep-ph/0204074]
71. Kubo, J., Mondragon, A., Mondragon, M., Rodriguez-Jauregui, E.: Prog. Theor. Phys. **109**, 795 (2003) [Erratum-ibid. **114**, 287 (2005)]. [arXiv:hep-ph/0302196]
72. Kobayashi, T., Kubo, J., Terao, H.: Phys. Lett. B **568**, 83 (2003). [arXiv:hep-ph/0303084]
73. Kubo, J.: Phys. Lett. B **578**, 156 (2004) [Erratum-ibid. B **619**, 387 (2005)]. [arXiv:hep-ph/0309167]
74. Chen, S.L., Frigerio, M., Ma, E.: Phys. Rev. D **70**, 073008 (2004) [Erratum-ibid. D **70**, 079905 (2004)]. [arXiv:hep-ph/0404084]
75. Choi, K.Y., Kajiyama, Y., Lee, H.M., Kubo, J. Phys. Rev. D **70**, 055004 (2004). [arXiv:hep-ph/0402026]
76. Kubo, J., Okada, H., Sakamaki, F.: Phys. Rev. D **70**, 036007 (2004). [arXiv:hep-ph/0402089]
77. Lavoura, L., Ma, E.: Mod. Phys. Lett. A **20**, 1217 (2005). [arXiv:hep-ph/0502181]
78. Grimus, W., Lavoura, L.: JHEP **0508**, 013 (2005). [arXiv:hep-ph/0504153]
79. Morisi, S., Picariello, M.: Int. J. Theor. Phys. **45**, 1267 (2006). [arXiv:hep-ph/0505113]
80. Teshima, T.: Phys. Rev. D **73**, 045019 (2006). [arXiv:hep-ph/0509094]
81. Koide, Y.: Phys. Rev. D **73**, 057901 (2006). [arXiv:hep-ph/0509214]
82. Kimura, T.: Prog. Theor. Phys. **114**, 329 (2005)
83. Araki, T., Kubo, J., Paschos, E.A.: Eur. Phys. J. C **45**, 465 (2006). [arXiv:hep-ph/0502164]
84. Kubo, J., Mondragon, A., Mondragon, M., Rodriguez-Jauregui, E., Felix-Beltran, O., Peinado, E.: J. Phys. Conf. Ser. **18**, 380 (2005)
85. Haba, N., Yoshioka, K.: Nucl. Phys. B **739**, 254 (2006). [arXiv:hep-ph/0511108]
86. Kim, J.E., Park, J.C.: JHEP **0605**, 017 (2006). [arXiv:hep-ph/0512130]
87. Kaneko, S., Sawanaka, H., Shingai, T., Tanimoto, M., Yoshioka, K.: Prog. Theor. Phys. **117**, 161 (2007). [arXiv:hep-ph/0609220]
88. Mohapatra, R.N., Nasri, S., Yu, H.B.: Phys. Lett. B **639**, 318 (2006). [arXiv:hep-ph/0605020]
89. Picariello, M.: Int. J. Mod. Phys. A **23**, 4435 (2008). [arXiv:hep-ph/0611189]
90. Koide, Y.: Eur. Phys. J. C **50**, 809 (2007). [arXiv:hep-ph/0612058]
91. Haba, N., Watanabe, A., Yoshioka, K.: Phys. Rev. Lett. **97**, 041601 (2006). [arXiv:hep-ph/0603116]
92. Morisi, S.: Int. J. Mod. Phys. A **22** (2007) 2921
93. Chen, C.Y., Wolfenstein, L.: Phys. Rev. D **77**, 093009 (2008). [arXiv:0709.3767 [hep-ph]]
94. Mitra, M., Choubey, S.: Phys. Rev. D **78**, 115014 (2008). [arXiv:0806.3254 [hep-ph]]
95. Feruglio, F., Lin, Y.: Nucl. Phys. B **800**, 77 (2008). [arXiv:0712.1528 [hep-ph]]
96. Jora, R., Schechter, J., Naeem Shahid, M.: Phys. Rev. D **80**, 093007 (2009). [arXiv:0909.4414 [hep-ph]]
97. Beltran, O.F., Mondragon, M., Rodriguez-Jauregui, E.: J. Phys. Conf. Ser. **171**, 012028 (2009)
98. Yamanaka, Y., Sugawara, H., Pakvasa, S.: Phys. Rev. D **25**, 1895 (1982) [Erratum-ibid. D **29**, 2135 (1984)]
99. Brown, T., Pakvasa, S., Sugawara, H., Yamanaka, Y.: Phys. Rev. D **30**, 255 (1984)
100. Brown, T., Deshpande, N., Pakvasa, S., Sugawara, H.: Phys. Lett. B **141**, 95 (1984)

101. Zhang, H.: Phys. Lett. B **655**, 132 (2007). [arXiv:hep-ph/0612214]
102. Hagedorn, C., Lindner, M., Mohapatra, R.N.: JHEP **0606**, 042 (2006). [arXiv:hep-ph/0602244]
103. Cai, Y., Yu, H.B.: Phys. Rev. D **74**, 115005 (2006). [arXiv:hep-ph/0608022]
104. Caravaglios, F., Morisi, S.: Int. J. Mod. Phys. A **22**, 2469 (2007). [arXiv:hep-ph/0611078]
105. Koide, Y.: JHEP **0708**, 086 (2007). [arXiv:0705.2275 [hep-ph]]
106. Parida, M.K.: Phys. Rev. D **78**, 053004 (2008). [arXiv:0804.4571 [hep-ph]]
107. Lam, C.S.: Phys. Rev. D **78**, 073015 (2008). [arXiv:0809.1185 [hep-ph]]
108. Bazzocchi, F., Morisi, S.: Phys. Rev. D **80**, 096005 (2009). [arXiv:0811.0345 [hep-ph]]
109. Ishimori, H., Shimizu, Y., Tanimoto, M.: Prog. Theor. Phys. **121**, 769 (2009). [arXiv:0812.5031 [hep-ph]]
110. Bazzocchi, F., Merlo, L., Morisi, S.: Nucl. Phys. B **816**, 204 (2009). [arXiv:0901.2086 [hep-ph]]
111. Bazzocchi, F., Merlo, L., Morisi, S.: Phys. Rev. D **80**, 053003 (2009). [arXiv:0902.2849 [hep-ph]]
112. Altarelli, G., Feruglio, F., Merlo, L.: JHEP **0905**, 020 (2009). [arXiv:0903.1940 [hep-ph]]
113. Grimus, W., Lavoura, L., Ludl, P.O.: J. Phys. G **36**, 115007 (2009). [arXiv:0906.2689 [hep-ph]]
114. Ding, G.J.: Nucl. Phys. B **827**, 82 (2010). [arXiv: 0909.2210 [hep-ph]]
115. Merlo, L.: AIP Conf. Proc. **1200**(1), 948–951 (2010). [arXiv:0909.2760 [hep-ph]]
116. Daikoku, Y., Okada, H.: arXiv:0910.3370 [hep-ph]
117. Meloni, D.: arXiv:0911.3591 [hep-ph]
118. Morisi, S., Peinado, E.: arXiv:1001.2265 [hep-ph]
119. de Adelhart Toorop, R., Bazzocchi, F., Merlo, L.: JHEP **1008**, 001 (2010). [arXiv:1003.4502 [hep-ph]]
120. Hagedorn, C., King, S.F., Luhn, C.: JHEP **1006**, 048 (2010). [arXiv:1003.4249 [hep-ph]]
121. Ishimori, H., Saga, K., Shimizu, Y., Tanimoto, M.: Phys. Rev. D **81**, 115009 (2010). [arXiv:1004.5004 [hep-ph]]
122. Ahn, Y.H., Kang, S.K., Kim, C.S., Nguyen, T.P.: Phys. Rev. D **82**, 093005 (2010). [arXiv:1004.3469 [hep-ph]]
123. Ding, G.-J.: Nucl. Phys. B **846**, 394–428 (2011). [arXiv:1006.4800 [hep-ph]]
124. Daikoku, Y., Okada, H.: [arXiv:1008.0914 [hep-ph]]
125. Dong, P.V., Long, H.N., Soa, D.V., Vien, V.V.: Eur. Phys. J. C **71**, 1544 (2011). [arXiv:1009.2328 [hep-ph]]
126. Ishimori, H., Shimizu, Y., Tanimoto, M., Watanabe, A.: Phys. Rev. D **83**, 033004 (2011). [arXiv:1010.3805 [hep-ph]]
127. Parida, M.K., Sahu, P.K., Bora, K.: Phys. Rev. D **83**, 093004 (2011). [arXiv:1011.4577 [hep-ph]]
128. Daikoku, Y., Okada, H., Toma, T.: [arXiv:1010.4963 [hep-ph]]
129. Ding, G.-J., Pan, D.-M.: Eur. Phys. J. C **71**, 1716 (2011). [arXiv:1011.5306 [hep-ph]]
130. Ishimori, H., Tanimoto, M.: Prog. Theor. Phys. **125**, 653–675 (2011). [arXiv:1012.2232 [hep-ph]]
131. Stech, B.: [arXiv:1012.6028 [hep-ph]]
132. Park, N.W., Nam, K.H., Siyeon, K.: Phys. Rev. D **83**, 056013 (2011). [arXiv:1101.4134 [hep-ph]]
133. Ishimori, H., Kajiyama, Y., Shimizu, Y., Tanimoto, M.: [arXiv:1103.5705 [hep-ph]]
134. Yang, R.-Z., Zhang, H.: Phys. Lett. B **700**, 316–321 (2011). [arXiv:1104.0380 [hep-ph]]
135. Morisi, S., Peinado, E.: Phys. Lett. B **701**, 451–457 (2011). [arXiv:1104.4961 [hep-ph]]
136. Zhao, Z.-H.: Phys. Lett. B **701**, 609–613 (2011). [arXiv:1106.2715 [hep-ph]]
137. Ishimori, H., Kobayashi, T.: [arXiv:1106.3604 [hep-ph]]
138. Hagedorn, C., Serone, M.: [arXiv:1106.4021 [hep-ph]]
139. Daikoku, Y., Okada, H., Toma, T.: [arXiv:1106.4717 [hep-ph]]
140. Meloni, D.: [arXiv:1107.0221 [hep-ph]]
141. Morisi, S., Patel, K.M., Peinado, E.: [arXiv:1107.0696 [hep-ph]]
142. Ma, E.: Phys. Lett. B **632**, 352 (2006). [arXiv:hep-ph/0508231]
143. Ma, E., Rajasekaran, G.: Phys. Rev. D **64**, 113012 (2001). [arXiv:hep-ph/0106291]

144. Everett, L.L., Stuart, A.J.: Phys. Rev. D **79**, 085005 (2009). [arXiv:0812.1057 [hep-ph]]
145. Luhn, C., Nasri, S., Ramond, P.: J. Math. Phys. **48**, 123519 (2007). [arXiv:0709.1447 [hep-th]]
146. Babu, K.S., Enkhbat, T., Gogoladze, I.: Phys. Lett. B **555**, 238 (2003). [arXiv:hep-ph/0204246]
147. Babu, K.S., Ma, E., Valle, J.W.F.: Phys. Lett. B **552**, 207 (2003). [arXiv:hep-ph/0206292]
148. Ma, E.: Mod. Phys. Lett. A **17**, 2361 (2002). [arXiv:hep-ph/0211393]
149. Babu, K.S., Kobayashi, T., Kubo, J.: Phys. Rev. D **67**, 075018 (2003). [arXiv:hep-ph/0212350]
150. Hirsch, M., Romao, J.C., Skadhauge, S., Valle, J.W.F., Villanova del Moral, A.: Phys. Rev. D **69**, 093006 (2004). [arXiv:hep-ph/0312265]
151. Ma, E.: Phys. Rev. D **70**, 031901 (2004). [arXiv:hep-ph/0404199]
152. Altarelli, G., Feruglio, F.: Nucl. Phys. B **720**, 64 (2005). [arXiv:hep-ph/0504165]
153. Chen, S.L., Frigerio, M., Ma, E.: Nucl. Phys. B **724**, 423 (2005). [arXiv:hep-ph/0504181]
154. Zee, A.: Phys. Lett. B **630**, 58 (2005). [arXiv:hep-ph/0508278]
155. Ma, E.: Mod. Phys. Lett. A **20**, 2601 (2005). [arXiv:hep-ph/0508099]
156. Ma, E.: Phys. Rev. D **73**, 057304 (2006). [arXiv:hep-ph/0511133]
157. Altarelli, G., Feruglio, F.: Nucl. Phys. B **741**, 215 (2006). [arXiv:hep-ph/0512103]
158. He, X.G., Keum, Y.Y., Volkas, R.R.: JHEP **0604**, 039 (2006). [arXiv:hep-ph/0601001]
159. Adhikary, B., Brahmachari, B., Ghosal, A., Ma, E., Parida, M.K.: Phys. Lett. B **638**, 345 (2006). [arXiv:hep-ph/0603059]
160. Ma, E., Sawanaka, H., Tanimoto, M.: Phys. Lett. B **641**, 301 (2006). [arXiv:hep-ph/0606103]
161. Valle, J.W.F.: J. Phys. Conf. Ser. **53**, 473 (2006). [arXiv:hep-ph/0608101]
162. Adhikary, B., Ghosal, A.: Phys. Rev. D **75**, 073020 (2007). [arXiv:hep-ph/0609193]
163. Lavoura, L., Kuhbock, H.: Mod. Phys. Lett. A **22**, 181 (2007). [arXiv:hep-ph/0610050]
164. King, S.F., Malinsky, M.: Phys. Lett. B **645**, 351 (2007). [arXiv:hep-ph/0610250]
165. Hirsch, M., Joshipura, A.S., Kaneko, S., Valle, J.W.F.: Phys. Rev. Lett. **99**, 151802 (2007). [arXiv:hep-ph/0703046]
166. Bazzocchi, F., Kaneko, S., Morisi, S.: JHEP **0803**, 063 (2008). [arXiv:0707.3032 [hep-ph]]
167. Grimus, W., Kuhbock, H.: Phys. Rev. D **77**, 055008 (2008). [arXiv:0710.1585 [hep-ph]]
168. Honda, M., Tanimoto, M.: Prog. Theor. Phys. **119**, 583 (2008). [arXiv:0801.0181 [hep-ph]]
169. Brahmachari, B., Choubey, S., Mitra, M.: Phys. Rev. D **77**, 073008 (2008) [Erratum-ibid. D **77**, 119901 (2008)]. [arXiv:0801.3554 [hep-ph]]
170. Adhikary, B., Ghosal, A.: Phys. Rev. D **78**, 073007 (2008). [arXiv:0803.3582 [hep-ph]]
171. Fukuyama, T.: [arXiv:0804.2107 [hep-ph]]
172. Lin, Y.: Nucl. Phys. B **813**, 91 (2009). [arXiv:0804.2867 [hep-ph]]
173. Frampton, P.H., Matsuzaki, S.: arXiv:0806.4592 [hep-ph]
174. Feruglio, F., Hagedorn, C., Lin, Y., Merlo, L.: Nucl. Phys. B **809**, 218 (2009). [arXiv:0807.3160 [hep-ph]]
175. Morisi, S.: Nuovo Cim. **123B**, 886 (2008). [arXiv:0807.4013 [hep-ph]]
176. Ishimori, H., Kobayashi, T., Omura, Y., Tanimoto, M.: JHEP **0812**, 082 (2008). [arXiv:0807.4625 [hep-ph]]
177. Ma, E.: Phys. Lett. B **671**, 366 (2009). [arXiv:0808.1729 [hep-ph]]
178. Bazzocchi, F., Frigerio, M., Morisi, S.: Phys. Rev. D **78**, 116018 (2008). [arXiv:0809.3573 [hep-ph]]
179. Hirsch, M., Morisi, S., Valle, J.W.F.: Phys. Rev. D **79**, 016001 (2009). [arXiv:0810.0121 [hep-ph]]
180. Merlo, L.: arXiv:0811.3512 [hep-ph]
181. Baek, S., Oh, M.C.: arXiv:0812.2704 [hep-ph]
182. Morisi, S.: Phys. Rev. D **79**, 033008 (2009). [arXiv:0901.1080 [hep-ph]]
183. Ciafaloni, P., Picariello, M., Torrente-Lujan, E., Urbano, A.: Phys. Rev. D **79**, 116010 (2009). [arXiv: 0901.2236 [hep-ph]]
184. Merlo, L.: J. Phys. Conf. Ser. **171**, 012083 (2009). [arXiv:0902.3067 [hep-ph]]
185. Chen, M.C., King, S.F.: JHEP **0906**, 072 (2009). [arXiv:0903.0125 [hep-ph]]
186. Branco, G.C., Gonzalez Felipe, R., Rebelo, M.N., Serodio, H.: Phys. Rev. D **79**, 093008 (2009). [arXiv:0904.3076 [hep-ph]]

187. Hayakawa, A., Ishimori, H., Shimizu, Y., Tanimoto, M.: Phys. Lett. B **680**, 334 (2009). [arXiv:0904.3820 [hep-ph]]
188. Altarelli, G., Meloni, D.: J. Phys. G **36**, 085005 (2009). [arXiv:0905.0620 [hep-ph]]
189. Urbano, A.: arXiv:0905.0863 [hep-ph]
190. Hirsch, M., Morisi, S., Valle, J.W.F.: Phys. Lett. B **679**, 454 (2009). [arXiv:0905.3056 [hep-ph]]
191. Lin, Y.: Nucl. Phys. B **824**, 95 (2010). [arXiv:0905.3534 [hep-ph]]
192. Hirsch, M.: Pramana **72**, 183 (2009)
193. Hagedorn, C., Molinaro, E., Petcov, S.T.: JHEP **0909**, 115 (2009). [arXiv:0908.0240 [hep-ph]]
194. Tamii, A., et al.: Mod. Phys. Lett. A **24**, 867 (2009)
195. Ma, E.: arXiv:0908.3165 [hep-ph]
196. Burrows, T.J., King, S.F.: arXiv:0909.1433 [hep-ph]
197. Albaid, A.: Phys. Rev. D **80**, 093002 (2009). [arXiv:0909.1762 [hep-ph]]
198. Ciafaloni, P., Picariello, M., Torrente-Lujan, E., Urbano, A.: arXiv:0909.2553 [hep-ph]
199. Feruglio, F., Hagedorn, C., Merlo, L.: arXiv:0910.4058 [hep-ph]
200. Morisi, S., Peinado, E.: arXiv:0910.4389 [hep-ph]
201. Berger, J., Grossman, Y.: arXiv:0910.4392 [hep-ph]
202. Hagedorn, C., Molinaro, E., Petcov, S.T.: arXiv:0911.3605 [hep-ph]
203. Feruglio, F., Hagedorn, C., Lin, Y., Merlo, L.: arXiv:0911.3874 [hep-ph]
204. Ding, G.J., Liu, J.F.: arXiv:0911.4799 [hep-ph]
205. Barry, J., Rodejohann, W.: arXiv:1003.2385 [hep-ph]
206. Cooper, I.K., King, S.F., Luhn, C.: Phys. Lett. B **690**, 396–402 (2010). [arXiv:1004.3243 [hep-ph]]
207. Albright, C.H., Dueck, A., Rodejohann, W.: Eur. Phys. J. C **70**, 1099–1110 (2010). [arXiv:1004.2798 [hep-ph]]
208. Riva, F.: Phys. Lett. B **690**, 443–450 (2010). [arXiv:1004.1177 [hep-ph]]
209. Kadosh, A., Pallante, E.: JHEP **1008**, 115 (2010). [arXiv:1004.0321 [hep-ph]]
210. Feruglio, F., Paris, A.: Nucl. Phys. B **840**, 405–423 (2010). [arXiv:1004.0321 [hep-ph]]
211. Fukuyama, T., Sugiyama, H., Tsumura, K.: Phys. Rev. D **82**, 036004 (2010). [arXiv:1005.5338 [hep-ph]]
212. Antusch, S., King, S.F., Spinrath, M.: Phys. Rev. D **83**, 013005 (2011). [arXiv:1005.0708 [hep-ph]]
213. Ahn, Y.H.: [arXiv:1006.2953 [hep-ph]]
214. Ma, E.: Phys. Rev. D **82**, 037301 (2010). [arXiv:1006.3524 [hep-ph]]
215. Hirsch, M., Morisi, S., Peinado, E., Valle, J.W.F.: Phys. Rev. D **82**, 116003 (2010). [arXiv:1007.0871 [hep-ph]]
216. Esteves, J.N., Joaquim, F.R., Joshipura, A.S., Romao, J.C., Tortola, M.A., Valle, J.W.F.: Phys. Rev. D **82**, 073008 (2010). [arXiv:1007.0898 [hep-ph]]
217. Burrows, T.J., King, S.F.: Nucl. Phys. B **842**, 107–121 (2011). [arXiv:1007.2310 [hep-ph]]
218. Haba, N., Kajiyama, Y., Matsumoto, S., Okada, H., Yoshioka, K.: Phys. Lett. B **695**, 476–481 (2011). [arXiv:1008.4777 [hep-ph]]
219. Araki, T., Mei, J., Xing, Z.-Z.: Phys. Lett. B **695**, 165–168 (2011). [arXiv:1010.3065 [hep-ph]]
220. Meloni, D., Morisi, S., Peinado, E.: Phys. Lett. B **697**, 339–342 (2011). [arXiv:1010.3065 [hep-ph]]
221. Machado, A.C.B., Montero, J.C., Pleitez, V.: Phys. Lett. B **697**, 318–322 (2011). [arXiv:1011.5855 [hep-ph]]
222. Carone, C.D., Lebed, R.F., Phys. Lett. B **696**, 454–458 (2011). [arXiv:1011.6379 [hep-ph]]
223. de Medeiros Varzielas, I., Merlo, L.: JHEP **1102**, 062 (2011). [arXiv:1011.6662 [hep-ph]]
224. de Adelhart Toorop, R., Bazzocchi, F., Merlo, L., Paris, A.: JHEP **1103**, 035 (2011). [arXiv:1012.1791 [hep-ph]]
225. de Adelhart Toorop, R., Bazzocchi, F., Merlo, L., Paris, A.: JHEP **1103**, 040 (2011). [arXiv:1012.2091 [hep-ph]]
226. Fukuyama, T., Sugiyama, H., Tsumura, K.: Phys. Rev. D **83**, 056016 (2011). [arXiv:1012.4886 [hep-ph]]

227. de Medeiros Varzielas, I., Gonzalez Felipe, R., Serodio, H.: Phys. Rev. D **83**, 033007 (2011). [arXiv:1101.0602 [hep-ph]]
228. Boucenna, M.S., Hirsch, M., Morisi, S., Peinado, E., Taoso, M., Valle, J.W.F.: JHEP **1105**, 037 (2011). [arXiv:1101.2874 [hep-ph]]
229. Kadosh, A., Pallante, E.: JHEP **1106**, 121 (2011). [arXiv:1101.5420 [hep-ph]]
230. Ahn, Y.H., Cheng, H.-Y., Oh, S.: Phys. Rev. D **83**, 076012 (2011). [arXiv:1102.0879 [hep-ph]]
231. Ahn, Y.H., Kim, C.S., Oh, S.: [arXiv:1103.0657 [hep-ph]]
232. Morisi, S., Peinado, E., Shimizu, Y., Valle, J.W.F.: Phys. Rev. D **84**, 036003 (2011). [arXiv:1104.1633 [hep-ph]]
233. Toorop, R.d.A., Bazzocchi, F., Morisi, S.: [arXiv:1104.5676 [hep-ph]]
234. Shimizu, Y., Tanimoto, M., Watanabe, A.: Prog. Theor. Phys. **126**, 81–90 (2011). [arXiv:1105.2929 [hep-ph]]
235. Barry, J., Rodejohann, W., Zhang, H.: JHEP **1107**, 091 (2011). [arXiv:1105.3911 [hep-ph]]
236. Ma, E., Wegman, D.: Phys. Rev. Lett. **107**, 061803 (2011). [arXiv:1106.4269 [hep-ph]]
237. Albaid, A.: [arXiv:1106.4070 [hep-ph]]
238. Adulpravitchai, A., Takahashi, R.: [arXiv:1107.3829 [hep-ph]]
239. King, S.F., Luhn, C.: JHEP **1109**, 042 (2011). [arXiv:1107.5332 [hep-ph]]
240. Machado, A.C.B., Montero, J.C., Pleitez, V.: [arXiv:1108.1767 [hep-ph]]
241. Ding, G.-J., Meloni, D.: [arXiv:1108.2733 [hep-ph]]
242. Aristizabal Sierra, D., Bazzocchi, F.: arXiv:1110.3781 [hep-ph]
243. Cooper, I.K., King, S.F., Luhn, C.: arXiv:1110.5676 [hep-ph]
244. Barry, J., Rodejohann, W., Zhang, H.: arXiv:1110.6382 [hep-ph]
245. Ferreira, P.M., Lavoura, L.: arXiv:1111.5859 [hep-ph]
246. King, S.F., Luhn, C.: arXiv:1112.1959 [hep-ph]
247. Ding, G.-J., Everett, L.L., Stuart, A.J.: arXiv:1110.1688 [hep-ph]
248. Chen, C.-S., Kephart, T.W., Yuan, T.-C.: JHEP **1104**, 015 (2011). [arXiv:1011.3199 [hep-ph]]
249. Feruglio, F., Paris, A.: JHEP **1103**, 101 (2011). [arXiv:1101.0393 [hep-ph]]
250. Shirai, K.: J. Phys. Soc. Jpn. **61**, 2735 (1992)
251. Luhn, C., Ramond, P.: J. Math. Phys. **49**, 053525 (2008). [arXiv:0803.0526 [hep-th]]
252. Merlo, L.: Nucl. Phys. Proc. Suppl. **188**, 345 (2009)
253. Aranda, A., Carone, C.D., Lebed, R.F.: Phys. Lett. B **474**, 170 (2000). [arXiv:hep-ph/9910392]
254. Eby, D.A., Frampton, P.H.: Phys. Rev. D **86**, 117304 (2012). [arXiv:1112.2675 [hep-ph]]
255. Aranda, A., Carone, C.D., Lebed, R.F.: Phys. Rev. D **62**, 016009 (2000). [arXiv:hep-ph/0002044]
256. Carr, P.D., Frampton, P.H.: arXiv:hep-ph/0701034
257. Feruglio, F., Hagedorn, C., Lin, Y., Merlo, L.: Nucl. Phys. B **775**, 120 (2007). [arXiv:hep-ph/0702194]
258. Chen, M.C., Mahanthappa, K.T.: Phys. Lett. B **652**, 34 (2007). [arXiv:0705.0714 [hep-ph]]
259. Frampton, P.H., Kephart, T.W.: JHEP **0709**, 110 (2007). [arXiv:0706.1186 [hep-ph]]
260. Aranda, A.: Phys. Rev. D **76**, 111301 (2007). [arXiv:0707.3661 [hep-ph]]
261. Ding, G.J.: Phys. Rev. D **78**, 036011 (2008). [arXiv:0803.2278 [hep-ph]]
262. Frampton, P.H., Kephart, T.W., Matsuzaki, S.: Phys. Rev. D **78**, 073004 (2008). [arXiv:0807.4713 [hep-ph]]
263. Eby, D.A., Frampton, P.H., Matsuzaki, S.: Phys. Lett. B **671**, 386 (2009). [arXiv:0810.4899 [hep-ph]]
264. Frampton, P.H., Matsuzaki, S.: Phys. Lett. B **679**, 347 (2009). [arXiv:0902.1140 [hep-ph]]
265. Chen, M.C., Mahanthappa, K.T., Yu, F.: arXiv:0909.547 [hep-ph]
266. Frampton, P.H., Ho, C.M., Kephart, T.W., Matsuzaki, S.: Phys. Rev. D **82**, 113007 (2010). [arXiv:1009.0307 [hep-ph]]
267. Aranda, A., Bonilla, C., Ramos, R., Rojas, A.D.: [arXiv:1011.6470 [hep-ph]]
268. BenTov, Y., Zee, A.: [arXiv:1101.1987 [hep-ph]]
269. Eby, D.A., Frampton, P.H., He, X.-G., Kephart, T.W.: Phys. Rev. D **84**, 037302 (2011). [arXiv:1103.5737 [hep-ph]]
270. Chen, M.-C., Mahanthappa, K.T.: [arXiv:1107.3856 [hep-ph]]

271. Merlo, L., Rigolin, S., Zaldivar, B.: JHEP **11**, 047 (2011). [arXiv:1108.1795 [hep-ph]]
272. Chen, M.-C., Mahanthappa, K.T., Meroni, A., Petcov, S.T.: [arXiv:1109.0731 [hep-ph]]
273. Bergshoeff, E., Janssen, B., Ortin, T.: Class. Quant. Grav. **13**, 321 (1996). [arXiv:hep-th/9506156]
274. Kajiyama, Y., Okada, H., Toma, T.: arXiv:1109.2722 [hep-ph]
275. Frampton, P.H., Kephart, T.W.: Phys. Rev. D **64**, 086007 (2001). [arXiv:hep-th/0011186]
276. Grimus, W., Lavoura, L.: Phys. Lett. B **572**, 189 (2003). [arXiv:hep-ph/0305046]
277. Grimus, W., Joshipura, A.S., Kaneko, S., Lavoura, L., Tanimoto, M.: JHEP **0407**, 078 (2004). [arXiv:hep-ph/0407112]
278. Grimus, W., Joshipura, A.S., Kaneko, S., Lavoura, L., Sawanaka, H., Tan-imoto, M.: Nucl. Phys. B **713**, 151 (2005). [arXiv:hep-ph/0408123]
279. Blum, A., Mohapatra, R.N., Rodejohann, W.: Phys. Rev. D **76**, 053003 (2007). [arXiv:0706.3801 [hep-ph]]
280. Ishimori, H., Kobayashi, T., Ohki, H., Omura, Y., Takahashi, R., Tanimoto, M.: Phys. Lett. B **662**, 178 (2008). [arXiv:0802.2310 [hep-ph]]
281. Ishimori, H., Kobayashi, T., Ohki, H., Omura, Y., Takahashi, R., Tanimoto, M.: Phys. Rev. D **77**, 115005 (2008). [arXiv:0803.0796 [hep-ph]]
282. Adulpravitchai, A., Blum, A., Hagedorn, C.: JHEP **0903**, 046 (2009). [arXiv:0812.3799 [hep-ph]]
283. Hagedorn, C., Ziegler, R.: Phys. Rev. D **82**, 053011 (2010). [arXiv:1007.1888 [hep-ph]]
284. Meloni, D., Morisi, S., Peinado, E.: Phys. Lett. B **703**, 281–287 (2011). [arXiv:1104.0178 [hep-ph]]
285. Hagedorn, C., Lindner, M., Plentinger, F.: Phys. Rev. D **74**, 025007 (2006). [arXiv:hep-ph/0604265]
286. Kajiyama, Y., Kubo, J., Okada, H.: Phys. Rev. D **75**, 033001 (2007). [arXiv:hep-ph/0610072]
287. Blum, A., Hagedorn, C., Lindner, M.: Phys. Rev. D **77**, 076004 (2008). [arXiv:0709.3450 [hep-ph]]
288. Hagedorn, C., Lindner, M., Plentinger, F.: Phys. Rev. D **74**, 025007 (2006). [arXiv:hep-ph/0604265]
289. Blum, A., Hagedorn, C., Hohenegger, A.: JHEP **0803**, 070 (2008). [arXiv:0710.5061 [hep-ph]]
290. Blum, A., Hagedorn, C.: Nucl. Phys. B **821**, 327 (2009). [arXiv:0902.4885 [hep-ph]]
291. Adulpravitchai, A., Blum, A., Rodejohann, W.: New J. Phys. **11**, 063026 (2009). [arXiv:0903.0531 [hep-ph]]
292. Kim, J.E., Seo, M.-S.: JHEP **1102**, 097 (2011). [arXiv:1005.4684 [hep-ph]]
293. Kajiyama, Y., Okada, H., Toma, T.: Eur. Phys. J. C **71**, 1688 (2011). [arXiv:1104.0367 [hep-ph]]
294. Kim, J.E., Seo, M.-S.: [arXiv:1106.6117 [hep-ph]]
295. Babu, K.S., Kubo, J.: Phys. Rev. D **71**, 056006 (2005). [arXiv:hep-ph/0411226]
296. Aranda, A., Bonilla, C., Ramos, R., Rojas, A.D.: Phys. Rev. D **84**, 016009 (2011). [arXiv:1105.6373 [hep-ph]]
297. Kajiyama, Y., Itou, E., Kubo, J.: Nucl. Phys. B **743**, 74 (2006). [arXiv:hep-ph/0511268]
298. Itou, E., Kajiyama, Y., Kubo, J.: AIP Conf. Proc. **903**, 389 (2007). [arXiv:hep-ph/0611052]
299. Kawashima, K., Kubo, J., Lenz, A.: Phys. Lett. B **681**, 60 (2009). [arXiv:0907.2302 [hep-ph]]
300. Volkov, G.: [arXiv:1006.5627 [math-ph]]
301. Kubo, J., Lenz, A.: Phys. Rev. D **82**, 075001 (2010). [arXiv:1007.0680 [hep-ph]]
302. Hackett, J., Kauffman, L.: [arXiv:1010.2979 [math-ph]]
303. Furui, S.: AIP Conf. Proc. **1343**, 533–535 (2011). [arXiv:1011.3086 [hep-ph]]
304. Kaburaki, Y., Konya, K., Kubo, J., Lenz, A.: Phys. Rev. D **84**, 016007 (2011). [arXiv:1012.2435 [hep-ph]]
305. Babu, K.S., Kawashima, K., Kubo, J.: Phys. Rev. D **83**, 095008 (2011). [arXiv:1103.1664 [hep-ph]]
306. Araki, T., Li, Y.F.: arXiv:1112.5819 [hep-ph]
307. Frigerio, M., Kaneko, S., Ma, E., Tanimoto, M.: Phys. Rev. D **71**, 011901 (2005). [arXiv:hep-ph/0409187]

308. Frigerio, M., Ma, E.: Phys. Rev. D **76**, 096007 (2007). [arXiv:0708.0166 [hep-ph]]
309. Dev, S., Verma, S.: Mod. Phys. Lett. A **25**, 2837–2848 (2010). [arXiv:1005.4521 [hep-ph]]
310. Ma, E.: arXiv:0705.0327 [hep-ph]
311. Bovier, A., Luling, M., Wyler, D.: J. Math. Phys. **22**, 1536 (1981)
312. Bovier, A., Luling, M., Wyler, D.: J. Math. Phys. **22**, 1543 (1981)
313. Fairbairn, W.M., Fulton, T.: J. Math. Phys. **23**, 1747 (1982)
314. Branco, G.C., Gerard, J.M., Grimus, W.: Phys. Lett. B **136**, 383 (1984)
315. Ma, E.: Mod. Phys. Lett. A **21**, 1917 (2006). [arXiv:hep-ph/0607056]
316. Luhn, C., Nasri, S., Ramond, P.: J. Math. Phys. **48**, 073501 (2007). [arXiv:hep-th/0701188]
317. de Medeiros Varzielas, I., King, S.F., Ross, G.G.: Phys. Lett. B **648**, 201 (2007). [arXiv:hep-ph/0607045]
318. Ma, E.: Phys. Lett. B **660**, 505 (2008). [arXiv:0709.0507 [hep-ph]]
319. Grimus, W., Lavoura, L.: JHEP **0809**, 106 (2008). [arXiv:0809.0226 [hep-ph]]
320. Howl, R., King, S.F.: JHEP **0805**, 008 (2008). [arXiv:0802.1909 [hep-ph]]
321. King, S.F., Luhn, C.: JHEP **0910**, 093 (2009). [arXiv:0908.1897 [hep-ph]]
322. King, S.F.: JHEP **1009**, 114 (2010). [arXiv:1006.5895 [hep-ph]]
323. Ding, G.J., Zhou, Y.L.: Chin. Phys. C **39**(2), 021001 (2015). [arXiv:1312.5222 [hep-ph]]
324. Ding, G.J., Zhou, Y.L.: JHEP **06**, 023 (2014). [arXiv:1404.0592 [hep-ph]]
325. Luhn, C., Nasri, S., Ramond, P.: Phys. Lett. B **652**, 27 (2007). [arXiv:0706.2341 [hep-ph]]
326. Hartmann, C.: [arXiv:1109.5143 [hep-ph]]
327. Hagedorn, C., Schmidt, M.A., Smirnov, A.Y.: Phys. Rev. D **79**, 036002 (2009). [arXiv:0811.2955 [hep-ph]]
328. Cao, Q.-H., Khalil, S., Ma, E., Okada, H.: Phys. Rev. Lett. **106**, 131801 (2011). [arXiv:1009.5415 [hep-ph]]
329. Ma, E.: Mod. Phys. Lett. A **26**, 377–385 (2011). [arXiv:1101.4972 [hep-ph]]
330. Cao, Q.-H., Khalil, S., Ma, E., Okada, H.: [arXiv:1108.0570 [hep-ph]]
331. Kajiyama, Y., Okada, H.: Nucl. Phys. B **848**, 303–313 (2011). [arXiv:1011.5753 [hep-ph]]
332. Parattu, K.M., Wingerter, A.: Phys. Rev. D **84**, 013011 (2011). [arXiv:1012.2842 [hep-ph]]
333. Ding, G.-J.: Nucl. Phys. B **853**, 635–662 (2011). [arXiv:1105.5879 [hep-ph]]
334. Hartmann, C., Zee, A.: Nucl. Phys. B **853**, 105–124 (2011). [arXiv:1106.0333 [hep-ph]]
335. Ishimori, H., Kobayashi, T.: Phys. Rev. D **85**, 125004 (2012). arXiv:1201.3429 [hep-ph]
336. Escobar, J.A., Luhn, C.: J. Math. Phys. **50**, 013524 (2009). [arXiv:0809.0639 [hep-th]]
337. de Adelhart Toorop, R., Feruglio, F., Hagedorn, C.: Phys. Lett. B **703**, 447–451 (2011). [arXiv:1107.3486 [hep-ph]]
338. King, S.F., Luhn, C., Stuart, A.J.: Nucl. Phys. B **867**, 203–235 (2013). [arXiv:1207.5741 [hep-ph]]
339. Ding, G.J., King, S.F.: Phys. Rev. D **89**(9), 093020 (2014). [arXiv:1403.5846 [hep-ph]]
340. Ishimori, H., Kobayashi, T., Okada, H., Shimizu, Y., Tanimoto, M.: JHEP **0904**, 011 (2009). [arXiv:0811.4683 [hep-ph]]
341. Ishimori, H., Kobayashi, T., Okada, H., Shimizu, Y., Tanimoto, M.: JHEP **0912**, 054 (2009). [arXiv:0907.2006 [hep-ph]]
342. Escobar, J.A.: [arXiv:1102.1649 [hep-ph]]
343. Varzielas, I.d.M., Emmanuel-Costa, D.: [arXiv:1106.5477 [hep-ph]]
344. King, S.F., Luhn, C.: Nucl. Phys. B **820**, 269 (2009). [arXiv:0905.1686 [hep-ph]]
345. King, S.F., Luhn, C.: Nucl. Phys. B **832**, 414 (2010). [arXiv:0912.1344 [hep-ph]]
346. Novichkov, P.P., Penedo, J.T., Petcov, S.T.: Nucl. Phys. B **963**, 115301 (2021). [arXiv:2006.03058 [hep-ph]]
347. Liu, X.G., Yao, C.Y., Ding, G.J.: Phys. Rev. D **103**(5), 056013 (2021). [arXiv:2006.10722 [hep-ph]]
348. Everett, L.L., Stuart, A.J.: Phys. Lett. B **698**, 131–139 (2011). [arXiv:1011.4928 [hep-ph]]
349. Chen, C.S., Kephart, T.W., Yuan, T.C.: PTEP **2013**(10), 103B01 (2013). [arXiv:1110.6233 [hep-ph]]
350. Hashimoto, K., Okada, H.: arXiv:1110.3640 [hep-ph]

351. de Medeiros Varzielas, I., King, S.F., Ross, G.G.: Phys. Lett. B **644**, 153 (2007). [arXiv:hep-ph/0512313]
352. Adulpravitchai, A., Blum, A., Lindner, M.: JHEP **0909**, 018 (2009). [arXiv:0907.2332 [hep-ph]]
353. Frampton, P.H., Kephart, T.W., Rohm, R.M.: Phys. Lett. B **679**, 478 (2009). [arXiv:0904.0420 [hep-ph]]
354. Ecker, G., Grimus, W., Konetschny, W.: Nucl. Phys. B **191**, 465–492 (1981)
355. Ecker, G., Grimus, W., Neufeld, H.: Nucl. Phys. B **247**, 70–82 (1984)
356. Ecker, G., Grimus, W., Neufeld, H.: J. Phys. A **20**, L807 (1987)
357. Neufeld, H., Grimus, W., Ecker, G.: Int. J. Mod. Phys. A **3**, 603–616 (1988)
358. Grimus, W., Rebelo, M.N.: Phys. Rept. **281**, 239–308 (1997). [arXiv:hep-ph/9506272 [hep-ph]]
359. Krauss, L.M., Wilczek, F.: Phys. Rev. Lett. **62**, 1221 (1989)
360. Luhn, C.: Phys. Lett. B **670**, 390 (2009). [arXiv:0807.1749 [hep-ph]]
361. Ibáñez, L.E., Ross, G.G.: Phys. Lett. B **260**, 291–295 (1991)
362. Banks, T., Dine, M.: Phys. Rev. D **45**, 1424–1427 (1992). [arXiv:hep-th/9109045]
363. Dine, M., Graesser, M.: JHEP **01**, 038 (2005). [arXiv:hep-th/0409209]
364. Csaki, C., Murayama, H.: Nucl. Phys. B **515** 114–162 (1998). [arXiv:hep-th/9710105]
365. Ibáñez, L.E., Ross, G.G.: Nucl. Phys. B **368**, 3–37 (1992)
366. Ibáñez, L.E.: Nucl. Phys. B **398**, 301–318 (1993). [arXiv:hep-ph/9210211]
367. Babu, K.S., Gogoladze, I., Wang, K.: Nucl. Phys. B **660**, 322–342 (2003). [arXiv:hep-ph/0212245]
368. Dreiner, H.K., Luhn, C., Thormeier, M.: Phys. Rev. D **73**, 075007 (2006). [arXiv:hep-ph/0512163]
369. Araki, T.: Prog. Theor. Phys. **117**, 1119–1138 (2007). [arXiv:hep-ph/0612306]
370. Araki, T., Choi, K.S., Kobayashi, T., Kubo, J., Ohki, H.: Phys. Rev. D **76**, 066006 (2007). [arXiv:0705.3075 [hep-ph]]
371. Araki, T., Kobayashi, T., Kubo, J., Ramos-Sanchez, S., Ratz, M., Vaudrevange, P.K.S.: Nucl. Phys. B **805**, 124 (2008). [arXiv:0805.0207 [hep-th]]
372. Luhn, C., Ramond, P.: JHEP **0807**, 085 (2008). [arXiv:0805.1736 [hep-ph]]

Basics of Finite Groups

2

We start with introducing basic aspects of group theory, in particular for finite groups. For the pedagogical purpose, we use several theorems without their proofs, but proofs of useful theorems are given in Appendix A. (See also e.g. Refs. [1–6].) On the other hand, we present several examples to understand these basic theorem clearly.

A group, G, is a set, where the multiplication is defined such that the following properties are satisfied:

1. **Closure**
   If a and b are elements of the group G, $c = ab$ is also its element.
2. **Associativity**
   $(ab)c = a(bc)$ for $a, b, c \in G$.
3. **Identity**
   The group G includes an identity element e, which satisfies $ae = ea = a$ for any element $a \in G$.
4. **Inverse**
   The group G includes an inverse element a^{-1} for any element $a \in G$ such that $aa^{-1} = a^{-1}a = e$.

Let us present simple examples.

Example: Cyclic Group Z_N
Discrete rotations of a complex plane form a group. Let us denote the $\exp[2\pi i/N]$ rotation by a. Then, the $\exp[2\pi i m/N]$ rotation for $m =$ integer can be written by a^m. The multiplication rule is defined such as $a^m a^n = a^{m+n}$. The operator a^N corresponds to the identity, $a^N = e$, and the inverse of a^m is obtained as a^{N-m}. Thus, the following set:

T. Kobayashi et al., *An Introduction to Non-Abelian Discrete Symmetries for Particle Physicists*, Lecture Notes in Physics 995, https://doi.org/10.1007/978-3-662-64679-3_2

$$\{e,\ a,\ a^2,\ \cdots,\ a^{N-1}\},\tag{2.1}$$

forms a group. Its closure and associativity would be obvious. This group is called the cyclic group Z_N.

Example: S_3 and S_N

All possible permutations among three objects, (x_1, x_2, x_3), form a group and it is denoted by S_3. There are six permutations as

$$\begin{aligned}
e &: (x_1, x_2, x_3) \to (x_1, x_2, x_3),\\
a_1 &: (x_1, x_2, x_3) \to (x_2, x_1, x_3),\\
a_2 &: (x_1, x_2, x_3) \to (x_3, x_2, x_1),\\
a_3 &: (x_1, x_2, x_3) \to (x_1, x_3, x_2),\\
a_4 &: (x_1, x_2, x_3) \to (x_3, x_1, x_2),\\
a_5 &: (x_1, x_2, x_3) \to (x_2, x_3, x_1).
\end{aligned}\tag{2.2}$$

Obviously, the element e is identity. Their multiplications form a closed algebra, e.g.

$$\begin{aligned}
a_1 a_2 &: (x_1, x_2, x_3) \to (x_2, x_3, x_1),\\
a_2 a_1 &: (x_1, x_2, x_3) \to (x_3, x_1, x_2),\\
a_4 a_2 &: (x_1, x_2, x_3) \to (x_1, x_3, x_2),
\end{aligned}\tag{2.3}$$

i.e.

$$a_1 a_2 = a_5, \quad a_2 a_1 = a_4, \quad a_4 a_2 = a_2 a_1 a_2 = a_3.\tag{2.4}$$

It would be straightforward to see the closure of other multiplications, associativity and the presence of inverse element for each element. By using their multiplication rules, one can write all of six elements in terms of two proper elements and their products. For example, by defining $a_1 = a, a_2 = b$, all of elements are written as

$$\{e, a, b, ab, ba, bab\}.\tag{2.5}$$

Note that $aba = bab$. The S_3 group is a symmetry of an equilateral triangle as shown in Fig. 2.1. The elements a and ab correspond to a reflection and the $2\pi/3$ rotation, respectively. Similarly, all possible permutations among N objects x_i with $i = 1, \ldots, N$,

$$(x_1, \ldots, x_N) \to (x_{i_1}, \ldots, x_{i_N}),\tag{2.6}$$

form a group. This is the so-called S_N with $N!$ elements, and S_N is often called as the symmetric group.

Fig. 2.1 The S_3 symmetry of an equilateral triangle

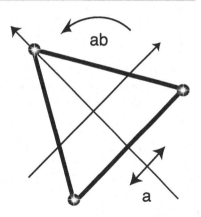

The **order** is the number of elements in G. Obviously, the order is finite in a finite group. For example, the order of the Z_N group is N, while the order of the S_N group is $N!$.

The group G is called **Abelian** if all of their elements are commutable each other, i.e. $ab = ba$ for any elements a and b in G. If all of elements do not satisfy the commutativity, the group is called **non-Abelian**. The Z_N group is Abelian, but the S_3 and S_N ($N \geq 3$) are non-Abelian. For example, for S_3 we see $a_1 a_2 \neq a_2 a_1$ in the above notation (2.3).

If a subset H of the group G is also a group, H is called the **subgroup** of G. The order of the subgroup H must be a divisor of the order of G. That is **Lagrange's theorem**. (See Appendix A.) If a subgroup N of G satisfies $g^{-1}Ng = N$ for any element $g \in G$, the subgroup N is called a **normal subgroup** or an **invariant subgroup**. The subgroup H and normal subgroup N of G satisfy $HN = NH$ and it is a subgroup of G, where HN denotes

$$\{h_i n_j | h_i \in H, n_j \in N\}, \tag{2.7}$$

and NH denotes a similar meaning.

Example

For example, the three elements, $\{e, ab, ba\}$, form a subgroup of S_3. Indeed, these elements correspond to even permutations, while the other elements, $\{a, b, bab\}$, correspond to odd permutations. This subgroup is nothing but the Z_3 group because of $(ab)^2 = ba$ and $(ab)^3 = e$. Lagrange's theorem implies that the order of subgroups in S_3 is equal to 1, 2 or 3, because the order of S_3 is $6(= 3 \times 2)$. The subgroup with order 3 corresponds to the above Z_3. In addition, the S_3 group includes three subgroups with order 2, $\{e, a\}$, $\{e, b\}$ and $\{e, bab\}$. These subgroups are Z_2 groups. Furthermore, we can find that the above Z_3 is a normal subgroup of S_3.

When $a^h = e$ for an element $a \in G$, the number h is called the **order** of a. The elements, $\{e, a, a^2, \ldots, a^{h-1}\}$, form a subgroup, which is the Abelian Z_h group with the order h.

The elements $g^{-1}ag$ for $g \in G$ are called elements conjugate to the element a. The set including all elements conjugate to an element a of G, $\{g^{-1}ag, \; \forall g \in G\}$, is called a **conjugacy class**. All of elements in a conjugacy class have the same order since

$$(gag^{-1})^h = ga(g^{-1}g)a(g^{-1}g) \cdots ag^{-1} = ga^h g^{-1} = geg^{-1} = e. \qquad (2.8)$$

The conjugacy class including the identity e consists of the single element e.

Example

All of the S_3 elements are classified to three conjugacy classes,

$$C_1 : \{e\}, \quad C_2 : \{ab, ba\}, \quad C_3 : \{a, b, bab\}. \qquad (2.9)$$

Here, the subscript of C_n, n, denotes the number of elements in the conjugacy class C_n.

We consider two groups, G and G', and a map f of G on G'. This map is **homomorphic** only if the map preserves the multiplication structure, that is,

$$f(a)f(b) = f(ab), \qquad (2.10)$$

for $a, b \in G$. Furthermore, the map is **isomorphic** when the map is one-to-one correspondence.

A **representation** of G is a homomorphic map of elements of G onto matrices, $D(g)$ for $g \in G$. The representation matrices should satisfy $D(a)D(b) = D(c)$ if $ab = c$ for $a, b, c \in G$. The vector space v_j, on which representation matrices act, is called a **representation space** such as $D(g)_{ij}v_j$ $(j = 1, \ldots, n)$. The dimension n of the vector space v_j $(j = 1, \ldots, n)$ is called as a **dimension** of the representation. A subspace in the representation space is called **invariant subspace** if $D(g)_{ij}v_j$ for any vector v_j in the subspace and any element $g \in G$ also corresponds to a vector in the same subspace. If a representation has an invariant subspace, such a representation is called **reducible**. A representation is **irreducible** if it has no invariant subspace. In particular, a representation is called **completely reducible** if $D(g)$ for $g \in G$ are written as the following block diagonal form,

$$\begin{pmatrix} D_1(g) & 0 & & \\ 0 & D_2(g) & & \\ & & \ddots & \\ & & & D_r(g) \end{pmatrix}, \qquad (2.11)$$

where each $D_\alpha(g)$ for $\alpha = 1, \ldots, r$ is irreducible. That implies that a reducible representation $D(g)$ is the direct sum of $D_\alpha(g)$,

$$\sum_{\alpha=1}^{r} \oplus D_\alpha(g). \tag{2.12}$$

Every (reducible) representation of a fine group is completely reducible. Furthermore, every representation of a fine group is equivalent to a unitary representation. (See Appendix A.) The simplest (irreducible) representation is found that $D(g) = 1$ for all elements g, that is, a trivial singlet. The matrix representations satisfy the following orthogonality relation,

$$\sum_{g \in G} D_\alpha(g)_{i\ell} D_\beta(g^{-1})_{mj} = \frac{N_G}{d_\alpha} \delta_{\alpha\beta} \delta_{ij} \delta_{\ell m}, \tag{2.13}$$

where N_G is the order of G and d_α is the dimension of the $D_\alpha(g)$. (See Appendix A.)

The **character** $\chi_D(g)$ of a representation $D(g)$ is the trace of the representation matrix,

$$\chi_D(g) = \text{tr } D(g) = \sum_{i=1}^{d_\alpha} D(g)_{ii}. \tag{2.14}$$

The elements conjugate to a have the same character because of the property of the trace,

$$\text{tr } D(g^{-1}ag) = \text{tr } \left(D(g^{-1})D(a)D(g) \right) = \text{tr } D(a), \tag{2.15}$$

that is, the characters are constant in a conjugacy class. The characters satisfy the following orthogonality relation,

$$\sum_{g \in G} \chi_{D_\alpha}(g)^* \chi_{D_\beta}(g) = N_G \delta_{\alpha\beta}, \tag{2.16}$$

where N_G denotes the order of a group G. (See Appendix A.) That is, the characters of different irreducible representations are orthogonal and different from each others. *Furthermore it is found that the number of irreducible representations must be equal to the number of conjugacy classes.* (See Appendix A.) In addition, they satisfy the following orthogonality relation,

$$\sum_{\alpha} \chi_{D_\alpha}(g_i)^* \chi_{D_\alpha}(g_j) = \frac{N_G}{n_i} \delta_{C_i C_j}, \tag{2.17}$$

where C_i denotes the conjugacy class of g_i and n_i denotes the number of elements in the conjugacy class C_i. (See Appendix A.) That is, the above equation means that the right hand side is equal $\frac{N_G}{n_i}$ if g_i and g_j belong to the same conjugacy class, and that otherwise it must vanish. A trivial singlet, $D(g) = 1$ for any $g \in G$, must always be included. Thus, the corresponding character satisfies $\chi_1(g) = 1$ for any $g \in G$.

Suppose that there are m_n n-dimensional irreducible representations, that is, $D(g)$ are represented by $(n \times n)$ matrices. The identity e is always represented by the $(n \times n)$ identity matrix. Obviously, the character $\chi_{D_\alpha}(C_1)$ for the conjugacy class $C_1 = \{e\}$ is found that $\chi_{D_\alpha}(C_1) = n$ for the n-dimensional representation. Then, the orthogonality relation (2.17) requires

$$\sum_\alpha [\chi_\alpha(C_1)]^2 = \sum_n m_n n^2 = m_1 + 4m_2 + 9m_3 + \cdots = N_G, \qquad (2.18)$$

where $m_n \geq 0$. Furthermore, m_n must satisfy

$$\sum_n m_n = \text{the number of conjugacy classes}, \qquad (2.19)$$

because the number of irreducible representations is equal to the number of conjugacy classes. Equations (2.18) and (2.19) as well as Eqs. (2.16) and (2.17) are often used in the following sections to determine characters.

Example
Let us study irreducible representations of S_3. The number of irreducible representations must be equal to three, because there are three conjugacy classes. We assume that there are m_n n-dimensional representations, that is, $D(g)$ are represented by $(n \times n)$ matrices. Here, m_n must satisfy $\sum_n m_n = 3$. Furthermore, the orthogonality relation (2.18) requires

$$\sum_\alpha [\chi_\alpha(C_1)]^2 = \sum_n m_n n^2 = m_1 + 4m_2 + 9m_3 + \cdots = 6, \qquad (2.20)$$

where $m_n \geq 0$. This equation has only two possible solutions, $(m_1, m_2) = (2, 1)$ and $(6, 0)$, but only the former $(m_1, m_2) = (2, 1)$ satisfies $m_1 + m_2 = 3$. Thus, irreducible representations of S_3 include two singlets $\mathbf{1}$ and $\mathbf{1}'$, and a doublet $\mathbf{2}$. We denote their characters by $\chi_1(g)$, $\chi_{1'}(g)$ and $\chi_2(g)$, respectively. Obviously, it is found that $\chi_1(C_1) = \chi_{1'}(C_1) = 1$ and $\chi_2(C_1) = 2$. Furthermore, one of singlet representations correspond to a trivial singlet, that is, $\chi_1(C_2) = \chi_1(C_3) = 1$. The characters, which are not fixed at this stage, are $\chi_{1'}(C_2)$, $\chi_{1'}(C_3)$, $\chi_2(C_2)$ and $\chi_2(C_3)$. Now let us determine them. For a non-trivial singlet $\mathbf{1}'$, representation matrices are nothing but characters, $\chi_{1'}(C_2)$ and $\chi_{1'}(C_3)$. They must satisfy

$$(\chi_{1'}(C_2))^3 = 1, \qquad (\chi_{1'}(C_3))^2 = 1. \qquad (2.21)$$

Thus, $\chi_{1'}(C_2)$ is one of 1, ω and ω^2, where $\omega = \exp[2\pi i/3]$, and $\chi_{1'}(C_3)$ is 1 or -1. On top of that, the orthogonality relation (2.16) requires

$$\sum_g \chi_1(g)\chi_{1'}(g) = 1 + 2\chi_{1'}(C_2) + 3\chi_{1'}(C_3) = 0. \qquad (2.22)$$

Table 2.1 Characters of S_3 representations

	h	χ_1	$\chi_{1'}$	χ_2
C_1	1	1	1	2
C_2	3	1	1	-1
C_3	2	1	-1	0

Its unique solution is obtained by $\chi_{1'}(C_2) = 1$ and $\chi_{1'}(C_3) = -1$. Furthermore, the orthogonality relations (2.16) and (2.17) require

$$\sum_g \chi_1(g)\chi_2(g) = 2 + 2\chi_2(C_2) + 3\chi_2(C_3) = 0, \qquad (2.23)$$

$$\sum_\alpha \chi_\alpha(C_1)^*\chi_\alpha(C_2) = 1 + \chi_{1'}(C_2) + 2\chi_2(C_2) = 0. \qquad (2.24)$$

Their solution is written by $\chi_2(C_2) = -1$ and $\chi_2(C_3) = 0$. These results are shown in Table 2.1.

Next, we figure out representation matrices $D(g)$ of S_3 by using the character Table 2.1. For singlets, their characters are nothing but representation matrices. Let us consider representation matrices $D(g)$ for the doublet, where $D(g)$ are (2×2) unitary matrices. Obviously, $D_2(e)$ is the (2×2) identity matrices. Because of $\chi_2(C_3) = 0$, one can diagonalize one element of the conjugacy class C_3. Here we choose e.g. a in C_3 as the diagonal element,

$$a = \begin{pmatrix} 1 & 0 \\ 0 & -1 \end{pmatrix}. \qquad (2.25)$$

The other elements in C_3 as well as C_2 are non-diagonal matrices. Recalling $b^2 = e$, we can write

$$b = \begin{pmatrix} \cos\theta & \sin\theta \\ \sin\theta & -\cos\theta \end{pmatrix}, \quad bab = \begin{pmatrix} \cos 2\theta & \sin 2\theta \\ \sin 2\theta & -\cos 2\theta \end{pmatrix}. \qquad (2.26)$$

Then, we can write elements in C_2 as

$$ab = \begin{pmatrix} \cos\theta & \sin\theta \\ -\sin\theta & \cos\theta \end{pmatrix}, \quad ba = \begin{pmatrix} \cos\theta & -\sin\theta \\ \sin\theta & \cos\theta \end{pmatrix}. \qquad (2.27)$$

Recall that the trace of elements in C_2 is equal to -1. Then, it is found that $\cos\theta = -1/2$, that is, $\theta = 2\pi/3, 4\pi/3$. When we choose $\theta = 4\pi/3$, we obtain the matrix representation of S_3 as

$$e = \begin{pmatrix} 1 & 0 \\ 0 & 1 \end{pmatrix}, \quad a = \begin{pmatrix} 1 & 0 \\ 0 & -1 \end{pmatrix}, \quad b = \begin{pmatrix} -\frac{1}{2} & -\frac{\sqrt{3}}{2} \\ -\frac{\sqrt{3}}{2} & \frac{1}{2} \end{pmatrix},$$

$$ab = \begin{pmatrix} -\frac{1}{2} & -\frac{\sqrt{3}}{2} \\ \frac{\sqrt{3}}{2} & -\frac{1}{2} \end{pmatrix}, \quad ba = \begin{pmatrix} -\frac{1}{2} & \frac{\sqrt{3}}{2} \\ -\frac{\sqrt{3}}{2} & -\frac{1}{2} \end{pmatrix}, \quad bab = \begin{pmatrix} -\frac{1}{2} & \frac{\sqrt{3}}{2} \\ \frac{\sqrt{3}}{2} & \frac{1}{2} \end{pmatrix}. \qquad (2.28)$$

The **automorphism** $\text{Aut}(G)$ of G is a one-to-one map from the element $a \in G$ to another $a' \in G$ preserving their algebraic relations, e.g. $ab = c$. One of such maps can be written by

$$a \longrightarrow a' = g^{-1}ag, \tag{2.29}$$

where g is an element of G. Then, we can realize $a'b' = c'$, where $b' = g^{-1}bg$ and $c' = g^{-1}cg$. In this map, both a and a' belong to the same conjugacy class. It is called the **inner automorphism**, $\text{Inn}(G)$. If an automorphism can not be written by conjugate maps, it is called the **outer automorphism** $\text{Out}(G)$,

$$\text{Out}(G) = \text{Aut}(G)/\text{Inn}(G). \tag{2.30}$$

That is, the outer automorphism is a map from a conjugacy class to another. One can find such symmetries in character tables. While the inner automorphism is a symmetry among elements in a multiplet, the outer automorphism is a symmetry among different multiplets.

Example: Z_3

The Z_3 group (2.1) is Abelian, and it satisfies $g^{-1}ag = a$ for any $g \in Z_3$. Thus, the inner automorphism is trivial. One can map $a = e^{2\pi i/3} \rightarrow a' = a^2 = e^{4\pi i/3}$ with preserving the multiplication rule, $a^m a^n = a^{m+n} \rightarrow a'^m a'^n = a'^{m+n}$. Thus, the Z_3 group has the Z_2 outer automorphism, $\text{Aut}(Z_3) = \text{Out}(Z_3) \simeq Z_2$.

Example: S_3

In the S_3 group, there are non-trivial maps (2.29), and all of g in S_3 lead to non-trivial maps, that is, $\text{Inn}(S_3) \simeq S_3$. In many non-Abelian groups, we have $\text{Inn}(G) \simeq G$. However, in some groups the inner automorphism $\text{Inn}(G)$ is smaller than G. If G has a center r, it commutes with all of the elements in G. Then, the above map by $g = r$ is trivial like Abelian symmetries for such a group G. Thus, the inner automorphism does not include the center $Z(G)$, i.e. $\text{Inn}(G) \simeq G/Z(G)$.

We can construct a lager group from more than two groups, G_i, by a certain product. A rather simple one is the **direct product**. We consider e.g. two groups G_1 and G_2. Their direct product is denoted as $G_1 \times G_2$, and its multiplication rule is defined as

$$(a_1, a_2)(b_1, b_2) = (a_1 b_1, a_2 b_2), \tag{2.31}$$

for $a_1, b_1 \in G_1$ and $a_2, b_2 \in G_2$.

The **semi-direct product** is more non-trivial product between two groups G_1 and G_2, and it is defined such as

$$(a_1, a_2)(b_1, b_2) = (a_1 f_{a_2}(b_1), a_2 b_2), \tag{2.32}$$

for $a_1, b_1 \in G_1$ and $a_2, b_2 \in G_2$, where $f_{a_2}(b_1)$ denotes a homomorphic map from G_2 to the automorphism of G_1, $\text{Aut}\, G_1$. This semi-direct product is denoted as

$G_1 \times_f G_2$. We consider the group G and its subgroup H and normal subgroup N, whose elements are h_i and n_j, respectively. When $G = NH = HN$ and $N \cap H = \{e\}$, the semi-direct product $N \times_f H$ is isomorphic to G, $G \simeq N \times_f H$, where we use the map f as

$$f_{h_i}(n_j) = h_i n_j (h_i)^{-1}. \tag{2.33}$$

For the notation of the semi-direct product, we will often omit f and denote it as $N \rtimes H$.

Example

Let us study the semi-direct product, $Z_3 \rtimes Z_2$. Here we denote the Z_3 and Z_2 generators by c and h, i.e., $c^3 = e$ and $h^2 = e$. In this case, Eq. (2.33) can be written by

$$hch^{-1} = c^m, \tag{2.34}$$

where $m \neq 0$, because all of the Z_3 elements can be written by c^m and the case with $m = 0$ is inconsistent. When $m = 1$, the above relation is trivial and that leads just to the direct product, $Z_3 \times Z_2$. Thus, only the case with $m = 2$ is non-trivial, i.e.,

$$hch^{-1} = c^2. \tag{2.35}$$

Indeed, this algebra is isomorphic to S_3, and h and c are identified as a and ab, respectively. Similarly, we can consider the $Z_n \rtimes Z_m$. When we denote the Z_n and Z_m generators by a and b, respectively, they satisfy

$$a^n = b^m = e, \qquad bab^{-1} = a^k, \tag{2.36}$$

where $k \neq 0$, although the case with $k = 1$ leads to the direct product $Z_n \times Z_m$.

References

1. Miller, G.A., Dickson, H.F., Blichfeldt, L.E.: Theory and Applications of Finite Groups. Wiley. New York (1916)
2. Hamermesh, M.: Group Theory and its Application to Physical Problems. Mass, Addison-Wesley, Reading (1962)
3. Georgi, H.: Lie algebras in particle physics. From isospin to unified theories. Front. Phys. **54**, 1 (1982)
4. Ramond, P.: Group Theory: A Physicist's Survey. Cambridge University Press (2010)
5. Ludl, P.O.: arXiv:0907.5587 [hep-ph]
6. Grimus, W., Ludl, P.O.: J. Phys. A **45**, 233001 (2012). [arXiv:1110.6376 [hep-ph]]

As introduced in the previous section, the symmetric group S_N consists of all possible permutations among N objects x_i with $i = 1, \cdots, N$,

$$(x_1, \cdots, x_N) \rightarrow (x_{i_1}, \cdots, x_{i_N}). \tag{3.1}$$

The group S_2 consists of two permutations,

$$(x_1, x_2) \rightarrow (x_1, x_2), \qquad (x_1, x_2) \rightarrow (x_2, x_1). \tag{3.2}$$

This group is nothing but the Abelian Z_2 group. Therefore, we study simple examples for $N = 3$ and 4, that is, S_3 and S_4.

3.1 S_3

We begin with S_3. Since some aspects of S_3 have been already discussed in Sect. 3.2, we summarize them briefly and study other aspects such as tensor products.

The S_3 group consists of all permutations among three objects, (x_1, x_2, x_3) and its order is equal to $3! = 6$. All of six elements are written as products of the element a and b

$$a : (x_1, x_2, x_3) \rightarrow (x_2, x_1, x_3),$$
$$b : (x_1, x_2, x_3) \rightarrow (x_3, x_2, x_1), \tag{3.3}$$

as well as the identity e, that is,

$$\{e, a, b, ab, ba, bab\}. \tag{3.4}$$

© The Author(s), under exclusive license to Springer-Verlag GmbH, DE,
part of Springer Nature 2022
T. Kobayashi et al., *An Introduction to Non-Abelian Discrete Symmetries for Particle Physicists*, Lecture Notes in Physics 995,
https://doi.org/10.1007/978-3-662-64679-3_3

• Conjugacy classes

As studied in Sect. 3.2, the S_3 group has the following three conjugacy classes:

$$C_1 : \{e\}, \quad C_2 : \{ab, ba\}, \quad C_3 : \{a, b, bab\}. \tag{3.5}$$

Their orders are found as

$$(ab)^3 = (ba)^3 = e, \quad a^2 = b^2 = (bab)^2 = e. \tag{3.6}$$

The elements $\{e, ab, ba\}$ correspond to even permutations, while the elements $\{a, b, bab\}$ are odd permutations.

• Characters and representations

The characters and representations are studied in Chap. 2. The group S_3 has two singlets **1** and **1'**, and a doublet **2**. Their characters are summarized in Table 2.1. Those characters for the singlets correspond to the representations on the singlets. In addition, the doublet representations are obtained in Eq. (2.28).

• Tensor products

Here, we consider tensor products of irreducible representations. Let us start with discussing the tensor products of two doublets, (x_1, x_2) and (y_1, y_2). For example, each element $x_i y_j$ is transformed under b as

$$
\begin{aligned}
x_1 y_1 &\rightarrow \frac{x_1 y_1 + 3 x_2 y_2 + \sqrt{3}(x_1 y_2 + x_2 y_1)}{4}, \\
x_1 y_2 &\rightarrow \frac{\sqrt{3} x_1 y_1 - \sqrt{3} x_2 y_2 - x_1 y_2 + 3 x_2 y_1}{4}, \\
x_2 y_1 &\rightarrow \frac{\sqrt{3} x_1 y_1 - \sqrt{3} x_2 y_2 - x_2 y_1 + 3 x_1 y_2}{4}, \\
x_2 y_2 &\rightarrow \frac{3 x_1 y_1 + x_2 y_2 - \sqrt{3}(x_1 y_2 + x_2 y_1)}{4}.
\end{aligned}
\tag{3.7}
$$

Thus, it is found that

$$b(x_1 y_1 + x_2 y_2) = (x_1 y_1 + x_2 y_2), \quad b(x_1 y_2 - x_2 y_1) = -(x_1 y_2 - x_2 y_1). \tag{3.8}$$

Therefore, these linear combinations correspond to the singlets,

$$\mathbf{1} : x_1 y_1 + x_2 y_2, \quad \mathbf{1'} : x_1 y_2 - x_2 y_1. \tag{3.9}$$

Furthermore, it is found that

$$
b \begin{pmatrix} x_2 y_2 - x_1 y_1 \\ x_1 y_2 + x_2 y_1 \end{pmatrix} = \begin{pmatrix} -\frac{1}{2} & -\frac{\sqrt{3}}{2} \\ -\frac{\sqrt{3}}{2} & \frac{1}{2} \end{pmatrix} \begin{pmatrix} x_2 y_2 - x_1 y_1 \\ x_1 y_2 + x_2 y_1 \end{pmatrix}. \tag{3.10}
$$

Hence, $(x_2 y_2 - x_2 y_2, x_1 y_2 + x_2 y_1)$ corresponds to the doublet, i.e.

$$\mathbf{2} = \begin{pmatrix} x_2 y_2 - x_1 y_1 \\ x_1 y_2 + x_2 y_1 \end{pmatrix}. \tag{3.11}$$

Similarly, we can study the tensor product of the doublet (x_1, x_2) and the singlet $\mathbf{1'}\, y'$. Their products $x_i y'$ transform under b as

$$x_1 y' \to \frac{1}{2} x_1 y' + \frac{\sqrt{3}}{2} x_2 y',$$
$$x_2 y' \to \frac{\sqrt{3}}{2} x_1 y' - \frac{1}{2} x_2 y'. \tag{3.12}$$

That is, those form a doublet,

$$\mathbf{2} : \begin{pmatrix} -x_2 y' \\ x_1 y' \end{pmatrix}. \tag{3.13}$$

These tensor products are summarized as follows,

$$\begin{pmatrix} x_1 \\ x_2 \end{pmatrix}_2 \otimes \begin{pmatrix} y_1 \\ y_2 \end{pmatrix}_2 = (x_1 y_1 + x_2 y_2)_1 + (x_1 y_2 - x_2 y_1)_{1'} + \begin{pmatrix} x_2 y_2 - x_1 y_1 \\ x_1 y_2 + x_2 y_1 \end{pmatrix}_2,$$
$$\begin{pmatrix} x_1 \\ x_2 \end{pmatrix}_2 \otimes (y')_1 = \begin{pmatrix} x_1 y' \\ x_2 y' \end{pmatrix}_2, \quad \begin{pmatrix} x_1 \\ x_2 \end{pmatrix}_2 \otimes (y')_{1'} = \begin{pmatrix} -x_2 y' \\ x_1 y' \end{pmatrix}_2, \tag{3.14}$$
$$(x)_1 \otimes (y)_{1'} = (xy)_{1'}, \quad (x)_{1'} \otimes (y)_{1'} = (xy)_1.$$

Obviously the tensor product of two trivial singlets corresponds to a trivial singlet.

Tensor products are important to applications for particle phenomenology. Matter and Higgs fields may be assigned to have certain representations of discrete symmetries. The Lagrangian must be invariant under discrete symmetries. That implies that n-point couplings corresponding to a trivial singlet can appear in Lagrangian.

In addition to the above (real) representation of S_3, another representation, i.e. the complex representation, is often used in the literature. Here, we mention about changing representation bases. All permutations of S_3 in Eq. (3.1) are represented on the reducible triplet (x_1, x_2, x_3) as

$$\begin{pmatrix} 1 & 0 & 0 \\ 0 & 1 & 0 \\ 0 & 0 & 1 \end{pmatrix}, \begin{pmatrix} 1 & 0 & 0 \\ 0 & 0 & 1 \\ 0 & 1 & 0 \end{pmatrix}, \begin{pmatrix} 0 & 1 & 0 \\ 1 & 0 & 0 \\ 0 & 0 & 1 \end{pmatrix},$$
$$\begin{pmatrix} 0 & 1 & 0 \\ 0 & 0 & 1 \\ 1 & 0 & 0 \end{pmatrix}, \begin{pmatrix} 0 & 0 & 1 \\ 0 & 1 & 0 \\ 1 & 0 & 0 \end{pmatrix}, \begin{pmatrix} 0 & 0 & 1 \\ 1 & 0 & 0 \\ 0 & 1 & 0 \end{pmatrix}. \tag{3.15}$$

We change the representation through the unitary transformation, $U^\dagger g U$, e.g. by using the following unitary matrix,

$$U = \begin{pmatrix} 1/\sqrt{3} & \sqrt{2/3} & 0 \\ 1/\sqrt{3} & -1/\sqrt{6} & -1/\sqrt{2} \\ 1/\sqrt{3} & -1/\sqrt{6} & 1/\sqrt{2} \end{pmatrix}. \tag{3.16}$$

Then, the six elements of S_3 are written as

$$\begin{pmatrix} 1 & 0 & 0 \\ 0 & 1 & 0 \\ 0 & 0 & 1 \end{pmatrix}, \begin{pmatrix} 1 & 0 & 0 \\ 0 & 1 & 0 \\ 0 & 0 & -1 \end{pmatrix}, \begin{pmatrix} 1 & 0 & 0 \\ 0 & -\frac{1}{2} & -\frac{\sqrt{3}}{2} \\ 0 & -\frac{\sqrt{3}}{2} & \frac{1}{2} \end{pmatrix},$$

$$\begin{pmatrix} 1 & 0 & 0 \\ 0 & -\frac{1}{2} & -\frac{\sqrt{3}}{2} \\ 0 & \frac{\sqrt{3}}{2} & -\frac{1}{2} \end{pmatrix}, \begin{pmatrix} 1 & 0 & 0 \\ 0 & -\frac{1}{2} & \frac{\sqrt{3}}{2} \\ 0 & \frac{\sqrt{3}}{2} & \frac{1}{2} \end{pmatrix}, \begin{pmatrix} 1 & 0 & 0 \\ 0 & -\frac{1}{2} & \frac{\sqrt{3}}{2} \\ 0 & -\frac{\sqrt{3}}{2} & -\frac{1}{2} \end{pmatrix}. \tag{3.17}$$

Note that this form is completely reducible and that the (right-bottom) (2×2) submatrices are exactly the same as those for the doublet representation (2.28).

We can use another unitary matrix U in order to obtain a completely reducible form from the reducible representation matrices (3.15). For example, let us use the following one for the unitary matrix,

$$U_w = \frac{1}{\sqrt{3}} \begin{pmatrix} 1 & 1 & 1 \\ 1 & w & w^2 \\ 1 & w^2 & w \end{pmatrix}, \tag{3.18}$$

which is called the magic matrix. Then, the six elements of S_3 are written as

$$\begin{pmatrix} 1 & 0 & 0 \\ 0 & 1 & 0 \\ 0 & 0 & 1 \end{pmatrix}, \begin{pmatrix} 1 & 0 & 0 \\ 0 & 0 & 1 \\ 0 & 1 & 0 \end{pmatrix}, \begin{pmatrix} 1 & 0 & 0 \\ 0 & 0 & w^2 \\ 0 & w & 0 \end{pmatrix},$$

$$\begin{pmatrix} 1 & 0 & 0 \\ 0 & w & 0 \\ 0 & 0 & w^2 \end{pmatrix}, \begin{pmatrix} 1 & 0 & 0 \\ 0 & 0 & w \\ 0 & w^2 & 0 \end{pmatrix}, \begin{pmatrix} 1 & 0 & 0 \\ 0 & w^2 & 0 \\ 0 & 0 & w \end{pmatrix}. \tag{3.19}$$

The (right-bottom) (2×2) submatrices correspond to the doublet representation in the different basis, that is, the complex representation. In different bases, the multiplication rule does not change. For example, we obtain $2 \times 2 = 1 + 1' + 2$ in both the real and complex bases. However, elements of doublets in the left hand side are written in a different way.

3.2 S_4

We now discuss the S_4 group, which consists of all permutations among four objects, (x_1, x_2, x_3, x_4),

$$(x_1, x_3, x_2, x_4), \quad \rightarrow \quad (x_i, x_j, x_k, x_l). \tag{3.20}$$

The order of S_4 is equal to $4! = 24$. We denote all of S_4 elements as

$$
\begin{aligned}
&a_1 : (x_1, x_3, x_2, x_4), \; a_2 : (x_2, x_1, x_4, x_3), \; a_3 : (x_3, x_4, x_1, x_2), \; a_4 : (x_4, x_3, x_2, x_1), \\
&b_1 : (x_1, x_4, x_2, x_3), \; b_2 : (x_4, x_1, x_3, x_2), \; b_3 : (x_2, x_3, x_1, x_4), \; b_4 : (x_3, x_2, x_4, x_1), \\
&c_1 : (x_1, x_3, x_4, x_2), \; c_2 : (x_3, x_1, x_2, x_4), \; c_3 : (x_4, x_2, x_1, x_3), \; c_4 : (x_2, x_4, x_3, x_1), \\
&d_1 : (x_1, x_2, x_4, x_3), \; d_2 : (x_2, x_1, x_3, x_4), \; d_3 : (x_4, x_3, x_1, x_2), \; d_4 : (x_3, x_4, x_2, x_1), \\
&e_1 : (x_1, x_3, x_2, x_4), \; e_2 : (x_3, x_1, x_4, x_2), \; e_3 : (x_2, x_4, x_1, x_3), \; e_4 : (x_4, x_2, x_3, x_1), \\
&f_1 : (x_1, x_4, x_3, x_2), \; f_2 : (x_4, x_1, x_2, x_3), \; f_3 : (x_3, x_2, x_1, x_4), \; f_4 : (x_2, x_3, x_4, x_1),
\end{aligned}
\tag{3.21}
$$

where we have shown the ordering of four objects after permutations. The S_4 is a symmetry of a cube as shown in Fig. 3.1.

It is obvious that $x_1 + x_2 + x_3 + x_4$ is invariant under any permutation of S_4, that is, a trivial singlet. Thus, we use the vector space, which is orthogonal to this singlet direction,

$$
3 : \begin{pmatrix} A_x \\ A_y \\ A_z \end{pmatrix} = \begin{pmatrix} x_1 + x_2 - x_3 - x_4 \\ x_1 - x_2 + x_3 - x_4 \\ x_1 - x_2 - x_3 + x_4 \end{pmatrix}, \tag{3.22}
$$

Fig. 3.1 The S_4 symmetry of a cube. This figure shows the transformations corresponding to the S_4 elements with $h = 2, 3$ and 4. Note that the group can be also considered as the regular octahedron in a way similar to a cube

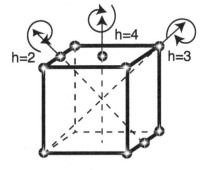

in order to construct matrix representations of S_4, that is, a triplet representation. In this triplet vector space, all of S_4 elements are represented by the following matrices,

$$
a_1 = \begin{pmatrix} 1 & 0 & 0 \\ 0 & 1 & 0 \\ 0 & 0 & 1 \end{pmatrix}, \quad a_2 = \begin{pmatrix} 1 & 0 & 0 \\ 0 & -1 & 0 \\ 0 & 0 & -1 \end{pmatrix},
$$

$$
a_3 = \begin{pmatrix} -1 & 0 & 0 \\ 0 & 1 & 0 \\ 0 & 0 & -1 \end{pmatrix}, \quad a_4 = \begin{pmatrix} -1 & 0 & 0 \\ 0 & -1 & 0 \\ 0 & 0 & 1 \end{pmatrix},
$$

$$
b_1 = \begin{pmatrix} 0 & 0 & 1 \\ 1 & 0 & 0 \\ 0 & 1 & 0 \end{pmatrix}, \quad b_2 = \begin{pmatrix} 0 & 0 & 1 \\ -1 & 0 & 0 \\ 0 & -1 & 0 \end{pmatrix},
$$

$$
b_3 = \begin{pmatrix} 0 & 0 & -1 \\ 1 & 0 & 0 \\ 0 & -1 & 0 \end{pmatrix}, \quad b_4 = \begin{pmatrix} 0 & 0 & -1 \\ -1 & 0 & 0 \\ 0 & 1 & 0 \end{pmatrix},
$$

$$
c_1 = \begin{pmatrix} 0 & 1 & 0 \\ 0 & 0 & 1 \\ 1 & 0 & 0 \end{pmatrix}, \quad c_2 = \begin{pmatrix} 0 & 1 & 0 \\ 0 & 0 & -1 \\ -1 & 0 & 0 \end{pmatrix},
$$

$$
c_3 = \begin{pmatrix} 0 & -1 & 0 \\ 0 & 0 & 1 \\ -1 & 0 & 0 \end{pmatrix}, \quad c_4 = \begin{pmatrix} 0 & -1 & 0 \\ 0 & 0 & -1 \\ 1 & 0 & 0 \end{pmatrix},
$$

$$
d_1 = \begin{pmatrix} 1 & 0 & 0 \\ 0 & 0 & 1 \\ 0 & 1 & 0 \end{pmatrix}, \quad d_2 = \begin{pmatrix} 1 & 0 & 0 \\ 0 & 0 & -1 \\ 0 & -1 & 0 \end{pmatrix}, \tag{3.23}
$$

$$
d_3 = \begin{pmatrix} -1 & 0 & 0 \\ 0 & 0 & 1 \\ 0 & -1 & 0 \end{pmatrix}, \quad d_4 = \begin{pmatrix} -1 & 0 & 0 \\ 0 & 0 & -1 \\ 0 & 1 & 0 \end{pmatrix},
$$

$$
e_1 = \begin{pmatrix} 0 & 1 & 0 \\ 1 & 0 & 0 \\ 0 & 0 & 1 \end{pmatrix}, \quad e_2 = \begin{pmatrix} 0 & 1 & 0 \\ -1 & 0 & 0 \\ 0 & 0 & -1 \end{pmatrix},
$$

$$
e_3 = \begin{pmatrix} 0 & -1 & 0 \\ 1 & 0 & 0 \\ 0 & 0 & -1 \end{pmatrix}, \quad e_4 = \begin{pmatrix} 0 & -1 & 0 \\ -1 & 0 & 0 \\ 0 & 0 & 1 \end{pmatrix},
$$

$$
f_1 = \begin{pmatrix} 0 & 0 & 1 \\ 0 & 1 & 0 \\ 1 & 0 & 0 \end{pmatrix}, \quad f_2 = \begin{pmatrix} 0 & 0 & 1 \\ 0 & -1 & 0 \\ -1 & 0 & 0 \end{pmatrix},
$$

$$
f_3 = \begin{pmatrix} 0 & 0 & -1 \\ 0 & 1 & 0 \\ -1 & 0 & 0 \end{pmatrix}, \quad f_4 = \begin{pmatrix} 0 & 0 & -1 \\ 0 & -1 & 0 \\ 1 & 0 & 0 \end{pmatrix}.
$$

• **Conjugacy classes**

The S_4 elements can be classified by the order h of each element, where $a^h = e$, as

$$
\begin{aligned}
h &= 1 &&: && \{a_1\}, \\
h &= 2 &&: && \{a_2, a_3, a_4, d_1, d_2, e_1, e_4, f_1, f_3\}, \\
h &= 3 &&: && \{b_1, b_2, b_3, b_4, c_1, c_2, c_3, c_4\}, \\
h &= 4 &&: && \{d_3, d_4, e_2, e_3, f_2, f_4\}.
\end{aligned}
\tag{3.24}
$$

Moreover, they are classified by the conjugacy classes as

$$
\begin{aligned}
C_1 &: & \{a_1\}, & & h &= 1, \\
C_3 &: & \{a_2, a_3, a_4\}, & & h &= 2, \\
C_6 &: & \{d_1, d_2, e_1, e_4, f_1, f_3\}, & & h &= 2, \\
C_8 &: & \{b_1, b_2, b_3, b_4, c_1, c_2, c_3, c_4\}, & & h &= 3, \\
C_{6'} &: & \{d_3, d_4, e_2, e_3, f_2, f_4\}, & & h &= 4.
\end{aligned}
\tag{3.25}
$$

• **Characters and representations**

The group S_4 includes five conjugacy classes, that is, there are five irreducible representations. For example, all of elements are written as multiplications of b_1 in C_8 and d_4 in $C_{6'}$, which satisfy

$$
(b_1)^3 = e, \quad (d_4)^4 = e, \quad d_4(b_1)^2 d_4 = b_1, \quad d_4 b_1 d_4 = b_1(d_4)^2 b_1. \tag{3.26}
$$

It is straightforward to write $a_i, b_i, c_i, d_i, e_i, f_i, (i = 1 - 4)$ in terms of b_1 and d_4. Then, the conjugacy classes can be reexpressed as

$$
\begin{aligned}
C_1 &: & \{e\}, & & h &= 1, \\
C_3 &: & \{d_4^2, d_4 b_1 d_4^2 b_1^2 d_4, d_4 b_1 d_4^2 b_1^2 d_4^3\}, & & h &= 2, \\
C_6 &: & \{d_4^2 b_1 d_4 b_1, b_1 d_4 b_1, d_4^2 b_1 d_4, b_1 d_4^3, d_4 b_1 d_4^2, b_1^2 d_4\}, & & h &= 2, \\
C_8 &: & \{b_1, d_4^2 b_1, b_1 d_4^2, d_4^2 b_1 d_4^2, b_1^2, d_4^2 b_1^2, d_4^2 b_1^2 d_4^2, b_1^2 d_4^2\}, & & h &= 3, \\
C_{6'} &: & \{d_4^3, d_4, b_1 d_4, d_4^3 b_1^2, b_1^2 d_4^3, d_4 b_1\}, & & h &= 4.
\end{aligned}
\tag{3.27}
$$

The orthogonality relation (2.18) requires

$$
\sum_\alpha [\chi_\alpha(C_1)]^2 = \sum_n m_n n^2 = m_1 + 4m_2 + 9m_3 + \cdots = 24, \tag{3.28}
$$

like Eq. (2.20), and m_n also satisfy $m_1 + m_2 + m_3 + \cdots = 5$, because there must be five irreducible representations. Then, we can easily find the unique solution as $(m_1, m_2, m_3) = (2, 1, 2)$. Therefore, irreducible representations of S_4 include two singlets **1** and **1'**, one doublet **2**, and two triplets **3** and **3'**, where **1** corresponds to a trivial singlet and **3** corresponds to (3.22) and (3.23). We can compute the character for each representation by an analysis similar to S_3. The characters are shown in Table 3.1.

Table 3.1 Characters of S_4 representations

	h	χ_1	$\chi_{1'}$	χ_2	χ_3	$\chi_{3'}$
C_1	1	1	1	2	3	3
C_3	2	1	1	2	-1	-1
C_6	2	1	-1	0	1	-1
$C_{6'}$	4	1	-1	0	-1	1
C_8	3	1	1	-1	0	0

For **2**, the representation matrices are written as e.g.

$$a_2(\mathbf{2}) = \begin{pmatrix} 1 & 0 \\ 0 & 1 \end{pmatrix}, \quad b_1(\mathbf{2}) = \begin{pmatrix} \omega & 0 \\ 0 & \omega^2 \end{pmatrix},$$

$$d_1(\mathbf{2}) = d_3(\mathbf{2}) = d_4(\mathbf{2}) = \begin{pmatrix} 0 & 1 \\ 1 & 0 \end{pmatrix}. \tag{3.29}$$

For **3′**, the representation matrices are written as e.g.

$$a_2(\mathbf{3'}) = \begin{pmatrix} 1 & 0 & 0 \\ 0 & -1 & 0 \\ 0 & 0 & -1 \end{pmatrix}, \quad b_1(\mathbf{3'}) = \begin{pmatrix} 0 & 0 & 1 \\ 1 & 0 & 0 \\ 0 & 1 & 0 \end{pmatrix}, \tag{3.30}$$

$$d_1(\mathbf{3'}) = \begin{pmatrix} -1 & 0 & 0 \\ 0 & 0 & -1 \\ 0 & -1 & 0 \end{pmatrix}, \quad d_3(\mathbf{3'}) = \begin{pmatrix} 1 & 0 & 0 \\ 0 & 0 & -1 \\ 0 & 1 & 0 \end{pmatrix}, \quad d_4(\mathbf{3'}) = \begin{pmatrix} 1 & 0 & 0 \\ 0 & 0 & 1 \\ 0 & -1 & 0 \end{pmatrix}.$$

It is noted that $a_2(\mathbf{3'}) = a_2(\mathbf{3})$ and $b_1(\mathbf{3'}) = b_1(\mathbf{3})$, but $d_1(\mathbf{3'}) = -d_1(\mathbf{3})$, $d_3(\mathbf{3'}) = -d_3(\mathbf{3})$ and $d_4(\mathbf{3'}) = -d_4(\mathbf{3})$. This aspect would be obvious from the above character table.

• **Tensor products**

Finally, we present the tensor products. The tensor products of $\mathbf{3} \times \mathbf{3}$ can be decomposed as

$$(\mathbf{A})_\mathbf{3} \times (\mathbf{B})_\mathbf{3} = (\mathbf{A} \cdot \mathbf{B})_\mathbf{1} + \begin{pmatrix} \mathbf{A} \cdot \Sigma \cdot \mathbf{B} \\ \mathbf{A} \cdot \Sigma^* \cdot \mathbf{B} \end{pmatrix}_\mathbf{2} + \begin{pmatrix} \{A_y B_z\} \\ \{A_z B_x\} \\ \{A_x B_y\} \end{pmatrix}_\mathbf{3} + \begin{pmatrix} [A_y B_z] \\ [A_z B_x] \\ [A_x B_y] \end{pmatrix}_\mathbf{3'},$$

$$\tag{3.31}$$

where

$$\mathbf{A} \cdot \mathbf{B} = A_x B_x + A_y B_y + A_z B_z,$$
$$\{A_i B_j\} = A_i B_j + A_j B_i,$$
$$[A_i B_j] = A_i B_j - A_j B_i, \tag{3.32}$$
$$\mathbf{A} \cdot \Sigma \cdot \mathbf{B} = A_x B_x + \omega A_y B_y + \omega^2 A_z B_z,$$
$$\mathbf{A} \cdot \Sigma^* \cdot \mathbf{B} = A_x B_x + \omega^2 A_y B_y + \omega A_z B_z.$$

The tensor products of other representations are also decomposed as e.g.

$$(\mathbf{A})_{3'} \times (\mathbf{B})_{3'} = (\mathbf{A} \cdot \mathbf{B})_1 + \begin{pmatrix} \mathbf{A} \cdot \Sigma \cdot \mathbf{B} \\ \mathbf{A} \cdot \Sigma^* \cdot \mathbf{B} \end{pmatrix}_2 + \begin{pmatrix} \{A_y B_z\} \\ \{A_z B_x\} \\ \{A_x B_y\} \end{pmatrix}_3 + \begin{pmatrix} [A_y B_z] \\ [A_z B_x] \\ [A_x B_y] \end{pmatrix}_{3'} , \qquad (3.33)$$

$$(\mathbf{A})_3 \times (\mathbf{B})_{3'} = (\mathbf{A} \cdot \mathbf{B})_{1'} + \begin{pmatrix} \mathbf{A} \cdot \Sigma \cdot \mathbf{B} \\ -\mathbf{A} \cdot \Sigma^* \cdot \mathbf{B} \end{pmatrix}_2 + \begin{pmatrix} \{A_y B_z\} \\ \{A_z B_x\} \\ \{A_x B_y\} \end{pmatrix}_{3'} + \begin{pmatrix} [A_y B_z] \\ [A_z B_x] \\ [A_x B_y] \end{pmatrix}_3 , \qquad (3.34)$$

and

$$(\mathbf{A})_2 \times (\mathbf{B})_2 = \{A_x B_y\}_1 + [A_x B_y]_{1'} + \begin{pmatrix} A_y B_y \\ A_x B_x \end{pmatrix}_2 , \qquad (3.35)$$

$$\begin{pmatrix} A_x \\ A_y \end{pmatrix}_2 \times \begin{pmatrix} B_x \\ B_y \\ B_z \end{pmatrix}_3 = \begin{pmatrix} (A_x + A_y)B_x \\ (\omega^2 A_x + \omega A_y)B_y \\ (\omega A_x + \omega^2 A_y)B_z \end{pmatrix}_3 + \begin{pmatrix} (A_x - A_y)B_x \\ (\omega^2 A_x - \omega A_y)B_y \\ (\omega A_x - \omega^2 A_y)B_z \end{pmatrix}_{3'} , \qquad (3.36)$$

$$\begin{pmatrix} A_x \\ A_y \end{pmatrix}_2 \times \begin{pmatrix} B_x \\ B_y \\ B_z \end{pmatrix}_{3'} = \begin{pmatrix} (A_x + A_y)B_x \\ (\omega^2 A_x + \omega A_y)B_y \\ (\omega A_x + \omega^2 A_y)B_z \end{pmatrix}_{3'} + \begin{pmatrix} (A_x - A_y)B_x \\ (\omega^2 A_x - \omega A_y)B_y \\ (\omega A_x - \omega^2 A_y)B_z \end{pmatrix}_3 . \qquad (3.37)$$

$$a_1 \otimes \begin{pmatrix} B_x \\ B_y \\ B_z \end{pmatrix}_{3,3'} = \begin{pmatrix} a B_x \\ a B_y \\ a B_z \end{pmatrix}_{3,3'} , \qquad a_{1'} \otimes \begin{pmatrix} B_x \\ B_y \\ B_z \end{pmatrix}_{3,3'} = \begin{pmatrix} a B_x \\ a B_y \\ a B_z \end{pmatrix}_{3',3} , \qquad (3.38)$$

$$a_1 \otimes \begin{pmatrix} B_x \\ B_y \end{pmatrix}_2 = \begin{pmatrix} a B_x \\ a B_y \end{pmatrix}_2 , \qquad a_{1'} \otimes \begin{pmatrix} B_x \\ B_y \end{pmatrix}_2 = \begin{pmatrix} a B_x \\ -a B_y \end{pmatrix}_2 . \qquad (3.39)$$

In the literature, several bases are used for S_4. The decomposition of tensor products, $\mathbf{r} \times \mathbf{r}' = \sum_m \mathbf{r}_m$, does not depend on the basis. For example, we obtain $\mathbf{3} \times \mathbf{3}' = \mathbf{1}' + \mathbf{2} + \mathbf{3} + \mathbf{3}'$ in any basis. However, the multiplication rules written by components depend on the basis, which we use. We have used the basis (3.29). In Appendix B, we show the relations between several bases and give explicitly the multiplication rules in terms of components.

Similarly, we can study the S_N group with $N > 4$. Here we give a brief comment on such groups. The S_N group with $N > 4$ has only one invariant subgroup, that is, the alternating group, A_N. The S_N group has two one-dimensional representations: one is trivial singlet, that is, invariant under all the elements (symmetric representation), the other is pseudo singlet, that is, symmetric under the even permutation-elements but antisymmetric under the odd permutation-elements. Group-theoretical aspects of S_5 are derived from those of S_4 by applying a theorem of Frobenuis (Frobenuis formula), graphical method (Young tableaux), recursion formulas for characters (branching laws). The details are given in, e.g., the text book of [1]. Such analysis would be extended recursively from S_N to S_{N+1}.

Reference

1. Hamermesh, M.: Group Theory and its Application to Physical Problems. Mass, Addison-Wesley, Reading (1962)

In this section, we study the A_N group. All even permutations among S_N form a group, which is A_N. It is also called the alternating group. Therefore, the order of this group is $(N!)/2$. Let us consider a simple example. In S_3 in Sect. 3.1 the even permutations include

$$
\begin{aligned}
e &: (x_1, x_2, x_3) \rightarrow (x_1, x_2, x_3), \\
a_4 &: (x_1, x_2, x_3) \rightarrow (x_3, x_1, x_2), \\
a_5 &: (x_1, x_2, x_3) \rightarrow (x_2, x_3, x_1),
\end{aligned}
\tag{4.1}
$$

while the odd permutations include

$$
\begin{aligned}
a_1 &: (x_1, x_2, x_3) \rightarrow (x_2, x_1, x_3), \\
a_2 &: (x_1, x_2, x_3) \rightarrow (x_3, x_2, x_1), \\
a_3 &: (x_1, x_2, x_3) \rightarrow (x_1, x_3, x_2).
\end{aligned}
\tag{4.2}
$$

The three elements of even permutations, $\{e, a_4, a_5\}$ form the group, which is A_3. Since $(a_4)^2 = a_5$ and $(a_4)^3 = e$, the group A_3 is nothing but Z_3. Therefore, we start with studying A_4, which is the smallest non-Abelian group with a triplet irreducible representation.

4.1 A_4

The A_4 group is formed by all even permutations of S_4. Thus, its order is equal to $(4!)/2 = 12$. The A_4 group is the symmetry of a tetrahedron as shown in Fig. 4.1. Thus, the A_4 group is often denoted as T. Using the notation in Sect. 3.2, all of 12 elements are denoted as

© The Author(s), under exclusive license to Springer-Verlag GmbH, DE,
part of Springer Nature 2022
T. Kobayashi et al., *An Introduction to Non-Abelian Discrete Symmetries for Particle Physicists*, Lecture Notes in Physics 995,
https://doi.org/10.1007/978-3-662-64679-3_4

Fig. 4.1 The A_4 symmetry of tetrahedron

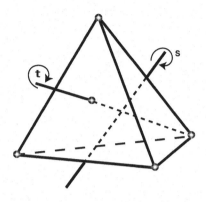

$$
a_1 = \begin{pmatrix} 1 & 0 & 0 \\ 0 & 1 & 0 \\ 0 & 0 & 1 \end{pmatrix}, \quad
a_2 = \begin{pmatrix} 1 & 0 & 0 \\ 0 & -1 & 0 \\ 0 & 0 & -1 \end{pmatrix},
$$

$$
a_3 = \begin{pmatrix} -1 & 0 & 0 \\ 0 & 1 & 0 \\ 0 & 0 & -1 \end{pmatrix}, \quad
a_4 = \begin{pmatrix} -1 & 0 & 0 \\ 0 & -1 & 0 \\ 0 & 0 & 1 \end{pmatrix},
$$

$$
b_1 = \begin{pmatrix} 0 & 0 & 1 \\ 1 & 0 & 0 \\ 0 & 1 & 0 \end{pmatrix}, \quad
b_2 = \begin{pmatrix} 0 & 0 & 1 \\ -1 & 0 & 0 \\ 0 & -1 & 0 \end{pmatrix},
$$

$$
b_3 = \begin{pmatrix} 0 & 0 & -1 \\ 1 & 0 & 0 \\ 0 & -1 & 0 \end{pmatrix}, \quad
b_4 = \begin{pmatrix} 0 & 0 & -1 \\ -1 & 0 & 0 \\ 0 & 1 & 0 \end{pmatrix}, \tag{4.3}
$$

$$
c_1 = \begin{pmatrix} 0 & 1 & 0 \\ 0 & 0 & 1 \\ 1 & 0 & 0 \end{pmatrix}, \quad
c_2 = \begin{pmatrix} 0 & 1 & 0 \\ 0 & 0 & -1 \\ -1 & 0 & 0 \end{pmatrix},
$$

$$
c_3 = \begin{pmatrix} 0 & -1 & 0 \\ 0 & 0 & 1 \\ -1 & 0 & 0 \end{pmatrix}, \quad
c_4 = \begin{pmatrix} 0 & -1 & 0 \\ 0 & 0 & -1 \\ 1 & 0 & 0 \end{pmatrix}.
$$

As seen in these forms, A_4 is obviously isomorphic to $\Delta(12) \simeq (Z_2 \times Z_2) \rtimes Z_3$, which is explained in Sect. 10.

They are classified by the conjugacy classes as

$$
\begin{aligned}
C_1 &: \quad \{a_1\}, & h &= 1, \\
C_3 &: \quad \{a_2, a_3, a_4\}, & h &= 2, \\
C_4 &: \{b_1, b_2, b_3, b_4, \}, & h &= 3, \\
C_{4'} &: \{c_1, c_2, c_3, c_4, \}, & h &= 3,
\end{aligned} \tag{4.4}
$$

where we have also shown the orders of each element in the conjugacy class by h. There are four conjugacy classes and there must be four irreducible representations, i.e. $m_1 + m_2 + m_3 + \cdots = 4$.

The orthogonality relation (2.17) requires

$$\sum_\alpha [\chi_\alpha(C_1)]^2 = \sum_n m_n n^2 = m_1 + 4m_2 + 9m_3 + \cdots = 12, \qquad (4.5)$$

for m_i, which satisfy $m_1 + m_2 + m_3 + \cdots = 4$. Then, we obtain a solution, $(m_1, m_2, m_3) = (3, 0, 1)$. That is, the A_4 group has three singlets, $\mathbf{1}$, $\mathbf{1}'$, and $\mathbf{1}''$, and a single triplet $\mathbf{3}$, where the triplet corresponds to (4.3).

Another algebraic definition of A_4 is often used in the literature. We denote $a_1 = e$, $a_2 = s$ and $b_1 = t$, that is,

$$s = \begin{pmatrix} 1 & 0 & 0 \\ 0 & -1 & 0 \\ 0 & 0 & -1 \end{pmatrix}, \qquad t = \begin{pmatrix} 0 & 0 & 1 \\ 1 & 0 & 0 \\ 0 & 1 & 0 \end{pmatrix}. \qquad (4.6)$$

They satisfy the following algebraic relations,

$$s^2 = t^3 = (st)^3 = e. \qquad (4.7)$$

The closed algebra of these elements, s and t, is defined as the A_4. It is straightforward to write all of a_i, b_i and c_i elements by s and t. Then, the conjugacy classes are rewritten as

$$\begin{aligned} C_1 &: & \{e\}, & & h &= 1, \\ C_3 &: & \{s, tst^2, t^2st\}, & & h &= 2, \\ C_4 &: & \{t, ts, st, sts\}, & & h &= 3, \\ C_{4'} &: & \{t^2, st^2, t^2s, tst\}, & & h &= 3. \end{aligned} \qquad (4.8)$$

Using them, we study characters. First, we consider characters of three singlets. Because $s^2 = e$, the characters of C_3 have two possibilities, $\chi_\alpha(C_3) = \pm 1$. However, the two elements, t and ts, belong to the same conjugacy class C_4. That means that $\chi_\alpha(C_3)$ should have the unique value, $\chi_\alpha(C_3) = 1$. Similarly, because of $t^3 = e$, the characters $\chi_\alpha(t)$ can correspond to three values, i.e. $\chi_\alpha(t) = \omega^n$, $n = 0, 1, 2$, and all of these three values are consistent with the above structure of conjugacy classes. Thus, all of three singlets, $\mathbf{1}$, $\mathbf{1}'$ and $\mathbf{1}''$ are classified by these three values of $\chi_\alpha(t) = 1$, ω and ω^2, respectively. Obviously, it is found that $\chi_\alpha(C_{4'}) = (\chi_\alpha(C_4))^2$. Thus, the generators such as $s = a_2, t = b_1, t^2 = c_1$ are represented on the non-trivial singlets $\mathbf{1}'$ and $\mathbf{1}''$ as

$$\begin{aligned} s(\mathbf{1}') &= a_2(\mathbf{1}') = 1, & t(\mathbf{1}') &= b_1(\mathbf{1}') = \omega, & t^2(\mathbf{1}') &= c_1(\mathbf{1}') = \omega^2, \\ s(\mathbf{1}'') &= a_2(\mathbf{1}'') = 1, & t(\mathbf{1}'') &= b_1(\mathbf{1}'') = \omega^2, & t^2(\mathbf{1}'') &= c_1(\mathbf{1}'') = \omega. \end{aligned} \qquad (4.9)$$

These characters are shown in Table 4.1.

Table 4.1 Characters of A_4 representations

	h	χ_1	$\chi_{1'}$	$\chi_{1''}$	χ_3
C_1	1	1	1	1	3
C_3	2	1	1	1	-1
C_4	3	1	ω	ω^2	0
$C_{4'}$	3	1	ω^2	ω	0

Next, we consider the characters for the triplet representation. Obviously, the matrices in Eq. (4.3) correspond to the triplet representation. Thus, we can obtain their characters. Its result is also shown in Table 4.1.

The tensor product of $\mathbf{3} \times \mathbf{3}$ can be decomposed as

$$
\begin{pmatrix} x_1 \\ x_2 \\ x_3 \end{pmatrix}_3 \otimes \begin{pmatrix} y_1 \\ y_2 \\ y_3 \end{pmatrix}_3 = (x_1 y_1 + x_2 y_2 + x_3 y_3)_1 \oplus \left(x_1 y_1 + \omega x_2 y_2 + \omega^2 x_3 y_3\right)_{1'}
$$

$$
\oplus \left(x_1 y_1 + \omega^2 x_2 y_2 + \omega x_3 y_3\right)_{1''}
$$

$$
\oplus \begin{pmatrix} x_2 y_3 + x_3 y_2 \\ x_3 y_1 + x_1 y_3 \\ x_1 y_2 + x_2 y_1 \end{pmatrix}_3 \oplus \begin{pmatrix} x_2 y_3 - x_3 y_2 \\ x_3 y_1 - x_1 y_3 \\ x_1 y_2 - x_2 y_1 \end{pmatrix}_3 . \tag{4.10}
$$

The products of singlets and the triplet $\mathbf{3}$ are

$$
\mathbf{1}' \otimes \mathbf{3} = \alpha \otimes \begin{pmatrix} x_1 \\ x_2 \\ x_3 \end{pmatrix}_3 = \alpha \begin{pmatrix} x_3 \\ x_1 \\ x_2 \end{pmatrix}_3 , \quad \mathbf{1}'' \otimes \mathbf{3} = \beta \otimes \begin{pmatrix} x_1 \\ x_2 \\ x_3 \end{pmatrix}_3 = \beta \begin{pmatrix} x_2 \\ x_3 \\ x_1 \end{pmatrix}_3 .
$$
$$\tag{4.11}$$

The products of singlets are given as

$$
\mathbf{1} \otimes \mathbf{1} = \mathbf{1} , \quad \mathbf{1}' \otimes \mathbf{1}' = \mathbf{1}'' , \quad \mathbf{1}'' \otimes \mathbf{1}'' = \mathbf{1}' , \quad \mathbf{1}' \otimes \mathbf{1}'' = \mathbf{1} . \tag{4.12}
$$

We show another basis for representations of the A_4 group, which is often used e.g. in [1], in Appendix C.

4.2 A_5

Next, we study the A_5 group. This group is isomorphic to the symmetry of a regular icosahedron. Thus, it is pedagogical to explain group-theoretical aspects of A_5 as the symmetry of a regular icosahedron [2]. As shown in Fig. 4.2, a regular icosahedron consists of 20 identical equilateral triangular faces, 30 edges and 12 vertices. The icosahedron is dual to a dodecahedron, whose symmetry is also isomorphic to A_5. The A_5 elements correspond to all the proper rotations of the icosahedron. Such

Fig. 4.2 The regular icosahedron

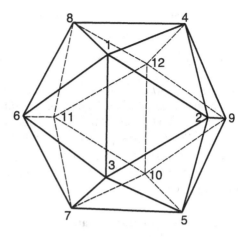

rotations are classified into five types, that is, the 0 rotation (identity), π rotations about the midpoint of each edge, rotations by $2\pi/3$ about axes through the center of each face, and rotations by $2\pi/5$ and $4\pi/5$ about an axis through each vertex. Following [2], we label the vertex by number $n = 1, \cdots, 12$ in Fig. 4.2. Here, we define two elements a and b such that a corresponds to the rotation by π about the midpoint of the edge between vertices 1 and 2 while b corresponds to the rotation by $2\pi/3$ about the axis through the center of the triangular face 10-11-12. That is, these two elements correspond to the transformations acting on the 12 vertices as follows,

$$a : (1, 2, 3, 4, 5, 6, 7, 8, 9, 10, 11, 12) \rightarrow (2, 1, 4, 3, 8, 9, 12, 5, 6, 11, 10, 7),$$
$$b : (1, 2, 3, 4, 5, 6, 7, 8, 9, 10, 11, 12) \rightarrow (2, 3, 1, 5, 6, 4, 8, 9, 7, 11, 12, 10).$$

Then the product ab is given by the following transformation as

$$ab : (1, 2, 3, 4, 5, 6, 7, 8, 9, 10, 11, 12) \rightarrow (3, 2, 5, 1, 9, 7, 10, 6, 4, 12, 11, 8),$$

which is the rotation by $2\pi/5$ about the axis through the vertex 2. All of the A_5 elements are written by products of these elements, which satisfy

$$a^2 = b^3 = (ab)^5 = e. \tag{4.13}$$

• Conjugacy classes
The order of A_5 is equal to $(5!)/2 = 60$. All of the A_5 elements, i.e. all the rotations of the icosahedron, are classified into five conjugacy classes as follows,

$$C_1 : \{e\},$$
$$C_{15} : \{a(12), a(13), a(14), a(16), a(18), a(23), a(24), a(25), a(29), a(35)\},$$
$$\qquad a(36), a(37), a(48), a(49), a(59), \},$$
$$C_{20} : \{b(123), b(124), b(126), b(136), b(168), b(235), b(249), \tag{4.14}$$

$b(259), b(357), b(367),$ and their inverse elements},

$C_{12} : \{c(1), c(2), c(3), c(4), c(5), c(6),$ and their inverse elements},

$C'_{12} : \{c^2(1), c^2(2), c^2(3), c^2(4), c^2(5), c^2(6),$ and their inverse elements},

where $a(km), b(kmn)$ and $c(k)$ denote the rotation by π about the midpoint of the edge $k - m$, the rotation by $2\pi/3$ about the axis through the center of the face $k - m - n$ and the rotation by $2\pi/5$ about the axis through the vertex k. The conjugacy classes, $C_1, C_{15}, C_{20}, C_{12}$ and C'_{12}, include 1, 15, 20, 12 and 12 elements, respectively. Since obviously $(a(km))^2 = (b(kmn))^3 = (c(k))^5 = e$, we find $h = 2$ in C_{15}, $h = 3$ in C_{20}, $h = 5$ in C_{12} and $h = 5$ in C'_{12}, where h denotes the order of each element in the conjugacy class, i.e. $g^h = e$.

• **Characters and representations**
The orthogonality relations (2.18) and (2.19) for A_5 is given as

$$m_1 + 4m_2 + 9m_3 + 16m_4 + 25m_5 + \cdots = 60, \tag{4.15}$$

$$m_1 + m_2 + m_3 + m_4 + m_5 + \cdots = 5. \tag{4.16}$$

By solving these equations, we have $(m_1, m_2, m_3, m_4, m_5) = (1, 0, 2, 1, 1)$. Therefore, the A_5 group has one trivial singlet, **1**, two triplets, **3** and **3′**, one quartet, **4**, and one quintet, **5**. The characters are shown in Table 4.2. Several ways are known to construct these representations, e.g., Cummins-Patera's basis [3], Shirai's basis [2], and Feruglio-Paris's basis [4]. Each of the relations is also summarized in Ref. [4]. Instead of a and b that is called Cummins-Patera's basis [3], we use the generators of Shirai's basis [2], that is, $s = a$ and $t = bab$, which satisfy

$$s^2 = t^5 = (t^2 s t^3 s t^{-1} s t s t^{-1})^3 = e. \tag{4.17}$$

It is straightforward to write sixty elements in Eq. (4.18) in terms of s and t. Then, the conjugacy classes can be reexpressed as [5]

$C_1 : \{e\},$

$C_{15} : \{s t^2 s t^3 s, t s t^4, t^4 (s t^2)^2, t^2 s t^3, t^2 s t^2 s t^3 s, s t^2 s t, s, t^3 s t^2 s t^3, t^3 \, s t^2 s t^3 s, t^3 s t^2,$

$\qquad t^4 s t^2 s t^3 s, t s t^2 s, s t^3 s t^2 s, t^4 s t, (t^2 s)^2 t^4\},$

$C_{20} : \{s t, t s, s t^4, t^4 s, t s t^3, t^2 s t^2, t^2 s t^4, t^3 s t, t^3 s t^3, t^4 s t^2, t s t^3 s, t^2 s t^3 s, t^3 s t^2 s, s t^2 s t^3$

$\qquad s t^3 s t, s t^3 s t^2, (t^2 s)^2 t^2, t^2 (t^2 s)^2, (s t^2)^2 s, (s t^2)^2 t^2\},$

$C_{12} : \{t, t^4, s t^2, t^2 s, s t^3, t^3 s, s t s, t s t, t s t^2, t^2 s t, t^3 s t^4, t^4 s t^3\},$

$C'_{12} : \{t^2, t^3, s t^2 s, s t^3 s, (s t^2)^2, (t^2 s)^2, (s t^3)^2, (t^3 s)^2, (t^2 s)^2 t^3, t^3 (s t^2)^2, t^3 s t^2 s t^4, t^4 s t^2 s t^3\}.$

$$\tag{4.18}$$

The generators, s and t, are represented as [2],

$$s = \frac{1}{2}\begin{pmatrix} -1 & \phi & \frac{1}{\phi} \\ \phi & \frac{1}{\phi} & 1 \\ \frac{1}{\phi} & 1 & -\phi \end{pmatrix}, \quad t = \frac{1}{2}\begin{pmatrix} 1 & \phi & \frac{1}{\phi} \\ -\phi & \frac{1}{\phi} & 1 \\ \frac{1}{\phi} & -1 & \phi \end{pmatrix}, \qquad \text{on } \mathbf{3}, \qquad (4.19)$$

$$s = \frac{1}{2}\begin{pmatrix} -\phi & \frac{1}{\phi} & 1 \\ \frac{1}{\phi} & -1 & \phi \\ 1 & \phi & \frac{1}{\phi} \end{pmatrix}, \quad t = \frac{1}{2}\begin{pmatrix} -\phi & -\frac{1}{\phi} & 1 \\ \frac{1}{\phi} & 1 & \phi \\ -1 & \phi & -\frac{1}{\phi} \end{pmatrix}, \qquad \text{on } \mathbf{3'}, \qquad (4.20)$$

$$s = \frac{1}{4}\begin{pmatrix} -1 & -1 & -3 & -\sqrt{5} \\ -1 & 3 & 1 & -\sqrt{5} \\ -3 & 1 & -1 & \sqrt{5} \\ -\sqrt{5} & -\sqrt{5} & \sqrt{5} & -1 \end{pmatrix},$$

$$t = \frac{1}{4}\begin{pmatrix} -1 & 1 & -3 & \sqrt{5} \\ -1 & -3 & 1 & \sqrt{5} \\ 3 & 1 & 1 & \sqrt{5} \\ \sqrt{5} & -\sqrt{5} & -\sqrt{5} & -1 \end{pmatrix}, \qquad \text{on } \mathbf{4}, \qquad (4.21)$$

$$s = \frac{1}{2}\begin{pmatrix} \frac{1-3\phi}{4} & \frac{\phi^2}{2} & -\frac{1}{2\phi^2} & \frac{\sqrt{5}}{2} & \frac{\sqrt{3}}{4\phi} \\ \frac{\phi^2}{2} & 1 & 1 & 0 & \frac{\sqrt{3}}{2\phi} \\ -\frac{1}{2\phi^2} & 1 & 0 & -1 & -\frac{\sqrt{3}\phi}{2} \\ \frac{\sqrt{5}}{2} & 0 & -1 & 1 & -\frac{\sqrt{3}}{2} \\ \frac{\sqrt{3}}{4\phi} & \frac{\sqrt{3}}{2\phi} & -\frac{\sqrt{3}\phi}{2} & -\frac{\sqrt{3}}{2} & \frac{3\phi-1}{4} \end{pmatrix},$$

$$t = \frac{1}{2}\begin{pmatrix} \frac{1-3\phi}{4} & -\frac{\phi^2}{2} & -\frac{1}{2\phi^2} & -\frac{\sqrt{5}}{2} & \frac{\sqrt{3}}{4\phi} \\ \frac{\phi^2}{2} & -1 & 1 & 0 & \frac{\sqrt{3}}{2\phi} \\ \frac{1}{2\phi^2} & 1 & 0 & -1 & \frac{\sqrt{3}\phi}{2} \\ -\frac{\sqrt{5}}{2} & 0 & 1 & 1 & \frac{\sqrt{3}}{2} \\ \frac{\sqrt{3}}{4\phi} & -\frac{\sqrt{3}}{2\phi} & -\frac{\sqrt{3}\phi}{2} & \frac{\sqrt{3}}{2} & \frac{3\phi-1}{4} \end{pmatrix}, \qquad \text{on } \mathbf{5}, \qquad (4.22)$$

where $\phi = \frac{1+\sqrt{5}}{2}$. Furthermore, the multiplication rules are also shown in Table 4.3.

Table 4.2 Characters of A_5 representations, where $\phi = \frac{1+\sqrt{5}}{2}$

	h	1	3	3′	4	5
C_1	1	1	3	3	4	5
C_{15}	2	1	-1	-1	0	1
C_{20}	3	1	0	0	1	-1
C_{12}	5	1	ϕ	$1-\phi$	-1	0
$C_{12'}$	5	1	$1-\phi$	ϕ	-1	0

Table 4.3 Multiplication rules for the A_5 group

$3 \otimes 3 = 1 \oplus 3 \oplus 5$
$3' \otimes 3' = 1 \oplus 3' \oplus 5$
$3 \otimes 3' = 4 \oplus 5$
$3 \otimes 4 = 3' \oplus 4 \oplus 5$
$3' \otimes 4 = 3 \oplus 4 \oplus 5$
$3 \otimes 5 = 3 \oplus 3' \oplus 4 \oplus 5$
$3' \otimes 5 = 3 \oplus 3' \oplus 4 \oplus 5$
$4 \otimes 4 = 1 \oplus 3 \oplus 3' \oplus 4 \oplus 5$
$4 \otimes 5 = 3 \oplus 3' \oplus 4 \oplus 5 \oplus 5$
$5 \otimes 5 = 1 \oplus 3 \oplus 3' \oplus 4 \oplus 4 \oplus 5 \oplus 5$

• Tensor products

Concrete tensor products have been partially completed in Ref. [6] based on Shirai basis. Here we show the complete set of the tensor products,

$$
\begin{pmatrix} x_1 \\ x_2 \\ x_3 \end{pmatrix}_3 \otimes \begin{pmatrix} y_1 \\ y_2 \\ y_3 \end{pmatrix}_3 = (x_1 y_1 + x_2 y_2 + x_3 y_3)_1 \oplus \begin{pmatrix} x_3 y_2 - x_2 y_3 \\ x_1 y_3 - x_3 y_1 \\ x_2 y_1 - x_1 y_2 \end{pmatrix}_3
$$
$$
\oplus \begin{pmatrix} x_2 y_2 - x_1 y_1 \\ x_2 y_1 + x_1 y_2 \\ x_3 y_2 + x_2 y_3 \\ x_1 y_3 + x_3 y_1 \\ -\frac{1}{\sqrt{3}}(x_1 y_1 + x_2 y_2 - 2x_3 y_3) \end{pmatrix}_5 , \tag{4.23}
$$

$$
\begin{pmatrix} x_1 \\ x_2 \\ x_3 \end{pmatrix}_{3'} \otimes \begin{pmatrix} y_1 \\ y_2 \\ y_3 \end{pmatrix}_{3'} = (x_1 y_1 + x_2 y_2 + x_3 y_3)_1 \oplus \begin{pmatrix} x_3 y_2 - x_2 y_3 \\ x_1 y_3 - x_3 y_1 \\ x_2 y_1 - x_1 y_2 \end{pmatrix}_{3'}
$$
$$
\oplus \begin{pmatrix} \frac{1}{2}(-\frac{1}{\phi}x_1 y_1 - \phi x_2 y_2 + \sqrt{5}x_3 y_3) \\ x_2 y_1 + x_1 y_2 \\ -(x_3 y_1 + x_1 y_3) \\ x_2 y_3 + x_3 y_2 \\ \frac{1}{2\sqrt{3}}((1-3\phi)x_1 y_1 + (3\phi-2)x_2 y_2 + x_3 y_3) \end{pmatrix}_5 , \tag{4.24}
$$

$$
\begin{pmatrix} x_1 \\ x_2 \\ x_3 \end{pmatrix}_3 \otimes \begin{pmatrix} y_1 \\ y_2 \\ y_3 \end{pmatrix}_{3'} = \begin{pmatrix} \frac{1}{\phi}x_3y_2 - \phi x_1y_3 \\ \phi x_3y_1 + \frac{1}{\phi}x_2y_3 \\ -\frac{1}{\phi}x_1y_1 + \phi x_2y_2 \\ x_2y_1 - x_1y_2 + x_3y_3 \end{pmatrix}_4 \oplus \begin{pmatrix} \frac{1}{2}(\phi^2 x_2y_1 + \frac{1}{\phi^2}x_1y_2 - \sqrt{5}x_3y_3) \\ -(\phi x_1y_1 + \frac{1}{\phi}x_2y_2) \\ \frac{1}{\phi}x_3y_1 - \phi x_2y_3 \\ \phi x_3y_2 + \frac{1}{\phi}x_1y_3 \\ \frac{\sqrt{3}}{2}(\frac{1}{\phi}x_2y_1 + \phi x_1y_2 + x_3y_3) \end{pmatrix}_5 ,
$$
$$(4.25)$$

$$
\begin{pmatrix} x_1 \\ x_2 \\ x_3 \end{pmatrix}_3 \otimes \begin{pmatrix} y_1 \\ y_2 \\ y_3 \\ y_4 \end{pmatrix}_4 = \begin{pmatrix} -\frac{1}{\phi^2}x_1y_3 + \frac{1}{\phi}x_2y_4 + x_3y_2 \\ -\frac{1}{\phi}x_1y_4 + x_2y_3 + \frac{1}{\phi^2}x_3y_1 \\ -x_1y_1 + \frac{1}{\phi^2}x_2y_2 + \frac{1}{\phi}x_3y_4 \end{pmatrix}_{3'} \oplus \begin{pmatrix} -x_1y_3 + x_2y_4 - x_3y_2 \\ -x_1y_4 - x_2y_3 + x_3y_1 \\ x_1y_1 + x_2y_2 + x_3y_4 \\ x_1y_2 - x_2y_1 - x_3y_3 \end{pmatrix}_4
$$
$$
\oplus \begin{pmatrix} \frac{1}{2}((6\phi+5)x_1y_2 + (3\phi+4)x_2y_1 + (3\phi+1)x_3y_3) \\ -x_1y_1 + (3\phi+2)x_2y_2 - (3\phi+1)x_3y_4 \\ -(3\phi+1)x_1y_4 - x_2y_3 - (3\phi+2)x_3y_1 \\ -(3\phi+2)x_1y_3 - (3\phi+1)x_2y_4 + x_3y_2 \\ \frac{\sqrt{3}}{2}(x_1y_2 - (3\phi+2)x_2y_1 + 3(\phi+1)x_3y_3) \end{pmatrix}_5 , \qquad (4.26)
$$

$$
\begin{pmatrix} x_1 \\ x_2 \\ x_3 \end{pmatrix}_{3'} \otimes \begin{pmatrix} y_1 \\ y_2 \\ y_3 \\ y_4 \end{pmatrix}_4 = \begin{pmatrix} x_1y_3 + \phi x_2y_4 + \phi^2 x_3y_1 \\ -\phi x_1y_4 - \phi^2 x_2y_3 - x_3y_2 \\ -\phi^2 x_1y_2 - x_2y_1 - \phi x_3y_4 \end{pmatrix}_3 \oplus \begin{pmatrix} x_1y_4 - x_2y_3 + x_3y_2 \\ x_1y_3 + x_2y_4 - x_3y_1 \\ -x_1y_2 + x_2y_1 + x_3y_4 \\ -(x_1y_1 + x_2y_2 + x_3y_3) \end{pmatrix}_4
$$
$$
\oplus \begin{pmatrix} x_1y_1 - \phi^4 x_2y_2 + \phi^2(2\phi-1)x_3y_3 \\ x_1y_2 - \phi^4 x_2y_1 + \phi^2(2\phi-1)x_3y_4 \\ \phi^4 x_1y_3 - \phi^2(2\phi-1)x_2y_4 + x_3y_1 \\ \phi^2(2\phi-1)x_1y_4 - x_2y_3 - \phi^4 x_3y_2 \\ -\sqrt{3}\phi\left(\phi^2 x_1y_1 - x_2y_2 - \phi x_3y_3\right) \end{pmatrix}_5 , \qquad (4.27)
$$

$$\begin{pmatrix} x_1 \\ x_2 \\ x_3 \end{pmatrix}_3 \otimes \begin{pmatrix} y_1 \\ y_2 \\ y_3 \\ y_4 \\ y_5 \end{pmatrix}_5 = \begin{pmatrix} x_1(y_1 + \frac{1}{\sqrt{3}}y_5) - x_2 y_2 - x_3 y_4 \\ -x_1 y_2 - x_2(y_1 - \frac{1}{\sqrt{3}}y_5) - x_3 y_3 \\ -x_1 y_4 - x_2 y_3 - \frac{2}{\sqrt{3}}x_3 y_5 \end{pmatrix}_3$$

$$\oplus \begin{pmatrix} x_1 y_2 - \frac{\phi}{2}x_2 y_1 - \frac{\sqrt{3}}{2\phi^2}x_2 y_5 - \frac{1}{\phi^2}x_3 y_3 \\ -\frac{\sqrt{3}}{2}x_1 y_5 - \frac{1}{2\phi^3}x_1 y_1 + \frac{1}{\phi^2}x_2 y_2 - x_3 y_4 \\ -\frac{1}{\phi^2}x_1 y_4 + x_2 y_3 + \frac{\sqrt{5}}{2\phi}x_3 y_1 - \frac{\sqrt{3}}{2\phi}x_3 y_5 \end{pmatrix}_{3'}$$

$$\oplus \begin{pmatrix} \frac{1}{\phi^2}x_1 y_2 + \frac{\phi^2-6}{2}x_2 y_1 + \frac{\sqrt{3}}{2}\phi^2 x_2 y_5 + \phi^2 x_3 y_3 \\ -\frac{\phi+4}{2}x_1 y_1 - \frac{\sqrt{3}}{2\phi^2}x_1 y_5 - \phi^2 x_2 y_2 - \frac{1}{\phi^2}x_3 y_4 \\ \phi^2 x_1 y_4 + \frac{1}{\phi^2}x_2 y_3 - \frac{\sqrt{5}}{2}x_3 y_1 - \frac{3\sqrt{3}}{2}x_3 y_5 \\ \sqrt{5}(x_1 y_3 + x_2 y_4 + x_3 y_2) \end{pmatrix}_4$$

$$\oplus \begin{pmatrix} x_1 y_3 + x_2 y_4 - 2x_3 y_2 \\ x_1 y_4 - x_2 y_3 + 2x_3 y_1 \\ -x_1 y_1 + x_2 y_2 - x_3 y_4 + \sqrt{3}x_1 y_5 \\ -x_1 y_2 - x_2 y_1 + x_3 y_3 - \sqrt{3}x_2 y_5 \\ -\sqrt{3}(x_1 y_3 - x_2 y_4) \end{pmatrix}_5 , \qquad (4.28)$$

$$\begin{pmatrix} x_1 \\ x_2 \\ x_3 \end{pmatrix}_{3'} \otimes \begin{pmatrix} y_1 \\ y_2 \\ y_3 \\ y_4 \\ y_5 \end{pmatrix}_5 = \begin{pmatrix} -\phi^2 x_1 y_2 + \frac{1}{2\phi}x_2 y_1 + \frac{\sqrt{3}}{2}\phi^2 x_2 y_5 + x_3 y_4 \\ \frac{2\phi+1}{2}x_1 y_1 + \frac{\sqrt{3}}{2}x_1 y_5 - x_2 y_2 - \phi^2 x_3 y_3 \\ x_1 y_3 + \phi^2 x_2 y_4 - \frac{\sqrt{5}}{2}\phi x_3 y_1 + \frac{\sqrt{3}}{2}\phi x_3 y_5 \end{pmatrix}_3$$

$$\oplus \begin{pmatrix} \frac{1}{2\phi}x_1 y_1 - x_2 y_2 + x_3 y_3 + \frac{3\phi-1}{2\sqrt{3}}x_1 y_5 \\ -x_1 y_2 + \frac{\phi}{2}x_2 y_1 - x_3 y_4 - \frac{3\phi-2}{2\sqrt{3}}x_2 y_5 \\ x_1 y_3 - x_2 y_4 - \frac{\sqrt{5}}{2}x_3 y_1 - \frac{1}{2\sqrt{3}}x_3 y_5 \end{pmatrix}_{3'}$$

$$\oplus \begin{pmatrix} \frac{1}{\sqrt{5}}\left(\frac{1}{\phi^2}x_1 y_1 + \phi^2 x_2 y_2 + \frac{1}{\phi^2}x_3 y_3 - \sqrt{3}\phi x_1 y_5\right) \\ \frac{1}{\sqrt{5}}\left(-\frac{1}{\phi^2}x_1 y_2 - \phi^2 x_2 y_1 - \frac{\sqrt{3}(\phi-3)}{\sqrt{5}}x_2 y_5 - \phi^2 x_3 y_4\right) \\ \frac{1}{\sqrt{5}}\left(\phi^2 x_1 y_3 - \frac{1}{\phi^2}x_2 y_4 + \sqrt{5}x_3 y_1 + \sqrt{3}x_3 y_5\right) \\ x_1 y_4 - x_2 y_3 + x_3 y_2 \end{pmatrix}_4$$

$$\oplus \begin{pmatrix} -(3\phi-1)x_1 y_4 + (2-3\phi)x_2 y_3 + x_3 y_2 \\ -2x_1 y_3 - 2x_2 y_4 - x_3 y_1 + \sqrt{15}x_3 y_5 \\ 2x_1 y_2 - (2-3\phi)x_2 y_1 - 2x_3 y_4 + \sqrt{3}\phi x_2 y_5 \\ (3\phi-1)x_1 y_1 + 2x_2 y_2 + 2x_3 y_3 - \frac{\sqrt{3}}{\phi}x_1 y_5 \\ \frac{\sqrt{3}}{\phi}x_1 y_4 - \phi\sqrt{3}x_2 y_3 - \sqrt{15}x_3 y_2 \end{pmatrix}_5 , \qquad (4.29)$$

$$\begin{pmatrix} x_1 \\ x_2 \\ x_3 \\ x_4 \end{pmatrix}_4 \otimes \begin{pmatrix} y_1 \\ y_2 \\ y_3 \\ y_4 \end{pmatrix}_4 = (x_1 y_1 + x_2 y_2 + x_3 y_3 + x_4 y_4)\mathbf{1} \oplus \begin{pmatrix} x_1 y_3 + x_2 y_4 - x_3 y_1 - x_4 y_2 \\ -x_1 y_4 + x_2 y_3 - x_3 y_2 + x_4 y_1 \\ x_1 y_2 - x_2 y_1 - x_3 y_4 + x_4 y_3 \end{pmatrix}_3$$

$$\oplus \begin{pmatrix} x_1 y_4 + x_2 y_3 - x_3 y_2 - x_4 y_1 \\ -x_1 y_3 + x_2 y_4 + x_3 y_1 - x_4 y_2 \\ x_1 y_2 - x_2 y_1 + x_3 y_4 - x_4 y_3 \end{pmatrix}_{3'}$$

$$\oplus \begin{pmatrix} x_1 y_4 - \sqrt{5}x_2 y_3 - \sqrt{5}x_3 y_2 + x_4 y_1 \\ -\sqrt{5}x_1 y_3 + x_2 y_4 - \sqrt{5}x_3 y_1 + x_4 y_2 \\ -\sqrt{5}x_1 y_2 - \sqrt{5}x_2 y_1 + x_3 y_4 + x_4 y_3 \\ x_1 y_1 + x_2 y_2 + x_3 y_3 - 3 x_4 y_4 \end{pmatrix}_4$$

$$\oplus \begin{pmatrix} -\frac{\phi^2}{\sqrt{5}}x_1 y_1 + \frac{1}{\sqrt{5}\phi^2}x_2 y_2 + x_3 y_3 \\ -\frac{1}{\sqrt{5}}x_1 y_2 - \frac{1}{\sqrt{5}}x_2 y_1 - x_3 y_4 - x_4 y_3 \\ \frac{1}{\sqrt{5}}x_1 y_3 + x_2 y_4 + \frac{1}{\sqrt{5}}x_3 y_1 + x_4 y_2 \\ -x_1 y_4 - \frac{1}{\sqrt{5}}x_2 y_3 - \frac{1}{\sqrt{5}}x_3 y_2 - x_4 y_1 \\ -\sqrt{\frac{3}{5}}\left(\frac{1}{\phi}x_1 y_1 - \phi x_2 y_2 + x_3 y_3\right) \end{pmatrix}_5, \tag{4.30}$$

$$\begin{pmatrix} x_1 \\ x_2 \\ x_3 \\ x_4 \end{pmatrix}_4 \otimes \begin{pmatrix} y_1 \\ y_2 \\ y_3 \\ y_4 \\ y_5 \end{pmatrix}_5 = \begin{pmatrix} \frac{2}{\phi^2}x_1 y_2 - (\phi+4)x_2 y_1 + 2\phi^2 x_3 y_4 + 2\sqrt{5}x_4 y_3 - \frac{\sqrt{3}}{\phi^2}x_2 y_5 \\ (\phi-5)x_1 y_1 + \sqrt{3}\phi^2 x_1 y_5 - 2\phi^2 x_2 y_2 + \frac{2}{\phi^2}x_3 y_3 + 2\sqrt{5}x_4 y_4 \\ 2\phi^2 x_1 y_3 - \frac{2}{\phi^2}x_2 y_4 - \sqrt{5}x_3 y_1 - 3\sqrt{3}x_3 y_5 + 2\sqrt{5}x_4 y_2 \end{pmatrix}_3$$

$$\oplus \begin{pmatrix} \frac{1}{\phi^2}x_1 y_1 - \sqrt{3}\phi x_1 y_5 - \frac{1}{\phi^2}x_2 y_2 + \phi^2 x_3 y_3 + \sqrt{5}x_4 y_4 \\ -\phi^2 x_1 y_2 - \phi^2 x_2 y_1 + \frac{\sqrt{3}}{\phi}x_2 y_5 - \frac{1}{\phi^2}x_3 y_4 - \sqrt{5}x_4 y_3 \\ \frac{1}{\phi^2}x_1 y_3 - \phi^2 x_2 y_4 + \sqrt{5}x_3 y_1 + \sqrt{3}x_3 y_5 + \sqrt{5}x_4 y_2 \end{pmatrix}_{3'}$$

$$\oplus \begin{pmatrix} -\frac{\phi^2}{\sqrt{5}}x_1 y_1 - \frac{1}{\sqrt{5}}x_2 y_2 + \frac{1}{\sqrt{5}}x_3 y_3 - x_4 y_4 - \frac{\sqrt{3}}{\sqrt{5}\phi}x_1 y_5 \\ -\frac{1}{\sqrt{5}}x_1 y_2 + \frac{1}{\sqrt{5}\phi^2}x_2 y_1 + \sqrt{\frac{3}{5}}\phi x_2 y_5 - \frac{1}{\sqrt{5}}x_3 y_4 + x_4 y_3 \\ \frac{1}{\sqrt{5}}x_1 y_3 - \frac{1}{\sqrt{5}}x_2 y_4 + x_3 y_1 - \sqrt{\frac{3}{5}}x_3 y_5 - x_4 y_2 \\ -x_1 y_4 + x_2 y_3 - x_3 y_2 \end{pmatrix}_4$$

$$\oplus \begin{pmatrix} \frac{1}{2}\left(\phi^2 x_1 y_4 - \frac{1}{\phi^2}x_2 y_3 - 3 x_3 y_2 + 3 x_4 y_1 + \sqrt{\frac{5}{3}}x_4 y_5\right) \\ \phi x_1 y_3 - \frac{1}{\phi}x_2 y_4 - x_3 y_1 - x_4 y_2 + \sqrt{\frac{5}{3}}x_3 y_5 \\ \frac{1}{\phi}x_1 y_2 + \frac{1}{\phi}x_2 y_1 + \frac{1}{\sqrt{3}}\phi^2 x_2 y_5 + \phi x_3 y_4 - x_4 y_3 \\ \phi x_1 y_1 + \frac{1}{\sqrt{3}\phi^2}x_1 y_5 - \phi x_2 y_2 + \frac{1}{\phi}x_3 y_3 - x_4 y_4 \\ \frac{1}{2\sqrt{3}}\left(-(\phi-5)x_1 y_4 + (\phi+4)x_2 y_3 + \sqrt{5}x_3 y_2 - \sqrt{5}x_4 y_1\right) + \frac{3}{2}x_4 y_5 \end{pmatrix}_5,$$

$$\oplus \begin{pmatrix} x_1 y_4 - x_2 y_3 - 2 x_3 y_2 + \frac{3}{2}x_4 y_1 + \frac{\sqrt{15}}{2}x_4 y_5 \\ \phi^2 x_1 y_3 + \frac{1}{\phi^2}x_2 y_4 - \frac{1}{2}x_3 y_1 + \frac{\sqrt{15}}{2}x_3 y_5 - x_4 y_2 \\ -\frac{1}{\phi^2}x_1 y_2 + \frac{3\phi-2}{2}x_2 y_1 + \frac{\sqrt{3}}{2}\phi x_2 y_5 + \phi^2 x_3 y_4 - x_4 y_3 \\ \frac{3\phi-1}{2}x_1 y_1 - \phi^2 x_2 y_2 - \frac{1}{\phi^2}x_3 y_3 - x_4 y_4 - \frac{\sqrt{3}}{2\phi}x_1 y_5 \\ \sqrt{3}x_1 y_4 + \sqrt{3}x_2 y_3 + \frac{3}{2}x_4 y_5 - \frac{\sqrt{15}}{2}x_4 y_1 \end{pmatrix}_5 \tag{4.31}$$

$$\begin{pmatrix} x_1 \\ x_2 \\ x_3 \\ x_4 \\ x_5 \end{pmatrix}_5 \otimes \begin{pmatrix} y_1 \\ y_2 \\ y_3 \\ y_4 \\ y_5 \end{pmatrix}_5$$

$$= (x_1 y_1 + x_2 y_2 + x_3 y_3 + x_4 y_4 + x_5 y_5)_1$$

$$\oplus \begin{pmatrix} x_1 y_3 - x_3 y_1 + x_2 y_4 - x_4 y_2 + \sqrt{3}(x_3 y_5 - x_5 y_3) \\ x_1 y_4 - x_4 y_1 - (x_2 y_3 - x_3 y_2) - \sqrt{3}(x_4 y_5 - x_5 y_4) \\ -2(x_1 y_2 - x_2 y_1) - (x_3 y_4 - x_4 y_3) \end{pmatrix}_3$$

$$\oplus \begin{pmatrix} (2\phi + 3)(x_1 y_4 - x_4 y_1) + 2\phi(x_2 y_3 - x_3 y_2) + \sqrt{3}(x_4 y_5 - x_5 y_4) \\ (\phi + 3)(x_1 y_3 - x_3 y_1) + 2\phi(x_2 y_4 - x_4 y_2) - \sqrt{3}\phi^2(x_3 y_5 - x_5 y_3) \\ -\phi(x_1 y_2 - x_2 y_1) - \sqrt{15}\phi(x_2 y_5 - x_5 y_2) + 2\phi(x_3 y_4 - x_4 y_3) \end{pmatrix}_{3'}$$

$$\oplus \begin{pmatrix} \sqrt{5}\left(\frac{\phi^2}{2}(x_1 y_4 + x_4 y_1) + x_2 y_3 + x_3 y_2 + \frac{\sqrt{3}}{2\phi}(x_4 y_5 + x_5 y_4)\right) \\ \sqrt{5}\left(\frac{1}{2\phi^2}(x_1 y_3 + x_3 y_1) - (x_2 y_4 + x_4 y_2) + \frac{\sqrt{3}}{2}\phi(x_3 y_5 + x_5 y_3)\right) \\ \frac{\sqrt{5}}{2}\left(-\sqrt{5}(x_1 y_2 + x_2 y_1) + \sqrt{3}(x_2 y_5 + x_5 y_2) + 2(x_3 y_4 + x_4 y_3)\right) \\ 3x_1 y_1 - 2(x_2 y_2 + x_3 y_3 + x_4 y_4) + 3x_5 y_5 \end{pmatrix}_4$$

$$\oplus \begin{pmatrix} \frac{1}{\sqrt{5}}\left(\frac{1}{2\phi}(x_1 y_4 - x_4 y_1) - (x_2 y_3 - x_3 y_2) + \frac{\phi^2}{2\sqrt{3}}(x_4 y_5 - x_5 y_4)\right) \\ \frac{1}{\sqrt{5}}\left(\frac{\phi}{2}(x_1 y_3 - x_3 y_1) - (x_2 y_4 - x_4 y_2) + \frac{1}{2\sqrt{3}\phi^2}(x_3 y_5 - x_5 y_3)\right) \\ \frac{1}{2\sqrt{5}}(x_1 y_2 - x_2 y_1) - \frac{1}{2\sqrt{3}}(x_2 y_5 - x_5 y_2) - \frac{1}{\sqrt{5}}(x_3 y_4 - x_4 y_3) \\ -\frac{1}{\sqrt{3}}(x_1 y_5 - x_5 y_1) \end{pmatrix}_4$$

$$\oplus \begin{pmatrix} -x_1 y_1 - \frac{11}{3\sqrt{15}}(x_1 y_5 + x_5 y_1) + \frac{4}{3}x_2 y_2 - \frac{4\sqrt{5}}{15}\left(\phi x_3 y_3 + \frac{1}{\phi}x_4 y_4\right) + x_5 y_5 \\ \frac{4}{3}\left(x_1 y_2 + x_2 y_1 + \frac{1}{\sqrt{15}}(x_2 y_5 + x_5 y_2) + \frac{2}{\sqrt{5}}(x_3 y_4 + x_4 y_3)\right) \\ \frac{4}{3\sqrt{5}}\left(-\phi(x_1 y_3 + x_3 y_1) + 2(x_2 y_4 + x_4 y_2) - \frac{2-3\phi}{\sqrt{3}}(x_3 y_5 + x_5 y_3)\right) \\ \frac{4\sqrt{5}}{15}\left(-\frac{1}{\phi}(x_1 y_4 + x_4 y_1) + 2(x_2 y_3 + x_3 y_2) - \frac{\sqrt{3}}{3}(3\phi - 1)(x_4 y_5 + x_5 y_4)\right) \\ x_1 y_5 + x_5 y_1 + \frac{1}{3\sqrt{15}}(-11 x_1 y_1 + 4 x_2 y_2 + 11 x_5 y_5) - \frac{4\sqrt{15}}{45}((2 - 3\phi)x_3 y_3 + (3\phi - 1)x_4 y_4) \end{pmatrix}_5$$

$$\oplus \begin{pmatrix} -\frac{3\sqrt{5}}{4}(x_1 y_1 - x_5 y_5) - \frac{\sqrt{3}}{4}(x_1 y_5 + x_5 y_1) + \sqrt{5}x_2 y_2 - \phi^2 x_3 y_3 + \frac{1}{\phi^2}x_4 y_4 \\ \sqrt{5}(x_1 y_2 + x_2 y_1) + \sqrt{3}(x_2 y_5 + x_5 y_2) + x_3 y_4 + x_4 y_3 \\ -\phi^2(x_1 y_3 + x_3 y_1) + x_2 y_4 + x_4 y_2 + \frac{\sqrt{3}}{\phi}(x_3 y_5 + x_5 y_3) \\ \frac{1}{\phi^2}(x_1 y_4 + x_4 y_1) + (x_2 y_3 + x_3 y_2) - \sqrt{3}\phi(x_4 y_5 + x_5 y_4) \\ -\frac{\sqrt{3}}{4}(x_1 y_1 - 4 x_2 y_2 - x_5 y_5) + \frac{3\sqrt{5}}{4}(x_1 y_5 + x_5 y_1) + \frac{\sqrt{3}}{\phi}x_3 y_3 - \sqrt{3}\phi x_4 y_4 \end{pmatrix}_5 \qquad (4.32)$$

References

1. Altarelli, G., Feruglio, F.: Nucl. Phys. B **741**, 215 (2006). [arXiv:hep-ph/0512103]
2. Shirai, K.: J. Phys. Soc. Jpn. **61**, 2735 (1992)
3. Cummins, C.J., Patera, J.: J. Math. Phys. **29**, 1736 (1988)
4. Feruglio, F., Paris, A.: JHEP **1103**, 101 (2011). [arXiv:1101.0393 [hep-ph]]
5. Ding, G.-J., Everett, L.L., Stuart, A.J.: arXiv:1110.1688 [hep-ph]
6. Everett, L.L., Stuart, A.J.: Phys. Rev. D **79**, 085005 (2009). [arXiv:0812.1057 [hep-ph]]

5

In this section, we study the T' group, which is the double covering group of $A_4 = T$. Instead of Eq. (4.7) for the case of A_4, we consider the following algebraic relations,

$$s^2 = r, \qquad r^2 = t^3 = (st)^3 = e, \qquad rt = tr. \tag{5.1}$$

The closed algebra including r, s and t forms the T' group. This group consists of 24 elements.

The algebraic relation $s^2 = e$ in A_4 is replaced by $s^2 = r$ and $r^2 = e$ in T'. For other groups G, we can construct covering groups, e.g. by replacing the algebraic relation $a^m = e$ in G by $a^m = r$ and $r^n = e$, where r must be commutable with any element in G. For example, when $n = 2$, we can construct double covering groups of G such as S_4', A_5', etc.

● **Conjugacy classes**

All of 24 elements in T' are classified by their orders as

$$
\begin{aligned}
h &= 1 &&: &&\{e\}, \\
h &= 2 &&: &&\{r\}, \\
h &= 3 &&: &&\{t, t^2, ts, st, rst^2, rt^2s, rtst, rsts\}, \\
h &= 4 &&: &&\{s, rs, tst^2, t^2st, rtst^2, rt^2st\}, \\
h &= 6 &&: &&\{rt, rst, rts, rt^2, sts, st^2, t^2s, tst\}.
\end{aligned} \tag{5.2}
$$

Moreover, these elements are classified into seven conjugacy classes as

T. Kobayashi et al., *An Introduction to Non-Abelian Discrete Symmetries for Particle Physicists*, Lecture Notes in Physics 995, https://doi.org/10.1007/978-3-662-64679-3_5

$$
\begin{aligned}
C_1: & & \{e\}, & & h=1, \\
C_{1'}: & & \{r\}, & & h=2, \\
C_4: & & \{t, rsts, st, ts\}, & & h=3, \\
C_{4'}: & & \{t^2, rtst, rt^2s, rst^2\}, & & h=3, \\
C_6: & \{s, rs, tst^2, t^2st, rtst^2, rt^2st\}, & & h=4, \\
C_{4''}: & & \{rt, sts, rst, rts\}, & & h=6, \\
C_{4'''}: & & \{rt^2, tst, t^2s, st^2\}, & & h=6.
\end{aligned}
\tag{5.3}
$$

• Characters and representations

The orthogonality relations (2.18) and (2.19) for T' lead to

$$
m_1 + 2^2 m_2 + 3^2 m_3 + \cdots = 24, \tag{5.4}
$$

$$
m_1 + m_2 + m_3 + \cdots = 7. \tag{5.5}
$$

The solution is given as $(m_1, m_2, m_3) = (3, 3, 1)$. Therefore, we find three singlets, three doublets and a triplet in T'.

Now, we study characters, which are obtained by the quite similar analysis on $A_4 = T$. Let us start with discussing singlets. Because of $s^4 = r^2 = e$, there are four possibilities for $\chi_\alpha(s) = (i)^n$ ($n = 0, 1, 2, 3$). However, since t and ts belong to the same conjugacy class, C_4, the character consistent with the structure of conjugacy classes is obtained as $\chi_\alpha(s) = 1$ for singlets. That also means $\chi_\alpha(r) = 1$. Then, similarly to $A_4 = T$, three singlets are classified by three possible values of $\chi_\alpha(t) = \omega^n$. That is, three singlets, $\mathbf{1}, \mathbf{1}'$, and $\mathbf{1}''$ are classified by these three values of $\chi_\alpha(t) = 1$, ω and ω^2, respectively. These are shown in Table 5.1.

Next, let us study three doublet representations $\mathbf{2}, \mathbf{2}'$ and $\mathbf{2}''$, and a triplet representation $\mathbf{3}$ for r. The element r commutes with all of elements. That implies by the Shur's lemma that r can be represented by

$$
\lambda_{2,2',2''} \begin{pmatrix} 1 & 0 \\ 0 & 1 \end{pmatrix}, \tag{5.6}
$$

on $\mathbf{2}, \mathbf{2}'$ and $\mathbf{2}''$ and

$$
\lambda_3 \begin{pmatrix} 1 & 0 & 0 \\ 0 & 1 & 0 \\ 0 & 0 & 1 \end{pmatrix}, \tag{5.7}
$$

on $\mathbf{3}$. In addition, possible values of $\lambda_{2,2',2''}$ and λ_3 must be equal to $\lambda_{2,2',2''} = \pm 1$ and $\lambda_3 = \pm 1$ because of $r^2 = e$. Thus, we obtain possible values of characters as $\chi_2(r), \chi_{2'}(r), \chi_{2''}(r) = \pm 2$ and $\chi_3(r) = \pm 3$. On the other hand, the second orthogonality relation between e and r gives

$$
\sum_\alpha \chi_{D_\alpha}(e)^* \chi_{D_\alpha}(r) = 3 + 2\chi_2(r) + 2\chi_{2'}(r) + 2\chi_{2''}(r) + 3\chi_3(r) = 0, \tag{5.8}
$$

where $\chi_{1,1',1''}(r) = 1$ is used. We obtain the solution as $\chi_2(r) = \chi_{2'}(r) = \chi_{2''}(r) = -2$ and $\chi_3(r) = 3$. These result are summarized in Table 5.1. Therefore, the r element is represented by

$$r = -\begin{pmatrix} 1 & 0 \\ 0 & 1 \end{pmatrix}, \tag{5.9}$$

on **2**, **2′** and **2″**, and

$$r = \begin{pmatrix} 1 & 0 & 0 \\ 0 & 1 & 0 \\ 0 & 0 & 1 \end{pmatrix}, \tag{5.10}$$

on **3**.

Now, we study doublet representation of t. We use the basis diagonalizing t. Because of $t^3 = e$, the element t could be written as

$$\begin{pmatrix} \omega^k & 0 \\ 0 & \omega^\ell \end{pmatrix}, \tag{5.11}$$

with $k, \ell = 0, 1, 2$. However, if $k = \ell$, the above matrix would become proportional to the (2×2) identity matrix, that is, the element t would also commute with all of elements. It is nothing but a singlet representation. Then, we should have the condition $k \neq \ell$. As a result, there are three possible values for the trace of the above values as $\omega^k + \omega^\ell = -\omega^n$ with $k, \ell, n = 0, 1, 2$ and $k \neq \ell, \ell \neq n, n \neq k$. That is, the characters of t for three doublets, **2**, **2′** and **2″** are classified as $\chi_2(t) = -1$, $\chi_{2'}(t) = -\omega$ and $\chi_{2''}(t) = -\omega^2$. These are shown in Table 5.1. Then, the element t is represented by

$$t = \begin{pmatrix} \omega^2 & 0 \\ 0 & \omega \end{pmatrix}, \qquad \text{on } \mathbf{2}, \tag{5.12}$$

$$t = \begin{pmatrix} 1 & 0 \\ 0 & \omega^2 \end{pmatrix}, \qquad \text{on } \mathbf{2'}, \tag{5.13}$$

$$t = \begin{pmatrix} \omega & 0 \\ 0 & 1 \end{pmatrix}, \qquad \text{on } \mathbf{2''}. \tag{5.14}$$

Since we have found the explicit (2×2) matrices for r and t on all of three doublets, it is straightforward to calculate the explicit forms of rt and rt^2, which belong to the conjugacy classes, C'_4 and C''_4, respectively. Then, it is also straightforward to compute the characters of C'_4 and C''_4 for doublets by such explicit forms of (2×2) matrices for rt and rt^2. They are shown in Table 5.1.

In order to determine the character of t for the triplet, $\chi_3(t)$, we remark the second orthogonality relation between e and t,

$$\sum_\alpha \chi_{D_\alpha}(e)^* \chi_{D_\alpha}(t) = 0. \tag{5.15}$$

Table 5.1 Characters of T' representations

	h	χ_1	$\chi_{1'}$	$\chi_{1''}$	χ_2	$\chi_{2'}$	$\chi_{2''}$	χ_3
C_1	1	1	1	1	2	2	2	3
$C_{1'}$	2	1	1	1	-2	-2	-2	3
C_4	3	1	ω	ω^2	-1	$-\omega$	$-\omega^2$	0
C_4'	3	1	ω^2	ω	-1	$-\omega^2$	$-\omega$	0
C_4''	6	1	ω	ω^2	1	ω	ω^2	0
C_4'''	6	1	ω^2	ω	1	ω^2	ω	0
C_6	4	1	1	1	0	0	0	-1

As all of the characters $\chi_\alpha(t)$ except $\chi_3(t)$ have been derived in the above, the above orthogonality relation (5.15) requires $\chi_3(t) = 0$, that is, $\chi_3(C_4) = 0$. Similarly, it is found that $\chi_3(C_4') = \chi_3(C_4'') = \chi_3(C_4''') = 0$ as shown in Table 5.1. Now, we study the explicit form of the (3×3) matrix for t on the triplet. We take the basis to diagonalize t. Since $t^3 = e$ and $\chi_3(t) = 0$, we can obtain

$$t = \begin{pmatrix} 1 & 0 & 0 \\ 0 & \omega & 0 \\ 0 & 0 & \omega^2 \end{pmatrix}, \quad \text{on } \mathbf{3}. \tag{5.16}$$

Finally, we study the characters of C_6 including s for the doublets and the triplet. Here, we use the first orthogonality relation between the trivial singlet representation and the doublet representation $\mathbf{2}$

$$\sum_{g \in G} \chi_1(g)^* \chi_2(g) = 0. \tag{5.17}$$

Since all of characters except $\chi_2(C_6)$ have been already given, this orthogonality relation (5.17) requires $\chi_2(C_6) = 0$. Similarly, we find that $\chi_{2'}(C_6) = \chi_{2''}(C_6) = 0$. In addition, the character of C_6 for the triplet $\chi_3(C_6)$ is also determined by using the orthogonality relation $\sum_{g \in G} \chi_1(g)^* \chi_2(g) = 0$ with the other known characters. As a result, we obtain $\chi_3(C_6) = -1$. Now, we have completed all of characters in the T' group, which are summarized in Table 5.1. Then, we study the explicit form of s on the doublets and triplet. On the doublets, the element must be the (2×2) unitary matrix, which satisfies $\text{tr}(s) = 0$ and $s^2 = r$. Recall that the doublet representation for r is already obtained in Eq. (5.9). Thus, the element s could be represented as

$$s = -\frac{1}{\sqrt{3}} \begin{pmatrix} i & \sqrt{2}p \\ \sqrt{2}\bar{p} & -i \end{pmatrix}, \quad p = e^{i\phi}, \tag{5.18}$$

on the doublet representations. For example, for $\mathbf{2}$ this representation of s satisfies

$$\text{tr}(st) = -\frac{i}{\sqrt{3}}(\omega^2 - \omega) = -1, \tag{5.19}$$

so the ambiguity of p cannot be removed. Similarly, we can study the explicit form of s on the tiplet.

Here, we summarize the doublet and triplet representations,

$$t = \begin{pmatrix} \omega^2 & 0 \\ 0 & \omega \end{pmatrix}, \quad r = \begin{pmatrix} -1 & 0 \\ 0 & -1 \end{pmatrix}, \quad s = -\frac{1}{\sqrt{3}} \begin{pmatrix} i & \sqrt{2}p \\ -\sqrt{2}\bar{p} & -i \end{pmatrix} \quad \text{on } \mathbf{2}, \quad (5.20)$$

$$t = \begin{pmatrix} 1 & 0 \\ 0 & \omega^2 \end{pmatrix}, \quad r = \begin{pmatrix} -1 & 0 \\ 0 & -1 \end{pmatrix}, \quad s = -\frac{1}{\sqrt{3}} \begin{pmatrix} i & \sqrt{2}p \\ -\sqrt{2}\bar{p} & -i \end{pmatrix} \quad \text{on } \mathbf{2'}, \quad (5.21)$$

$$t = \begin{pmatrix} \omega & 0 \\ 0 & 1 \end{pmatrix}, \quad r = \begin{pmatrix} -1 & 0 \\ 0 & -1 \end{pmatrix}, \quad s = -\frac{1}{\sqrt{3}} \begin{pmatrix} i & \sqrt{2}p \\ -\sqrt{2}\bar{p} & -i \end{pmatrix} \quad \text{on } \mathbf{2''}, \quad (5.22)$$

$$t = \begin{pmatrix} 1 & 0 & 0 \\ 0 & \omega & 0 \\ 0 & 0 & \omega^2 \end{pmatrix}, \quad r = \begin{pmatrix} 1 & 0 & 0 \\ 0 & 1 & 0 \\ 0 & 0 & 1 \end{pmatrix}, \quad s = \frac{1}{3} \begin{pmatrix} -1 & 2p_1 & 2p_1p_2 \\ 2\bar{p}_1 & -1 & 2p_2 \\ 2\bar{p}_1\bar{p}_2 & 2\bar{p}_2 & -1 \end{pmatrix} \quad \text{on } \mathbf{3}, \quad (5.23)$$

where $p_1 = e^{i\phi_1}$ and $p_2 = e^{i\phi_2}$.

• Tensor products

Now we can deterimine tensor products. First, we study the tensor product of $\mathbf{2}$ and $\mathbf{2}$, i.e.

$$\begin{pmatrix} x_1 \\ x_2 \end{pmatrix}_\mathbf{2} \otimes \begin{pmatrix} y_1 \\ y_2 \end{pmatrix}_\mathbf{2}. \quad (5.24)$$

Let us investigate the transformation property of elements $x_i y_j$ for $i, j = 1, 2$ under t, r and s. It is easily found that

$$\begin{pmatrix} x_1 \\ x_2 \end{pmatrix}_{\mathbf{2}(\mathbf{2'})} \otimes \begin{pmatrix} y_1 \\ y_2 \end{pmatrix}_{\mathbf{2}(\mathbf{2''})} = \begin{pmatrix} \frac{x_1 y_2 - x_2 y_1}{\sqrt{2}} \end{pmatrix}_\mathbf{1} \oplus \begin{pmatrix} \frac{i}{\sqrt{2}} p_1 p_2 \bar{p}(x_1 y_2 + x_2 y_1) \\ p_2 \bar{p}^2 x_1 y_1 \\ x_2 y_2 \end{pmatrix}_\mathbf{3}. \quad (5.25)$$

Similarly, we can obtain

$$\begin{pmatrix} x_1 \\ x_2 \end{pmatrix}_{\mathbf{2'}(\mathbf{2})} \otimes \begin{pmatrix} y_1 \\ y_2 \end{pmatrix}_{\mathbf{2'}(\mathbf{2''})} = \begin{pmatrix} \frac{x_1 y_2 - x_2 y_1}{\sqrt{2}} \end{pmatrix}_{\mathbf{1''}} \oplus \begin{pmatrix} p_1 \bar{p}^2 x_1 y_1 \\ x_2 y_2 \\ \frac{i}{\sqrt{2}} \bar{p} \bar{p}_2 (x_1 y_2 + x_2 y_1) \end{pmatrix}_\mathbf{3}, \quad (5.26)$$

$$\begin{pmatrix} x_1 \\ x_2 \end{pmatrix}_{\mathbf{2''}(\mathbf{2})} \otimes \begin{pmatrix} y_1 \\ y_2 \end{pmatrix}_{\mathbf{2''}(\mathbf{2'})} = \begin{pmatrix} \frac{x_1 y_2 - x_2 y_1}{\sqrt{2}} \end{pmatrix}_{\mathbf{1'}} \oplus \begin{pmatrix} x_2 y_2 \\ \frac{i}{\sqrt{2}} \bar{p} \bar{p}_1 (x_1 y_2 + x_2 y_1) \\ \bar{p}^2 \bar{p}_1 \bar{p}_2 x_1 y_1 \end{pmatrix}_\mathbf{3}. \quad (5.27)$$

We can also compute other products such as $\mathbf{2} \times \mathbf{2'}, \mathbf{2} \times \mathbf{2''}$ and $\mathbf{2'} \times \mathbf{2''}$. It is found that

$$\mathbf{2} \times \mathbf{2'} = \mathbf{2''} \times \mathbf{2''}, \qquad \mathbf{2} \times \mathbf{2''} = \mathbf{2'} \times \mathbf{2'}, \qquad \mathbf{2'} \times \mathbf{2''} = \mathbf{2} \times \mathbf{2}. \quad (5.28)$$

Moreover, a similar analysis leads to

$$\begin{pmatrix} x_1 \\ x_2 \\ x_3 \end{pmatrix}_3 \otimes \begin{pmatrix} y_1 \\ y_2 \\ y_3 \end{pmatrix}_3 = [x_1 y_1 + p_1^2 p_2 (x_2 y_3 + x_3 y_2)]_1$$

$$\oplus [x_3 y_3 + \bar{p}_1 \bar{p}_2^2 (x_1 y_2 + x_2 y_1)]_{1'} \oplus [(x_2 y_2 + \bar{p}_1 p_2 (x_1 y_3 + x_3 y_1)]_{1''}$$

$$\oplus \begin{pmatrix} 2x_1 y_1 - p_1^2 p_2 (x_2 y_3 + x_3 y_2) \\ 2p_1 p_2^2 x_3 y_3 - x_1 y_2 - x_2 y_1 \\ 2p_1 \bar{p}_2 x_2 y_2 - x_1 y_3 - x_3 y_1 \end{pmatrix}_3$$

$$\oplus \begin{pmatrix} x_2 y_3 - x_3 y_2 \\ \bar{p}_1^2 \bar{p}_2 (x_1 y_2 - x_2 y_1) \\ \bar{p}_1^2 \bar{p}_2 (x_3 y_1 - x_1 y_3) \end{pmatrix}_3 , \tag{5.29}$$

$$\begin{pmatrix} x_1 \\ x_2 \end{pmatrix}_{2,2',2''} \otimes \begin{pmatrix} y_1 \\ y_2 \\ y_3 \end{pmatrix}_3 = \begin{pmatrix} -i\sqrt{2} p p_1 x_2 y_2 + x_1 y_1 \\ i\sqrt{2} \bar{p} p_1 p_2 x_1 y_3 - x_2 y_1 \end{pmatrix}_{2,2',2''}$$

$$\oplus \begin{pmatrix} -i\sqrt{2} p p_2 x_2 y_3 + x_1 y_2 \\ i\sqrt{2} \bar{p} \bar{p}_1 x_1 y_1 - x_2 y_2 \end{pmatrix}_{2',2'',2}$$

$$\oplus \begin{pmatrix} -i\sqrt{2} p \bar{p}_1 \bar{p}_2 x_2 y_1 + x_1 y_3 \\ i\sqrt{2} \bar{p} \bar{p}_2 x_1 y_2 - x_2 y_3 \end{pmatrix}_{2'',2,2'} , \tag{5.30}$$

$$(x)_{1'(1'')} \otimes \begin{pmatrix} y_1 \\ y_2 \end{pmatrix}_{2,2',2''} = \begin{pmatrix} x y_1 \\ x y_2 \end{pmatrix}_{2'(2''),2''(2),2(2')} , \tag{5.31}$$

$$(x)_{1'} \otimes \begin{pmatrix} y_1 \\ y_2 \\ y_3 \end{pmatrix}_3 = \begin{pmatrix} x y_3 \\ \bar{p}_1^2 \bar{p}_2 x y_1 \\ \bar{p}_1 \bar{p}_2^2 x y_2 \end{pmatrix}_3 , \quad (x)_{1''} \otimes \begin{pmatrix} y_1 \\ y_2 \\ y_3 \end{pmatrix}_3 = \begin{pmatrix} x y_2 \\ \bar{p}_1 p_2 x y_3 \\ \bar{p}_1^2 \bar{p}_2 x y_1 \end{pmatrix}_3 . \tag{5.32}$$

The representations for p' can be in general obtained by transforming p as follows,

$$\Phi_2(p') = \begin{pmatrix} 1 & 0 \\ 0 & e^{-i\gamma} \end{pmatrix} \Phi_2(p), \quad p' = p e^{i\gamma}, \tag{5.33}$$

$$\Phi_3(p') = \begin{pmatrix} 1 & 0 & 0 \\ 0 & e^{i\alpha} & 0 \\ 0 & 0 & e^{-i(\alpha+\beta)} \end{pmatrix} \Phi_3(p), \quad p_1' = p_1 e^{i\alpha}, \quad p_2' = p_2 e^{-i\beta}. \tag{5.34}$$

If one takes the parameters $p = i$ and $p_1 = p_2 = 1$, then the generator s is simplified as

$$s = -\frac{i}{\sqrt{3}} \begin{pmatrix} 1 & \sqrt{2} \\ \sqrt{2} & -1 \end{pmatrix}, \quad \text{on } \mathbf{2}, \tag{5.35}$$

$$s = \frac{1}{3} \begin{pmatrix} -1 & 2 & 2 \\ 2 & -1 & 2 \\ 2 & 2 & -1 \end{pmatrix}, \quad \text{on } \mathbf{3}.$$

These tensor products can be also simplified as

$$\begin{pmatrix} x_1 \\ x_2 \end{pmatrix}_{2(2')} \otimes \begin{pmatrix} y_1 \\ y_2 \end{pmatrix}_{2(2'')} = \begin{pmatrix} \frac{x_1 y_2 - x_2 y_1}{\sqrt{2}} \end{pmatrix}_1 \oplus \begin{pmatrix} \frac{x_1 y_2 + x_2 y_1}{\sqrt{2}} \\ -x_1 y_1 \\ x_2 y_2 \end{pmatrix}_3 , \quad (5.36)$$

$$\begin{pmatrix} x_1 \\ x_2 \end{pmatrix}_{2'(2)} \otimes \begin{pmatrix} y_1 \\ y_2 \end{pmatrix}_{2'(2'')} = \begin{pmatrix} \frac{x_1 y_2 - x_2 y_1}{\sqrt{2}} \end{pmatrix}_{1''} \oplus \begin{pmatrix} -x_1 y_1 \\ x_2 y_2 \\ \frac{x_1 y_2 + x_2 y_1}{\sqrt{2}} \end{pmatrix}_3 , \quad (5.37)$$

$$\begin{pmatrix} x_1 \\ x_2 \end{pmatrix}_{2''(2)} \otimes \begin{pmatrix} y_1 \\ y_2 \end{pmatrix}_{2''(2')} = \begin{pmatrix} \frac{x_1 y_2 - x_2 y_1}{\sqrt{2}} \end{pmatrix}_{1'} \oplus \begin{pmatrix} x_2 y_2 \\ \frac{x_1 y_2 + x_2 y_1}{\sqrt{2}} \\ -x_1 y_1 \end{pmatrix}_3 , \quad (5.38)$$

$$\begin{pmatrix} x_1 \\ x_2 \\ x_3 \end{pmatrix}_3 \otimes \begin{pmatrix} y_1 \\ y_2 \\ y_3 \end{pmatrix}_3 = [x_1 y_1 + x_2 y_3 + x_3 y_2]_1$$

$$\oplus [x_3 y_3 + x_1 y_2 + x_2 y_1]_{1'} \oplus (x_2 y_2 + x_1 y_3 + x_3 y_1)_{1''}$$

$$\oplus \begin{pmatrix} 2x_1 y_1 - x_2 y_3 - x_3 y_3 \\ 2x_3 y_3 - x_1 y_2 - x_2 y_1 \\ 2x_2 y_2 - x_1 y_3 - x_3 y_1 \end{pmatrix}_3$$

$$\oplus \begin{pmatrix} x_2 y_3 - x_3 y_2 \\ x_1 y_2 - x_2 y_1 \\ x_3 y_1 - x_1 y_3 \end{pmatrix}_3 , \quad (5.39)$$

$$\begin{pmatrix} x_1 \\ x_2 \end{pmatrix}_{2,2',2''} \otimes \begin{pmatrix} y_1 \\ y_2 \\ y_3 \end{pmatrix}_3 = \begin{pmatrix} \sqrt{2}x_2 y_2 + x_1 y_1 \\ \sqrt{2}x_1 y_3 - x_2 y_1 \end{pmatrix}_{2,2',2''} \oplus \begin{pmatrix} \sqrt{2}x_2 y_3 + x_1 y_2 \\ \sqrt{2}x_1 y_1 - x_2 y_2 \end{pmatrix}_{2',2'',2}$$

$$\oplus \begin{pmatrix} \sqrt{2}x_2 y_1 + x_1 y_3 \\ \sqrt{2}x_1 y_2 - x_2 y_3 \end{pmatrix}_{2'',2,2'} , \quad (5.40)$$

$$(x)_{1'(1'')} \otimes \begin{pmatrix} y_1 \\ y_2 \end{pmatrix}_{2,2',2''} = \begin{pmatrix} x y_1 \\ x y_2 \end{pmatrix}_{2'(2''),2''(2),2(2')} , \quad (5.41)$$

$$(x)_{1'} \otimes \begin{pmatrix} y_1 \\ y_2 \\ y_3 \end{pmatrix}_3 = \begin{pmatrix} x y_3 \\ x y_1 \\ x y_2 \end{pmatrix}_3 , \quad (x)_{1''} \otimes \begin{pmatrix} y_1 \\ y_2 \\ y_3 \end{pmatrix}_3 = \begin{pmatrix} x y_2 \\ x y_3 \\ x y_1 \end{pmatrix}_3 . \quad (5.42)$$

When $p = e^{i\pi/12}$ and $p_1 = p_2 = \omega$, the representation and their tensor products are given in Appendix E.2.

In this section, we discuss dihedral group, which is denoted by D_N. It is a symmetry of a regular polygon with N sides. This group is isomorphic to $Z_N \rtimes Z_2$ and is also denoted by $\Delta(2N)$. It consists of cyclic rotation, Z_N and reflection. That is, it is generated by two generators a and b, which act on N edges x_i ($i = 1, \ldots, N$) of N-polygon as

$$a : (x_1, x_2 \ldots, x_N) \rightarrow (x_N, x_1 \ldots, x_{N-1}), \tag{6.1}$$
$$b : (x_1, x_2 \ldots, x_N) \rightarrow (x_1, x_N \ldots, x_2). \tag{6.2}$$

These two generators satisfy

$$a^N = e, \quad b^2 = e, \quad bab = a^{-1}, \tag{6.3}$$

where the third equation is equivalent to $aba = b$. The order of D_N is equal to $2N$, and all of $2N$ elements are written as $a^m b^k$ with $m = 0, \ldots, N - 1$ and $k = 0, 1$. The third equation in (6.3) implies that the Z_N subgroup including a^m is a normal subgroup of D_N. Thus, D_N corresponds to a semi-direct product between Z_N including a^m and Z_2 including b^k, i.e. $Z_N \rtimes Z_2$. Equation (6.1) corresponds to the (reducible) N-dimensional representation. The simple doublet representation is written as

$$a = \begin{pmatrix} \cos 2\pi/N & -\sin 2\pi/N \\ \sin 2\pi/N & \cos 2\pi/N \end{pmatrix}, \quad b = \begin{pmatrix} 1 & 0 \\ 0 & -1 \end{pmatrix}. \tag{6.4}$$

T. Kobayashi et al., *An Introduction to Non-Abelian Discrete Symmetries for Particle Physicists*, Lecture Notes in Physics 995, https://doi.org/10.1007/978-3-662-64679-3_6

6.1 D_N with $N =$ Even

D_N groups have different features for $N =$ even and odd. We begin to study D_N with $N =$ even.

• Conjugacy classes
The algebraic relations (6.3) tell us that a^m and a^{N-m} belong to the same conjugacy class and also b and $a^{2m}b$ belong to the same conjugacy class. When N is even, D_N has the following $3 + N/2$ conjugacy classes,

$$
\begin{array}{lll}
C_1 : & \{e\}, & h = 1, \\
C_2^{(1)} : & \{a, a^{N-1}\}, & h = N, \\
\quad\vdots & \quad\vdots & \quad\vdots, \\
C_2^{(N/2-1)} : & \{a^{N/2-1}, a^{N/2+1}\}, & h = N/gcd(N, N/2 - 1), \\
C_1' : & \{a^{N/2}\}, & h = 2, \\
C_{N/2} : & \{b, a^2b, \dots, a^{N-2}b\}, & h = 2, \\
C_{N/2}' : & \{ab, a^3b, \dots, a^{N-1}b\}, & h = 2,
\end{array}
\tag{6.5}
$$

where we have also shown the orders of each element in the conjugacy class by h. That implies that there are $3 + N/2$ irreducible representations. Furthermore, the orthogonality relation (2.18) requires

$$
\sum_\alpha [\chi_\alpha(C_1)]^2 = \sum_n m_n n^2 = m_1 + 4m_2 + 9m_3 + \cdots = 2N, \tag{6.6}
$$

for m_i, which satisfies $m_1 + m_2 + m_3 + \cdots = 3 + N/2$. The solution is given as $(m_1, m_2) = (4, N/2 - 1)$. Therefore, it is noticed there are four singlets and $(N/2 - 1)$ doublets.

• Characters and representations
We start with studying on singlets in the case of $N =$ even, where there are four singlets. Since generators satisfy $b^2 = e$ in $C_{N/2}$ and $(ab)^2 = e$ in $C_{N/2}'$, the characters $\chi_\alpha(g)$ for four singlets should be $\chi_\alpha(C_{N/2}) = \pm 1$ and $\chi_\alpha(C_{N/2}') = \pm 1$. Therefore, there are four possible combinations of $\chi_\alpha(C_{N/2}) = \pm 1$ and $\chi_\alpha(C_{N/2}') = \pm 1$ and they correspond to four singlets, $\mathbf{1}_{\pm\pm}$, which are presented in Table 6.1.

Now, let us study doublet representations, that is, (2×2) matrix representations. As seen in Eq. (6.4), those (2×2) matrices correspond to one of doublet representations. The (2×2) matrix representations for generic doublet $\mathbf{2}_k$ are obtained by replacing

$$
a \to a^k. \tag{6.7}
$$

Thus, a and b are represented for the doublet $\mathbf{2}_k$ as

$$
a = \begin{pmatrix} \cos 2\pi k/N & -\sin 2\pi k/N \\ \sin 2\pi k/N & \cos 2\pi k/N \end{pmatrix}, \quad b = \begin{pmatrix} 1 & 0 \\ 0 & -1 \end{pmatrix}, \tag{6.8}
$$

Table 6.1 Characters of $D_{N=\text{even}}$ representations

	h	χ_{1++}	χ_{1+-}	χ_{1-+}	χ_{1--}	χ_{2k}
C_1	1	1	1	1	1	2
C_2^1	N	1	-1	-1	1	$2\cos(2\pi k/N)$
$\vdots$						
$C_2^{N/2-1}$	$N/gcd(N,N/2-1)$	1	$(-1)^{(N/2-1)}$	$(-1)^{(N/2-1)}$	1	$2\cos(2\pi k(N/2-1)/N)$
C_1'	2	1	$(-1)^{N/2}$	$(-1)^{N/2}$	1	-2
$C_{N/2}$	2	1	1	-1	-1	0
$C_{N/2}'$	2	1	-1	1	-1	0

where $k = 1, \ldots, N/2 - 1$ for $N =$ even and $k = 1, \ldots, (N-1)/2$ for $N =$ odd. In the expression of the doublet $\mathbf{2}_k$ as

$$\mathbf{2}_k = \begin{pmatrix} x_k \\ y_k \end{pmatrix}, \tag{6.9}$$

the generator a is the Z_N rotation on the two-dimensional real coordinates (x_k, y_k) and the generator b is the reflection along y_k, i.e. $y_k \to -y_k$. These transformations can be represented on the complex coordinate z_k and its conjugate $\bar{z}_{-k}$. These bases are transformed as

$$\begin{pmatrix} z_k \\ \bar{z}_{-k} \end{pmatrix} = U \begin{pmatrix} x_k \\ y_k \end{pmatrix}, \qquad U = \frac{1}{\sqrt{2}} \begin{pmatrix} 1 & i \\ 1 & -i \end{pmatrix}. \tag{6.10}$$

In the complex basis, the generators, a and b, can be given as $\tilde{a} = UaU^{-1}$ and $\tilde{b} = UbU^{-1}$,

$$\tilde{a} = \begin{pmatrix} \exp 2\pi ik/N & 0 \\ 0 & \exp -2\pi ik/N \end{pmatrix}, \qquad \tilde{b} = \begin{pmatrix} 0 & 1 \\ 1 & 0 \end{pmatrix}. \tag{6.11}$$

This complex basis may be useful. Actually, the generator $\tilde{a}$ is the diagonal matrix. That implies that in the doublet $\mathbf{2}_k$, which is denoted by

$$\mathbf{2}_k = \begin{pmatrix} z_k \\ \bar{z}_{-k} \end{pmatrix}, \tag{6.12}$$

each of up and down components, z_k and $\bar{z}_{-k}$, has the definite Z_N charge. That is, Z_N charges of z_k and $\bar{z}_{-k}$ are equal to k and $-k$, respectively. The characters of these matrices for the doublets $\mathbf{2}_k$ are presented in Table 6.1. It is easily found that these characters satisfy the orthogonality relations (2.16) and (2.17).

• **Tensor products**

In the next step, we discuss the tensor products of the D_N group with $N =$ even. Let us start with $\mathbf{2}_k \times \mathbf{2}_{k'}$, i.e.

$$\begin{pmatrix} z_k \\ \bar{z}_{-k} \end{pmatrix}_{\mathbf{2}_k} \otimes \begin{pmatrix} z_{k'} \\ \bar{z}_{-k'} \end{pmatrix}_{\mathbf{2}_{k'}}, \tag{6.13}$$

where $k, k' = 1, \ldots, N/2 - 1$. It is noted that $z_k z_{k'}, z_k \bar{z}_{-k'}, \bar{z}_{-k} z_{k'}$ and $\bar{z}_{-k} \bar{z}_{-k'}$ have define Z_N changes, i.e. $k + k'$, $k - k'$, $-k + k'$ and $-k - k'$, respectively. For the case with $k + k' \neq N/2$ and $k - k' \neq 0$, they are decomposed into two doublets as

$$\begin{pmatrix} z_k \\ \bar{z}_{-k} \end{pmatrix}_{\mathbf{2}_k} \otimes \begin{pmatrix} z_{k'} \\ \bar{z}_{-k'} \end{pmatrix}_{\mathbf{2}_{k'}} = \begin{pmatrix} z_k z_{k'} \\ \bar{z}_{-k} \bar{z}_{-k'} \end{pmatrix}_{\mathbf{2}_{k+k'}} \oplus \begin{pmatrix} z_k \bar{z}_{-k'} \\ \bar{z}_{-k} z_{k'} \end{pmatrix}_{\mathbf{2}_{k-k'}}. \tag{6.14}$$

In the case of $k + k' = N/2$, the matrix a is represented on the above (reducible) doublet $(z_k z_{k'}, \bar{z}_{-k} \bar{z}_{-k'})$ as

$$a \begin{pmatrix} z_k z_{k'} \\ \bar{z}_{-k} \bar{z}_{-k'} \end{pmatrix} = \begin{pmatrix} -1 & 0 \\ 0 & -1 \end{pmatrix} \begin{pmatrix} z_k z_{k'} \\ \bar{z}_{-k} \bar{z}_{-k'} \end{pmatrix}. \tag{6.15}$$

Since a is proportional to the (2×2) identity matrix for $(z_k z_{k'}, \bar{z}_{-k} \bar{z}_{-k'})$ with $k + k' = N/2$, we can diagonalize another matrix b in this vector space $(z_k z_{k'}, \bar{z}_{-k} \bar{z}_{-k'})$. Such a basis is given as $(z_k z_{k'} + \bar{z}_{-k} \bar{z}_{-k'}, z_k z_{k'} - \bar{z}_{-k} \bar{z}_{-k'})$ and their eigenvalues of b are given as

$$b \begin{pmatrix} z_k z_{k'} + \bar{z}_{-k} \bar{z}_{-k'} \\ z_k z_{k'} - \bar{z}_{-k} \bar{z}_{-k'} \end{pmatrix} = \begin{pmatrix} 1 & 0 \\ 0 & -1 \end{pmatrix} \begin{pmatrix} z_k z_{k'} + \bar{z}_{-k} \bar{z}_{-k'} \\ z_k z_{k'} - \bar{z}_{-k} \bar{z}_{-k'} \end{pmatrix}. \tag{6.16}$$

Thus, $z_k z_{k'} + \bar{z}_{-k} \bar{z}_{-k'}$ and $z_k z_{k'} - \bar{z}_{-k} \bar{z}_{-k'}$ correspond to $\mathbf{1}_{+-}$ and $\mathbf{1}_{-+}$, respectively.

In the case of $k - k' = 0$, a similar decomposition is obtained for the (reducible) doublet $(z_k \bar{z}_{-k'}, \bar{z}_{-k} z_{k'})$. The generator a is the (2×2) identity matrix on the vector space $(z_k \bar{z}_{-k'}, \bar{z}_{-k} z_{k'})$ with $k - k' = 0$. Therefore, we can take the basis $(z_k \bar{z}_{-k'} + \bar{z}_{-k} z_{k'}, z_k \bar{z}_{-k'} - \bar{z}_{-k} z_{k'})$, where b is diagonalized. That is, $z_k \bar{z}_{-k'} + \bar{z}_{-k} z_{k'}$ and $z_k \bar{z}_{-k'} - \bar{z}_{-k} z_{k'}$ correspond to $\mathbf{1}_{++}$ and $\mathbf{1}_{--}$, respectively.

Now, we study the tensor products of the doublets $\mathbf{2}_k$ and singlets, e.g. $\mathbf{1}_{--} \times \mathbf{2}_k$. Here we denote the vector space for the singlet $\mathbf{1}_{--}$ by w, where $aw = w$ and $bw = -w$. It is easily found that $(wz_k, -w\bar{z}_k)$ is nothing but the doublet $\mathbf{2}_k$, that is, $\mathbf{1}_{--} \times \mathbf{2}_k = \mathbf{2}_k$. Similar results are obtained for other singlets. Furthermore, it is straightforward to study the tensor products among singlets.

Hence, the tensor products of D_N irreducible representations with $N =$ even are summarized as

$$\begin{pmatrix} z_k \\ \bar{z}_{-k} \end{pmatrix}_{\mathbf{2}_k} \otimes \begin{pmatrix} z_{k'} \\ \bar{z}_{-k'} \end{pmatrix}_{\mathbf{2}_{k'}} = \begin{pmatrix} z_k z_{k'} \\ \bar{z}_{-k} \bar{z}_{-k'} \end{pmatrix}_{\mathbf{2}_{k+k'}} \oplus \begin{pmatrix} z_k \bar{z}_{-k'} \\ \bar{z}_{-k} z_{k'} \end{pmatrix}_{\mathbf{2}_{k-k'}}, \tag{6.17}$$

for $k + k' \neq N/2$ and $k - k' \neq 0$,

$$\begin{pmatrix} z_k \\ \bar{z}_{-k} \end{pmatrix}_{2_k} \otimes \begin{pmatrix} z_{k'} \\ \bar{z}_{-k'} \end{pmatrix}_{2_{k'}} = \left(z_k z_{k'} + \bar{z}_{-k}\bar{z}_{-k'}\right)_{1_{+-}} \oplus \left(z_k z_{k'} - \bar{z}_{-k}\bar{z}_{-k'}\right)_{1_{-+}}$$
$$\oplus \begin{pmatrix} z_k \bar{z}_{-k'} \\ \bar{z}_{-k} z_{k'} \end{pmatrix}_{2_{k-k'}}, \tag{6.18}$$

for $k + k' = N/2$ and $k - k' \neq 0$,

$$\begin{pmatrix} z_k \\ \bar{z}_{-k} \end{pmatrix}_{2_k} \otimes \begin{pmatrix} z_{k'} \\ \bar{z}_{-k'} \end{pmatrix}_{2_{k'}} = \left(z_k \bar{z}_{-k'} + \bar{z}_{-k} z_{k'}\right)_{1_{++}} \oplus \left(z_k \bar{z}_{-k'} - \bar{z}_{-k} z_{k'}\right)_{1_{--}}$$
$$\oplus \begin{pmatrix} z_k z_{k'} \\ \bar{z}_{-k}\bar{z}_{-k'} \end{pmatrix}_{2_{k+k'}}, \tag{6.19}$$

for $k + k' \neq N/2$ and $k - k' = 0$,

$$\begin{pmatrix} z_k \\ \bar{z}_{-k} \end{pmatrix}_{2_k} \otimes \begin{pmatrix} z_{k'} \\ \bar{z}_{-k'} \end{pmatrix}_{2_{k'}} = \left(z_k \bar{z}_{-k'} + \bar{z}_{-k} z_{k'}\right)_{1_{++}} \oplus \left(z_k \bar{z}_{-k'} - \bar{z}_{-k} z_{k'}\right)_{1_{--}}$$
$$\oplus \left(z_k z_{k'} + \bar{z}_{-k}\bar{z}_{-k'}\right)_{1_{+-}} \oplus \left(z_k z_{k'} - \bar{z}_{-k}\bar{z}_{-k'}\right)_{1_{-+}}, \tag{6.20}$$

for $k + k' = N/2$ and $k - k' = 0$, and

$$(w)_{1_{++}} \otimes \begin{pmatrix} z_k \\ \bar{z}_{-k} \end{pmatrix}_{2_k} = \begin{pmatrix} w z_k \\ w\bar{z}_{-k} \end{pmatrix}_{2_k},$$

$$(w)_{1_{--}} \otimes \begin{pmatrix} z_k \\ \bar{z}_{-k} \end{pmatrix}_{2_k} = \begin{pmatrix} w z_k \\ -w\bar{z}_{-k} \end{pmatrix}_{2_k},$$

$$(w)_{1_{+-}} \otimes \begin{pmatrix} z_k \\ \bar{z}_{-k} \end{pmatrix}_{2_k} = \begin{pmatrix} w\bar{z}_{-k} \\ w z_k \end{pmatrix}_{2_{N/2-k}},$$

$$(w)_{1_{-+}} \otimes \begin{pmatrix} z_k \\ \bar{z}_{-k} \end{pmatrix}_{2_k} = \begin{pmatrix} w\bar{z}_{-k} \\ -w z_k \end{pmatrix}_{2_{N/2-k}}. \tag{6.21}$$

$$\mathbf{1}_{s_1 s_2} \otimes \mathbf{1}_{s_1' s_2'} = \mathbf{1}_{s_1'' s_2''}, \tag{6.22}$$

with $s_i, s_i', s_i'' = \pm$ $(i = 1, 2)$, where $s_i'' = +$ for $(s_i, s_i') = (+, +)$ and $(-, -)$, and $s_i'' = -$ for $(s_i, s_i') = (+, -)$ and $(-, +)$. Hereafter, this sign rule for s_i'' is denoted by $s_i'' = s_i s_i'$ $(i = 1, 2)$ for simplicity.

Note that the above multiplication rules are the same between the complex basis and the real basis. For example, in both bases we get

$$\mathbf{2}_k \otimes \mathbf{2}_{k'} = \mathbf{2}_{k+k'} + \mathbf{2}_{k-k'},$$

for $k + k' \neq N/2$ and $k - k' \neq 0$. On the other hand, elements of doublets are written in a different way, although those transform as (6.10).

6.2 D_N with $N = $ Odd

In the case of D_N with $N = $ odd, a similar study is given for conjugacy classes, characters, representations and tensor products.

• Conjugacy classes
The D_N group with $N = $ odd has the following $2 + (N-1)/2$ conjugacy classes,

$$
\begin{aligned}
C_1 : & \quad \{e\}, & h &= 1, \\
C_2^{(1)} : & \quad \{a, a^{N-1}\}, & h &= N, \\
\vdots & \quad \quad \vdots & \vdots& \\
C_2^{(N-1)/2} : & \quad \{a^{(N-1)/2}, a^{(N+1)/2}\}, & h &= N/gcd(N, (N-1)/2), \\
C_N : & \quad \{b, ab, \dots, a^{N-1}b\}, & h &= 2.
\end{aligned}
\tag{6.23}
$$

That is, there are $2 + (N-1)/2$ irreducible representations. Furthermore, the orthogonality relation (2.18) requires the same equation as (6.6) for m_i, which satisfies $m_1 + m_2 + m_3 + \cdots = 2 + (N-1)/2$. The solution is found as $(m_1, m_2) = (2, (N-1)/2)$. Thus, it is found that there are two singlets and $(N-1)/2$ doublets.

• Characters and representations
We study two singlets of D_N with $N = $ odd. Since $b^2 = e$ in C_N is satisfied, the characters $\chi_\alpha(g)$ for two singlets should be $\chi_\alpha(C_N) = \pm 1$. Since both b and ab belong to the same conjugacy class C_N, the characters $\chi_\alpha(a)$ for two singlets must always satisfy $\chi_\alpha(a) = 1$. That is, there are two singlets, $\mathbf{1}_+$ and $\mathbf{1}_-$. Their characters are determined by whether the conjugacy class includes b or not as shown in Table 6.2.

The doublet representations of D_N with $N = $ odd are the same as those in D_N with $N = $ even. Their characters are shown in Table 6.2.

• Tensor products
Let us discuss the tensor products of D_N irreducible representations with $N = $ odd. We can analyze them in the similar way of D_N with $N = $ even. Results are summa-

Table 6.2 Characters of $D_{N=\text{odd}}$ representations

	h	χ_{1_+}	χ_{1_-}	χ_{2k}
C_1	1	1	1	2
C_2^1	N	1	1	$2\cos(2\pi k/N)$
$\vdots$				
$C_2^{(N-1)/2}$	$N/gcd(N, (N-1)/2)$	1	1	$2\cos(2\pi k(N-1)/2N)$
C_N	2	1	-1	0

rized as follows,

$$
\begin{pmatrix} z_k \\ \bar{z}_{-k} \end{pmatrix}_{2_k} \otimes \begin{pmatrix} z_{k'} \\ \bar{z}_{-k'} \end{pmatrix}_{2_{k'}} = \begin{pmatrix} z_k z_{k'} \\ \bar{z}_{-k}\bar{z}_{-k'} \end{pmatrix}_{2_{k+k'}} \oplus \begin{pmatrix} z_k \bar{z}_{-k'} \\ \bar{z}_{-k} z_{k'} \end{pmatrix}_{2_{k-k'}}, \tag{6.24}
$$

for $k - k' \neq 0$, where $k, k' = 1, \ldots, (N-1)/2$,

$$
\begin{pmatrix} z_k \\ \bar{z}_{-k} \end{pmatrix}_{2_k} \otimes \begin{pmatrix} z_{k'} \\ \bar{z}_{-k'} \end{pmatrix}_{2_{k'}} = \left(z_k \bar{z}_{-k'} + \bar{z}_{-k} z_{k'} \right)_{1_+} \oplus \left(z_k \bar{z}_{-k'} - \bar{z}_{-k} z_{k'} \right)_{1_-}
$$
$$
\oplus \begin{pmatrix} z_k z_{k'} \\ \bar{z}_{-k}\bar{z}_{-k'} \end{pmatrix}_{2_{k+k'}}, \tag{6.25}
$$

for $k - k' = 0$, and

$$
(w)_{1_+} \otimes \begin{pmatrix} z_k \\ \bar{z}_{-k} \end{pmatrix}_{2_k} = \begin{pmatrix} w z_k \\ w\bar{z}_{-k} \end{pmatrix}_{2_k}, \qquad (w)_{1_-} \otimes \begin{pmatrix} z_k \\ \bar{z}_{-k} \end{pmatrix}_{2_k} = \begin{pmatrix} w z_k \\ -w\bar{z}_{-k} \end{pmatrix}_{2_k}, \tag{6.26}
$$

$$
1_s \otimes 1_{s'} = 1_{s''}, \tag{6.27}
$$

where $s'' = ss'$.

6.3 D_4

In this subsection, we present simple examples of D_N. The smallest non-Abelian group in D_N is D_3. However, D_3 corresponds to a group of all possible permutations of three objects, that is, S_3. Thus, we show D_4 and D_5 as simple examples.

The D_4 is the symmetry of a square, which is generated by the $\pi/2$ rotation a and the reflection b, where they satisfy $a^4 = e$, $b^2 = e$ and $bab = a^{-1}$. (See Fig. 6.1.) The D_4 consists of the eight elements, $a^m b^k$ with $m = 0, 1, 2, 3$ and $k = 0, 1$. The D_4 has the following five conjugacy classes,

$$
\begin{aligned}
C_1 &: \quad \{e\}, \quad h = 1, \\
C_2 &: \quad \{a, a^3\}, \quad h = 4, \\
C_1' &: \quad \{a^2\}, \quad h = 2, \\
C_2' &: \quad \{b, a^2 b\}, \quad h = 2, \\
C_2'' &: \quad \{ab, a^3 b\}, \quad h = 2,
\end{aligned} \tag{6.28}
$$

where the orders of each element in the conjugacy class are given by h.

Fig. 6.1 The D_4 symmetry of a square

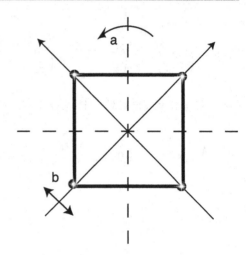

Table 6.3 Characters of D_4 representations

	h	$\chi_{1_{++}}$	$\chi_{1_{+-}}$	$\chi_{1_{-+}}$	$\chi_{1_{--}}$	χ_2
C_1	1	1	1	1	1	2
C_2	4	1	-1	-1	1	0
C_1'	2	1	1	1	1	-2
C_2'	2	1	1	-1	-1	0
C_2''	2	1	-1	1	-1	0

The D_4 has four singlets, 1_{++}, 1_{+-}, 1_{-+} and 1_{--}, and one doublet 2. The characters are shown in Table 6.3. The tensor products are given as

$$
\begin{pmatrix} z \\ \bar{z} \end{pmatrix}_2 \otimes \begin{pmatrix} z' \\ \bar{z}' \end{pmatrix}_2 = (z\bar{z}' + \bar{z}z')_{1_{++}} \oplus (z\bar{z}' - \bar{z}z')_{1_{--}}
$$
$$
\oplus (zz' + \bar{z}\bar{z}')_{1_{+-}} \oplus (zz' - \bar{z}\bar{z}')_{1_{-+}} , \qquad (6.29)
$$

$$
(w)_{1_{++}} \otimes \begin{pmatrix} z \\ \bar{z} \end{pmatrix}_2 = \begin{pmatrix} wz \\ w\bar{z} \end{pmatrix}_2, \qquad (w)_{1_{--}} \otimes \begin{pmatrix} z \\ \bar{z} \end{pmatrix}_2 = \begin{pmatrix} wz \\ -w\bar{z} \end{pmatrix}_2,
$$
$$
(w)_{1_{+-}} \otimes \begin{pmatrix} z \\ \bar{z} \end{pmatrix}_2 = \begin{pmatrix} w\bar{z} \\ wz \end{pmatrix}_2, \qquad (w)_{1_{-+}} \otimes \begin{pmatrix} z \\ \bar{z} \end{pmatrix}_2 = \begin{pmatrix} w\bar{z} \\ -wz \end{pmatrix}_2, \qquad (6.30)
$$

$$
1_{s_1 s_2} \otimes 1_{s_1' s_2'} = 1_{s_1'' s_2''}, \qquad (6.31)
$$

where $s_1'' = s_1 s_1'$ and $s_2'' = s_2 s_2'$.

6.4 D_5

Let us show the D_5 group, which is the symmetry of a regular pentagon. This is generated by the $2\pi/5$ rotation a and the reflection b. See Fig. 6.2. Generators satisfy that $a^5 = e$, $b^2 = e$ and $bab = a^{-1}$. The D_5 includes the 10 elements, $a^m b^k$ with $m = 0, 1, 2, 3, 4$ and $k = 0, 1$. They are classified into the following four conjugacy classes,

$$
\begin{array}{lll}
C_1: & \{e\}, & h = 1, \\
C_2^{(1)}: & \{a, a^4\}, & h = 5, \\
C_2^{(2)}: & \{a^2, a^3\}, & h = 5, \\
C_5: & \{b, ab, a^2b, a^3b, a^4b\}, & h = 2.
\end{array}
\tag{6.32}
$$

The D_5 has two singlets, $\mathbf{1}_+$ and $\mathbf{1}_-$, and two doublets, $\mathbf{2}_1$ and $\mathbf{2}_2$. Their characters are shown in Table 6.4.

The tensor products are given as

$$
\begin{pmatrix} z \\ \bar{z} \end{pmatrix}_{\mathbf{2}_2} \otimes \begin{pmatrix} z' \\ \bar{z}' \end{pmatrix}_{\mathbf{2}_1} = \begin{pmatrix} zz' \\ \bar{z}\bar{z}' \end{pmatrix}_{\mathbf{2}_2} \oplus \begin{pmatrix} z\bar{z}' \\ \bar{z}z' \end{pmatrix}_{\mathbf{2}_1},
\tag{6.33}
$$

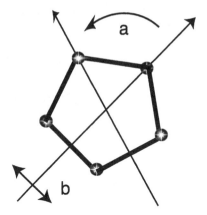

Fig. 6.2 The D_5 symmetry of a regular pentagon

Table 6.4 Characters of D_5 representations

	h	χ_{1_+}	χ_{1_-}	χ_{2_1}	χ_{2_2}
C_1	1	1	1	2	2
C_2^1	5	1	1	$2\cos(2\pi/5)$	$2\cos(4\pi/5)$
C_2^2	5	1	1	$2\cos(4\pi/5)$	$2\cos(8\pi/5)$
C_5	2	1	-1	0	0

$$\begin{pmatrix} z_k \\ \bar{z}_{-k} \end{pmatrix}_{2_k} \otimes \begin{pmatrix} z'_k \\ \bar{z}'_{-k} \end{pmatrix}_{2_k} = (z_k \bar{z}'_{-k} + \bar{z}_{-k} z'_k)_{1_+} \oplus (z_k \bar{z}'_{-k} - \bar{z}_{-k} z'_k)_{1_-} \oplus \begin{pmatrix} z_k z'_k \\ \bar{z}_{-k} \bar{z}'_{-k} \end{pmatrix}_{2_{2k}},$$
$$\tag{6.34}$$

$$(w)_{1_+} \otimes \begin{pmatrix} z_k \\ \bar{z}_{-k} \end{pmatrix}_{2_k} = \begin{pmatrix} w z_k \\ w \bar{z}_{-k} \end{pmatrix}_{2_k}, \qquad (w)_{1_-} \otimes \begin{pmatrix} z_k \\ \bar{z}_{-k} \end{pmatrix}_{2_k} = \begin{pmatrix} w z_k \\ -w \bar{z}_{-k} \end{pmatrix}_{2_k},$$
$$\tag{6.35}$$

$$\mathbf{1}_s \otimes \mathbf{1}_{s'} = \mathbf{1}_{s''}, \tag{6.36}$$

where $s'' = ss'$.

The binary dihedral group is called as Q_N, where N is even. It consists of the elements, $a^m b^k$ with $m = 0, \ldots, N - 1$ and $k = 0, 1$, where the generators a and b satisfy

$$a^N = e, \qquad b^2 = (a^{N/2}), \qquad b^{-1}ab = a^{-1}. \tag{7.1}$$

The order of Q_N is equal to $2N$. The generator a can be represented by the same (2×2) matrices as D_N, i.e.

$$a = \begin{pmatrix} \exp 2\pi i k/N & 0 \\ 0 & \exp -2\pi i k/N \end{pmatrix}. \tag{7.2}$$

It is noted that $a^{N/2} = e$ for $k =$ even and $a^{N/2} = -e$ for $k =$ odd. That leads to that $b^2 = e$ for $k =$ even and $b^2 = -e$ for $k =$ odd. Thus, the generators a and b are represented by (2×2) matrices, e.g. as

$$a = \begin{pmatrix} \exp 2\pi i k/N & 0 \\ 0 & \exp -2\pi i k/N \end{pmatrix}, \qquad b = \begin{pmatrix} 0 & i \\ i & 0 \end{pmatrix}, \tag{7.3}$$

for $k =$ odd,

$$a = \begin{pmatrix} \exp 2\pi i k/N & 0 \\ 0 & \exp -2\pi i k/N \end{pmatrix}, \qquad b = \begin{pmatrix} 0 & 1 \\ 1 & 0 \end{pmatrix}, \tag{7.4}$$

for $k =$ even.

© The Author(s), under exclusive license to Springer-Verlag GmbH, DE,
part of Springer Nature 2022
T. Kobayashi et al., *An Introduction to Non-Abelian Discrete Symmetries for Particle Physicists*, Lecture Notes in Physics 995,
https://doi.org/10.1007/978-3-662-64679-3_7

7.1 Q_N with $N = 4n$

The Q_N groups have different features between $N = 4n$ and $4n + 2$. At first, we study Q_N with $N = 4n$.

• Conjugacy classes
The conjugacy classes are given by the algebraic relations (7.1). The elements are classified into the $(3 + N/2)$ conjugacy classes as

$$
\begin{array}{lll}
C_1 : & \{e\}, & h = 1, \\
C_2^{(1)} : & \{a, a^{N-1}\}, & h = N, \\
\vdots & \vdots & \vdots, \\
C_2^{(N/2-1)} : & \{a^{N/2-1}, a^{N/2+1}\}, & h = N/\gcd(N, N/2 - 1), \\
C_1' : & \{a^{N/2}\}, & h = 2, \\
C_{N/2} : & \{b, a^2 b, \ldots, a^{N-2} B\}, & h = 4, \\
C_{N/2}' : & \{ab, a^3 b, \ldots, a^{N-1} B\}, & h = 4,
\end{array}
\tag{7.5}
$$

where h is the orders of each element in the conjugacy class. These are almost the same as the conjugacy classes of D_N with $N = $ even. There must be the $(3 + N/2)$ irreducible representations, and similarly to $D_{N=\text{even}}$ there are four singlets and $(N/2 - 1)$ doublets.

• Characters and representations
The characters of Q_N for doublets are the same as those of $D_{N=\text{even}}$, and are shown in Table 7.1. We study the characters of the singlets of Q_N with $N = 4n$. Then, we have the following relation,

$$
b^2 = a^{2n}. \tag{7.6}
$$

Since $b^4 = e$ in $C_{N/2}$ is satisfied, the characters $\chi_\alpha(b)$ for four singlets must be $\chi_\alpha(b) = e^{\pi i n/2}$ with $n = 0, 1, 2, 3$. Furthermore, it is noticed that the element ba^2 belongs to the same conjugacy class as b. That implies $\chi_\alpha(a^2) = 1$ for four singlets. By using Eq. (7.6), we have $\chi_\alpha(b^2) = 1$, that is, $\chi_\alpha(b) = \pm 1$. Thus, the characters of Q_N with $N = 4n$ for singlets are the same as those of $D_{N=\text{even}}$ and are shown in Table 7.1.

• Tensor products
The tensor products of Q_N irreducible representations can be analyzed similarly to D_N with $N = $ even. The results for Q_N with $N = 4n$ are obtained as

$$
\begin{pmatrix} z_k \\ \bar{z}_{-k} \end{pmatrix}_{2_k} \otimes \begin{pmatrix} z_{k'} \\ \bar{z}_{-k'} \end{pmatrix}_{2_{k'}} = \begin{pmatrix} z_k z_{k'} \\ (-1)^{kk'} \bar{z}_{-k} \bar{z}_{-k'} \end{pmatrix}_{2_{k+k'}} \oplus \begin{pmatrix} z_k \bar{z}_{-k'} \\ (-1)^{kk'} \bar{z}_{-k} z_{k'} \end{pmatrix}_{2_{k-k'}},
\tag{7.7}
$$

Table 7.1 Characters of Q_N representations for $N = 4n$

	h	χ_{1++}	χ_{1+-}	χ_{1-+}	χ_{1--}	χ_{2k}
C_1	1	1	1	1	1	2
C_2^1	N	1	-1	-1	1	$2\cos(2\pi k/N)$
$\vdots$						
$C_2^{N/2-1}$	$N/gcd(N, N/2-1)$	1	$(-1)^{(N/2-1)}$	$(-1)^{(N/2-1)}$	1	$2\cos(2\pi k(N/2-1)/N)$
C_1'	2	1	$(-1)^{N/2}$	$(-1)^{N/2}$	1	-2
$C_{N/2}$	4	1	1	-1	-1	0
$C_{N/2}'$	4	1	-1	1	-1	0

for $k + k' \neq N/2$ and $k - k' \neq 0$,

$$
\begin{pmatrix} z_k \\ \bar{z}_{-k} \end{pmatrix}_{2_k} \otimes \begin{pmatrix} z_{k'} \\ \bar{z}_{-k'} \end{pmatrix}_{2_{k'}} = \left(z_k z_{k'} + (-1)^{kk'} \bar{z}_{-k} \bar{z}_{-k'} \right)_{1_{+-}} \oplus \left(z_k z_{k'} - (-1)^{kk'} \bar{z}_{-k} \bar{z}_{-k'} \right)_{1_{-+}}
$$
$$
\oplus \begin{pmatrix} z_k \bar{z}_{-k'} \\ (-1)^{kk'} \bar{z}_{-k} z_{k'} \end{pmatrix}_{2_{k-k'}}, \tag{7.8}
$$

for $k + k' = N/2$ and $k - k' \neq 0$,

$$
\begin{pmatrix} z_k \\ \bar{z}_{-k} \end{pmatrix}_{2_k} \otimes \begin{pmatrix} z_{k'} \\ \bar{z}_{-k'} \end{pmatrix}_{2_{k'}} = \left(z_k \bar{z}_{-k'} + (-1)^{kk'} \bar{z}_{-k} z_{k'} \right)_{1_{++}} \oplus \left(z_k \bar{z}_{-k'} - (-1)^{kk'} \bar{z}_{-k} z_{k'} \right)_{1_{--}}
$$
$$
\oplus \begin{pmatrix} z_k z_{k'} \\ (-1)^{kk'} \bar{z}_{-k} \bar{z}_{-k'} \end{pmatrix}_{2_{k+k'}}, \tag{7.9}
$$

for $k + k' \neq N/2$ and $k - k' = 0$,

$$
\begin{pmatrix} z_k \\ \bar{z}_{-k} \end{pmatrix}_{2_k} \otimes \begin{pmatrix} z_{k'} \\ \bar{z}_{-k'} \end{pmatrix}_{2_{k'}} = \left(z_k \bar{z}_{-k'} + (-1)^{kk'} \bar{z}_{-k} z_{k'} \right)_{1_{++}} \oplus \left(z_k \bar{z}_{-k'} - (-1)^{kk'} \bar{z}_{-k} z_{k'} \right)_{1_{--}}
$$
$$
\oplus \left(z_k z_{k'} + (-1)^{kk'} \bar{z}_{-k} \bar{z}_{-k'} \right)_{1_{+-}} \oplus \left(z_k z_{k'} - (-1)^{kk'} \bar{z}_{-k} \bar{z}_{-k'} \right)_{1_{-+}}, \tag{7.10}
$$

for $k + k' = N/2$ and $k - k' = 0$,

$$
(w)_{1_{++}} \otimes \begin{pmatrix} z_k \\ \bar{z}_{-k} \end{pmatrix}_{2_k} = \begin{pmatrix} w z_k \\ w \bar{z}_{-k} \end{pmatrix}_{2_k},
$$
$$
(w)_{1_{--}} \otimes \begin{pmatrix} z_k \\ \bar{z}_{-k} \end{pmatrix}_{2_k} = \begin{pmatrix} w z_k \\ -w \bar{z}_{-k} \end{pmatrix}_{2_k},
$$
$$
(w)_{1_{+-}} \otimes \begin{pmatrix} z_k \\ \bar{z}_{-k} \end{pmatrix}_{2_k} = \begin{pmatrix} w \bar{z}_{-k} \\ w z_k \end{pmatrix}_{2_{N/2-k}},
$$
$$
(w)_{1_{-+}} \otimes \begin{pmatrix} z_k \\ \bar{z}_{-k} \end{pmatrix}_{2_k} = \begin{pmatrix} w \bar{z}_{-k} \\ -w z_k \end{pmatrix}_{2_{N/2-k}}, \tag{7.11}
$$

$$1_{s_1 s_2} \otimes 1_{s_1' s_2'} = 1_{s_1'' s_2''}, \tag{7.12}$$

where $s_1'' = s_1 s_1'$ and $s_2'' = s_2 s_2'$. It should be noticed that some minus signs are different from the tensor products of D_N.

7.2 Q_N with $N = 4n + 2$

Similar to Q_N with $N = 4n$, we now investigate Q_N with $N = 4n + 2$.

• **Conjugacy classes**

The conjugacy classes of Q_N with $N = 4n + 2$ are exactly the same as those of Q_N with $N = 4n$.

• **Characters and representations**

The characters of Q_N for doublets are the same as those of $D_{N=\text{even}}$, and are shown in Table 7.2.

Let us consider four singlets of Q_N for $N = 4n + 2$. In this case we have the relation,

$$b^2 = a^{2n+1}. \tag{7.13}$$

Since b and $a^2 b$ are included in the same conjugacy class, the characters $\chi_\alpha(a^2)$ for four singlets must be $\chi_\alpha(a^2) = 1$. Thus, we obtain $\chi_\alpha(a) = \pm 1$. In the case of $\chi_\alpha(a) = 1$, the relation (7.13) leads to the two possibilities $\chi_\alpha(b) = \pm 1$. On the other hand, in the case of $\chi_\alpha(a) = -1$, the relation (7.13) gives other possibilities $\chi_\alpha(b) = \pm i$. Thus, there are four possibilities corresponding to the four singlets. It is noted that $\chi_\alpha(a) = \chi_\alpha(b^2)$ is satisfied for all of singlets.

Table 7.2 Characters of Q_N representations for $N = 4n + 2$

	h	$\chi_{1_{++}}$	$\chi_{1_{+-}}$	$\chi_{1_{-+}}$	$\chi_{1_{--}}$	χ_{2k}
C_1	1	1	1	1	1	2
C_2^1	N	1	-1	-1	1	$2\cos(2\pi k/N)$
$\vdots$						
$C_2^{N/2-1}$	$N/gcd(N, N/2-1)$	1	$(-1)^{(N/2-1)}$	$(-1)^{(N/2-1)}$	1	$2\cos(2\pi k(N/2-1)/N)$
C_1'	2	1	$(-1)^{N/2}$	$(-1)^{N/2}$	1	-2
$C_{N/2}$	4	1	i	$-i$	-1	0
$C_{N/2}'$	4	1	$-i$	i	-1	0

● **Tensor products**

We can obtain the tensor products of Q_N with $N = 4n + 2$ similarly to tensor products of Q_N with $N = 4n$ as follows:

$$\begin{pmatrix} z_k \\ \bar{z}_{-k} \end{pmatrix}_{2_k} \otimes \begin{pmatrix} z_{k'} \\ \bar{z}_{-k'} \end{pmatrix}_{2_{k'}} = \begin{pmatrix} z_k z_{k'} \\ (-1)^{kk'} \bar{z}_{-k} \bar{z}_{-k'} \end{pmatrix}_{2_{k+k'}} \oplus \begin{pmatrix} z_k \bar{z}_{-k'} \\ (-1)^{kk'} \bar{z}_{-k} z_{k'} \end{pmatrix}_{2_{k-k'}},$$

(7.14)

for $k + k' \neq N/2$ and $k - k' \neq 0$,

$$\begin{pmatrix} z_k \\ \bar{z}_{-k} \end{pmatrix}_{2_k} \otimes \begin{pmatrix} z_{k'} \\ \bar{z}_{-k'} \end{pmatrix}_{2_{k'}} = \left(z_k z_{k'} + \bar{z}_{-k} \bar{z}_{-k'} \right)_{1_{+-}} \oplus \left(z_k z_{k'} - \bar{z}_{-k} \bar{z}_{-k'} \right)_{1_{-+}}$$

$$\oplus \begin{pmatrix} z_k \bar{z}_{-k'} \\ (-1)^{kk'} \bar{z}_{-k} z_{k'} \end{pmatrix}_{2_{k-k'}},$$

(7.15)

for $k + k' = N/2$ and $k - k' \neq 0$,

$$\begin{pmatrix} z_k \\ \bar{z}_{-k} \end{pmatrix}_{2_k} \otimes \begin{pmatrix} z_{k'} \\ \bar{z}_{-k'} \end{pmatrix}_{2_{k'}} = \left(z_k \bar{z}_{-k'} + (-1)^{kk'} \bar{z}_{-k} z_{k'} \right)_{1_{++}} \oplus \left(z_k \bar{z}_{-k'} - (-1)^{kk'} \bar{z}_{-k} z_{k'} \right)_{1_{--}}$$

$$\oplus \begin{pmatrix} z_k z_{k'} \\ (-1)^{kk'} \bar{z}_{-k} \bar{z}_{-k'} \end{pmatrix}_{2_{k+k'}},$$

(7.16)

for $k + k' \neq N/2$ and $k - k' = 0$,

$$\begin{pmatrix} z_k \\ \bar{z}_{-k} \end{pmatrix}_{2_k} \otimes \begin{pmatrix} z_{k'} \\ \bar{z}_{-k'} \end{pmatrix}_{2_{k'}} = \left(z_k \bar{z}_{-k'} + (-1)^{kk'} \bar{z}_{-k} z_{k'} \right)_{1_{++}} \oplus \left(z_k \bar{z}_{-k'} - (-1)^{kk'} \bar{z}_{-k} z_{k'} \right)_{1_{--}}$$

$$\oplus \left(z_k z_{k'} + \bar{z}_{-k} \bar{z}_{-k'} \right)_{1_{+-}} \oplus \left(z_k z_{k'} - \bar{z}_{-k} \bar{z}_{-k'} \right)_{1_{-+}},$$

(7.17)

for $k + k' = N/2$ and $k - k' = 0$,

$$(w)_{1_{++}} \otimes \begin{pmatrix} z_k \\ \bar{z}_{-k} \end{pmatrix}_{2_k} = \begin{pmatrix} w z_k \\ w \bar{z}_{-k} \end{pmatrix}_{2_k},$$

$$(w)_{1_{--}} \otimes \begin{pmatrix} z_k \\ \bar{z}_{-k} \end{pmatrix}_{2_k} = \begin{pmatrix} w z_k \\ -w \bar{z}_{-k} \end{pmatrix}_{2_k},$$

$$(w)_{1_{+-}} \otimes \begin{pmatrix} z_k \\ \bar{z}_{-k} \end{pmatrix}_{2_k} = \begin{pmatrix} w \bar{z}_{-k} \\ w z_k \end{pmatrix}_{2_{N/2-k}},$$

$$(w)_{1_{-+}} \otimes \begin{pmatrix} z_k \\ \bar{z}_{-k} \end{pmatrix}_{2_k} = \begin{pmatrix} w \bar{z}_{-k} \\ -w z_k \end{pmatrix}_{2_{N/2-k}},$$

(7.18)

$$\mathbf{1}_{s_1 s_2} \otimes \mathbf{1}_{s_1' s_2'} = \mathbf{1}_{s_1'' s_2''},$$

(7.19)

where $s_1'' = s_1 s_1'$ and $s_2'' = s_2 s_2'$.

7.3 Q_4

Simple examples of Q_N are useful for applications. At first, we present the results on Q_4, and in the next subsection we show Q_6.

The group Q_4 is called Quaternion group. It has the eight elements: $a^m b^k$, for $m = 0, 1, 2, 3$ and $k = 0, 1$, where a and b satisfy $a^4 = e$, $b^2 = a^2$ and $b^{-1}ab = a^{-1}$. These elements are classified into the five conjugacy classes,

$$
\begin{aligned}
C_1 &: \quad \{e\}, \quad h = 1, \\
C_2 &: \quad \{a, a^3\}, \quad h = 4, \\
C_1' &: \quad \{a^2\}, \quad h = 2, \\
C_2' &: \quad \{b, a^2 b\}, \quad h = 4, \\
C_2'' &: \quad \{ab, a^3 b\}, \quad h = 4.
\end{aligned}
\tag{7.20}
$$

The Q_4 has four singlets, $\mathbf{1}_{++}$, $\mathbf{1}_{+-}$, $\mathbf{1}_{-+}$ and $\mathbf{1}_{--}$, and one doublet $\mathbf{2}$. The characters are shown in Table 7.3. The tensor products are given as

$$
\begin{aligned}
\begin{pmatrix} z \\ \bar{z} \end{pmatrix}_2 \otimes \begin{pmatrix} z' \\ \bar{z}' \end{pmatrix}_2 &= \left(z\bar{z}' - \bar{z}z' \right)_{\mathbf{1}_{++}} \oplus \left(z\bar{z}' + \bar{z}z' \right)_{\mathbf{1}_{--}} \\
&\oplus \left(zz' - \bar{z}\bar{z}' \right)_{\mathbf{1}_{+-}} \oplus \left(zz' + \bar{z}\bar{z}' \right)_{\mathbf{1}_{-+}},
\end{aligned}
\tag{7.21}
$$

$$
(w)_{\mathbf{1}_{++}} \otimes \begin{pmatrix} z \\ \bar{z} \end{pmatrix}_2 = \begin{pmatrix} wz \\ w\bar{z} \end{pmatrix}_2, \quad (w)_{\mathbf{1}_{--}} \otimes \begin{pmatrix} z \\ \bar{z} \end{pmatrix}_2 = \begin{pmatrix} wz \\ -w\bar{z} \end{pmatrix}_2,
$$

$$
(w)_{\mathbf{1}_{+-}} \otimes \begin{pmatrix} z \\ \bar{z} \end{pmatrix}_2 = \begin{pmatrix} w\bar{z} \\ wz \end{pmatrix}_2, \quad (w)_{\mathbf{1}_{-+}} \otimes \begin{pmatrix} z \\ \bar{z} \end{pmatrix}_2 = \begin{pmatrix} w\bar{z} \\ -wz \end{pmatrix}_2,
\tag{7.22}
$$

$$
\mathbf{1}_{s_1 s_2} \otimes \mathbf{1}_{s_1' s_2'} = \mathbf{1}_{s_1'' s_2''},
\tag{7.23}
$$

where $s_1'' = s_1 s_1'$ and $s_2'' = s_2 s_2'$. Some minus signs are different from the tensor products of D_4.

Table 7.3 Characters of Q_4 representations

	h	$\chi_{1_{++}}$	$\chi_{1_{+-}}$	$\chi_{1_{-+}}$	$\chi_{1_{--}}$	χ_2
C_1	1	1	1	1	1	2
C_2	4	1	-1	-1	1	0
C_1'	2	1	1	1	1	-2
C_2'	4	1	1	-1	-1	0
C_2''	4	1	-1	1	-1	0

7.4 Q_6

The Q_6 has the 12 elements, $a^m b^k$, for $m = 0, 1, 2, 3, 4, 5$ and $k = 0, 1$, where a and b satisfy $a^6 = e$, $b^2 = a^3$ and $b^{-1}ab = a^{-1}$. These elements are classified into the six conjugacy classes,

$$
\begin{aligned}
C_1 &: & \{e\}, & \quad h = 1, \\
C_2^{(1)} &: & \{a, a^5\}, & \quad h = 6, \\
C_2^{(2)} &: & \{a^2, a^4\}, & \quad h = 3, \\
C_1' &: & \{a^3\}, & \quad h = 2, \\
C_3 &: & \{b, a^2b, a^4b\}, & \quad h = 4, \\
C_3' &: & \{ab, a^3b, a^5b\}, & \quad h = 4.
\end{aligned}
\tag{7.24}
$$

The Q_6 has four singlets, $\mathbf{1}_{++}$, $\mathbf{1}_{+-}$, $\mathbf{1}_{-+}$, and $\mathbf{1}_{--}$, and two doublets, $\mathbf{2}_1$ and $\mathbf{2}_2$. The characters are shown in Table 7.4. The tensor products are given as

$$
\begin{pmatrix} z \\ \bar{z} \end{pmatrix}_{\mathbf{2}_2} \otimes \begin{pmatrix} z' \\ \bar{z}' \end{pmatrix}_{\mathbf{2}_1} = (zz' - \bar{z}\bar{z}')_{\mathbf{1}_{+-}} \oplus (zz' + \bar{z}\bar{z}')_{\mathbf{1}_{-+}} \oplus \begin{pmatrix} z\bar{z}' \\ \bar{z}z' \end{pmatrix}_{\mathbf{2}_1}, \tag{7.25}
$$

$$
\begin{pmatrix} z \\ \bar{z} \end{pmatrix}_{\mathbf{2}_k} \otimes \begin{pmatrix} z' \\ \bar{z}' \end{pmatrix}_{\mathbf{2}_k} = (z\bar{z}' - \bar{z}z')_{\mathbf{1}_{++}} \oplus (z\bar{z}' + \bar{z}z')_{\mathbf{1}_{--}} \oplus \begin{pmatrix} zz' \\ -\bar{z}\bar{z}' \end{pmatrix}_{\mathbf{2}_{k'}}, \tag{7.26}
$$

for $k, k' = 1, 2$ and $k' \neq k$, and

$$
\begin{aligned}
(w)_{\mathbf{1}_{++}} \otimes \begin{pmatrix} z_k \\ \bar{z}_{-k} \end{pmatrix}_{\mathbf{2}_k} = \begin{pmatrix} wz_k \\ w\bar{z}_{-k} \end{pmatrix}_{\mathbf{2}_k}, & \quad (w)_{\mathbf{1}_{--}} \otimes \begin{pmatrix} z_k \\ \bar{z}_{-k} \end{pmatrix}_{\mathbf{2}_k} = \begin{pmatrix} wz_k \\ -w\bar{z}_{-k} \end{pmatrix}_{\mathbf{2}_k}, \\
(w)_{\mathbf{1}_{+-}} \otimes \begin{pmatrix} z_k \\ \bar{z}_{-k} \end{pmatrix}_{\mathbf{2}_k} = \begin{pmatrix} w\bar{z}_{-k} \\ wz_k \end{pmatrix}_{\mathbf{2}_{3-k}}, & \quad (w)_{\mathbf{1}_{-+}} \otimes \begin{pmatrix} z_k \\ \bar{z}_{-k} \end{pmatrix}_{\mathbf{2}_k} = \begin{pmatrix} w\bar{z}_{-k} \\ -wz_k \end{pmatrix}_{\mathbf{2}_{3-k}},
\end{aligned}
\tag{7.27}
$$

Table 7.4 Characters of Q_6 representations

	h	$\chi_{1_{++}}$	$\chi_{1_{+-}}$	$\chi_{1_{-+}}$	$\chi_{1_{--}}$	χ_{2_1}	χ_{2_2}
C_1	1	1	1	1	1	2	2
C_2^1	6	1	-1	-1	1	$2\cos(2\pi/6)$	$2\cos(4\pi/6)$
C_2^2	3	1	1	1	1	$2\cos(4\pi/6)$	$2\cos(8\pi/6)$
C_1'	2	1	1	1	1	-2	2
C_3	4	1	i	$-i$	-1	0	0
C_3'	4	1	$-i$	i	-1	0	0

for $k = 1, 2$.

$$\mathbf{1}_{s_1 s_2} \otimes \mathbf{1}_{s_1' s_2'} = \mathbf{1}_{s_1'' s_2''}, \tag{7.28}$$

where $s_1'' = s_1 s_1'$ and $s_2'' = s_2 s_2'$.

8.1 Generic Aspects

Here, we briefly study generic aspects of QD_{2N}. Let us start with a semi-direct product, $Z_{2^{N'-1}} \rtimes Z_2$, whose order is $2^{N'}$. The Abelian groups, $Z_{2^{N'-1}}$ and Z_2, are generated by two generators a and b, respectively, and they satisfy,

$$a^{2^{N'-1}} = 1, \qquad b^2 = 1. \tag{8.1}$$

Because of the semi-direct product, we require

$$bab^{-1} = a^m. \tag{8.2}$$

If $m = 1 \bmod 2^{N'-1}$, this corresponds to the direct product. Thus, we require $m \neq 1 \bmod 2^{N'-1}$. Furthermore, by use of $a = ba^m b$, it is found that

$$a^m = (ba^m b) \dots (ba^m b) = ba^{m^2} b. \tag{8.3}$$

Note that $a^m = bab$ because of $a = ba^m b$. Then, the consistency requires $a^{m^2} = a$. That implies

$$m^2 = 1 \bmod 2^{N'-1}. \tag{8.4}$$

It is generally known that the solutions of the above equation are given as $m = \pm 1$ for $N' \leq 3$ and $m = \pm 1$, $2^{N'-2} \pm 1$ for $N' \geq 4$.

For $m = 1$, it is nothing but a direct product group. For $m = -1$, one finds that it is identified as a dihedral group. For the non-trivial solution, $m = 2^{N'-2} - 1$, the

T. Kobayashi et al., *An Introduction to Non-Abelian Discrete Symmetries for Particle Physicists*, Lecture Notes in Physics 995, https://doi.org/10.1007/978-3-662-64679-3_8

group $Z_{2^{N'-1}} \rtimes Z_2$ is called *"Quasi Dihedral group"*.[1] Therefore, we define $QD_{2^{N'}}$ for $N' \geq 4$ such as QD_{16}, QD_{32}, etc.

Hereafter we denote $N \equiv 2^{N'-1}$ for the convenience. Then, the QD_{2N} group is isomorphic to $Z_N \rtimes Z_2$ and the generators of Z_N and Z_2, a and b, satisfy

$$a^N = 1, \quad b^2 = 1, \quad bab = a^{N/2-1}. \tag{8.5}$$

All of the QD_{2N} elements are written as $a^k b^\ell$ for $k = 0, \dots, N-1$ and $\ell = 0, 1$. The generators, a and b, are represented, e.g. as

$$a = \begin{pmatrix} \rho & 0 \\ 0 & \rho^{N/2-1} \end{pmatrix}, \quad b = \begin{pmatrix} 0 & 1 \\ 1 & 0 \end{pmatrix}, \tag{8.6}$$

where $\rho = e^{2\pi i/N}$.

• **Conjugacy classes**

The algebraic relations (8.5) tell us that a^k and $a^{k(N/2-1)}$ belong to the same conjugacy class and also b and $a^{m(N/2-2)}b$ belong to the same conjugacy class. The QD_{2N} group has the following $(3 + N/2)$ conjugacy classes,

$$
\begin{array}{lll}
C_1: & \{e\}, & h = 1, \\
C_2^{[k]}: & \{a^k, a^{k(N/2-1)}\}, & h = N/gcd(N, k), \\
C_1': & \{a^{N/2}\}, & h = 2, \\
C_{N/2}: & \{b, a^2b, \dots, a^{N-2}b\}, & h = 2, \\
C_{N/2}': & \{ab, a^3b, \dots, a^{N-1}b\}, & h = 4,
\end{array} \tag{8.7}
$$

where $[k] = k$ or $k(N/2 - 1)$ mod N with $k = 1, \dots, N-1$ except $N/2$. We have also shown the orders of each element in the conjugacy class by h. That implies that there are $(3 + N/2)$ irreducible representations. Furthermore, the orthogonality relation (2.18) requires

$$\sum_\alpha [\chi_\alpha(C_1)]^2 = \sum_n m_n n^2 = m_1 + 4m_2 + 9m_3 + \cdots = 2N, \tag{8.8}$$

for m_i, which satisfies $m_1 + m_2 + m_3 + \cdots = 3 + N/2$. The solution is given as $(m_1, m_2) = (4, N/2 - 1)$. Therefore, it is found that there are four singlets and $(N/2 - 1)$ doublets.

• **Characters and representations**

Here, we study characters and representations. The QD_{2N} group has four singlets and $(N/2 - 1)$ doublets. We denote the four singlets $\mathbf{1}_{ss'}$ with $s, s' = \pm$. Characters of a and b are obtained as $\chi_{\mathbf{1}_{\pm s'}}(a) = \pm 1$ for any s' and $\chi_{\mathbf{1}_{s\pm}}(b) = \pm 1$ for any s.

---

[1] For the non-trivial solution; $m = 2^{N'-2} + 1$, no one gives a name.

Table 8.1 Characters of QD_{2N} representations, where $\rho = e^{2\pi i/N}$. Notice that $\rho^{mk} + \rho^{mk(N/2-1)} = 2\cos(2\pi mk/N)$ when mk is even, and $\rho^{mk} + \rho^{mk(N/2-1)} = 2i\sin(2\pi mk/N)$ when mk is odd

	h	χ_{1++}	χ_{1-+}	χ_{1--}	χ_{1+-}	$\chi_{2_{[k]}}$
C_1	1	1	1	1	1	2
$C_2^{[m]}$	$N/gcd(N,m)$	1	$(-1)^m$	$(-1)^m$	1	$\rho^{mk} + \rho^{mk(N/2-1)}$
C_1'	2	1	$(-1)^{N/2}=1$	$(-1)^{N/2}=1$	1	-2
$C_{N/2}$	2	1	1	-1	-1	0
$C_{N/2}'$	4	1	-1	1	-1	0

Next, we study doublets. The two generators, a and b, are represented, e.g., as

$$a = \begin{pmatrix} \rho^k & 0 \\ 0 & \rho^{k(N/2-1)} \end{pmatrix}, \qquad b = \begin{pmatrix} 0 & 1 \\ 1 & 0 \end{pmatrix}, \tag{8.9}$$

on the doublet $2_{[k]}$. We also denote the vector $2_{[k]}$ as

$$2_{[k]} = \begin{pmatrix} x_k \\ x_{(N/2-1)k} \end{pmatrix}, \tag{8.10}$$

where each of up and down components, x_k and $x_{(N/2-1)k}$, has the definite Z_N charge. The characters are shown in Table 8.1.

• Tensor products
Here, we can discuss the tensor products of the QD_{2N} group. Let us start with $2_k \times 2_{k'}$, i.e.

$$\begin{pmatrix} x_k \\ x_{(N/2-1)k} \end{pmatrix}_{2_k} \otimes \begin{pmatrix} y_{k'} \\ y_{(N/2-1)k'} \end{pmatrix}_{2_{k'}}, \tag{8.11}$$

where $k, k' = 1, \ldots, N-1$ except $N/2$.

Hence, the tensor products of QD_{2N} irreducible representations are generally given by

$$\begin{pmatrix} x_k \\ x_{(N/2-1)k} \end{pmatrix}_{2_{[k]}} \otimes \begin{pmatrix} y_{k'} \\ y_{(N/2-1)k'} \end{pmatrix}_{2_{[k']}} = \begin{pmatrix} x_k y_{k'} \\ x_{(N/2-1)k} y_{(N/2-1)k'} \end{pmatrix}_{2_{[k+k']}} \oplus \begin{pmatrix} x_k y_{(N/2-1)k'} \\ x_{N/2-1k} y_{k'} \end{pmatrix}_{2_{[k+(N/2-1)k']}}, \tag{8.12}$$

for $k + k', k + (N/2 - 1)k' \neq 0, N/2$. In a certain case, the above representation becomes reducible. For example, if $k + k' = 0 \mod N$, the doublet $2_{[k+k']}$ can be reduced as

$$\begin{pmatrix} x_k y_{k'} \\ x_{(N/2-1)k} y_{(N/2-1)k'} \end{pmatrix}_{2_{[k+k']}} = (x_k y_{k'} + x_{(N/2-1)k} y_{(N/2-1)k'})_{1_{++}} \oplus (x_k y_{k'} - x_{(N/2-1)k} y_{(N/2-1)k'})_{1_{+-}}. \tag{8.13}$$

Similarly, when $k + k' = N/2 \mod N$, the doublet $2_{[k+k']}$ can be reduced as

$$
\begin{pmatrix} x_k y_{k'} \\ x_{(N/2-1)k} y_{(N/2-1)k'} \end{pmatrix}_{2_{[k+k']}} = \left(x_k y_{k'} + x_{(N/2-1)k} y_{(N/2-1)k'} \right)_{1_{-+}} \oplus \left(x_k y_{k'} - x_{(N/2-1)k} y_{(N/2-1)k'} \right)_{1_{--}}.
$$

(8.14)

In addition, the tensor products between singlets and doublets are obtained as

$$
(w)_{1_{++}} \otimes \begin{pmatrix} x_k \\ x_{(N/2-1)k} \end{pmatrix}_{2_{[k]}} = \begin{pmatrix} w x_k \\ w x_{(N/2-1)k} \end{pmatrix}_{2_{[k]}},
\tag{8.15}
$$

$$
(w)_{1_{--}} \otimes \begin{pmatrix} x_k \\ x_{(N/2-1)k} \end{pmatrix}_{2_{[k]}} = \begin{pmatrix} w x_{(N/2-1)k} \\ -w x_k \end{pmatrix}_{2_{[k+N/2]}},
\tag{8.16}
$$

$$
(w)_{1_{+-}} \otimes \begin{pmatrix} x_k \\ x_{(N/2-1)k} \end{pmatrix}_{2_{[k]}} = \begin{pmatrix} w x_k \\ -w x_{(N/2-1)k} \end{pmatrix}_{2_{[k]}},
\tag{8.17}
$$

$$
(w)_{1_{-+}} \otimes \begin{pmatrix} x_k \\ x_{(N/2-1)k} \end{pmatrix}_{2_{[k]}} = \begin{pmatrix} w x_{(N/2-1)k} \\ w x_k \end{pmatrix}_{2_{[k+N/2]}}.
\tag{8.18}
$$

Furthermore, the tensor products among singlets are obtained as

$$
1_{s_1 s_2} \otimes 1_{s_1' s_2'} = 1_{s_1'' s_2''},
\tag{8.19}
$$

where $s_i'' = s_i s_i'$ $(i = 1, 2)$.

8.2 QD_{16}

We show the simple example for $N' = 4$ and $N = 8$, i.e. QD_{16}. This group is generated by the $2\pi/8$ rotation a and the reflection b. These generators satisfy

$$
a^8 = 1, \qquad b^2 = 1, \qquad bab = a^3.
\tag{8.20}
$$

The QD_{16} includes the 16 elements, $a^m b^k$ with $m = 0, \ldots, 7$ and $k = 0, 1$. They are classified into the following seven conjugacy classes,

$$
\begin{array}{lll}
C_1 : & \{e\}, & h = 1, \\
C_2^{[1]} : & \{a, a^3\}, & h = 8, \\
C_2^{[2]} : & \{a^2, a^6\}, & h = 4, \\
C_2^{[5]} : & \{a^5, a^7\}, & h = 8, \\
C_1' : & \{a^4\}, & h = 2, \\
C_4 : & \{b, a^2 b, a^4 b, a^6 b\}, & h = 2, \\
C_4' : & \{ab, a^3 b, a^5 b, a^7 b\}, & h = 4.
\end{array}
\tag{8.21}
$$

Table 8.2 Characters of QD_{16} representations

	h	$\chi_{1_{++}}$	$\chi_{1_{-+}}$	$\chi_{1_{--}}$	$\chi_{1_{+-}}$	χ_{2_1}	χ_{2_2}	χ_{2_5}
C_1	1	1	1	1	1	2	2	2
$C_2^{[1]}$	8	1	-1	-1	1	$\sqrt{2}i$	0	$-\sqrt{2}i$
$C_2^{[2]}$	4	1	1	1	1	0	-2	0
$C_2^{[5]}$	8	1	-1	-1	1	$-\sqrt{2}i$	0	$\sqrt{2}i$
C_1'	2	1	1	1	1	-2	2	-2
C_4	2	1	1	-1	-1	0	0	0
C_4'	2	1	-1	1	-1	0	0	0

The QD_{16} has four singlets, $\mathbf{1}_{\pm\pm}$, and three doublets, $\mathbf{2}_1$, $\mathbf{2}_2$, and $\mathbf{2}_3$. Their characters are shown in Table 8.2.

We define three doublets as

$$\mathbf{2}_1 = \begin{pmatrix} x_1 \\ x_3 \end{pmatrix}, \quad \mathbf{2}_2 = \begin{pmatrix} x_2 \\ x_6 \end{pmatrix}, \quad \mathbf{2}_5 = \begin{pmatrix} x_5 \\ x_7 \end{pmatrix}. \tag{8.22}$$

Then the tensor products are given as

$$\begin{pmatrix} x_1 \\ x_3 \end{pmatrix}_{\mathbf{2}_1} \otimes \begin{pmatrix} y_1 \\ y_3 \end{pmatrix}_{\mathbf{2}_1} = \begin{pmatrix} x_1 y_1 \\ x_3 y_3 \end{pmatrix}_{\mathbf{2}_2} \oplus (x_1 y_3 \pm x_3 y_1)_{\mathbf{1}_{-\pm}}, \tag{8.23}$$

$$\begin{pmatrix} x_1 \\ x_3 \end{pmatrix}_{\mathbf{2}_1} \otimes \begin{pmatrix} y_2 \\ y_6 \end{pmatrix}_{\mathbf{2}_2} = \begin{pmatrix} x_3 y_2 \\ x_1 y_6 \end{pmatrix}_{\mathbf{2}_5} \oplus \begin{pmatrix} x_3 y_6 \\ x_1 y_2 \end{pmatrix}_{\mathbf{2}_1}, \tag{8.24}$$

$$\begin{pmatrix} x_1 \\ x_3 \end{pmatrix}_{\mathbf{2}_1} \otimes \begin{pmatrix} y_5 \\ y_7 \end{pmatrix}_{\mathbf{2}_5} = \begin{pmatrix} x_3 y_7 \\ x_1 y_5 \end{pmatrix}_{\mathbf{2}_2} \oplus (x_1 y_7 \pm x_3 y_5)_{\mathbf{1}_{+\pm}}, \tag{8.25}$$

$$\begin{pmatrix} x_2 \\ x_6 \end{pmatrix}_{\mathbf{2}_2} \otimes \begin{pmatrix} y_2 \\ y_6 \end{pmatrix}_{\mathbf{2}_2} = (x_2 y_2 \pm x_6 y_6)_{\mathbf{1}_{-\pm}} \oplus (x_2 y_6 \pm x_6 y_2)_{\mathbf{1}_{+\pm}}, \tag{8.26}$$

$$\begin{pmatrix} x_2 \\ x_6 \end{pmatrix}_{\mathbf{2}_2} \otimes \begin{pmatrix} y_5 \\ y_7 \end{pmatrix}_{\mathbf{2}_5} = \begin{pmatrix} x_6 y_7 \\ x_2 y_5 \end{pmatrix}_{\mathbf{2}_5} \oplus \begin{pmatrix} x_2 y_7 \\ x_6 y_5 \end{pmatrix}_{\mathbf{2}_1}, \tag{8.27}$$

$$\begin{pmatrix} x_5 \\ x_7 \end{pmatrix}_{\mathbf{2}_5} \otimes \begin{pmatrix} y_5 \\ y_7 \end{pmatrix}_{\mathbf{2}_5} = \begin{pmatrix} x_5 y_5 \\ x_7 y_7 \end{pmatrix}_{\mathbf{2}_2} \oplus (x_5 y_7 \pm x_7 y_5)_{\mathbf{1}_{-\pm}}, \tag{8.28}$$

$$(w)_{\mathbf{1}_{-\pm}} \otimes \begin{pmatrix} y_1 \\ y_3 \end{pmatrix}_{\mathbf{2}_1} = \begin{pmatrix} w y_3 \\ \pm w y_1 \end{pmatrix}_{\mathbf{2}_5}, \quad (w)_{\mathbf{1}_{-\pm}} \otimes \begin{pmatrix} y_2 \\ y_6 \end{pmatrix}_{\mathbf{2}_2} = \begin{pmatrix} w y_6 \\ \pm w y_2 \end{pmatrix}_{\mathbf{2}_2}, \tag{8.29}$$

$$(w)_{\mathbf{1}_{-\pm}} \otimes \begin{pmatrix} y_5 \\ y_7 \end{pmatrix}_{\mathbf{2}_5} = \begin{pmatrix} wy_7 \\ \pm wy_5 \end{pmatrix}_{\mathbf{2}_1}, \tag{8.30}$$

$$(w)_{\mathbf{1}_{+\pm}} \otimes \begin{pmatrix} y_1 \\ y_3 \end{pmatrix}_{\mathbf{2}_1} = \begin{pmatrix} wy_1 \\ \pm wy_3 \end{pmatrix}_{\mathbf{2}_5}, \quad (w)_{\mathbf{1}_{+\pm}} \otimes \begin{pmatrix} y_2 \\ y_6 \end{pmatrix}_{\mathbf{2}_2} = \begin{pmatrix} wy_2 \\ \pm wy_6 \end{pmatrix}_{\mathbf{2}_2}, \tag{8.31}$$

$$(w)_{\mathbf{1}_{+\pm}} \otimes \begin{pmatrix} y_5 \\ y_7 \end{pmatrix}_{\mathbf{2}_5} = \begin{pmatrix} wy_5 \\ \pm wy_7 \end{pmatrix}_{\mathbf{2}_5}, \tag{8.32}$$

$$\mathbf{1}_{+\pm} \otimes \mathbf{1}_{+\pm} = \mathbf{1}_{++}, \ \mathbf{1}_{-\pm} \otimes \mathbf{1}_{+\pm} = \mathbf{1}_{-+}, \ \mathbf{1}_{-\pm} \otimes \mathbf{1}_{-\pm} = \mathbf{1}_{++}, \tag{8.33}$$

$$\mathbf{1}_{+\pm} \otimes \mathbf{1}_{+\mp} = \mathbf{1}_{+-}, \ \mathbf{1}_{-\pm} \otimes \mathbf{1}_{+\mp} = \mathbf{1}_{--}, \ \mathbf{1}_{-\pm} \otimes \mathbf{1}_{-\mp} = \mathbf{1}_{+-}. \tag{8.34}$$

$\Sigma(2N^2)$

9.1 Generic Aspects

In this section, we study the discrete group $\Sigma(2N^2)$, which is isomorphic to $(Z_N \times Z_N') \rtimes Z_2$. Let us denote the generators of Z_N and Z_N' by a and a', respectively, and the Z_2 generator by b. These generators satisfy

$$a^N = a'^N = b^2 = e,$$
$$aa' = a'a, \quad bab = a'. \tag{9.1}$$

Therefore, all of $\Sigma(2N^2)$ elements are given as

$$g = b^k a^m a'^n, \tag{9.2}$$

for $k = 0, 1$ and $m, n = 0, 1, \ldots, N - 1$.

Since these generators, a, a' and b, are represented, e.g. as

$$a = \begin{pmatrix} 1 & 0 \\ 0 & \rho \end{pmatrix}, \quad a' = \begin{pmatrix} \rho & 0 \\ 0 & 1 \end{pmatrix}, \quad b = \begin{pmatrix} 0 & 1 \\ 1 & 0 \end{pmatrix}, \tag{9.3}$$

where $\rho = e^{2\pi i/N}$, all of $\Sigma(2N^2)$ elements are expressed by the 2×2 matrices as

$$\begin{pmatrix} \rho^m & 0 \\ 0 & \rho^n \end{pmatrix}, \quad \begin{pmatrix} 0 & \rho^m \\ \rho^n & 0 \end{pmatrix}. \tag{9.4}$$

© The Author(s), under exclusive license to Springer-Verlag GmbH, DE, part of Springer Nature 2022
T. Kobayashi et al., *An Introduction to Non-Abelian Discrete Symmetries for Particle Physicists*, Lecture Notes in Physics 995, https://doi.org/10.1007/978-3-662-64679-3_9

● **Conjugacy classes**

The conjugacy classes of $\Sigma(2N^2)$ are found easily by using following algebraic relations,

$$b(a^l a'^m)b^{-1} = a^m a'^l, \qquad b(ba^l a'^m)b^{-1} = ba^m a'^l,$$
$$a^k(ba^l a'^m)a^{-k} = ba^{l-k}a'^{m+k}, \qquad a'^k(ba^l a'^m)a'^{-k} = ba^{l+k}a'^{m-k}, \quad (9.5)$$

which are given from Eq. (9.3). Now, we find that the $\Sigma(2N^2)$ group has the following conjugacy classes,

$$
\begin{array}{lll}
C_1: & \{e\}, & h = 1, \\
C_1^{(1)}: & \{aa'\}, & h = N, \\
\vdots & \vdots & \vdots \\
C_1^{(k)}: & \{a^k a'^k\}, & h = N/gcd(N, k), \\
\vdots & \vdots & \vdots \\
C_1^{(N-1)}: & \{a^{N-1}a'^{N-1}\}, & h = N/gcd(N, N-1), \\
C_N'^{(k)}: & \{ba^k, ba^{k-1}a', \ldots, ba'^k, \ldots, ba^{k+1}a'^{N-1}\}, & h = 2N/gcd(N, k), \\
C_2^{(l,m)}: & \{a^l a'^m, \; a'^l a^m\}, & h = N/gcd(N, l, m),
\end{array}
\tag{9.6}
$$

where $l > m$ for $l, m = 0, \ldots, N - 1$. The number of conjugacy classes $C_2^{(l,m)}$ is $N(N - 1)/2$. The total number of conjugacy classes of $\Sigma(2N^2)$ is given as $N(N - 1)/2 + N + N = (N^2 + 3N)/2$.

● **Characters and representations**

The orthogonality relations (2.18) and (2.19) for $\Sigma(2N^2)$ give

$$m_1 + 2^2 m_2 + \cdots = 2N^2, \tag{9.7}$$
$$m_1 + m_2 + \cdots = (N^2 + 3N)/2. \tag{9.8}$$

Then, we have solution, $(m_1, m_2) = (2N, N(N - 1)/2)$. That is, there are $2N$ singlets and $N(N - 1)/2$ doublets.

Let us discuss on singlets. Since a and a' belong to the same conjugacy class $C_2^{(1,0)}$, the characters $\chi_\alpha(g)$ for singlets should satisfy $\chi_\alpha(a) = \chi_\alpha(a')$. Because of $b^2 = e$ and $a^N = e$, possible values of $\chi_\alpha(g)$ for singlets are obtained as $\chi_\alpha(a) = \rho^n$ and $\chi_\alpha(b) = \pm 1$. Then, totally we have $2N$ combinations, which correspond to $2N$ singlets, $\mathbf{1}_{\pm n}$ for $n = 0, 1, \ldots, N - 1$. These characters are summarized in Table 9.1.

Next, we discuss on doublet representations. As seen in (9.3), generators are represented in the doublet representation. Similarly, (2×2) matrix representations for generic doublets $\mathbf{2}_{q,p}$ are written by replacing

$$a \to a^p a'^q \quad \text{and} \quad a' \to a^q a'^p. \tag{9.9}$$

Table 9.1 Characters of $\Sigma(2N^2)$ representations

	h	$\chi_{1_{+n}}$	$\chi_{1_{-n}}$	$\chi_{2_{p,q}}$
C_1	1	1	1	2
$C_1^{(1)}$	N	ρ^{2n}	ρ^{2n}	$2\rho^{p+q}$
$\vdots$				
$C_1^{(N-1)}$	$N/gcd(N, N-1)$	$\rho^{2n(N-1)}$	$\rho^{2n(N-1)}$	$2\rho^{(N-1)(p+q)}$
$C_N'^{(k)}$	$2N/gcd(N, k)$	ρ^{kn}	$-\rho^{kn}$	0
$C_2^{(l,m)}$	$N/gcd(N, l, m)$	$\rho^{(l+m)n}$	$\rho^{(l+m)n}$	$\rho^{lq+mp} + \rho^{lp+mq}$

That is, for doublets $2_{p,q}$, the generators a and a' as well as b are given as

$$a = \begin{pmatrix} \rho^q & 0 \\ 0 & \rho^p \end{pmatrix}, \qquad a' = \begin{pmatrix} \rho^p & 0 \\ 0 & \rho^q \end{pmatrix}, \qquad b = \begin{pmatrix} 0 & 1 \\ 1 & 0 \end{pmatrix}. \tag{9.10}$$

Let us denote the doublet $2_{q,p}$ as

$$2_{q,p} = \begin{pmatrix} x_q \\ x_p \end{pmatrix}, \tag{9.11}$$

where we take $q > p$ and $p = 0, 1, \ldots, N-2$. Then, each of up and down components, x_q and x_p, has definite $Z_N \times Z_N'$ charges. The characters for doublets are also summarized in Table 9.1.

• Tensor products
We show tensor products of doublets $2_{q,p}$ in $\Sigma(2N^2)$. Taking account of $Z_N \times Z_N'$ charges, their tensor products are given as

$$\begin{pmatrix} x_q \\ x_p \end{pmatrix}_{2_{q,p}} \otimes \begin{pmatrix} y_{q'} \\ y_{p'} \end{pmatrix}_{2_{q',p'}} = \begin{pmatrix} x_q y_{q'} \\ x_p y_{p'} \end{pmatrix}_{2_{q+q',p+p'}} \oplus \begin{pmatrix} x_p y_{q'} \\ x_q y_{p'} \end{pmatrix}_{2_{q'+p,q+p'}}, \tag{9.12}$$

for $q + q' \neq p + p' \bmod(N)$ and $q + p' \neq p + q' \bmod(N)$,

$$\begin{pmatrix} x_q \\ x_p \end{pmatrix}_{2_{q,p}} \otimes \begin{pmatrix} y_{q'} \\ y_{p'} \end{pmatrix}_{2_{q',p'}} = (x_q y_{q'} + x_p y_{p'})_{1_{+,q+q'}} \oplus (x_q y_{q'} - x_p y_{p'})_{1_{-,q+q'}}$$

$$\oplus \begin{pmatrix} x_p y_{q'} \\ x_q y_{p'} \end{pmatrix}_{2_{q'+p,q+p'}}, \tag{9.13}$$

for $q + q' = p + p' \bmod(N)$ and $q + p' \neq p + q' \bmod(N)$,

$$\begin{pmatrix} x_q \\ x_p \end{pmatrix}_{2_{q,p}} \otimes \begin{pmatrix} y_{q'} \\ y_{p'} \end{pmatrix}_{2_{q',p'}} = (x_p y_{q'} + x_q y_{p'})_{1_{+,q+p'}} \oplus (x_p y_{q'} - x_q y_{p'})_{1_{-,q+p'}}$$

$$\oplus \begin{pmatrix} x_q y_{q'} \\ x_p y_{p'} \end{pmatrix}_{2_{q+q',p+p'}}, \tag{9.14}$$

for $q + q' \neq p + p' \bmod(N)$ and $q + p' = p + q' \bmod(N)$,

$$\begin{pmatrix} x_q \\ x_p \end{pmatrix}_{2_{q,p}} \otimes \begin{pmatrix} y_{q'} \\ y_{p'} \end{pmatrix}_{2_{q',p'}} = (x_q y_{q'} + x_p y_{p'})_{1_{+,q+q'}} \oplus (x_q y_{q'} - x_p y_{p'})_{1_{-,q+q'}}$$

$$\oplus (x_p y_{q'} + x_q y_{p'})_{1_{+,q+p'}} \oplus (x_p y_{q'} - x_q y_{p'})_{1_{-,q+p'}}, \quad (9.15)$$

for $q + q' = p + p' \bmod(N)$ and $q + p' = p + q' \bmod(N)$.

Furthermore, we obtain the tensor products between singlets and doublets as

$$(y)_{1_{s,n}} \otimes \begin{pmatrix} x_q \\ x_p \end{pmatrix}_{2_{q,p}} = \begin{pmatrix} y x_q \\ s y x_p \end{pmatrix}_{2_{q+n,p+n}} \quad \text{for } q + n > p + n \bmod(N), \quad (9.16)$$

$$(y)_{1_{s,n}} \otimes \begin{pmatrix} x_q \\ x_p \end{pmatrix}_{2_{q,p}} = \begin{pmatrix} y x_p \\ s y x_q \end{pmatrix}_{2_{p+n,q+n}} \quad \text{for } p + n > q + n \bmod(N). \quad (9.17)$$

These tensor products are independent of $s = \pm$. The tensor products of singlets are simply given as

$$\mathbf{1}_{sn} \otimes \mathbf{1}_{s'n'} = \mathbf{1}_{ss',n+n'}. \quad (9.18)$$

9.2 $\Sigma(18)$

Let us present simple examples of $\Sigma(2N^2)$. The simplest group of $\Sigma(2N^2)$ is $\Sigma(2)$ which is nothing but the Abelian Z_2 group. The next one is the $\Sigma(8)$ group, which is isomorphic to D_4. Consequently, the simple and non-trivial example is $\Sigma(18)$.

In the $\Sigma(18)$, there are eighteen elements $b^k a^m a'^n$ for $k = 0, 1$ and $m, n = 0, 1, 2$, where a, a' and b satisfy $b^2 = e$, $a^3 = a'^3 = e$, $aa' = a'a$ and $bab = a'$. These elements are classified into nine conjugacy classes,

$$
\begin{array}{llll}
C_1 : & \{e\}, & h = 1, \\
C_1^{(1)} : & \{aa'\}, & h = 3, \\
C_1^{(2)} : & \{a^2 a'^2\}, & h = 3, \\
C_3^{'(0)} : & \{b, ba'^2 a, ba'a^2\}, & h = 2, \\
C_3^{'(1)} : & \{ba', ba, ba'^2 a^2\}, & h = 6, & (9.19) \\
C_3^{'(2)} : & \{ba'^2, ba'a, ba^2\}, & h = 6, \\
C_2^{(1,0)} : & \{a, a'\}, & h = 3, \\
C_2^{(2,0)} : & \{a^2, a'^2\}, & h = 3, \\
C_2^{(2,1)} : & \{a^2 a', aa'^2\}, & h = 3,
\end{array}
$$

where h is the order of each element in the conjugacy class.

The $\Sigma(18)$ has six singlets $\mathbf{1}_{\pm,n}$ with $n = 0, 1, 2$ and three doublets $\mathbf{2}_{q,p}$ with $(q, p) = (1, 0), (2, 0), (2, 1)$. The characters are shown in Table 9.2.

Table 9.2 Characters of $\Sigma(18)$ representations

	h	$\chi_{1_{+0}}$	$\chi_{1_{+1}}$	$\chi_{1_{+2}}$	$\chi_{1_{-0}}$	$\chi_{1_{-1}}$	$\chi_{1_{-2}}$	$\chi_{2_{1,0}}$	$\chi_{2_{2,0}}$	$\chi_{2_{2,1}}$
C_1	1	1	1	1	1	1	1	2	2	2
$C_1^{(1)}$	3	1	ρ^2	ρ	1	ρ^2	ρ	2ρ	$2\rho^2$	2
$C_1^{(2)}$	3	1	ρ	ρ^2	1	ρ	ρ^2	$2\rho^2$	2ρ	2
$C_3'^{(0)}$	2	1	1	1	-1	-1	-1	0	0	0
$C_3'^{(1)}$	6	1	ρ	ρ^2	-1	$-\rho$	$-\rho^2$	0	0	0
$C_3'^{(2)}$	6	1	ρ^2	ρ	-1	$-\rho^2$	$-\rho$	0	0	0
$C_2^{(1,0)}$	3	1	ρ	ρ^2	1	ρ	ρ^2	$-\rho^2$	$-\rho$	-1
$C_1^{(2,0)}$	3	1	ρ^2	ρ	1	ρ^2	ρ	$-\rho$	$-\rho^2$	-1
$C_1^{(3,0)}$	3	1	1	1	1	1	1	-1	-1	-1

We show the tensor products between doublets as follows:

$$
\begin{pmatrix} x_2 \\ x_1 \end{pmatrix}_{2,1} \otimes \begin{pmatrix} y_2 \\ y_1 \end{pmatrix}_{2,1} = (x_1 y_2 + x_2 y_1)_{1_{+,0}} \oplus (x_1 y_2 - x_2 y_1)_{1_{-,0}} \oplus \begin{pmatrix} x_1 y_1 \\ x_2 y_2 \end{pmatrix}_{2,1} ,
$$

$$
\begin{pmatrix} x_2 \\ x_0 \end{pmatrix}_{2,0} \otimes \begin{pmatrix} y_2 \\ y_0 \end{pmatrix}_{2,0} = (x_0 y_2 + x_2 y_0)_{1_{+,2}} \oplus (x_0 y_2 - x_2 y_0)_{1_{-,2}} \oplus \begin{pmatrix} x_2 y_2 \\ x_0 y_0 \end{pmatrix}_{2_{1,0}} ,
$$

$$
\begin{pmatrix} x_1 \\ x_0 \end{pmatrix}_{2_{1,0}} \otimes \begin{pmatrix} y_1 \\ y_0 \end{pmatrix}_{2_{1,0}} = (x_0 y_1 + x_1 y_0)_{1_{+,1}} \oplus (x_0 y_1 - x_1 y_0)_{1_{-,1}} \oplus \begin{pmatrix} x_1 y_1 \\ x_0 y_0 \end{pmatrix}_{2,0} ,
$$

$$
\begin{pmatrix} x_2 \\ x_1 \end{pmatrix}_{2,1} \otimes \begin{pmatrix} y_2 \\ y_0 \end{pmatrix}_{2,0} = (x_2 y_2 + x_1 y_0)_{1_{+,1}} \oplus (x_2 y_2 - x_1 y_0)_{1_{-,1}} \oplus \begin{pmatrix} x_2 y_0 \\ x_1 y_2 \end{pmatrix}_{2,0} ,
$$

$$
\begin{pmatrix} x_2 \\ x_1 \end{pmatrix}_{2,1} \otimes \begin{pmatrix} y_1 \\ y_0 \end{pmatrix}_{2_{1,0}} = (x_1 y_1 + x_2 y_0)_{1_{+,2}} \oplus (x_1 y_1 - x_2 y_0)_{1_{-,2}} \oplus \begin{pmatrix} x_1 y_0 \\ x_2 y_1 \end{pmatrix}_{2_{1,0}} ,
$$

$$
\begin{pmatrix} x_2 \\ x_0 \end{pmatrix}_{2,0} \otimes \begin{pmatrix} y_1 \\ y_0 \end{pmatrix}_{2_{1,0}} = (x_2 y_1 + x_0 y_0)_{1_{+,0}} \oplus (x_2 y_1 - x_0 y_0)_{1_{-,0}} \oplus \begin{pmatrix} x_2 y_0 \\ x_0 y_1 \end{pmatrix}_{2,1} .
$$

$$(9.20)$$

The tensor products between singlets are given as

$$
1_{\pm,0} \otimes 1_{\pm,0} = 1_{+,0} , \ 1_{\pm,1} \otimes 1_{\pm,1} = 1_{+,2} , \ 1_{\pm,2} \otimes 1_{\pm,2} = 1_{+,1} , \ 1_{\pm,1} \otimes 1_{\pm,0} = 1_{+,1} ,
$$

$$
1_{\pm,2} \otimes 1_{\pm,0} = 1_{+,2} , \ 1_{\pm,2} \otimes 1_{\pm,1} = 1_{+,0} , \ 1_{\pm,0} \otimes 1_{\mp,0} = 1_{-,0} , \ 1_{\pm,1} \otimes 1_{\mp,1} = 1_{-,2} ,
$$

$$
1_{\pm,2} \otimes 1_{\mp,2} = 1_{-,1} , \ 1_{\pm,1} \otimes 1_{\mp,0} = 1_{-,1} , \ 1_{\pm,2} \otimes 1_{\pm,0} = 1_{-,2} , \ 1_{\pm,2} \otimes 1_{\mp,1} = 1_{-,0} .
$$

$$(9.21)$$

We also present the tensor products between singlets and doublets as

$$
(y)_{1_{\pm,0}} \otimes \begin{pmatrix} x_2 \\ x_1 \end{pmatrix}_{2,1} = \begin{pmatrix} y x_2 \\ \pm y x_1 \end{pmatrix}_{2,1} , \quad (y)_{1_{\pm,1}} \otimes \begin{pmatrix} x_2 \\ x_1 \end{pmatrix}_{2,1} = \begin{pmatrix} y x_1 \\ \pm y x_2 \end{pmatrix}_{2,0} ,
$$

$$
(y)_{1_{\pm,2}} \otimes \begin{pmatrix} x_2 \\ x_1 \end{pmatrix}_{2,1} = \begin{pmatrix} y x_2 \\ \pm y x_1 \end{pmatrix}_{2_{1,0}} , \quad (y)_{1_{\pm,0}} \otimes \begin{pmatrix} x_2 \\ x_0 \end{pmatrix}_{2,0} = \begin{pmatrix} y x_2 \\ \pm y x_0 \end{pmatrix}_{2,0} ,
$$

$$(y)\mathbf{1}_{\pm,1} \otimes \begin{pmatrix} x_2 \\ x_0 \end{pmatrix}_{2_{2,0}} = \begin{pmatrix} yx_0 \\ \pm yx_2 \end{pmatrix}_{2_{1,0}}, \quad (y)\mathbf{1}_{\pm,2} \otimes \begin{pmatrix} x_2 \\ x_0 \end{pmatrix}_{2_{2,0}} = \begin{pmatrix} yx_0 \\ \pm yx_2 \end{pmatrix}_{2_{2,1}},$$

$$(y)\mathbf{1}_{\pm,0} \otimes \begin{pmatrix} x_1 \\ x_0 \end{pmatrix}_{2_{1,0}} = \begin{pmatrix} yx_1 \\ \pm yx_0 \end{pmatrix}_{2_{1,0}}, \quad (y)\mathbf{1}_{\pm,1} \otimes \begin{pmatrix} x_1 \\ x_0 \end{pmatrix}_{2_{1,0}} = \begin{pmatrix} yx_1 \\ \pm yx_0 \end{pmatrix}_{2_{2,1}},$$

$$(y)\mathbf{1}_{\pm,2} \otimes \begin{pmatrix} x_1 \\ x_0 \end{pmatrix}_{2_{1,0}} = \begin{pmatrix} yx_0 \\ \pm yx_1 \end{pmatrix}_{2_{2,0}}. \tag{9.22}$$

9.3 $\Sigma(32)$

The next example is the $\Sigma(32)$ group, which has thirty-two elements, $b^k a^m a'^n$ for $k = 0, 1$ and $m, n = 0, 1, 2, 3$, where a, a' and b satisfy $b^2 = e, a^4 = a'^4 = e, aa' = a'a$ and $bab = a'$. These elements are classified into fourteen conjugacy classes,

$$
\begin{array}{llll}
C_1 : & \{e\}, & h = 1, \\
C_1^{(1)} : & \{aa'\}, & h = 4, \\
C_1^{(2)} : & \{a^2 a'^2\}, & h = 2, \\
C_1^{(3)} : & \{a^3 a'^3\}, & h = 4, \\
C_4'^{(0)} : & \{b, ba'a^3, ba'^2a^2, ba'^3a\}, & h = 2, \\
C_4'^{(1)} : & \{ba', ba, ba'^2a^3, ba'^3a^2\}, & h = 8, \\
C_4'^{(2)} : & \{ba'^2, ba'a, ba^2, ba'^3a^3\}, & h = 4, \\
C_4'^{(3)} : & \{ba'^3, ba'^2a, ba'a^2, ba^3\}, & h = 8, \\
C_2^{(1,0)} : & \{a, a'\}, & h = 4, \\
C_2^{(2,0)} : & \{a^2, a'^2\}, & h = 2, \\
C_2^{(2,1)} : & \{a^2a', aa'^2\}, & h = 4, \\
C_2^{(3,0)} : & \{a^3, a'^3\}, & h = 4, \\
C_2^{(3,1)} : & \{a^3a', aa'^3\}, & h = 4, \\
C_2^{(3,2)} : & \{a^3a'^2, a^2a'^3\}, & h = 4, \\
\end{array}
\tag{9.23}
$$

where h is the order of each element in the conjugacy class.

The $\Sigma(32)$ has eight singlets $\mathbf{1}_{\pm,n}$ with $n = 0, 1, 2, 3$ and six doublets $\mathbf{2}_{q,p}$ with $(q, p) = (1, 0), (2, 0), (3, 0), (2, 1), (3, 1), (3, 2)$. The characters are shown in Table 9.3.

We present the tensor products between doublets as follows:

$$\begin{pmatrix} x_3 \\ x_2 \end{pmatrix}_{2_{3,2}} \otimes \begin{pmatrix} y_3 \\ y_2 \end{pmatrix}_{2_{3,2}} = (x_2 y_3 + x_3 y_2)\mathbf{1}_{+,1} \oplus (x_2 y_3 - x_3 y_2)\mathbf{1}_{-,1} \oplus \begin{pmatrix} x_3 y_3 \\ x_2 y_2 \end{pmatrix}_{2_{2,0}}. \tag{9.24}$$

$$\begin{pmatrix} x_3 \\ x_1 \end{pmatrix}_{2_{3,1}} \otimes \begin{pmatrix} y_3 \\ y_1 \end{pmatrix}_{2_{3,1}} = (x_1 y_3 + x_3 y_1)\mathbf{1}_{+,0} \oplus (x_1 y_3 - x_3 y_1)\mathbf{1}_{-,0}$$

$$\oplus (x_3 y_3 + x_1 y_1)\mathbf{1}_{+,2} \oplus (x_3 y_3 - x_1 y_1)\mathbf{1}_{-,2}, \tag{9.25}$$

Table 9.3 Characters of $\Sigma(32)$ representations

	h	$\chi1_{+0}$	$\chi1_{+1}$	$\chi1_{+2}$	$\chi1_{+3}$	$\chi1_{-0}$	$\chi1_{-1}$	$\chi1_{-2}$	$\chi1_{-3}$	$\chi2_{1,0}$	$\chi2_{2,0}$	$\chi2_{2,1}$	$\chi2_{3,0}$	$\chi2_{3,1}$	$\chi2_{3,2}$
C_1	1	1	1	1	1	1	1	1	1	2	2	2	2	2	2
$C_1^{(1)}$	4	1	-1	1	-1	1	-1	1	-1	$2i$	-2	$-2i$	$-2i$	2	$2i$
$C_1^{(2)}$	2	1	1	1	1	1	1	1	1	-2	2	-2	-2	2	-2
$C_1^{(3)}$	4	1	-1	1	-1	1	-1	1	-1	$-2i$	-2	$2i$	$2i$	2	$-2i$
$C_4'^{(0)}$	2	1	1	1	1	-1	-1	-1	-1	0	0	0	0	0	0
$C_4'^{(1)}$	8	1	i	-1	$-i$	-1	$-i$	1	i	0	0	0	0	0	0
$C_4'^{(2)}$	4	1	-1	1	-1	-1	1	-1	1	0	0	0	0	0	0
$C_4'^{(3)}$	8	1	$-i$	-1	i	-1	i	1	$-i$	0	0	0	0	0	0
$C_2^{(1,0)}$	4	1	i	-1	$-i$	1	i	-1	$-i$	$1+i$	0	$-1+i$	$1-i$	0	$-1-i$
$C_2^{(2,0)}$	2	1	-1	1	-1	1	-1	1	-1	0	2	0	0	-2	0
$C_2^{(2,1)}$	4	1	$-i$	-1	i	1	$-i$	-1	i	$-1+i$	0	$1+i$	$-1-i$	0	$1-i$
$C_2^{(3,0)}$	4	1	$-i$	-1	i	1	$-i$	-1	i	$1-i$	0	$-1-i$	$1+i$	0	$-1+i$
$C_2^{(3,1)}$	4	1	1	1	1	1	1	1	1	0	-2	0	0	-2	0
$C_2^{(3,2)}$	4	1	i	-1	$-i$	1	i	-1	$-i$	$-1-i$	0	$1-i$	$-1+i$	0	$1+i$

$$\begin{pmatrix} x_3 \\ x_0 \end{pmatrix}_{\mathbf{2}_{3,0}} \otimes \begin{pmatrix} y_3 \\ y_0 \end{pmatrix}_{\mathbf{2}_{3,0}} = (x_0 y_3 + x_3 y_0)_{\mathbf{1}_{+,3}} \oplus (x_0 y_3 - x_3 y_0)_{\mathbf{1}_{-,3}} \oplus \begin{pmatrix} x_3 y_3 \\ x_0 y_0 \end{pmatrix}_{\mathbf{2}_{2,0}},$$
$$(9.26)$$

$$\begin{pmatrix} x_2 \\ x_1 \end{pmatrix}_{\mathbf{2}_{2,1}} \otimes \begin{pmatrix} y_2 \\ y_1 \end{pmatrix}_{\mathbf{2}_{2,1}} = (x_1 y_2 + x_2 y_1)_{\mathbf{1}_{+,3}} \oplus (x_1 y_2 - x_2 y_1)_{\mathbf{1}_{-,3}} \oplus \begin{pmatrix} x_1 y_1 \\ x_2 y_2 \end{pmatrix}_{\mathbf{2}_{2,0}},$$
$$(9.27)$$

$$\begin{pmatrix} x_2 \\ x_0 \end{pmatrix}_{\mathbf{2}_{2,0}} \otimes \begin{pmatrix} y_2 \\ y_0 \end{pmatrix}_{\mathbf{2}_{2,0}} = (x_0 y_2 + x_2 y_0)_{\mathbf{1}_{+,2}} \oplus (x_0 y_2 - x_2 y_0)_{\mathbf{1}_{-,2}}$$
$$\oplus (x_2 y_2 + x_0 y_0)_{\mathbf{1}_{+,0}} \oplus (x_2 y_2 - x_0 y_0)_{\mathbf{1}_{-,0}}, \quad (9.28)$$

$$\begin{pmatrix} x_1 \\ x_0 \end{pmatrix}_{\mathbf{2}_{1,0}} \otimes \begin{pmatrix} y_1 \\ y_0 \end{pmatrix}_{\mathbf{2}_{1,0}} = (x_0 y_1 + x_1 y_0)_{\mathbf{1}_{+,1}} \oplus (x_0 y_1 - x_1 y_0)_{\mathbf{1}_{-,1}} \oplus \begin{pmatrix} x_1 y_1 \\ x_0 y_0 \end{pmatrix}_{\mathbf{2}_{2,0}},$$
$$(9.29)$$

$$\begin{pmatrix} x_3 \\ x_2 \end{pmatrix}_{\mathbf{2}_{3,2}} \otimes \begin{pmatrix} y_3 \\ y_1 \end{pmatrix}_{\mathbf{2}_{3,1}} = \begin{pmatrix} x_2 y_3 \\ x_3 y_1 \end{pmatrix}_{\mathbf{2}_{1,0}} \oplus \begin{pmatrix} x_2 y_1 \\ x_3 y_3 \end{pmatrix}_{\mathbf{2}_{3,2}}, \quad (9.30)$$

$$\begin{pmatrix} x_3 \\ x_2 \end{pmatrix}_{\mathbf{2}_{3,2}} \otimes \begin{pmatrix} y_3 \\ y_0 \end{pmatrix}_{\mathbf{2}_{3,0}} = (x_3 y_3 + x_2 y_0)_{\mathbf{1}_{+,2}} \oplus (x_3 y_3 - x_2 y_0)_{\mathbf{1}_{-,2}} \oplus \begin{pmatrix} x_3 y_0 \\ x_2 y_3 \end{pmatrix}_{\mathbf{2}_{3,1}},$$
$$(9.31)$$

$$\begin{pmatrix} x_3 \\ x_2 \end{pmatrix}_{\mathbf{2}_{3,2}} \otimes \begin{pmatrix} y_2 \\ y_1 \end{pmatrix}_{\mathbf{2}_{2,1}} = (x_2 y_2 + x_3 y_1)_{\mathbf{1}_{+,0}} \oplus (x_2 y_2 - x_3 y_1)_{\mathbf{1}_{-,0}} \oplus \begin{pmatrix} x_2 y_1 \\ x_3 y_2 \end{pmatrix}_{\mathbf{2}_{3,1}},$$
$$(9.32)$$

$$\begin{pmatrix} x_3 \\ x_2 \end{pmatrix}_{\mathbf{2}_{3,2}} \otimes \begin{pmatrix} y_2 \\ y_0 \end{pmatrix}_{\mathbf{2}_{2,0}} = \begin{pmatrix} x_3 y_0 \\ x_2 y_2 \end{pmatrix}_{\mathbf{2}_{3,1}} \oplus \begin{pmatrix} x_2 y_0 \\ x_3 y_2 \end{pmatrix}_{\mathbf{2}_{2,1}}, \qquad (9.33)$$

$$\begin{pmatrix} x_3 \\ x_2 \end{pmatrix}_{\mathbf{2}_{3,2}} \otimes \begin{pmatrix} y_1 \\ y_0 \end{pmatrix}_{\mathbf{2}_{1,0}} = (x_2 y_1 + x_3 y_0)_{\mathbf{1}_{+,3}} \oplus (x_2 y_1 - x_3 y_0)_{\mathbf{1}_{-,3}} \oplus \begin{pmatrix} x_2 y_0 \\ x_3 y_1 \end{pmatrix}_{\mathbf{2}_{2,0}},$$
$$(9.34)$$

$$\begin{pmatrix} x_3 \\ x_1 \end{pmatrix}_{\mathbf{2}_{3,1}} \otimes \begin{pmatrix} y_3 \\ y_0 \end{pmatrix}_{\mathbf{2}_{3,0}} = \begin{pmatrix} x_3 y_0 \\ x_1 y_3 \end{pmatrix}_{\mathbf{2}_{3,0}} \oplus \begin{pmatrix} x_3 y_3 \\ x_1 y_0 \end{pmatrix}_{\mathbf{2}_{2,1}}, \qquad (9.35)$$

$$\begin{pmatrix} x_3 \\ x_1 \end{pmatrix}_{\mathbf{2}_{3,1}} \otimes \begin{pmatrix} y_2 \\ y_1 \end{pmatrix}_{\mathbf{2}_{2,1}} = \begin{pmatrix} x_1 y_2 \\ x_3 y_1 \end{pmatrix}_{\mathbf{2}_{3,0}} \oplus \begin{pmatrix} x_1 y_1 \\ x_3 y_2 \end{pmatrix}_{\mathbf{2}_{2,1}}, \qquad (9.36)$$

$$\begin{pmatrix} x_3 \\ x_1 \end{pmatrix}_{\mathbf{2}_{3,1}} \otimes \begin{pmatrix} y_2 \\ y_0 \end{pmatrix}_{\mathbf{2}_{2,0}} = (x_1 y_2 + x_3 y_0)_{\mathbf{1}_{+,3}} \oplus (x_1 y_2 - x_3 y_0)_{\mathbf{1}_{-,3}}$$
$$\oplus (x_3 y_2 + x_1 y_0)_{\mathbf{1}_{+,1}} \oplus (x_3 y_2 - x_1 y_0)_{\mathbf{1}_{-,1}}, \qquad (9.37)$$

$$\begin{pmatrix} x_3 \\ x_1 \end{pmatrix}_{\mathbf{2}_{3,1}} \otimes \begin{pmatrix} y_1 \\ y_0 \end{pmatrix}_{\mathbf{2}_{1,0}} = \begin{pmatrix} x_3 y_0 \\ x_1 y_1 \end{pmatrix}_{\mathbf{2}_{3,2}} \oplus \begin{pmatrix} x_1 y_0 \\ x_3 y_1 \end{pmatrix}_{\mathbf{2}_{1,0}}, \qquad (9.38)$$

$$\begin{pmatrix} x_3 \\ x_0 \end{pmatrix}_{\mathbf{2}_{3,0}} \otimes \begin{pmatrix} y_2 \\ y_1 \end{pmatrix}_{\mathbf{2}_{2,1}} = (x_3 y_2 + x_0 y_1)_{\mathbf{1}_{+,1}} \oplus (x_3 y_2 - x_0 y_1)_{\mathbf{1}_{-,1}} \oplus \begin{pmatrix} x_0 y_2 \\ x_3 y_1 \end{pmatrix}_{\mathbf{2}_{2,0}},$$
$$(9.39)$$

$$\begin{pmatrix} x_3 \\ x_0 \end{pmatrix}_{\mathbf{2}_{3,0}} \otimes \begin{pmatrix} y_2 \\ y_0 \end{pmatrix}_{\mathbf{2}_{2,0}} = \begin{pmatrix} x_3 y_0 \\ x_0 y_2 \end{pmatrix}_{\mathbf{2}_{3,2}} \oplus \begin{pmatrix} x_3 y_2 \\ x_0 y_0 \end{pmatrix}_{\mathbf{2}_{1,0}}, \qquad (9.40)$$

$$\begin{pmatrix} x_3 \\ x_0 \end{pmatrix}_{\mathbf{2}_{3,0}} \otimes \begin{pmatrix} y_1 \\ y_0 \end{pmatrix}_{\mathbf{2}_{1,0}} = (x_3 y_1 + x_0 y_0)_{\mathbf{1}_{+,0}} \oplus (x_3 y_1 - x_0 y_0)_{\mathbf{1}_{-,0}} \oplus \begin{pmatrix} x_3 y_0 \\ x_0 y_1 \end{pmatrix}_{\mathbf{2}_{3,1}},$$
$$(9.41)$$

$$\begin{pmatrix} x_2 \\ x_1 \end{pmatrix}_{\mathbf{2}_{2,1}} \otimes \begin{pmatrix} y_2 \\ y_0 \end{pmatrix}_{\mathbf{2}_{2,0}} = \begin{pmatrix} x_1 y_0 \\ x_2 y_2 \end{pmatrix}_{\mathbf{2}_{1,0}} \oplus \begin{pmatrix} x_1 y_2 \\ x_2 y_0 \end{pmatrix}_{\mathbf{2}_{3,2}}, \qquad (9.42)$$

$$\begin{pmatrix} x_2 \\ x_1 \end{pmatrix}_{\mathbf{2}_{2,1}} \otimes \begin{pmatrix} y_1 \\ y_0 \end{pmatrix}_{\mathbf{2}_{1,0}} = (x_1 y_1 + x_2 y_0)_{\mathbf{1}_{+,2}} \oplus (x_1 y_1 - x_2 y_0)_{\mathbf{1}_{-,2}} \oplus \begin{pmatrix} x_2 y_1 \\ x_1 y_0 \end{pmatrix}_{\mathbf{2}_{3,1}},$$
$$(9.43)$$

$$\begin{pmatrix} x_2 \\ x_0 \end{pmatrix}_{\mathbf{2}_{2,0}} \otimes \begin{pmatrix} y_1 \\ y_0 \end{pmatrix}_{\mathbf{2}_{1,0}} = \begin{pmatrix} x_2 y_1 \\ x_0 y_0 \end{pmatrix}_{\mathbf{2}_{3,0}} \oplus \begin{pmatrix} x_2 y_0 \\ x_0 y_1 \end{pmatrix}_{\mathbf{2}_{2,1}}. \qquad (9.44)$$

The tensor products between singlets are obtained as

$$
\begin{aligned}
&\mathbf{1}_{\pm,0} \otimes \mathbf{1}_{\pm,0} = \mathbf{1}_{+,0} \, , \ \mathbf{1}_{\pm,1} \otimes \mathbf{1}_{\pm,1} = \mathbf{1}_{+,2} \, , \ \mathbf{1}_{\pm,2} \otimes \mathbf{1}_{\pm,2} = \mathbf{1}_{+,0} \, , \ \mathbf{1}_{\pm,3} \otimes \mathbf{1}_{\pm,3} = \mathbf{1}_{+,2} , \\
&\mathbf{1}_{\pm,3} \otimes \mathbf{1}_{\pm,2} = \mathbf{1}_{+,1} \, , \ \mathbf{1}_{\pm,3} \otimes \mathbf{1}_{\pm,1} = \mathbf{1}_{+,0} \, , \ \mathbf{1}_{\pm,3} \otimes \mathbf{1}_{\pm,0} = \mathbf{1}_{+,3} \, , \ \mathbf{1}_{\pm,2} \otimes \mathbf{1}_{\pm,1} = \mathbf{1}_{+,3} , \\
&\mathbf{1}_{\pm,2} \otimes \mathbf{1}_{\pm,0} = \mathbf{1}_{+,2} \, , \ \mathbf{1}_{\pm,1} \otimes \mathbf{1}_{\pm,0} = \mathbf{1}_{+,1} . \\
&\mathbf{1}_{\mp,0} \otimes \mathbf{1}_{\pm,0} = \mathbf{1}_{-,0} \, , \ \mathbf{1}_{\mp,1} \otimes \mathbf{1}_{\pm,1} = \mathbf{1}_{-,2} \, , \ \mathbf{1}_{\mp,2} \otimes \mathbf{1}_{\pm,2} = \mathbf{1}_{-,0} \, , \ \mathbf{1}_{\mp,3} \otimes \mathbf{1}_{\pm,3} = \mathbf{1}_{-,2} , \\
&\mathbf{1}_{\mp,3} \otimes \mathbf{1}_{\pm,2} = \mathbf{1}_{-,1} \, , \ \mathbf{1}_{\mp,3} \otimes \mathbf{1}_{\pm,1} = \mathbf{1}_{-,0} \, , \ \mathbf{1}_{\mp,3} \otimes \mathbf{1}_{\pm,0} = \mathbf{1}_{-,3} \, , \ \mathbf{1}_{\mp,2} \otimes \mathbf{1}_{\pm,1} = \mathbf{1}_{-,3} , \\
&\mathbf{1}_{\mp,2} \otimes \mathbf{1}_{\pm,0} = \mathbf{1}_{-,2} \, , \ \mathbf{1}_{\mp,1} \otimes \mathbf{1}_{\pm,0} = \mathbf{1}_{-,1} .
\end{aligned}
\tag{9.45}
$$

We also present the tensor products between singlets and doublets as

$$
\begin{aligned}
&(y)\mathbf{1}_{\pm,0} \otimes \begin{pmatrix} x_3 \\ x_2 \end{pmatrix}_{\mathbf{2}_{3,2}} = \begin{pmatrix} yx_3 \\ \pm yx_2 \end{pmatrix}_{\mathbf{2}_{3,2}} , \ (y)\mathbf{1}_{\pm,1} \otimes \begin{pmatrix} x_3 \\ x_2 \end{pmatrix}_{\mathbf{2}_{3,2}} = \begin{pmatrix} yx_2 \\ \pm yx_3 \end{pmatrix}_{\mathbf{2}_{3,0}} , \\
&(y)\mathbf{1}_{\pm,2} \otimes \begin{pmatrix} x_3 \\ x_2 \end{pmatrix}_{\mathbf{2}_{3,2}} = \begin{pmatrix} yx_3 \\ \pm yx_2 \end{pmatrix}_{\mathbf{2}_{1,0}} , \ (y)\mathbf{1}_{\pm,3} \otimes \begin{pmatrix} x_3 \\ x_2 \end{pmatrix}_{\mathbf{2}_{3,2}} = \begin{pmatrix} yx_3 \\ \pm yx_2 \end{pmatrix}_{\mathbf{2}_{2,1}} , \\
&(y)\mathbf{1}_{\pm,0} \otimes \begin{pmatrix} x_3 \\ x_1 \end{pmatrix}_{\mathbf{2}_{3,1}} = \begin{pmatrix} yx_3 \\ \pm yx_1 \end{pmatrix}_{\mathbf{2}_{3,1}} , \ (y)\mathbf{1}_{\pm,1} \otimes \begin{pmatrix} x_3 \\ x_1 \end{pmatrix}_{\mathbf{2}_{3,1}} = \begin{pmatrix} yx_1 \\ \pm yx_3 \end{pmatrix}_{\mathbf{2}_{2,0}} , \\
&(y)\mathbf{1}_{\pm,2} \otimes \begin{pmatrix} x_3 \\ x_1 \end{pmatrix}_{\mathbf{2}_{3,1}} = \begin{pmatrix} yx_1 \\ \pm yx_3 \end{pmatrix}_{\mathbf{2}_{3,1}} , \ (y)\mathbf{1}_{\pm,3} \otimes \begin{pmatrix} x_3 \\ x_1 \end{pmatrix}_{\mathbf{2}_{3,1}} = \begin{pmatrix} yx_3 \\ \pm yx_1 \end{pmatrix}_{\mathbf{2}_{2,0}} , \\
&(y)\mathbf{1}_{\pm,0} \otimes \begin{pmatrix} x_3 \\ x_0 \end{pmatrix}_{\mathbf{2}_{3,0}} = \begin{pmatrix} yx_3 \\ \pm yx_0 \end{pmatrix}_{\mathbf{2}_{3,0}} , \ (y)\mathbf{1}_{\pm,1} \otimes \begin{pmatrix} x_3 \\ x_0 \end{pmatrix}_{\mathbf{2}_{3,0}} = \begin{pmatrix} yx_0 \\ \pm yx_3 \end{pmatrix}_{\mathbf{2}_{1,0}} , \\
&(y)\mathbf{1}_{\pm,2} \otimes \begin{pmatrix} x_3 \\ x_0 \end{pmatrix}_{\mathbf{2}_{3,0}} = \begin{pmatrix} yx_0 \\ \pm yx_3 \end{pmatrix}_{\mathbf{2}_{2,1}} , \ (y)\mathbf{1}_{\pm,3} \otimes \begin{pmatrix} x_3 \\ x_0 \end{pmatrix}_{\mathbf{2}_{3,0}} = \begin{pmatrix} yx_0 \\ \pm yx_3 \end{pmatrix}_{\mathbf{2}_{3,2}} , \\
&(y)\mathbf{1}_{\pm,0} \otimes \begin{pmatrix} x_2 \\ x_1 \end{pmatrix}_{\mathbf{2}_{2,1}} = \begin{pmatrix} yx_2 \\ \pm yx_1 \end{pmatrix}_{\mathbf{2}_{2,1}} , \ (y)\mathbf{1}_{\pm,1} \otimes \begin{pmatrix} x_2 \\ x_1 \end{pmatrix}_{\mathbf{2}_{2,1}} = \begin{pmatrix} yx_2 \\ \pm yx_1 \end{pmatrix}_{\mathbf{2}_{3,2}} , \\
&(y)\mathbf{1}_{\pm,2} \otimes \begin{pmatrix} x_2 \\ x_1 \end{pmatrix}_{\mathbf{2}_{2,1}} = \begin{pmatrix} yx_1 \\ \pm yx_2 \end{pmatrix}_{\mathbf{2}_{3,0}} , \ (y)\mathbf{1}_{\pm,3} \otimes \begin{pmatrix} x_2 \\ x_1 \end{pmatrix}_{\mathbf{2}_{2,1}} = \begin{pmatrix} yx_2 \\ \pm yx_1 \end{pmatrix}_{\mathbf{2}_{1,0}} , \\
&(y)\mathbf{1}_{\pm,0} \otimes \begin{pmatrix} x_2 \\ x_0 \end{pmatrix}_{\mathbf{2}_{2,0}} = \begin{pmatrix} yx_2 \\ \pm yx_0 \end{pmatrix}_{\mathbf{2}_{2,0}} , \ (y)\mathbf{1}_{\pm,1} \otimes \begin{pmatrix} x_2 \\ x_0 \end{pmatrix}_{\mathbf{2}_{2,0}} = \begin{pmatrix} yx_2 \\ \pm yx_0 \end{pmatrix}_{\mathbf{2}_{3,1}} , \\
&(y)\mathbf{1}_{\pm,2} \otimes \begin{pmatrix} x_2 \\ x_0 \end{pmatrix}_{\mathbf{2}_{2,0}} = \begin{pmatrix} yx_0 \\ \pm yx_2 \end{pmatrix}_{\mathbf{2}_{2,0}} , \ (y)\mathbf{1}_{\pm,3} \otimes \begin{pmatrix} x_2 \\ x_0 \end{pmatrix}_{\mathbf{2}_{2,0}} = \begin{pmatrix} yx_0 \\ \pm yx_2 \end{pmatrix}_{\mathbf{2}_{3,1}} , \\
&(y)\mathbf{1}_{\pm,0} \otimes \begin{pmatrix} x_1 \\ x_0 \end{pmatrix}_{\mathbf{2}_{1,0}} = \begin{pmatrix} yx_1 \\ \pm yx_0 \end{pmatrix}_{\mathbf{2}_{1,0}} , \ (y)\mathbf{1}_{\pm,1} \otimes \begin{pmatrix} x_1 \\ x_0 \end{pmatrix}_{\mathbf{2}_{1,0}} = \begin{pmatrix} yx_1 \\ \pm yx_0 \end{pmatrix}_{\mathbf{2}_{2,1}} , \\
&(y)\mathbf{1}_{\pm,2} \otimes \begin{pmatrix} x_1 \\ x_0 \end{pmatrix}_{\mathbf{2}_{1,0}} = \begin{pmatrix} yx_1 \\ \pm yx_0 \end{pmatrix}_{\mathbf{2}_{3,2}} , \ (y)\mathbf{1}_{\pm,3} \otimes \begin{pmatrix} x_1 \\ x_0 \end{pmatrix}_{\mathbf{2}_{1,0}} = \begin{pmatrix} yx_0 \\ \pm yx_1 \end{pmatrix}_{\mathbf{2}_{3,0}} .
\end{aligned}
\tag{9.46}
$$

9.4 $\Sigma(50)$

We present an example, $\Sigma(50)$. This group has fifty elements, $b^k a^m a'^n$ for $k = 0, 1$ and m, $n = 0, 1, 2, 3, 4$ where a, a' and b satisfy the same conditions as Eq. (9.1) in the case of $N = 5$. These elements are classified into twenty conjugacy classes,

$$
\begin{aligned}
&C_1: &&\{e\}, &&h = 1,\\
&C_1^{(1)}: &&\{aa'\}, &&h = 5,\\
&C_1^{(2)}: &&\{a^2 a'^2\}, &&h = 5,\\
&C_1^{(3)}: &&\{a^3 a'^3\}, &&h = 5,\\
&C_1^{(4)}: &&\{a^4 a'^4\}, &&h = 5,\\
&C_5'^{(0)}: &&\{b, ba'^2 a^3, ba'^3 a^2, ba'^4 a, ba'a^4\}, &&h = 2,\\
&C_5'^{(1)}: &&\{ba', ba, ba'^3 a^3, ba'^4 a^2, ba'^2 a^4\}, &&h = 10,\\
&C_5'^{(2)}: &&\{ba'^2, ba'a, ba^2, ba'^4 a^3, ba'^3 a^4\}, &&h = 10,\\
&C_5'^{(3)}: &&\{ba'^3, ba'^2 a, ba'a^2, ba^3, ba'^4 a^4\}, &&h = 10,\\
&C_5'^{(4)}: &&\{ba'^4, ba'^2 a^2, ba'a^3, ba'^3 a, ba^4\}, &&h = 10,\\
&C_2^{(1,0)}: &&\{a, a'\}, &&h = 5,\\
&C_2^{(2,0)}: &&\{a^2, a'^2\}, &&h = 5,\\
&C_2^{(2,1)}: &&\{a^2 a', aa'^2\}, &&h = 5,\\
&C_2^{(3,0)}: &&\{a^3, a'^3\}, &&h = 5,\\
&C_2^{(3,1)}: &&\{a^3 a', aa'^3\}, &&h = 5,\\
&C_2^{(3,2)}: &&\{a^3 a'^2, a^2 a'^3\}, &&h = 5,\\
&C_2^{(4,0)}: &&\{a^4, a'^4\}, &&h = 5,\\
&C_2^{(4,1)}: &&\{a^4 a', aa'^4\}, &&h = 5,\\
&C_2^{(4,2)}: &&\{a^4 a'^2, a^2 a'^4\}, &&h = 5,\\
&C_2^{(4,3)}: &&\{a^4 a'^3, a^3 a'^4\}, &&h = 5,
\end{aligned}
\tag{9.47}
$$

where h is the order of each element in the conjugacy class.

The $\Sigma(50)$ has ten singlets $\mathbf{1}_{\pm,n}$ with $n = 0, 1, 2, 3, 4$ and ten doublets $\mathbf{2}_{q,p}$ with $(q, p) = (1, 0), (2, 0), (3, 0), (4, 0), (2, 1), (3, 1), (4, 1), (3, 1), (3, 2), (4, 3)$.

Table 9.4 Characters of $\Sigma(50)$ representations, where $\rho = e^{2i\pi/5}$

	h	$\chi_{1_{\pm 0}}$	$\chi_{1_{\pm 1}}$	$\chi_{1_{\pm 2}}$	$\chi_{1_{\pm 3}}$	$\chi_{1_{\pm 4}}$
C_1	1	1	1	1	1	1
$C_1^{(1)}$	5	1	ρ^2	ρ^4	ρ	ρ^3
$C_1^{(2)}$	5	1	ρ^4	ρ^3	ρ^2	ρ
$C_1^{(3)}$	5	1	ρ	ρ^2	ρ^3	ρ^4
$C_1^{(4)}$	5	1	ρ^3	ρ	ρ^4	ρ^2
$C_5'^{(0)}$	2	± 1	± 1	± 1	± 1	± 1
$C_5'^{(1)}$	10	± 1	$\pm\rho$	$\pm\rho^2$	$\pm\rho^3$	$\pm\rho^4$
$C_5'^{(2)}$	10	± 1	$\pm\rho^2$	$\pm\rho^4$	$\pm\rho$	$\pm\rho^3$
$C_5'^{(3)}$	10	± 1	$\pm\rho^3$	$\pm\rho$	$\pm\rho^4$	$\pm\rho^2$
$C_5'^{(4)}$	10	± 1	$\pm\rho^4$	$\pm\rho^3$	$\pm\rho^2$	$\pm\rho$
$C_2^{(1,0)}$	5	1	ρ	ρ^2	ρ^3	ρ^3
$C_2^{(2,0)}$	5	1	ρ^2	ρ^4	ρ	ρ^3
$C_2^{(2,1)}$	5	1	ρ^3	ρ	ρ^4	ρ^2
$C_2^{(3,0)}$	5	1	ρ^3	ρ	ρ^4	ρ^2
$C_2^{(3,1)}$	5	1	ρ^4	ρ^3	ρ^2	ρ
$C_2^{(3,2)}$	5	1	1	1	1	1
$C_2^{(4,0)}$	5	1	ρ^4	ρ^3	ρ^2	ρ
$C_2^{(4,1)}$	5	1	1	1	1	1
$C_2^{(4,2)}$	5	1	ρ	ρ^2	ρ^3	ρ^4
$C_2^{(4,3)}$	5	1	ρ^2	ρ^4	ρ	ρ^3

Table 9.5 Characters of $\Sigma(50)$ representations, where $\rho = e^{2i\pi/5}$

	h	$\chi_{2_{1,0}}$	$\chi_{2_{2,0}}$	$\chi_{2_{2,1}}$	$\chi_{2_{3,0}}$	$\chi_{2_{3,1}}$	$\chi_{2_{3,2}}$	$\chi_{2_{4,0}}$	$\chi_{2_{4,1}}$	$\chi_{2_{4,2}}$	$\chi_{2_{4,3}}$
C_1	1	2	2	2	2	2	2	2	2	2	2
$C_1^{(1)}$	5	2ρ	$2\rho^2$	$2\rho^3$	$2\rho^3$	$2\rho^4$	2	$2\rho^4$	2	2ρ	$2\rho^2$
$C_1^{(2)}$	5	$2\rho^2$	$2\rho^4$	2ρ	2ρ	$2\rho^3$	2	$2\rho^3$	2	$2\rho^2$	$2\rho^4$
$C_1^{(3)}$	5	$2\rho^3$	2ρ	$2\rho^4$	$2\rho^4$	$2\rho^2$	2	$2\rho^2$	2	$2\rho^3$	2ρ
$C_1^{(4)}$	5	$2\rho^4$	$2\rho^3$	$2\rho^2$	$2\rho^2$	2ρ	2	2ρ	2	$2\rho^4$	$2\rho^3$
$C_5'^{(0)}$	2	0	0	0	0	0	0	0	0	0	0
$C_5'^{(1-4)}$	10	0	0	0	0	0	0	0	0	0	0
$C_2^{(1,0)}$	5	$1+\rho$	$1+\rho^2$	$\rho+\rho^2$	$1+\rho^3$	$\rho+\rho^3$	$\rho^2+\rho^3$	$1+\rho^4$	$\rho+\rho^4$	$\rho^2+\rho^4$	$\rho^3+\rho^4$
$C_2^{(2,0)}$	5	$1+\rho^2$	$1+\rho^4$	$\rho^2+\rho^4$	$1+\rho$	$\rho+\rho^2$	$\rho+\rho^4$	$1+\rho^3$	$\rho^2+\rho^3$	$\rho^3+\rho^4$	$\rho+\rho^3$
$C_2^{(2,1)}$	5	$\rho+\rho^2$	$\rho^2+\rho^4$	$1+\rho^4$	$\rho+\rho^3$	$1+\rho^2$	$\rho^2+\rho^3$	$\rho^3+\rho^4$	$\rho+\rho^4$	$1+\rho^3$	$1+\rho$
$C_2^{(3,0)}$	5	$1+\rho^3$	$1+\rho$	$\rho+\rho^3$	$1+\rho^4$	$\rho^3+\rho^4$	$\rho+\rho^4$	$1+\rho^2$	$\rho^2+\rho^3$	$\rho+\rho^2$	$\rho^2+\rho^4$
$C_2^{(3,1)}$	5	$\rho+\rho^3$	$\rho+\rho^2$	$1+\rho^2$	$\rho^3+\rho^4$	$1+\rho$	$\rho+\rho^4$	$\rho^2+\rho^4$	$\rho^2+\rho^3$	$1+\rho^4$	$1+\rho^3$
$C_2^{(3,2)}$	5	$\rho^2+\rho^3$	$\rho+\rho^4$	$\rho^2+\rho^3$	$\rho+\rho^4$	$\rho+\rho^4$	$\rho^2+\rho^3$	$\rho^2+\rho^3$	$\rho+\rho^4$	$\rho+\rho^4$	$\rho^2+\rho^3$
$C_2^{(4,0)}$	5	$1+\rho^4$	$1+\rho^3$	$\rho^3+\rho^4$	$1+\rho^2$	$\rho^2+\rho^4$	$\rho^2+\rho^3$	$1+\rho$	$\rho+\rho^4$	$\rho+\rho^3$	$\rho+\rho^2$
$C_2^{(4,1)}$	5	$\rho+\rho^4$	$\rho^2+\rho^3$	$\rho+\rho^4$	$\rho^2+\rho^3$	$\rho^2+\rho^3$	$\rho+\rho^4$	$\rho+\rho^4$	$\rho^2+\rho^3$	$\rho^2+\rho^3$	$\rho+\rho^4$
$C_2^{(4,2)}$	5	$\rho^2+\rho^4$	$\rho^3+\rho^4$	$1+\rho^3$	$\rho+\rho^2$	$1+\rho^4$	$1+\rho$	$\rho+\rho^3$	$\rho^2+\rho^3$	$1+\rho$	$1+\rho^2$
$C_2^{(4,3)}$	5	$\rho^3+\rho^4$	$\rho+\rho^3$	$1+\rho$	$\rho^2+\rho^4$	$1+\rho^3$	$\rho^2+\rho^3$	$\rho+\rho^2$	$\rho+\rho^4$	$1+\rho^2$	$1+\rho^4$

The characters are shown in Tables 9.4 and 9.5. We omit the explicit expressions of the tensor products since they can be obtained in the same ways as the cases of $\Sigma(18)$ and $\Sigma(32)$.

In this section, we study the discrete group $\Delta(3N^2)$, which is isomorphic to $(Z_N \times Z'_N) \rtimes Z_3$. (See also Ref. [1].) The generators of Z_N and Z'_N are denoted by a and a', respectively, and the Z_3 generator is written by b. These generators satisfy

$$a^N = a'^N = b^3 = e, \qquad aa' = a'a,$$
$$bab^{-1} = a^{-1}(a')^{-1}, \qquad ba'b^{-1} = a. \tag{10.1}$$

Therefore, all of $\Delta(3N^2)$ elements are written as

$$g = b^k a^m a'^n, \tag{10.2}$$

for $k = 0, 1, 2$ and $m, n = 0, 1, 2, \ldots, N - 1$.

Since the generators, a, a' and b, are represented, e.g. as

$$b = \begin{pmatrix} 0 & 1 & 0 \\ 0 & 0 & 1 \\ 1 & 0 & 0 \end{pmatrix}, \quad a = \begin{pmatrix} \rho & 0 & 0 \\ 0 & 1 & 0 \\ 0 & 0 & \rho^{-1} \end{pmatrix}, \quad a' = \begin{pmatrix} \rho^{-1} & 0 & 0 \\ 0 & \rho & 0 \\ 0 & 0 & 1 \end{pmatrix}, \tag{10.3}$$

where $\rho = e^{2\pi i/N}$, all elements of $\Delta(3N^2)$ are written as

$$\begin{pmatrix} \rho^m & 0 & 0 \\ 0 & p^n & 0 \\ 0 & 0 & \rho^{-m-n} \end{pmatrix}, \quad \begin{pmatrix} 0 & \rho^m & 0 \\ 0 & 0 & \rho^n \\ \rho^{-m-n} & 0 & 0 \end{pmatrix}, \quad \begin{pmatrix} 0 & 0 & \rho^m \\ \rho^n & 0 & 0 \\ 0 & \rho^{-m-n} & 0 \end{pmatrix}, \tag{10.4}$$

for $m, n = 0, 1, 2, \ldots, N - 1$.

T. Kobayashi et al., *An Introduction to Non-Abelian Discrete Symmetries
for Particle Physicists*, Lecture Notes in Physics 995,
https://doi.org/10.1007/978-3-662-64679-3_10

10.1 $\Delta(3N^2)$ with $N/3 \neq$ Integer

The $\Delta(3N^2)$ groups have different features between $N/3 \neq$ integer and $N/3 =$ integer. First, we study $\Delta(3N^2)$ groups with $N/3 \neq$ integer.

• Conjugacy classes

The conjugacy classes of $\Delta(3N^2)$ with $N/3 \neq$ integer are found to be

$$ba^\ell a'^m b^{-1} = a^{-\ell+m} a'^{-\ell}, \qquad b^2 a^\ell a'^m b^{-2} = a^{-m} a'^{\ell-m}. \tag{10.5}$$

Thus, these elements, $a^\ell a'^m$, $a^{-\ell+m} a'^{-\ell}$, $a^{-m} a'^{\ell-m}$, must belong to the same conjugacy class. These are independent elements of $\Delta(3N^2)$ unless $N/3 =$ integer and $3\ell = \ell + m = 0 \pmod N$. As a result, the elements $a^\ell a'^m$ are classified into the following conjugacy classes,

$$C_3^{(\ell,m)} = \{a^\ell a'^m,\ a^{-\ell+m} a'^{-\ell},\ a^{-m} a'^{\ell-m}\}, \tag{10.6}$$

for $N/3 \neq$ integer.

In the same way, we can obtain conjugacy classes including $ba^\ell a'^m$. Let us consider the conjugates of the simplest element b among $ba^\ell a'^m$. We find

$$a^p a'^q (b) a^{-p} a'^{-q} = ba^{-p-q} a'^{p-2q} = ba^{-n+3q} a'^n, \tag{10.7}$$

where we have written for convenience by using $n \equiv p - 2q$. We also get

$$b(ba^{-n+3q} a'^n)b^{-1} = ba^{2n-3q} a'^{n-3q}, \tag{10.8}$$

$$b^2(ba^{-n+3q} a'^n)b^{-2} = ba^{-n} a'^{-2n+3q}. \tag{10.9}$$

The important property is that q appears only in the form of $3q$. Thus, if $N/3 \neq$ integer, the element of b is conjugate to all of $ba^\ell a'^m$. That is, all of them belong to the same conjugacy class $C_{N^2}^1$. Similarly, all of $b^2 a^\ell a'^m$ belong to the same conjugacy class $C_{N^2}^2$ for $N/3 \neq$ integer.

We summarize the conjugacy classes of $\Delta(3N^2)$ for $N/3 \neq$ integer. It has the following conjugacy classes,

$$\begin{array}{lll}
C_1: & \{e\}, & h = 1, \\
C_3^{(\ell,m)}: & \{a^\ell a'^m,\ a^{-\ell+m} a'^{-\ell},\ a^{-\sigma} a'^{\ell-m}\}, & h = N/\gcd(N, \ell, m), \\
C_{N^2}^1: & \{ba^\ell a'^m | \ell, m = 0, 1, \ldots, N-1\}, & h = 3, \\
C_{N^2}^2: & \{b^2 a^\ell a'^m | \ell, m = 0, 1, \ldots, N-1\}, & h = 3.
\end{array} \tag{10.10}$$

The number of the conjugacy classes $C_3^{(\ell,m)}$ is $(N^2 - 1)/3$. Then, the total number of conjugacy classes is $3 + (N^2 - 1)/3$. The relations (2.18) and (2.19) for $\Delta(3n^2)$ with $N/3 \neq$ integer give

$$m_1 + 2^2 m_2 + 3^2 m_3 + \cdots = 3N^2, \tag{10.11}$$

$$m_1 + m_2 + m_3 + \cdots = 3 + (N^2 - 1)/3, \tag{10.12}$$

which leads to $(m_1, m_3) = (3, (N^2 - 1)/3)$. Therefore, we find three singlets and $(N^2 - 1)/3$ triplets.

• **Characters and representations**

There are 3 singlets in the $\Delta(3N^2)$ group with $N/3 \neq$ integer. Since $b^3 = e$ is satisfied in this group, characters of three singlets have three possible values $\chi_{1k}(b) = \omega^k$ with $k = 0, 1, 2$, which correspond to three singlets, $\mathbf{1}_k$. Because of $\chi_{1k}(b) = \chi_{1k}(ba) = \chi_{1k}(ba')$, it is found to be $\chi_{1k}(a) = \chi_{1k}(a') = 1$. These characters are given in Table 10.1.

Now, we discuss triplet representations. As seen in (10.3), we have 3×3 matrices correspond to one of triplet representations. Therefore, (3×3) matrix representations for generic triplets are given by replacing

$$a \to a^\ell a'^m, \qquad a' \to b^2 ab^{-2} = a^{-m} a'^{\ell-m}. \tag{10.13}$$

However, it is noticed that the following two types of replacing

$$\begin{aligned}
(a, a') &\to (a^{-\ell+m} a'^{-\ell}, a^\ell a'^m), \\
(a, a') &\to (a^{-m} a'^{\ell-m}, a^{-\ell+m} a'^{-\ell}),
\end{aligned} \tag{10.14}$$

also lead to the representation equivalent to the above (10.13), because the three elements, $a^\ell a'^m$, $a^{-\ell+m} a'^{-\ell}$ and $a^{-m} a'^{\ell-m}$, belong to the same conjugacy class, $C_3^{(\ell,m)}$. Thus, the generators of $\Delta(3N^2)$ group with $N/3 \neq$ integer is represented as

$$b = \begin{pmatrix} 0 & 1 & 0 \\ 0 & 0 & 1 \\ 1 & 0 & 0 \end{pmatrix}, \quad a = \begin{pmatrix} \rho^\ell & 0 & 0 \\ 0 & \rho^k & 0 \\ 0 & 0 & \rho^{-k-\ell} \end{pmatrix}, \quad a' = \begin{pmatrix} \rho^{-k-\ell} & 0 & 0 \\ 0 & \rho^\ell & 0 \\ 0 & 0 & \rho^k \end{pmatrix}, \tag{10.15}$$

on the triplet $\mathbf{3}_{[k][\ell]}$, where $[k][\ell]$ denotes[1]

$$[k][\ell] = (k, \ell), \ (-k - \ell, k) \text{ or } (\ell, -k - \ell). \tag{10.16}$$

We denote the vector of $\mathbf{3}_{[k][\ell]}$ as

$$\mathbf{3}_{[k][\ell]} = \begin{pmatrix} x_{\ell, -k-\ell} \\ x_{k, \ell} \\ x_{-k-\ell, k} \end{pmatrix}, \tag{10.17}$$

for $k, \ell = 0, 1, \ldots, N - 1$, where k and ℓ correspond to Z_N and Z'_N charges, respectively. When $(k, \ell) = (0, 0)$, the matrices a and a' are the identity matrices. Thus, we exclude the case with $(k, \ell) = (0, 0)$. The characters are given in Table 10.1.

---

[1] The notation $[k][\ell]$ corresponds to $\widetilde{(k, \ell)}$ in Ref. [1].

Table 10.1 Characters of $\Delta(3N^2)$ for $N/3 \neq$ integer

	h	χ_{1_0}	χ_{1_1}	χ_{1_2}	$\chi_{3_{[n][m]}}$
C_1	1	1	1	1	3
$C_3^{(k,\ell)}$	$\dfrac{N}{\gcd(N,k,\ell)}$	1	1	1	$\rho^{mk-n\ell-m\ell}$ $+\rho^{nk+m\ell}+$ $\rho^{-nk-mk+n\ell}$
$C_{N^2}^1$	3	1	ω	ω^2	0
$C_{N^2}^2$	3	1	ω^2	ω	0

• Tensor products

Let us study tensor products of triplets. Taking account of $Z_N \times Z'_N$ charges, those are given as

$$
\begin{pmatrix} x_{\ell,-k-\ell} \\ x_{k,\ell} \\ x_{-k-\ell,k} \end{pmatrix}_{3_{[k][\ell]}} \otimes \begin{pmatrix} y_{\ell',-k'-\ell'} \\ y_{k',\ell'} \\ y_{-k'-\ell',k'} \end{pmatrix}_{3_{[k'][\ell']}} = \begin{pmatrix} x_{\ell,-k-\ell}y_{\ell',-k'-\ell'} \\ x_{k,\ell}y_{k',\ell'} \\ x_{-k-\ell,k}y_{-k'-\ell',k'} \end{pmatrix}_{3_{[k+k'][\ell+\ell']}}
$$

$$
\oplus \begin{pmatrix} x_{k,\ell}y_{-k'-\ell',k'} \\ x_{-k-\ell,k}y_{\ell',-k'-\ell'} \\ x_{\ell,-k-\ell}y_{k',\ell'} \end{pmatrix}_{3_{[-k-\ell+\ell'][k-k'-\ell']}} \oplus \begin{pmatrix} x_{-k-\ell,k}y_{k',\ell'} \\ x_{\ell,-k-\ell}y_{-k'-\ell',k'} \\ x_{k,\ell}y_{\ell',-k'-\ell'} \end{pmatrix}_{3_{[\ell-k'-\ell'][-k-\ell+k']}} \tag{10.18}
$$

for $-(k,\ell) \neq [k'][\ell']$,

$$
\begin{pmatrix} x_{\ell,-k-\ell} \\ x_{k,\ell} \\ x_{-k-\ell,k} \end{pmatrix}_{3_{[k][\ell]}} \otimes \begin{pmatrix} y_{-\ell,k+\ell} \\ y_{-k,-\ell} \\ y_{k+\ell,-k} \end{pmatrix}_{3_{-[k][\ell]}} = \left(x_{\ell,-k-\ell}y_{-\ell,k+\ell} + x_{k,\ell}y_{-k,-\ell} + x_{-k-\ell,k}y_{k+\ell,-k} \right)_{1_0}
$$

$$
\oplus \left(x_{\ell,-k-\ell}y_{-\ell,k+\ell} + \omega^2 x_{k,\ell}y_{-k,-\ell} + \omega x_{-k-\ell,k}y_{k+\ell,-k} \right)_{1_1}
$$

$$
\oplus \left(x_{\ell,-k-\ell}y_{-\ell,k+\ell} + \omega x_{k,\ell}y_{-k,-\ell} + \omega^2 x_{-k-\ell,k}y_{k+\ell,-k} \right)_{1_2}
$$

$$
\oplus \begin{pmatrix} x_{k,\ell}y_{k+\ell,-k} \\ x_{-k-\ell,k}y_{-\ell,k+\ell} \\ x_{\ell,-k-\ell}y_{-k,-\ell} \end{pmatrix}_{3_{[-k-2\ell][2k+\ell]}} \oplus \begin{pmatrix} x_{-k-\ell,k}y_{-k,-\ell} \\ x_{\ell,-k-\ell}y_{k+\ell,-k} \\ x_{k,\ell}y_{-\ell,k+\ell} \end{pmatrix}_{3_{[k+2\ell][-2k-\ell]}} \tag{10.19}
$$

We present a product of $3_{[k][\ell]}$ and 1_r as

$$
\begin{pmatrix} x_{\ell,-k-\ell} \\ x_{k,\ell} \\ x_{-k-\ell,k} \end{pmatrix}_{[k][\ell]} \otimes (z)_{1_r} = \begin{pmatrix} x_{\ell,-k-\ell}z \\ \omega^r x_{k,\ell}z \\ \omega^{2r} x_{-k-\ell,k}z \end{pmatrix}_{[k][\ell]} . \tag{10.20}
$$

The tensor products of singlets 1_k and $1_{k'}$ are given simply as

$$
1_k \otimes 1_{k'} = 1_{k+k' \bmod 3}. \tag{10.21}
$$

10.2 $\Delta(3N^2)$ with $N/3 =$ Integer

In this subsection, we study $\Delta(3N^2)$ groups with $N/3 =$ integer.

• **Conjugacy classes**

The algebraic relation (10.5) hold true for the case of $N/3 =$ integer, as well as for the case of $N/3 \neq$ integer. Thus, these elements, $a^\ell a'^m$, $a^{-\ell+m}a'^{-\ell}$, $a^{-m}a'^{\ell-m}$, must belong to the same conjugacy class. In the case of $N/3 =$ integer and $3\ell = \ell + m = 0$ (mod N), the above elements are the same, i.e. $a^\ell a'^{-\ell}$. As a result, the elements $a^\ell a'^m$ are classified into the following conjugacy classes,

$$C_1^\ell = \{a^\ell a'^{-\ell}\}, \qquad \ell = \frac{N}{3}, \frac{2N}{3},$$
$$C_3^{(\ell,m)} = \{a^\ell a'^m,\ a^{-\ell+m}a'^{-\ell},\ a^{-m}a'^{\ell-m}\}, \quad (\ell,m) \neq \left(\frac{N}{3}, \frac{2N}{3}\right), \left(\frac{2N}{3}, \frac{N}{3}\right),$$

$$(10.22)$$

for $N/3 =$ integer.

Similarly, we obtain conjugacy classes including $ba^\ell a'^m$. One can obtain the conjugates of b among $ba^\ell a'^m$ by using Eqs. (10.7), (10.8) and (10.9). When $N/3 \neq$ integer, the element of b is conjugate to all of $ba^\ell a'^m$ as shown in the previous subsection. However, the situation is different when $N/3 =$ integer. The elements conjugate to b do not include ba. The conjugates to ba are also obtained as

$$a^p a'^q (ba) a^{-p} a'^{-q} = ba^{1-p-q} a'^{p-2q} = ba^{1-n+3q} a'^n, \qquad (10.23)$$
$$b(ba^{1-n+3q} a'^n)b^{-1} = ba^{-1+2n-3q} a'^{-1+n-3q}, \qquad (10.24)$$
$$b^2(ba^{1-n+3q} a'^n)b^{-2} = ba^{-n} a'^{1-2n+3q}, \qquad (10.25)$$

where it is noticed that these elements conjugate to ba as well as conjugates of b do not include ba^2 when $N/3 =$ integer. Therefore, for $N/3 =$ integer, the elements $ba^\ell a'^m$ are classified into three conjugacy classes, $C_{N^2/3}^{(\ell)}$ for $\ell = 0, 1, 2$, i.e.,

$$C_{N^2/3}^{(\ell)} = \{ba^{\ell-n-3m} a'^n | m = 0, 1, \ldots, \frac{N-3}{3};\ n = 0, \ldots, N-1\}. \quad (10.26)$$

In the same way, the $b^2 a^\ell a'^m$ are classified into three conjugacy classes, $C_{N^2/3}^{(\ell)}$ for $\ell = 0, 1, 2$, i.e.,

$$C_{N^2/3}^{(\ell)} = \{b^2 a^{\ell-n-3m} a'^n | m = 0, 1, \ldots, \frac{N-3}{3};\ n = 0, \ldots, N-1\}. \quad (10.27)$$

we summarize the conjugacy classes of $\Delta(3N^2)$ for $N/3 =$ integer as follows:

$$C_1 : \qquad\qquad\qquad\qquad\qquad \{e\},$$
$$C_1^{(k)} : \qquad\qquad\qquad \{a^k a'^{-k}\}, \; k = \tfrac{N}{3}, \tfrac{2N}{3},$$
$$C_3^{(\ell,m)} : \quad \{a^\ell a'^m, a^{-\ell+m} a'^{-\ell}, a^{-m} a'^{\ell-m}\}, \; (\ell, m) \neq \left(\tfrac{N}{3}, \tfrac{2N}{3}\right), \left(\tfrac{2N}{3}, \tfrac{N}{3}\right), \quad (10.28)$$
$$C_{N^2/3}^{(1,p)} : \{b a^{p-n-3m} a'^n | m = 0, 1, \ldots, \tfrac{N-3}{3}, n = 0, 1, \ldots, N-1\}, \quad p = 0, 1, 2,$$
$$C_{N^2/3}^{(2,p)} : \{b^2 a^{p-n-3m} a'^n | m = 0, 1, \ldots, \tfrac{N-3}{3}, n = 0, 1, \ldots, N-1\}, \quad p = 0, 1, 2.$$

The orders of each element in the conjugacy classes, i.e. $g^h = e$ are given as follows,

$$\begin{aligned}
C_1 : &\qquad h = 1, \\
C_1^{(k)} : &\qquad h = 3, \\
C_3^{(\ell,m)} : &\; h = N/\gcd(N, \ell, m), \\
C_{N^2/3}^{(1,p)} : &\qquad h = 3, \\
C_{N^2/3}^{(2,p)} : &\qquad h = 3.
\end{aligned} \qquad (10.29)$$

The numbers of the conjugacy classes $C_1^{(k)}$, $C_3^{(\ell,m)}$, $C_{N^2/3}^{(1,p)}$ and $C_{N^2/3}^{(2,p)}$, are 2, $(N^2 - 3)/3$, 3 and 3, respectively. The total number of conjugacy classes is equal to $9 + (N^2 - 3)/3$. The relations (2.18) and (2.19) for $\Delta(3N^2)$ with $N/3 =$ integer lead to

$$m_1 + 2^2 m_2 + 3^2 m_3 + \cdots = 3N^2, \qquad (10.30)$$
$$m_1 + m_2 + m_3 + \cdots = 9 + (N^2 - 3)/3, \qquad (10.31)$$

which give a solution $(m_1, m_3) = (9, (N^2 - 3)/3)$. That is, there are nine singlets and $(N^2 - 3)/3$ triplets.

• Characters and representations

There are nine singlets in the $\Delta(3N^2)$ group with $N/3 =$ integer. Their characters must satisfy $\chi_\alpha(b) = \omega^k$ $(k = 0, 1, 2)$ similarly to the case of $N/3 \neq$ integer. In addition, it is found that $\chi_\alpha(a) = \chi_\alpha(a') = \omega^\ell$ $(\ell = 0, 1, 2)$. Thus, nine singlets can be specified by combinations of $\chi_\alpha(b)$ and $\chi_\alpha(a)$, i.e. $\mathbf{1}_{k,\ell}$ $(k, \ell = 0, 1, 2)$ with $\chi_\alpha(b) = \omega^k$ and $\chi_\alpha(a) = \chi_\alpha(a') = \omega^\ell$. These characters are shown in Table 10.2.

The triplet representations are also given similarly to the case with $N/3 \neq$ integer. That is, the generators of the $\Delta(3N^2)$ group with $N/3 =$ integer is represented as

$$b = \begin{pmatrix} 0 & 1 & 0 \\ 0 & 0 & 1 \\ 1 & 0 & 0 \end{pmatrix}, \quad a = \begin{pmatrix} \rho^\ell & 0 & 0 \\ 0 & \rho^k & 0 \\ 0 & 0 & \rho^{-k-\ell} \end{pmatrix}, \quad a' = \begin{pmatrix} \rho^{-k-\ell} & 0 & 0 \\ 0 & \rho^\ell & 0 \\ 0 & 0 & \rho^k \end{pmatrix}, \quad (10.32)$$

on the triplet $\mathbf{3}_{[k][\ell]}$. It should be noticed that the matrices, a and a' are trivial for the case of $(k, \ell) = (0, 0)$, $(N/3, N/3)$, $(2N/3, 2N/3)$. Thus, we exclude such values of (k, ℓ). These characters are shown in Table 10.2.

Table 10.2 Characters of $\Delta(3N^2)$ for $N/3 =$ integer

	h	$\chi_{1_{r,s}}$	$\chi_{3_{[n][m]}}$
C_1	1	1	3
$C_1^{(k)}$	3	1	$\rho^{nk+2mk} + \rho^{nk+mk} + \rho^{-2nk-mk}$
$C_3^{(k,\ell)}$	$\dfrac{N}{\gcd(N,k,\ell)}$	$\omega^{s(\ell+m)}$	$\rho^{mk-n\ell-m\ell}$ $+\rho^{nk-m\ell} + \rho^{-nk-mk+n\ell}$
$C_{N^2/3}^{(1,p)}$	3	ω^{r+sp}	0
$C_{N^2/3}^{(2,p)}$	3	ω^{2r+sp}	0

- **Tensor products**

For $N/3 =$ integer, we present the tensor products of two triplets,

$$\begin{pmatrix} x_{\ell,-k-\ell} \\ x_{k,\ell} \\ x_{-k-\ell,k} \end{pmatrix}_{3_{[k][\ell]}} \quad \text{and} \quad \begin{pmatrix} y_{\ell',-k'-\ell'} \\ y_{k',\ell'} \\ y_{-k'-\ell',k'} \end{pmatrix}_{3_{[k'][\ell']}}. \tag{10.33}$$

Unless $(k', \ell') = [k + mN/3][\ell + mN/3]$ for $m = 0, 1, 2$, their tensor products are the same as (10.18).

For $(k', \ell') = [-k + mN/3][-\ell + mN/3]$ ($m = 0, 1$ or 2), tensor products of the above triplets are given as

$$\begin{pmatrix} x_{\ell,-k-\ell} \\ x_{k,\ell} \\ x_{-k-\ell,k} \end{pmatrix}_{3_{[k][\ell]}} \otimes \begin{pmatrix} y_{-\ell+mN/3,k+\ell-2mN/3} \\ y_{-k+mN/3,-\ell+mN/3} \\ y_{k+\ell-2mN/3,-k+mN/3} \end{pmatrix}_{3_{[-k+mN/3][-\ell+mN/3]}}$$

$$= \left(x_{\ell,-k-\ell} y_{-\ell+mN/3,k+\ell-2mN/3} + x_{k,\ell} y_{-k+mN/3,-\ell+mN/3} + x_{-k-\ell,k} y_{k+\ell-2mN/3,-k+mN/3} \right)_{1_{0,m}}$$

$$\oplus \left(x_{\ell,-k-\ell} y_{-\ell+mN/3,k+\ell-2mN/3} + \omega^2 x_{k,\ell} y_{-k+mN/3,-\ell+mN/3} + \omega x_{-k-\ell,k} y_{k+\ell-2mN/3,-k+mN/3} \right)_{1_{1,m}}$$

$$\oplus \left(x_{\ell,-k-\ell} y_{-\ell+mN/3,k+\ell-2mN/3} + \omega x_{k,\ell} y_{-k+mN/3,-\ell+mN/3} + \omega^2 x_{-k-\ell,k} y_{k+\ell-2mN/3,-k+mN/3} \right)_{1_{2,m}}$$

$$\oplus \begin{pmatrix} x_{k,\ell} y_{k+\ell-2mN/3,-k+mN/3} \\ x_{-k-\ell,k} y_{-\ell+mN/3,k+\ell-2mN/3} \\ x_{\ell,-k-\ell} y_{-k+mN/3,-\ell+mN/3} \end{pmatrix}_{3_{[-k-2\ell+mN/3][2k+\ell-2mN/3]}}$$

$$\oplus \begin{pmatrix} x_{-k-\ell,k} y_{-k+mN/3,-\ell+mN/3} \\ x_{\ell,-k-\ell} y_{k+\ell-2mN/3,-k+mN/3} \\ x_{k,\ell} y_{-\ell+mN/3,k+\ell-2mN/3} \end{pmatrix}_{3_{[k+2\ell-2mN/3][-2k-\ell+mN/3]}} \tag{10.34}$$

A product of $3_{[k][\ell]}$ and $1_{r,s}$ is

$$\begin{pmatrix} x_{\ell,-k-\ell} \\ x_{k,\ell} \\ x_{-k-\ell,k} \end{pmatrix}_{3_{[k][\ell]}} \otimes (z)_{1_{r,s}} = \begin{pmatrix} x_{\ell,-k-\ell} z \\ \omega^r x_{k,\ell} z \\ \omega^{2r} x_{-k-\ell,k} z \end{pmatrix}_{3_{[k+sN/3][\ell+sN/3]}}. \tag{10.35}$$

We find the tensor products of singlets $\mathbf{1}_{k,\ell}$ and $\mathbf{1}_{k',\ell'}$ as follow:

$$\mathbf{1}_{k,\ell} \otimes \mathbf{1}_{k',\ell'} = \mathbf{1}_{k+k' \bmod 3, \ell+\ell' \bmod 3}. \tag{10.36}$$

10.3 $\Delta(27)$

We present a simple example of the $\Delta(3N^2)$ group. The simplest group $\Delta(3)$ is nothing but the Z_3 group. The next one $\Delta(12)$ is isomorphic to A_4. Thus, the simple and non-trivial example is $\Delta(27)$.

The conjugacy classes of $\Delta(27)$ are given as

$$
\begin{aligned}
C_1 &: & \{e\}, & & h = 1, \\
C_1^{(1)} &: & \{aa'^2\}, & & h = 3, \\
C_1^{(2)} &: & \{a^2a'\}, & & h = 3, \\
C_3^{(0,1)} &: & \{a', a, a^2a'^2\}, & & h = 3, \\
C_3^{(0,2)} &: & \{a'^2, a^2, aa'\}, & & h = 3, \\
C_3^{(1,p)} &: & \{ba^p, ba^{p-1}a', ba^{p-2}a'^2\}, & & h = 3, \\
C_3^{(2,p)} &: & \{b^2a^p, b^2a^{p-1}a', b^2a^{p-2}a'^2\}, & & h = 3.
\end{aligned}
\tag{10.37}
$$

The $\Delta(27)$ has nine singlets $\mathbf{1}_{k,\ell}$ ($k, \ell = 0, 1, 2$) and two triplets, $\mathbf{3}_{[0][1]}$ and $\mathbf{3}_{[0][2]}$. The characters are shown in Table 10.3.

We show tensor products between triplets as follows:

$$
\begin{pmatrix} x_{1,-1} \\ x_{0,1} \\ x_{-1,0} \end{pmatrix}_{\mathbf{3}_{[0][1]}} \otimes \begin{pmatrix} y_{1,-1} \\ y_{0,1} \\ y_{-1,0} \end{pmatrix}_{\mathbf{3}_{[0][1]}} = \begin{pmatrix} x_{1,-1}y_{1,-1} \\ x_{0,1}y_{0,1} \\ x_{-1,0}y_{-1,0} \end{pmatrix}_{\mathbf{3}_{[0][2]}} \oplus \begin{pmatrix} x_{0,1}y_{-1,0} \\ x_{-1,0}y_{1,-1} \\ x_{1,-1}y_{0,1} \end{pmatrix}_{\mathbf{3}_{[0][2]}} \oplus \begin{pmatrix} x_{-1,0}y_{0,1} \\ x_{1,-1}y_{-1,0} \\ x_{0,1}y_{1,-1} \end{pmatrix}_{\mathbf{3}_{[0][2]}},
\tag{10.38}
$$

$$
\begin{pmatrix} x_{2,-2} \\ x_{0,2} \\ x_{-2,0} \end{pmatrix}_{\mathbf{3}_{[0][2]}} \otimes \begin{pmatrix} y_{2,-2} \\ y_{0,2} \\ y_{-2,0} \end{pmatrix}_{\mathbf{3}_{[0][2]}} = \begin{pmatrix} x_{2,-2}y_{2,-2} \\ x_{0,2}y_{0,2} \\ x_{-2,0}y_{-2,0} \end{pmatrix}_{\mathbf{3}_{[0][1]}} \oplus \begin{pmatrix} x_{0,2}y_{-2,0} \\ x_{-2,0}y_{2,-2} \\ x_{2,-2}y_{0,2} \end{pmatrix}_{\mathbf{3}_{[0][1]}} \oplus \begin{pmatrix} x_{-2,0}y_{0,2} \\ x_{2,-2}y_{-2,0} \\ x_{0,2}y_{2,-2} \end{pmatrix}_{\mathbf{3}_{[0][1]}},
\tag{10.39}
$$

Table 10.3 Characters of $\Delta(27)$

	h	$\chi_{\mathbf{1}_{k,\ell}}$	$\chi_{\mathbf{3}_{[0,1]}}$	$\chi_{\mathbf{3}_{[0,2]}}$
C_1	1	1	3	3
$C_1^{(1)}$	1	1	$3\omega^2$	3ω
$C_1^{(2)}$	1	1	3ω	$3\omega^2$
$3C_1^{(0,1)}$	3	ω^ℓ	0	0
$3C_1^{(0,2)}$	3	$\omega^{2\ell}$	0	0
$C_3^{(1,p)}$	3	$\omega^{k+\ell p}$	0	0
$C_3^{(2,p)}$	3	$\omega^{2k+\ell p}$	0	0

$$\begin{pmatrix} x_{1,-1} \\ x_{0,1} \\ x_{-1,0} \end{pmatrix}_{3_{[0][1]}} \otimes \begin{pmatrix} y_{-1,1} \\ y_{0,-1} \\ y_{1,0} \end{pmatrix}_{3_{[0][2]}} = \sum_r (x_{1,-1}y_{-1,1} + \omega^{2r}x_{0,1}y_{0,-1} + \omega^r x_{-1,0}y_{1,0})\mathbf{1}_{(r,0)}$$

$$\oplus \sum_r (x_{1,-1}y_{0,-1} + \omega^{2r}x_{0,1}y_{1,0} + \omega^r x_{-1,0}y_{-1,1})\mathbf{1}_{(r,1)}$$

$$\oplus \sum_r (x_{1,-1}y_{1,0} + \omega^{2r}x_{0,1}y_{-1,1} + \omega^r x_{-1,0}y_{0,-1})\mathbf{1}_{(r,2)}.$$

$$(10.40)$$

The tensor products between singlets and triplets are given as

$$\begin{pmatrix} x_{(1,-1)} \\ x_{(0,1)} \\ x_{(-1,0)} \end{pmatrix}_{3_{[0][1]}} \otimes (z)_{1_{k,\ell}} = \begin{pmatrix} x_{(1,-1)}z \\ \omega^r x_{(0,1)}z \\ \omega^{2r} x_{(-1,0)}z \end{pmatrix}_{3_{[\ell][1+\ell]}},$$

$$\begin{pmatrix} x_{(2,-2)} \\ x_{(0,2)} \\ x_{(-2,0)} \end{pmatrix}_{3_{[0][2]}} \otimes (z)_{1_{k,\ell}} = \begin{pmatrix} x_{(2,-2)}z \\ \omega^r x_{(0,2)}z \\ \omega^{2r} x_{(-2,0)}z \end{pmatrix}_{3_{[\ell][2+\ell]}}.$$

$$(10.41)$$

The tensor products of singlets are easily obtained from Eq. (10.36).

10.4 $\Delta(48)$

We present another simple example of $\Delta(3N^2)$ in case of $N \neq 3$, that is $\Delta(48)$. It is isomorphic to $(Z_4 \times Z_4') \rtimes Z_3$. (See also Refs. [2,3].) Let us denote the generators of Z_4 and Z_4' by a and a', respectively. We denote Z_3 generators by b, where b is Z_3 generator of Z_3. These generators satisfy

$$a^4 = a'^4 = b^3 = e, \quad aa' = a'a,$$
$$bab^{-1} = a^{-1}(a')^{-1}, \quad ba'b^{-1} = a. \tag{10.42}$$

Using them, all of $\Delta(48)$ elements are written as

$$g = b^k a^m a'^n, \tag{10.43}$$

for $k = 0, 1, 2$, and $m, n = 0, 1, 2, 3$.

10.4.1 Conjugacy Classes and Tensor Products

Here, we summarize the conjugacy classes of $\Delta(48)$ as follows: conjugacy classes,

$$
\begin{aligned}
C_1 : & \quad \{e\}, \quad h = 1, \\
C_3^{(0,1)} : & \quad \{a, a', a^3 a'^3\}, \quad h = 4, \\
C_3^{(0,2)} : & \quad \{a^2, a'^2, a^2 a'^2\}, \quad h = 4, \\
C_3^{(0,3)} : & \quad \{a^3, a'^3, aa'\}, \quad h = 4, \\
C_3^{(1,2)} : & \quad \{aa'^2, aa'^3, a^2 a'^3\}, \quad h = 4, \\
C_3^{(2,1)} : & \quad \{a^2 a', a^3 a'^2, a^3 a'\}, \quad h = 4, \\
C_{16}^{(1)} : & \quad \{b a^\ell a'^m \mid \ell, m = 0, 1, 2, 3\}, \quad h = 3, \\
C_{16}^{(2)} : & \quad \{b^2 a^\ell a'^m \mid \ell, m = 0, 1, 2, 3\}, \quad h = 3,
\end{aligned}
\tag{10.44}
$$

$\Delta(48)$ has three singlets $1_{0,1,2}$, and five triplets $3_{[1][0]}$, $3_{[3][0]}$, $3_{[1][2]}$, $3_{[3][2]}$, $3_{[0][2]}$. The characters are shown in Table 10.4.

The tensor products among singlets are given as follows:

$$
1_k \otimes 1_{k'} = 1_{k+k' \bmod 4}.
\tag{10.45}
$$

The tensor products between singlets and triplets are given as follows:

$$
(x)_{1_r} \otimes \begin{pmatrix} y_1 \\ y_2 \\ y_3 \end{pmatrix}_{3_{[k][\ell]}} = \begin{pmatrix} x y_1 \\ \omega^r x y_2 \\ \omega^{2r} x y_3 \end{pmatrix}_{3_{[k][\ell]}},
\tag{10.46}
$$

Table 10.4 Characters of $\Delta(48)$

	h	χ_{1_0}	χ_{1_1}	χ_{1_2}	$\chi_{3_{[1][0]}}$	$\chi_{3_{[3][0]}}$	$\chi_{3_{[1][2]}}$	$\chi_{3_{[3][2]}}$	$\chi_{3_{[0][2]}}$
C_1	1	1	1	1	3	3	3	3	3
$C_3^{(0,2)}$	2	1	1	1	-1	-1	-1	-1	3
$C_3^{(0,1)}$	4	1	1	1	1	1	$-1 + 2i$	$-1 - 2i$	-1
$C_3^{(0,3)}$	4	1	1	1	1	1	$-1 - 2i$	$-1 + 2i$	-1
$C_3^{(1,2)}$	4	1	1	1	$-1 + 2i$	$-1 - 2i$	1	1	-1
$C_3^{(2,1)}$	4	1	1	1	$-1 - 2i$	$-1 + 2i$	1	1	-1
$C_{16}^{(0)}$	3	1	ω	ω^2	0	0	0	0	0
$C_{16}^{(1)}$	3	1	ω^2	ω	0	0	0	0	0

The tensor products among triplets are given as follows:

$$
\begin{pmatrix} x_1 \\ x_2 \\ x_3 \end{pmatrix}_{3_{[1][0]}} \otimes \begin{pmatrix} y_1 \\ y_2 \\ y_3 \end{pmatrix}_{3_{[1][0]}} = \begin{pmatrix} x_3 y_2 + x_2 y_3 \\ x_3 y_1 + x_1 y_3 \\ x_2 y_1 + x_1 y_2 \end{pmatrix}_{3_{[3][0]}} \oplus \begin{pmatrix} x_3 y_2 - x_2 y_3 \\ -x_3 y_1 + x_1 y_3 \\ x_2 y_1 - x_1 y_2 \end{pmatrix}_{3_{[3][0]}} \oplus \begin{pmatrix} x_3 y_3 \\ x_1 y_1 \\ x_2 y_2 \end{pmatrix}_{3_{[0][2]}} ,
$$

(10.47)

$$
\begin{pmatrix} x_1 \\ x_2 \\ x_3 \end{pmatrix}_{3_{[1][2]}} \otimes \begin{pmatrix} y_1 \\ y_2 \\ y_3 \end{pmatrix}_{3_{[1][2]}} = \begin{pmatrix} x_3 y_2 + x_2 y_3 \\ x_3 y_1 + x_1 y_3 \\ x_2 y_1 + x_1 y_2 \end{pmatrix}_{3_{[3][2]}} \oplus \begin{pmatrix} x_3 y_2 - x_2 y_3 \\ -x_3 y_1 + x_1 y_3 \\ x_2 y_1 - x_1 y_2 \end{pmatrix}_{3_{[3][2]}} \oplus \begin{pmatrix} x_3 y_3 \\ x_1 y_1 \\ x_2 y_2 \end{pmatrix}_{3_{[0][2]}} ,
$$

(10.48)

$$
\begin{pmatrix} x_1 \\ x_2 \\ x_3 \end{pmatrix}_{3_{[3][0]}} \otimes \begin{pmatrix} y_1 \\ y_2 \\ y_3 \end{pmatrix}_{3_{[3][0]}} = \begin{pmatrix} x_3 y_2 + x_2 y_3 \\ x_3 y_1 + x_1 y_3 \\ x_2 y_1 + x_1 y_2 \end{pmatrix}_{3_{[1][0]}} \oplus \begin{pmatrix} x_3 y_2 - x_2 y_3 \\ -x_3 y_1 + x_1 y_3 \\ x_2 y_1 - x_1 y_2 \end{pmatrix}_{3_{[1][0]}} \oplus \begin{pmatrix} x_3 y_3 \\ x_1 y_1 \\ x_2 y_2 \end{pmatrix}_{3_{[0][2]}} ,
$$

(10.49)

$$
\begin{pmatrix} x_1 \\ x_2 \\ x_3 \end{pmatrix}_{3_{[3][2]}} \otimes \begin{pmatrix} y_1 \\ y_2 \\ y_3 \end{pmatrix}_{3_{[3][2]}} = \begin{pmatrix} x_3 y_2 + x_2 y_3 \\ x_3 y_1 + x_1 y_3 \\ x_2 y_1 + x_1 y_2 \end{pmatrix}_{3_{[1][2]}} \oplus \begin{pmatrix} -x_3 y_2 + x_2 y_3 \\ x_3 y_1 - x_1 y_3 \\ -x_2 y_1 + x_1 y_2 \end{pmatrix}_{3_{[1][2]}} \oplus \begin{pmatrix} x_3 y_3 \\ x_1 y_1 \\ x_2 y_2 \end{pmatrix}_{3_{[0][2]}} ,
$$

(10.50)

$$
\begin{pmatrix} x_1 \\ x_2 \\ x_3 \end{pmatrix}_{3_{[1][0]}} \otimes \begin{pmatrix} y_1 \\ y_2 \\ y_3 \end{pmatrix}_{3_{[3][0]}}
$$
$$
= (x_1 y_1 + x_2 y_2 + x_3 y_3)_{1_0} \oplus (x_1 y_1 + \omega^2 x_2 y_2 + \omega x_3 y_3)_{1_1} \oplus (x_1 y_1 + \omega x_2 y_2 + \omega^2 x_3 y_3)_{1_2}
$$
$$
\oplus \begin{pmatrix} x_3 y_2 \\ x_1 y_3 \\ x_2 y_1 \end{pmatrix}_{3_{[1][2]}} \oplus \begin{pmatrix} x_2 y_3 \\ x_3 y_1 \\ x_1 y_2 \end{pmatrix}_{3_{[3][2]}} ,
$$

(10.51)

$$
\begin{pmatrix} x_1 \\ x_2 \\ x_3 \end{pmatrix}_{3_{[1][2]}} \otimes \begin{pmatrix} y_1 \\ y_2 \\ y_3 \end{pmatrix}_{3_{[3][2]}}
$$
$$
= (x_1 y_1 + x_2 y_2 + x_3 y_3)_{1_0} \oplus (x_1 y_1 + \omega^2 x_2 y_2 + \omega x_3 y_3)_{1_1} \oplus (x_1 y_1 + \omega x_2 y_2 + \omega^2 x_3 y_3)_{1_2}
$$
$$
\oplus \begin{pmatrix} x_3 y_2 \\ x_1 y_3 \\ x_2 y_1 \end{pmatrix}_{3_{[1][0]}} \oplus \begin{pmatrix} x_2 y_3 \\ x_3 y_1 \\ x_1 y_2 \end{pmatrix}_{3_{[3][0]}} ,
$$

(10.52)

$$
\begin{pmatrix} x_1 \\ x_2 \\ x_3 \end{pmatrix}_{3_{[0][2]}} \otimes \begin{pmatrix} y_1 \\ y_2 \\ y_3 \end{pmatrix}_{3_{[0][2]}}
$$
$$
= (x_1 y_1 + x_2 y_2 + x_3 y_3)_{1_0} \oplus (x_2 y_2 + \omega^2 x_3 y_3 + \omega x_1 y_1)_{1_1} \oplus (x_2 y_2 + \omega x_3 y_3 + \omega^2 x_1 y_1)_{1_2}
$$
$$
\oplus \begin{pmatrix} x_3 y_2 + x_2 y_3 \\ x_3 y_1 + x_1 y_3 \\ x_2 y_1 + x_1 y_2 \end{pmatrix}_{3_{[0][2]}} \oplus \begin{pmatrix} -x_3 y_2 + x_2 y_3 \\ x_3 y_1 - x_1 y_3 \\ -x_2 y_1 + x_1 y_2 \end{pmatrix}_{3_{[0][2]}} ,
$$

(10.53)

$$
\begin{pmatrix} x_1 \\ x_2 \\ x_3 \end{pmatrix}_{\mathbf{3}_{[1][0]}} \otimes \begin{pmatrix} y_1 \\ y_2 \\ y_3 \end{pmatrix}_{\mathbf{3}_{[1][2]}} = \begin{pmatrix} x_3 y_2 \\ x_1 y_3 \\ x_2 y_1 \end{pmatrix}_{\mathbf{3}_{[1][0]}} \oplus \begin{pmatrix} x_2 y_3 \\ x_3 y_1 \\ x_1 y_2 \end{pmatrix}_{\mathbf{3}_{[1][2]}} \oplus \begin{pmatrix} x_2 y_2 \\ x_3 y_3 \\ x_1 y_1 \end{pmatrix}_{\mathbf{3}_{[0][2]}} \quad,
$$
$$(10.54)$$

$$
\begin{pmatrix} x_1 \\ x_2 \\ x_3 \end{pmatrix}_{\mathbf{3}_{[1][0]}} \otimes \begin{pmatrix} y_1 \\ y_2 \\ y_3 \end{pmatrix}_{\mathbf{3}_{[3][2]}} = \begin{pmatrix} x_2 y_3 \\ x_3 y_1 \\ x_1 y_2 \end{pmatrix}_{\mathbf{3}_{[1][0]}} \oplus \begin{pmatrix} x_3 y_2 \\ x_1 y_3 \\ x_2 y_1 \end{pmatrix}_{\mathbf{3}_{[3][2]}} \oplus \begin{pmatrix} x_1 y_1 \\ x_2 y_2 \\ x_3 y_3 \end{pmatrix}_{\mathbf{3}_{[0][2]}} \quad,
$$
$$(10.55)$$

$$
\begin{pmatrix} x_1 \\ x_2 \\ x_3 \end{pmatrix}_{\mathbf{3}_{[3][0]}} \otimes \begin{pmatrix} y_1 \\ y_2 \\ y_3 \end{pmatrix}_{\mathbf{3}_{[3][2]}} = \begin{pmatrix} x_2 y_2 \\ x_1 y_3 \\ x_3 y_1 \end{pmatrix}_{\mathbf{3}_{[3][0]}} \oplus \begin{pmatrix} x_3 y_3 \\ x_2 y_1 \\ x_1 y_2 \end{pmatrix}_{\mathbf{3}_{[3][2]}} \oplus \begin{pmatrix} x_3 y_2 \\ x_2 y_3 \\ x_1 y_1 \end{pmatrix}_{\mathbf{3}_{[0][2]}} \quad,
$$
$$(10.56)$$

$$
\begin{pmatrix} x_1 \\ x_2 \\ x_3 \end{pmatrix}_{\mathbf{3}_{[3][0]}} \otimes \begin{pmatrix} y_1 \\ y_2 \\ y_3 \end{pmatrix}_{\mathbf{3}_{[1][2]}} = \begin{pmatrix} x_2 y_3 \\ x_3 y_1 \\ x_1 y_2 \end{pmatrix}_{\mathbf{3}_{[3][0]}} \oplus \begin{pmatrix} x_3 y_2 \\ x_1 y_3 \\ x_2 y_1 \end{pmatrix}_{\mathbf{3}_{[1][2]}} \oplus \begin{pmatrix} x_1 y_1 \\ x_2 y_2 \\ x_3 y_3 \end{pmatrix}_{\mathbf{3}_{[0][2]}} \quad,
$$
$$(10.57)$$

$$
\begin{pmatrix} x_1 \\ x_2 \\ x_3 \end{pmatrix}_{\mathbf{3}_{[1][0]}} \otimes \begin{pmatrix} y_1 \\ y_2 \\ y_3 \end{pmatrix}_{\mathbf{3}_{[0][2]}} = \begin{pmatrix} x_1 y_2 \\ x_2 y_3 \\ x_3 y_1 \end{pmatrix}_{\mathbf{3}_{[3][0]}} \oplus \begin{pmatrix} x_1 y_1 \\ x_2 y_2 \\ x_3 y_3 \end{pmatrix}_{\mathbf{3}_{[1][2]}} \oplus \begin{pmatrix} x_1 y_3 \\ x_2 y_1 \\ x_3 y_2 \end{pmatrix}_{\mathbf{3}_{[3][2]}} \quad,
$$
$$(10.58)$$

$$
\begin{pmatrix} x_1 \\ x_2 \\ x_3 \end{pmatrix}_{\mathbf{3}_{[1][2]}} \otimes \begin{pmatrix} y_1 \\ y_2 \\ y_3 \end{pmatrix}_{\mathbf{3}_{[0][2]}} = \begin{pmatrix} x_1 y_1 \\ x_2 y_2 \\ x_3 y_3 \end{pmatrix}_{\mathbf{3}_{[1][0]}} \oplus \begin{pmatrix} x_1 y_3 \\ x_2 y_1 \\ x_3 y_2 \end{pmatrix}_{\mathbf{3}_{[3][0]}} \oplus \begin{pmatrix} x_1 y_2 \\ x_2 y_3 \\ x_3 y_1 \end{pmatrix}_{\mathbf{3}_{[3][2]}} \quad,
$$
$$(10.59)$$

$$
\begin{pmatrix} x_1 \\ x_2 \\ x_3 \end{pmatrix}_{\mathbf{3}_{[3][0]}} \otimes \begin{pmatrix} y_1 \\ y_2 \\ y_3 \end{pmatrix}_{\mathbf{3}_{[0][2]}} = \begin{pmatrix} x_1 y_2 \\ x_2 y_3 \\ x_3 y_1 \end{pmatrix}_{\mathbf{3}_{[1][0]}} \oplus \begin{pmatrix} x_1 y_3 \\ x_2 y_1 \\ x_3 y_2 \end{pmatrix}_{\mathbf{3}_{[1][2]}} \oplus \begin{pmatrix} x_1 y_1 \\ x_2 y_2 \\ x_3 y_3 \end{pmatrix}_{\mathbf{3}_{[3][2]}} \quad,
$$
$$(10.60)$$

$$
\begin{pmatrix} x_1 \\ x_2 \\ x_3 \end{pmatrix}_{\mathbf{3}_{[3][2]}} \otimes \begin{pmatrix} y_1 \\ y_2 \\ y_3 \end{pmatrix}_{\mathbf{3}_{[0][2]}} = \begin{pmatrix} x_1 y_3 \\ x_2 y_1 \\ x_3 y_2 \end{pmatrix}_{\mathbf{3}_{[1][0]}} \oplus \begin{pmatrix} x_1 y_1 \\ x_2 y_2 \\ x_3 y_3 \end{pmatrix}_{\mathbf{3}_{[3][0]}} \oplus \begin{pmatrix} x_1 y_2 \\ x_2 y_3 \\ x_3 y_1 \end{pmatrix}_{\mathbf{3}_{[1][2]}} \quad.
$$
$$(10.61)$$

10.4.2 The Different Bases of $\Delta(48)$

Our bases b and a are written by

$$b = \begin{pmatrix} 0 & 1 & 0 \\ 0 & 0 & 1 \\ 1 & 0 & 0 \end{pmatrix}, \quad a = \begin{pmatrix} \rho^\ell & 0 & 0 \\ 0 & \rho^k & 0 \\ 0 & 0 & \rho^{-k-\ell} \end{pmatrix}, \tag{10.62}$$

where the representation matrix of a' is given by $a' = b^{-1}ab$, and subscripts for triplets $3_{[k][\ell]}$ correspond to each of representations of b, a. On the other hand, another bases are given by Ref. [3] as follows:

$$3(\equiv 3_{[1][0]}): \quad b' = \begin{pmatrix} 1 & 0 & 0 \\ 0 & \omega & 0 \\ 0 & 0 & \omega^2 \end{pmatrix}, \quad a' = \frac{1}{3}\begin{pmatrix} 1 & 1-\sqrt{3} & 1+\sqrt{3} \\ 1+\sqrt{3} & 1 & 1-\sqrt{3} \\ 1-\sqrt{3} & 1+\sqrt{3} & 1 \end{pmatrix}, \tag{10.63}$$

$$\bar{3}(\equiv 3_{[3][0]}): \quad b' = \begin{pmatrix} 1 & 0 & 0 \\ 0 & \omega^2 & 0 \\ 0 & 0 & \omega \end{pmatrix}, \quad a' = \frac{1}{3}\begin{pmatrix} 1 & 1-\sqrt{3} & 1+\sqrt{3} \\ 1+\sqrt{3} & 1 & 1-\sqrt{3} \\ 1-\sqrt{3} & 1+\sqrt{3} & 1 \end{pmatrix}, \tag{10.64}$$

$$3'(\equiv 3_{[1][2]}): \quad b' = \begin{pmatrix} 1 & 0 & 0 \\ 0 & \omega & 0 \\ 0 & 0 & \omega^2 \end{pmatrix}, \quad a' = \frac{1}{3}\begin{pmatrix} -1+2i & -1-i & -1-i \\ -1-i & -1+2i & -1-i \\ -1-i & -1-i & -1+2i \end{pmatrix},$$
$$\tag{10.65}$$

$$\bar{3}'(\equiv 3_{[3][2]}): \quad b' = \begin{pmatrix} 1 & 0 & 0 \\ 0 & \omega^2 & 0 \\ 0 & 0 & \omega \end{pmatrix}, \quad a' = \frac{1}{3}\begin{pmatrix} -1-2i & -1+i & -1+i \\ -1+i & -1-2i & -1+i \\ -1+i & -1+i & -1-2i \end{pmatrix},$$
$$\tag{10.66}$$

$$\tilde{3}(\equiv 3_{[0][2]}): \quad b' = \begin{pmatrix} 1 & 0 & 0 \\ 0 & \omega & 0 \\ 0 & 0 & \omega^2 \end{pmatrix}, \quad a' = \frac{1}{3}\begin{pmatrix} -1 & 2 & 2 \\ 2 & -1 & 2 \\ 2 & 2 & -1 \end{pmatrix}, \tag{10.67}$$

where we neglect singlets since they are trivial.

These new basis are constructed in terms of our basis b, a as follows:

$$a' = V_R^\dagger a V_R, \quad b' = V_R^\dagger b V_R, \tag{10.68}$$

where the unitary matrix V_R depends on a kind of representation. Then, all the representations $\mathbf{R'}$ in [3] are uniquely transformed to our representations $\mathbf{R}$ except the overall coefficients by V_R as follows:

$$\mathbf{R} \propto V_R \cdot \mathbf{R'}. \tag{10.69}$$

On the triplet representations, there are five unitary matrices, and each of them are represented as

$$V_{3_{[1][0]}} = V_{3_{[1][2]}} = \frac{1}{\sqrt{3}} \begin{pmatrix} 1 & 1 & 1 \\ 1 & \omega & \omega^2 \\ 1 & \omega^2 & \omega \end{pmatrix}, \tag{10.70}$$

$$V_{3_{[3][0]}} = \frac{1}{\sqrt{3}} \begin{pmatrix} 1 & 1 & 1 \\ 1 & \omega & \omega^2 \\ 1 & \omega^2 & \omega \end{pmatrix} \begin{pmatrix} 1 & 0 & 0 \\ 0 & 0 & 1 \\ 0 & 1 & 0 \end{pmatrix}, \tag{10.71}$$

$$V_{3_{[3][2]}} = \frac{1}{\sqrt{3}} \begin{pmatrix} 1 & 0 & 0 \\ 0 & 0 & 1 \\ 0 & 1 & 0 \end{pmatrix} \begin{pmatrix} 1 & 1 & 1 \\ 1 & \omega & \omega^2 \\ 1 & \omega^2 & \omega \end{pmatrix}, \tag{10.72}$$

$$V_{3_{[0][2]}} = \frac{1}{\sqrt{3}} \begin{pmatrix} 0 & 1 & 0 \\ 1 & 0 & 0 \\ 0 & 0 & 1 \end{pmatrix} \begin{pmatrix} 1 & 1 & 1 \\ 1 & \omega & \omega^2 \\ 1 & \omega^2 & \omega \end{pmatrix} \begin{pmatrix} 1 & 0 & 0 \\ 0 & 0 & 1 \\ 0 & 1 & 0 \end{pmatrix}. \tag{10.73}$$

References

1. Luhn, C., Nasri, S., Ramond, P.: J. Math. Phys. **48**, 073501 (2007). arXiv:hep-th/0701188
2. Ding, G.J., Zhou, Y.L.: Chin. Phys. C **39**(2), 021001 (2015). arXiv:1312.5222 [hep-ph]
3. Ding, G.J., Zhou, Y.L.: JHEP **06**, 023 (2014). arXiv:1404.0592 [hep-ph]

11.1 Generic Aspects

We study the group T_N, which is isomorphic to $Z_N \rtimes Z_3$ (see e.g. Refs. [1–3]). Here we focus on the case that N is a prime number or its power except 3, i.e. $N = p^q$ with a prime number p ($p \neq 3$) and a positive number q. We denote the generators of Z_N and Z_3 by a and b, respectively, which satisfy

$$a^N = e, \qquad b^3 = e. \tag{11.1}$$

Because of the semi-direct product structure, we impose

$$ba = a^m b, \tag{11.2}$$

with $m \neq 0$. When $m = 1 \bmod N$, a and b are commutable each other and the group is just the direct product, $Z_N \times Z_3$. Thus, we impose $m \neq 1 \bmod N$. For example, the case with $N = 2$ is excluded because we have only $m = 1 \bmod N$ for $N = 2$ except $m = 0$.

It is found

$$a = b^2 a^m b, \quad a^m = (b^2 a^m b)\ldots(b^2 a^m b) = b^2 a^{m^2} b = bab^2, \tag{11.3}$$
$$a = ba^{m^2}b^2, \quad a^m = (ba^{m^2}b^2)\ldots(ba^{m^2}b^2) = ba^{m^3}b^2 = bab^2. \tag{11.4}$$

Due to the consistency of these equations, we have

$$a^{m^3} = a. \tag{11.5}$$

That implies

$$m^3 - 1 = (m - 1)(m^2 + m + 1) = 0 \quad \bmod \quad N. \tag{11.6}$$

Table 11.1 m for $N \le 50$

N	7	13	19	31	43	49(= 7 × 7)
m	2	3	7	5	6	18

Here, we are focusing on the case that the following condition,

$$m^2 + m + 1 = 0 \mod N, \tag{11.7}$$

with $1 < m < N$, is satisfied.

Suppose that $m = 3\ell$ with integer ℓ. Thus, it is found

$$m^2 + m + 1 = 3\ell(3\ell + 1) + 1, \tag{11.8}$$

and $3\ell(3\ell + 1)$ is always a multiple of 6, i.e. $3\ell(3\ell + 1) = 6k$ and $m^2 + m + 1 = 6k + 1$. Similarly, when $m = 3\ell + 2$, we obtain

$$m^2 + m + 1 = 3\ell(3\ell + 5) + 7. \tag{11.9}$$

Then, it is found that $m^2 + m + 1 = 6k + 1$. These results show possible values of N must be $N = 6k + 1$.

On the other hand, when $m = 3\ell + 1$, we find

$$m^2 + m + 1 = 3(3\ell(\ell + 1) + 1). \tag{11.10}$$

Here, $\ell(\ell + 1)$ is always even, i.e. $\ell(\ell + 1) = 2\ell'$, and we can write

$$m^2 + m + 1 = 3(6\ell' + 1). \tag{11.11}$$

This implies that for $N = p^q$ ($p \ne 3$) the possible value of N would be $N = 6\ell' + 1$. If we took $N = 3$, one could not find non-trivial m, because $m^3 - 1 = 7$ for $m = 2$. Then, we find that the possible values of N is $N = 6k + 1$ for $m = 3\ell + 1$, too.

Indeed, explicit computation on Eq. (11.7) leads to the possible values of N

$$N = (7, \ 13, \ 19, \ 31, \ 43, \ 49(= 7 \times 7)) \text{ for } N \ \le 50. \tag{11.12}$$

These values and the corresponding values of m are shown in Table 11.1.

All of T_N elements are written by the two generators, a and b, as

$$g = b^k a^\ell, \tag{11.13}$$

for $k = 0, 1, 2$ and $\ell = 0, \ldots, N - 1$. The generators, b and a are represented, e.g., as

$$b = \begin{pmatrix} 0 & 1 & 0 \\ 0 & 0 & 1 \\ 1 & 0 & 0 \end{pmatrix}, \quad a = \begin{pmatrix} \rho & 0 & 0 \\ 0 & \rho^m & 0 \\ 0 & 0 & \rho^{m^2} \end{pmatrix}, \tag{11.14}$$

where we define $\rho = e^{2\pi i/N}$.

• Conjugacy classes

All of T_N elements, $b^k a^\ell$, are classified into $3 + (N-1)/3$ conjugacy classes,

$$
\begin{array}{lll}
C_1: & \{e\}, & h = 1, \\
C_N^{(1)}: & \{b, ba, \ldots, ba^{N-2}, ba^{N-1}\}, & h = 3, \\
C_N^{(2)}: & \{b^2, b^2 a, \ldots, b^2 a^{N-2}, b^2 a^{N-1}\}, & h = 3, \\
C_{3_{[k]}}: & \{a^k, a^{km}, a^{km^2}\}, & h = N/\mathrm{g.c.d}(N, k),
\end{array}
\tag{11.15}
$$

where $\mathrm{g.c.d}(N, k) = N$ when N is a prime number.

The relations (2.18) and (2.19) lead to

$$
m_1 + m_2 + m_3 = 3 + (N - 1)/3, \tag{11.16}
$$
$$
m_1 + 4m_2 + 9m_3 = 3N. \tag{11.17}
$$

For specific values of N in Eq. (11.12), their solutions are found as $m_1 = 3$, $m_2 = 0$, and $m_3 = (N - 1)/3$.

• Characters and representations

Here, we study characters and representations. The T_N group has 3 singlets. Because $b^3 = e$, characters of three singlets have three possible values $\chi_{1k}(b) = \omega^k$ with $k = 0, 1, 2$ and they correspond to three singlets, $\mathbf{1}_k$. Note that $\chi_{1k}(a) = 1$, because $\chi_{1k}(b) = \chi_{1k}(ba)$. These characters are shown in Table 11.2.

Next, let us study triplets. We use the notation $[k]$ which denotes

$$
[k] = k, \quad km, \quad \text{or} \quad km^2, \quad (\text{mod } N). \tag{11.18}
$$

We also use $\xi_{[k]}$, which is $\xi_{[k]} = \rho^k + \rho^{km} + \rho^{km^2}$. The notation $\bar{\xi}_{[k]}$ is defined as the complex conjugate of $\bar{\xi}_{[k]} (= \xi_{[N-k]})$. The two generators, b and a, are represented, e.g., as

$$
b = \begin{pmatrix} 0 & 1 & 0 \\ 0 & 0 & 1 \\ 1 & 0 & 0 \end{pmatrix}, \quad a = \begin{pmatrix} \rho^k & 0 & 0 \\ 0 & \rho^{km} & 0 \\ 0 & 0 & \rho^{km^2} \end{pmatrix}, \tag{11.19}
$$

on the triplets $\mathbf{3}_{[k]}$. We also denote the vector of $\mathbf{3}_{[k]}$ as

$$
\mathbf{3}_{[k]} \equiv \begin{pmatrix} x_k \\ x_{km} \\ x_{km^2} \end{pmatrix}, \tag{11.20}
$$

for $k \in N - 1$, where k corresponds to Z_N charge. Thus we have $(N - 1)/3$ different triplets. The characters are shown in Table 11.2. We denote $\bar{\mathbf{3}}_{[k]} = \mathbf{3}_{[N-k]}$.

Table 11.2 Characters of T_N

	n	h	χ_{1_0}	χ_{1_1}	χ_{1_2}	$\chi_{3_{[\ell]}}$
$C_1^{(0)}$	1	1	1	1	1	3
$C_N^{(1)}$	N	3	1	ω	ω^2	0
$C_N^{(2)}$	N	3	1	ω^2	ω	0
$C_{3_{[k]}}$	3	$N/\text{g.c.d.}(N,k)$	1	1	1	$\xi_{[k\ell]}$

• Tensor products

Now we study tensor products. The tensor products between triplets are in general obtained as

$$
\begin{pmatrix} x_k \\ x_{km} \\ x_{km^2} \end{pmatrix}_{3_{[k]}}
\otimes
\begin{pmatrix} y_\ell \\ y_{\ell m} \\ y_{\ell m^2} \end{pmatrix}_{3_{[\ell]}}
=
\begin{pmatrix} x_k y_\ell \\ x_{km} y_{\ell m} \\ x_{km^2} y_{\ell m^2} \end{pmatrix}_{3_{[k+\ell]}}
$$
$$
\oplus
\begin{pmatrix} x_k y_{\ell m} \\ x_{km} y_{\ell m^2} \\ x_{km^2} y_\ell \end{pmatrix}_{3_{[k+\ell m]}}
$$
$$
\oplus
\begin{pmatrix} x_k y_{\ell m^2} \\ x_{km} y_\ell \\ x_{km^2} y_{\ell m} \end{pmatrix}_{3_{[k+\ell m^2]}} ,
\tag{11.21}
$$

for $[k] \neq [N - \ell]$. When $[k] = [N - \ell]$, one of $3_{[k+\ell]}$, $3_{[k+\ell m]}$ or $3_{[k+\ell m^2]}$ can be reduced to three singlets. For example, if $k + \ell = 0 \bmod N$, the triplet $3_{[k+\ell]}$ can be reduced as

$$
\begin{pmatrix} x_k y_\ell \\ x_{km} y_{\ell m} \\ x_{km^2} y_{\ell m^2} \end{pmatrix}_{3_{[k+\ell]}}
= \oplus \sum_{k'=0,1,2} \left(x_k y_\ell + \omega^{2k'} x_{km} y_{\ell m} + \omega^{k'} x_{km^2} y_{\ell m^2} \right)_{1_{k'}} .
\tag{11.22}
$$

Similarly, when $k + \ell m^n = 0 \bmod N$ for $n = 1, 2$, $3_{[k+\ell m^n]}$ can be reduced to three singlets.

In addition, the tensor products including singlets are written simply as

$$
1_k \otimes 3_{[\ell]} = 3_{[\ell]}, \qquad 1_k \otimes 1_{k'} = 1_{k+k'},
\tag{11.23}
$$

where

$$
1_{k'} \otimes
\begin{pmatrix} x_k \\ x_{km} \\ x_{km^2} \end{pmatrix}_{3_{[k]}}
=
\begin{pmatrix} x_k \\ \omega^{k'} x_{km} \\ \omega^{2k'} x_{km^2} \end{pmatrix}_{3_{[k]}}
, \quad
1_{k'} \otimes
\begin{pmatrix} \bar{x}_{(N-k)} \\ \bar{x}_{(N-k)m} \\ \bar{x}_{(N-k)m^2} \end{pmatrix}_{\bar{3}_{[(N-k)]}}
=
\begin{pmatrix} \bar{x}_{(N-k)} \\ \omega^{k'} \bar{x}_{(N-k)m} \\ \omega^{2k'} \bar{x}_{(N-k)m^2} \end{pmatrix}_{\bar{3}_{[(N-k)]}}
, \tag{11.24}
$$

where $k' = 0, 1, 2$.

11.2 T_7

Here we study the smallest T_N group, that is, T_7. We denote the generators of Z_7 by a and Z_3 generator is written by b. They satisfy

$$a^7 = 1, \quad ba = a^2 b. \tag{11.25}$$

Using them, all of T_7 elements are written as

$$g = b^k a^\ell, \tag{11.26}$$

with $k = 0, 1, 2$ and $\ell = 0, \ldots, 6$.

The generators, a and b, are represented e.g. as

$$b = \begin{pmatrix} 0 & 1 & 0 \\ 0 & 0 & 1 \\ 1 & 0 & 0 \end{pmatrix}, \quad a = \begin{pmatrix} \rho & 0 & 0 \\ 0 & \rho^2 & 0 \\ 0 & 0 & \rho^4 \end{pmatrix}, \tag{11.27}$$

where $\rho = e^{2i\pi/7}$. These elements are classified into five conjugacy classes,

$$
\begin{array}{llll}
C_1: & \{e\}, & h = 1, \\
C_7^{(1)}: & \{b, ba, ba^2, ba^3, ba^4, ba^5, ba^6\}, & h = 3, \\
C_7^{(2)}: & \{b^2, b^2a, b^2a^2, b^2a^3, b^2a^4, b^2a^5, b^2a^6\}, & h = 3, & (11.28) \\
C_3: & \{a, a^2, a^4\}, & h = 7, \\
C_{\bar{3}}: & \{a^3, a^5, a^6\}, & h = 7.
\end{array}
$$

The T_7 group has three singlets $\mathbf{1}_k$ with $k = 0, 1, 2$ and two triplets $\mathbf{3}$ and $\bar{\mathbf{3}}$. The characters are shown in Table 11.3, where $\xi = \frac{-1+i\sqrt{7}}{2}$.

Using the order of ρ in a, we define the triplets $\mathbf{3}$ and $\bar{\mathbf{3}}$ as

$$\mathbf{3} \equiv \begin{pmatrix} x_1 \\ x_2 \\ x_4 \end{pmatrix}, \quad \bar{\mathbf{3}} \equiv \begin{pmatrix} x_{-1} \\ x_{-2} \\ x_{-4} \end{pmatrix} = \begin{pmatrix} x_6 \\ x_5 \\ x_3 \end{pmatrix}. \tag{11.29}$$

Table 11.3 Characters of T_7

	n	h	$\chi_{\mathbf{1}_0}$	$\chi_{\mathbf{1}_1}$	$\chi_{\mathbf{1}_2}$	$\chi_{\mathbf{3}}$	$\chi_{\bar{\mathbf{3}}}$
$C_1^{(0)}$	1	1	1	1	1	3	3
$C_7^{(1)}$	7	3	1	ω	ω^2	0	0
$C_7^{(2)}$	7	3	1	ω^2	ω	0	0
C_3	3	7	1	1	1	ξ	$\bar{\xi}$
$C_{\bar{3}}$	3	7	1	1	1	$\bar{\xi}$	ξ

The tensor products between triplets are obtained as

$$\begin{pmatrix} x_1 \\ x_2 \\ x_4 \end{pmatrix}_3 \otimes \begin{pmatrix} y_1 \\ y_2 \\ y_4 \end{pmatrix}_3 = \begin{pmatrix} x_2 y_4 \\ x_4 y_1 \\ x_1 y_2 \end{pmatrix}_{\bar{3}} \oplus \begin{pmatrix} x_4 y_2 \\ x_1 y_4 \\ x_2 y_1 \end{pmatrix}_{\bar{3}} \oplus \begin{pmatrix} x_4 y_4 \\ x_1 y_1 \\ x_2 y_2 \end{pmatrix}_3 , \quad (11.30)$$

$$\begin{pmatrix} x_6 \\ x_5 \\ x_3 \end{pmatrix}_{\bar{3}} \otimes \begin{pmatrix} y_6 \\ y_5 \\ y_3 \end{pmatrix}_{\bar{3}} = \begin{pmatrix} x_5 y_3 \\ x_3 y_6 \\ x_6 y_5 \end{pmatrix}_3 \oplus \begin{pmatrix} x_3 y_5 \\ x_6 y_3 \\ x_5 y_6 \end{pmatrix}_3 \oplus \begin{pmatrix} x_3 y_3 \\ x_6 y_6 \\ x_5 y_5 \end{pmatrix}_{\bar{3}} , \quad (11.31)$$

$$\begin{pmatrix} x_1 \\ x_2 \\ x_4 \end{pmatrix}_3 \otimes \begin{pmatrix} y_6 \\ y_5 \\ y_3 \end{pmatrix}_{\bar{3}} = \begin{pmatrix} x_2 y_6 \\ x_4 y_5 \\ x_1 y_3 \end{pmatrix}_3 \oplus \begin{pmatrix} x_1 y_5 \\ x_2 y_3 \\ x_4 y_6 \end{pmatrix}_{\bar{3}}$$
$$\oplus \sum_{k=0,1,2} (x_1 y_6 + \omega^{2k} x_2 y_5 + \omega^k x_4 y_3)_{1_k}. \quad (11.32)$$

The tensor products between singlets are obtained as

$$(x)_{1_0}(y)_{1_0} = (x)_{1_1}(y)_{1_2} = (x)_{1_2}(y)_{1_1} = (xy)_{1_0},$$
$$(x)_{1_1}(y)_{1_1} = (xy)_{1_2}, \quad (x)_{1_2}(y)_{1_2} = (xy)_{1_1}. \quad (11.33)$$

The tensor products between triplets and singlets are found as

$$1_{k'} \otimes \begin{pmatrix} x_k \\ x_{km} \\ x_{km^2} \end{pmatrix}_{3_{[k]}} = \begin{pmatrix} x_k \\ \omega^{k'} x_{km} \\ \omega^{2k'} x_{km^2} \end{pmatrix}_{3_{[k]}}, \quad 1_{k'} \otimes \begin{pmatrix} \bar{x}_{7-k} \\ \bar{x}_{(7-k)m} \\ \bar{x}_{(7-k)m^2} \end{pmatrix}_{\bar{3}_{[7-k]}} = \begin{pmatrix} \bar{x}_{7-k} \\ \omega^{k'} \bar{x}_{(7-k)m} \\ \omega^{2k'} \bar{x}_{(7-k)m^2} \end{pmatrix}_{\bar{3}_{[7-k]}} ,$$
$$(11.34)$$

where $k' = 0, 1, 2$ and $k = 1$.

11.3 T_{13} Group Theory

The non-Abelian discrete group T_{13} is isomorphic to $Z_{13} \rtimes Z_3$ [3,4]. The T_{13} group is a subgroup of $SU(3)$, and known as the minimal non-Abelian discrete group having two complex triplets as the irreducible representations. We denote the generators of Z_{13} and Z_3 by a and b, respectively. They satisfy

$$a^{13} = 1, \quad ba = a^3 b. \quad (11.35)$$

Using them, all of T_{13} elements are written as

$$g = b^k a^\ell, \quad (11.36)$$

with $k = 0, 1, 2$ and $\ell = 0, \ldots, 12$.

Table 11.4 Characters of T_{13}. $\bar{\xi}_i$ is defined as the complex conjugate of ξ_i

	n	h	χ_{1_0}	χ_{1_1}	χ_{1_2}	χ_{3_1}	$\chi_{\bar{3}_1}$	χ_{3_2}	$\chi_{\bar{3}_2}$
$C_1^{(0)}$	1	1	1	1	1	3	3	3	3
$C_{13}^{(1)}$	13	3	1	ω	ω^2	0	0	0	0
$C_{13}^{(2)}$	13	3	1	ω^2	ω	0	0	0	0
C_{3_1}	3	13	1	1	1	ξ_1	$\bar{\xi}_1$	ξ_2	$\bar{\xi}_2$
$C_{\bar{3}_1}$	3	13	1	1	1	$\bar{\xi}_1$	ξ_1	$\bar{\xi}_2$	ξ_2
C_{3_2}	3	13	1	1	1	ξ_2	$\bar{\xi}_2$	ξ_1	$\bar{\xi}_1$
$C_{\bar{3}_2}$	3	13	1	1	1	$\bar{\xi}_2$	ξ_2	$\bar{\xi}_1$	ξ_1

The generators, a and b, are represented e.g. as

$$b = \begin{pmatrix} 0 & 1 & 0 \\ 0 & 0 & 1 \\ 1 & 0 & 0 \end{pmatrix}, \quad a = \begin{pmatrix} \rho & 0 & 0 \\ 0 & \rho^3 & 0 \\ 0 & 0 & \rho^9 \end{pmatrix}, \tag{11.37}$$

where $\rho = e^{2i\pi/13}$. These elements are classified into seven conjugacy classes,

$$\begin{aligned}
C_1 : & \qquad\qquad \{e\}, & h &= 1, \\
C_{13}^{(1)} : & \quad \{b, ba, ba^2, \quad \ldots, ba^{10}, ba^{11}, ba^{12}\}, & h &= 3, \\
C_{13}^{(2)} : & \{b^2, b^2 a, b^2 a^2, \quad \ldots, b^2 a^{10}, b^2 a^{11}, b^2 a^{12}\}, & h &= 3, \\
C_{3_1} : & \qquad\qquad \{a, a^3, a^9\}, & h &= 13, \\
C_{\bar{3}_1} : & \qquad\qquad \{a^4, a^{10}, a^{12}\}, & h &= 13, \\
C_{3_2} : & \qquad\qquad \{a^2, a^5, a^6\}, & h &= 13, \\
C_{\bar{3}_2} : & \qquad\qquad \{a^7, a^8, a^{11}\}, & h &= 13.
\end{aligned} \tag{11.38}$$

The T_{13} group has three singlets $\mathbf{1}_k$ with $k = 0, 1, 2$ and two complex triplets, $\mathbf{3}_1$ and $\mathbf{3}_2$, and their conjugates as irreducible representations. The characters are shown in Table 11.4, where $\xi_1 \equiv \rho + \rho^3 + \rho^9$, $\xi_2 \equiv \rho^2 + \rho^5 + \rho^6$, and $\omega \equiv e^{2i\pi/3}$.

Next we show the multiplication rules of the T_{13} group. We define the triplets as

$$\mathbf{3}_1 \equiv \begin{pmatrix} x_1 \\ x_3 \\ x_9 \end{pmatrix}, \quad \bar{\mathbf{3}}_1 \equiv \begin{pmatrix} \bar{x}_{12} \\ \bar{x}_{10} \\ \bar{x}_4 \end{pmatrix}, \quad \mathbf{3}_2 = \begin{pmatrix} y_2 \\ y_6 \\ y_5 \end{pmatrix}, \quad \bar{\mathbf{3}}_2 \equiv \begin{pmatrix} \bar{y}_{11} \\ \bar{y}_7 \\ \bar{y}_8 \end{pmatrix}, \tag{11.39}$$

where the subscripts denote Z_{13} charge of each element.

The tensor products between triplets are obtained as

$$
\begin{pmatrix} x_1 \\ x_3 \\ x_9 \end{pmatrix}_{3_1} \otimes \begin{pmatrix} y_1 \\ y_3 \\ y_9 \end{pmatrix}_{3_1} = \begin{pmatrix} x_3 y_9 \\ x_9 y_1 \\ x_1 y_3 \end{pmatrix}_{\bar{3}_1} \oplus \begin{pmatrix} x_9 y_3 \\ x_1 y_9 \\ x_3 y_1 \end{pmatrix}_{\bar{3}_1} \oplus \begin{pmatrix} x_1 y_1 \\ x_3 y_3 \\ x_9 y_9 \end{pmatrix}_{3_2} , \tag{11.40}
$$

$$
\begin{pmatrix} \bar{x}_{12} \\ \bar{x}_{10} \\ \bar{x}_4 \end{pmatrix}_{\bar{3}_1} \otimes \begin{pmatrix} \bar{y}_{12} \\ \bar{y}_{10} \\ \bar{y}_4 \end{pmatrix}_{\bar{3}_1} = \begin{pmatrix} \bar{x}_{10}\bar{y}_4 \\ \bar{x}_4\bar{y}_{12} \\ \bar{x}_{12}\bar{y}_{10} \end{pmatrix}_{3_1} \oplus \begin{pmatrix} \bar{x}_4\bar{y}_{10} \\ \bar{x}_{12}\bar{y}_4 \\ \bar{x}_{10}\bar{y}_{12} \end{pmatrix}_{3_1} \oplus \begin{pmatrix} \bar{x}_{12}\bar{y}_{12} \\ \bar{x}_{10}\bar{y}_{10} \\ \bar{x}_4\bar{y}_4 \end{pmatrix}_{\bar{3}_2} , \tag{11.41}
$$

$$
\begin{pmatrix} x_1 \\ x_3 \\ x_9 \end{pmatrix}_{3_1} \otimes \begin{pmatrix} \bar{y}_{12} \\ \bar{y}_{10} \\ \bar{y}_4 \end{pmatrix}_{\bar{3}_1} = \sum_{k=0,1,2} (x_1\bar{y}_{12} + \omega^{2k}x_3\bar{y}_{10} + \omega^k x_9\bar{y}_4)_{1_k}
$$
$$
\oplus \begin{pmatrix} x_3\bar{y}_{12} \\ x_9\bar{y}_{10} \\ x_1\bar{y}_4 \end{pmatrix}_{3_2} \oplus \begin{pmatrix} x_1\bar{y}_{10} \\ x_3\bar{y}_4 \\ x_9\bar{y}_{12} \end{pmatrix}_{\bar{3}_2} , \tag{11.42}
$$

$$
\begin{pmatrix} x_2 \\ x_6 \\ x_5 \end{pmatrix}_{3_2} \otimes \begin{pmatrix} y_2 \\ y_6 \\ y_5 \end{pmatrix}_{3_2} = \begin{pmatrix} x_5 y_6 \\ x_2 y_5 \\ x_6 y_2 \end{pmatrix}_{\bar{3}_2} \oplus \begin{pmatrix} x_6 y_5 \\ x_5 y_2 \\ x_2 y_6 \end{pmatrix}_{\bar{3}_2} \oplus \begin{pmatrix} x_6 y_6 \\ x_5 y_5 \\ x_2 y_2 \end{pmatrix}_{3_1} , \tag{11.43}
$$

$$
\begin{pmatrix} \bar{x}_{11} \\ \bar{x}_7 \\ \bar{x}_8 \end{pmatrix}_{\bar{3}_2} \otimes \begin{pmatrix} \bar{y}_{11} \\ \bar{y}_7 \\ \bar{y}_8 \end{pmatrix}_{\bar{3}_2} = \begin{pmatrix} \bar{x}_8\bar{y}_7 \\ \bar{x}_{11}\bar{y}_8 \\ \bar{x}_7\bar{y}_{11} \end{pmatrix}_{3_2} \oplus \begin{pmatrix} \bar{x}_7\bar{y}_8 \\ \bar{x}_8\bar{y}_{11} \\ \bar{x}_{11}\bar{y}_7 \end{pmatrix}_{3_2} \oplus \begin{pmatrix} \bar{x}_7\bar{y}_7 \\ \bar{x}_8\bar{y}_8 \\ \bar{x}_{11}\bar{y}_{11} \end{pmatrix}_{3_1} , \tag{11.44}
$$

$$
\begin{pmatrix} x_2 \\ x_6 \\ x_5 \end{pmatrix}_{3_2} \otimes \begin{pmatrix} \bar{y}_{11} \\ \bar{y}_7 \\ \bar{y}_8 \end{pmatrix}_{\bar{3}_2} = \sum_{k=0,1,2} (x_2\bar{y}_{11} + \omega^{2k}x_6\bar{y}_7 + \omega^k x_5\bar{y}_8)_{1_k}
$$
$$
\oplus \begin{pmatrix} x_6\bar{y}_8 \\ x_5\bar{y}_{11} \\ x_2\bar{y}_7 \end{pmatrix}_{3_1} \oplus \begin{pmatrix} x_5\bar{y}_7 \\ x_2\bar{y}_8 \\ x_6\bar{y}_{11} \end{pmatrix}_{\bar{3}_1} , \tag{11.45}
$$

$$
\begin{pmatrix} x_1 \\ x_3 \\ x_9 \end{pmatrix}_{3_1} \otimes \begin{pmatrix} y_2 \\ y_6 \\ y_5 \end{pmatrix}_{3_2} = \begin{pmatrix} x_9 y_6 \\ x_1 y_5 \\ x_3 y_2 \end{pmatrix}_{3_2} \oplus \begin{pmatrix} x_9 y_2 \\ x_1 y_6 \\ x_3 y_5 \end{pmatrix}_{\bar{3}_2} \oplus \begin{pmatrix} x_9 y_5 \\ x_1 y_2 \\ x_3 y_6 \end{pmatrix}_{3_1} , \tag{11.46}
$$

$$
\begin{pmatrix} x_1 \\ x_3 \\ x_9 \end{pmatrix}_{3_1} \otimes \begin{pmatrix} \bar{y}_{11} \\ \bar{y}_7 \\ \bar{y}_8 \end{pmatrix}_{\bar{3}_2} = \begin{pmatrix} x_1\bar{y}_{11} \\ x_3\bar{y}_7 \\ x_9\bar{y}_8 \end{pmatrix}_{\bar{3}_1} \oplus \begin{pmatrix} x_3\bar{y}_8 \\ x_9\bar{y}_{11} \\ x_1\bar{y}_7 \end{pmatrix}_{3_2} \oplus \begin{pmatrix} x_3\bar{y}_{11} \\ x_9\bar{y}_7 \\ x_1\bar{y}_8 \end{pmatrix}_{3_1} , \tag{11.47}
$$

$$
\begin{pmatrix} x_2 \\ x_6 \\ x_5 \end{pmatrix}_{3_2} \otimes \begin{pmatrix} \bar{y}_{12} \\ \bar{y}_{10} \\ \bar{y}_4 \end{pmatrix}_{\bar{3}_1} = \begin{pmatrix} x_2\bar{y}_{12} \\ x_6\bar{y}_{10} \\ x_5\bar{y}_4 \end{pmatrix}_{3_1} \oplus \begin{pmatrix} x_2\bar{y}_{10} \\ x_6\bar{y}_4 \\ x_5\bar{y}_{12} \end{pmatrix}_{\bar{3}_1} \oplus \begin{pmatrix} x_5\bar{y}_{10} \\ x_2\bar{y}_4 \\ x_6\bar{y}_{12} \end{pmatrix}_{3_2} , \tag{11.48}
$$

$$
\begin{pmatrix} \bar{x}_{12} \\ \bar{x}_{10} \\ \bar{x}_4 \end{pmatrix}_{\bar{3}_1} \otimes \begin{pmatrix} \bar{y}_{11} \\ \bar{y}_7 \\ \bar{y}_8 \end{pmatrix}_{\bar{3}_2} = \begin{pmatrix} \bar{x}_4\bar{y}_8 \\ \bar{x}_{12}\bar{y}_{11} \\ \bar{x}_{10}\bar{y}_7 \end{pmatrix}_{3_1} \oplus \begin{pmatrix} \bar{x}_4\bar{y}_7 \\ \bar{x}_{12}\bar{y}_8 \\ \bar{x}_{10}\bar{y}_{11} \end{pmatrix}_{3_2} \oplus \begin{pmatrix} \bar{x}_4\bar{y}_{11} \\ \bar{x}_{12}\bar{y}_7 \\ \bar{x}_{10}\bar{y}_8 \end{pmatrix}_{3_2} . \tag{11.49}
$$

The tensor products between singlets are obtained as

$$
(x)_{1_0}(y)_{1_0} = (x)_{1_1}(y)_{1_2} = (x)_{1_2}(y)_{1_1} = (xy)_{1_0},
$$
$$
(x)_{1_1}(y)_{1_1} = (xy)_{1_2}, \quad (x)_{1_2}(y)_{1_2} = (xy)_{1_1}. \tag{11.50}
$$

The tensor products between triplets and singlets are found as

$$
1_{k'} \otimes \begin{pmatrix} x_k \\ x_{km} \\ x_{km^2} \end{pmatrix}_{3_{[k]}} = \begin{pmatrix} x_k \\ \omega^{k'} x_{km} \\ \omega^{2k'} x_{km^2} \end{pmatrix}_{3_{[k]}}, \quad 1_{k'} \otimes \begin{pmatrix} \bar{x}_{13-k} \\ \bar{x}_{(13-k)m} \\ \bar{x}_{(13-k)m^2} \end{pmatrix}_{\bar{3}_{[13-k]}} = \begin{pmatrix} \bar{x}_{13-k} \\ \omega^{k'} \bar{x}_{(13-k)m} \\ \omega^{2k'} \bar{x}_{(13-k)m^2} \end{pmatrix}_{\bar{3}_{[13-k]}},
$$
$$\tag{11.51}$$

where $k' = 0, 1, 2$ and $k = 1, 2$.

11.4 T_{19} Group Theory

The non-Abelian discrete group T_{19} is isomorphic to $Z_{19} \rtimes Z_3$. The T_{19} group is a subgroup of $SU(3)$, and known as the minimal non-Abelian discrete group having three complex triplets as the irreducible representations. We denote the generators of Z_{19} and Z_3 by a and b, respectively. They satisfy

$$
a^{19} = 1, \quad ba = a^7 b. \tag{11.52}
$$

Using them, all of T_{19} elements are written as

$$
g = b^k a^\ell, \tag{11.53}
$$

with $k = 0, 1, 2$ and $\ell = 0, \ldots, 18$.
The generators, a and b, are represented e.g. as

$$
b = \begin{pmatrix} 0 & 1 & 0 \\ 0 & 0 & 1 \\ 1 & 0 & 0 \end{pmatrix}, \quad a = \begin{pmatrix} \rho & 0 & 0 \\ 0 & \rho^7 & 0 \\ 0 & 0 & \rho^{11} \end{pmatrix}, \tag{11.54}
$$

where $\rho = e^{2i\pi/19}$. These elements are classified into nine conjugacy classes,

$$
\begin{array}{lll}
C_1: & \{e\}, & h = 1, \\
C_{19}^{(1)}: & \{b, ba, ba^2, \ldots, ba^{16}, ba^{17}, ba^{18}\}, & h = 3, \\
C_{19}^{(2)}: & \{b^2, b^2 a, b^2 a^2, \ldots, b^2 a^{16}, b^2 a^{17}, b^2 a^{18}\}, & h = 3, \\
C_{3_1}: & \{a, a^7, a^{11}\}, & h = 19, \\
C_{\bar{3}_1}: & \{a^8, a^{12}, a^{18}\}, & h = 19, \quad (11.55) \\
C_{3_2}: & \{a^2, a^3, a^{14}\}, & h = 19, \\
C_{\bar{3}_2}: & \{a^5, a^{16}, a^{17}\}, & h = 19, \\
C_{3_3}: & \{a^4, a^6, a^9\}, & h = 19, \\
C_{\bar{3}_3}: & \{a^{10}, a^{13}, a^{15}\}, & h = 19.
\end{array}
$$

Table 11.5 Characters of T_{19}. $\bar{\xi}_i$ is defined as the complex conjugate of ξ_i

	n	h	χ_{1_0}	χ_{1_1}	χ_{1_2}	χ_{3_1}	$\chi_{\bar{3}_1}$	χ_{3_2}	$\chi_{\bar{3}_2}$	χ_{3_3}	$\chi_{\bar{3}_3}$
$C_1^{(0)}$	1	1	1	1	1	3	3	3	3	3	3
$C_{19}^{(1)}$	19	3	1	ω	ω^2	0	0	0	0	0	0
$C_{19}^{(2)}$	19	3	1	ω^2	ω	0	0	0	0	0	0
C_{3_1}	3	19	1	1	1	ξ_1	$\bar{\xi}_1$	ξ_2	$\bar{\xi}_2$	ξ_3	$\bar{\xi}_3$
$C_{\bar{3}_1}$	3	19	1	1	1	$\bar{\xi}_1$	ξ_1	$\bar{\xi}_2$	ξ_2	$\bar{\xi}_3$	ξ_3
C_{3_2}	3	19	1	1	1	ξ_2	$\bar{\xi}_2$	ξ_3	$\bar{\xi}_3$	ξ_1	$\bar{\xi}_1$
$C_{\bar{3}_2}$	3	19	1	1	1	$\bar{\xi}_2$	ξ_2	$\bar{\xi}_3$	ξ_3	$\bar{\xi}_1$	ξ_1
C_{3_3}	3	19	1	1	1	ξ_3	$\bar{\xi}_3$	ξ_1	$\bar{\xi}_1$	ξ_2	$\bar{\xi}_2$
$C_{\bar{3}_3}$	3	19	1	1	1	$\bar{\xi}_3$	ξ_3	$\bar{\xi}_1$	ξ_1	$\bar{\xi}_2$	ξ_2

The T_{19} group has three singlets $\mathbf{1}_k$ with $k = 0$, 1, 2 and three complex triplets, $\mathbf{3}_1$, $\mathbf{3}_2$ and $\mathbf{3}_3$, and their conjugates as irreducible representations. The characters are shown in Table 11.5, where $\xi_1 \equiv \rho + \rho^7 + \rho^{11}$, $\xi_2 \equiv \rho^2 + \rho^3 + \rho^{14}$, $\xi_3 \equiv \rho^4 + \rho^6 + \rho^9$, and $\omega \equiv e^{2i\pi/3}$.

Next we show the multiplication rules of the T_{19} group. We define the triplets as

$$\mathbf{3}_1 \equiv \begin{pmatrix} x_1 \\ x_7 \\ x_{11} \end{pmatrix}, \ \bar{\mathbf{3}}_1 \equiv \begin{pmatrix} \bar{x}_{18} \\ \bar{x}_{12} \\ \bar{x}_8 \end{pmatrix}, \ \mathbf{3}_2 = \begin{pmatrix} y_2 \\ y_{14} \\ y_3 \end{pmatrix}, \ \bar{\mathbf{3}}_2 = \begin{pmatrix} \bar{y}_{17} \\ \bar{y}_5 \\ \bar{y}_{16} \end{pmatrix}, \ \mathbf{3}_3 = \begin{pmatrix} z_4 \\ z_9 \\ z_6 \end{pmatrix}, \ \bar{\mathbf{3}}_3 \equiv \begin{pmatrix} \bar{z}_{15} \\ \bar{z}_{10} \\ \bar{z}_{13} \end{pmatrix},$$

$$\tag{11.56}$$

where the subscripts denote Z_{19} charge of each element.

The tensor products between triplets are obtained as

$$\begin{pmatrix} x_1 \\ x_7 \\ x_{11} \end{pmatrix}_{3_1} \otimes \begin{pmatrix} y_1 \\ y_7 \\ y_{11} \end{pmatrix}_{3_1} = \begin{pmatrix} x_7 y_{11} \\ x_{11} y_1 \\ x_1 y_7 \end{pmatrix}_{\bar{3}_1} \oplus \begin{pmatrix} x_{11} y_7 \\ x_1 y_{11} \\ x_7 y_1 \end{pmatrix}_{\bar{3}_1} \oplus \begin{pmatrix} x_1 y_1 \\ x_7 y_7 \\ x_{11} y_{11} \end{pmatrix}_{3_2}, \tag{11.57}$$

$$\begin{pmatrix} \bar{x}_{18} \\ \bar{x}_{12} \\ \bar{x}_8 \end{pmatrix}_{\bar{3}_1} \otimes \begin{pmatrix} \bar{y}_{18} \\ \bar{y}_{12} \\ \bar{y}_8 \end{pmatrix}_{\bar{3}_1} = \begin{pmatrix} \bar{x}_8 \bar{y}_{12} \\ \bar{x}_{18} \bar{y}_8 \\ \bar{x}_{12} \bar{y}_{18} \end{pmatrix}_{3_1} \oplus \begin{pmatrix} \bar{x}_{12} \bar{y}_8 \\ \bar{x}_8 \bar{y}_{18} \\ \bar{x}_{18} \bar{y}_{12} \end{pmatrix}_{3_1} \oplus \begin{pmatrix} \bar{x}_{18} \bar{y}_{18} \\ \bar{x}_{12} \bar{y}_{12} \\ \bar{x}_8 \bar{y}_8 \end{pmatrix}_{\bar{3}_2}, \tag{11.58}$$

$$\begin{pmatrix} x_1 \\ x_7 \\ x_{11} \end{pmatrix}_{3_1} \otimes \begin{pmatrix} \bar{y}_{18} \\ \bar{y}_{12} \\ \bar{y}_8 \end{pmatrix}_{\bar{3}_1} = \sum_{k=0,1,2} (x_1 \bar{y}_{18} + \omega^{2k} x_7 \bar{y}_{12} + \omega^k x_{11} \bar{y}_8)_{1_k} \oplus \begin{pmatrix} x_{11} \bar{y}_{12} \\ x_1 \bar{y}_8 \\ x_7 \bar{y}_{18} \end{pmatrix}_{3_3} \oplus \begin{pmatrix} x_7 \bar{y}_8 \\ x_{11} \bar{y}_{18} \\ x_1 \bar{y}_{12} \end{pmatrix}_{\bar{3}_3}, $$

$$\tag{11.59}$$

$$\begin{pmatrix} x_2 \\ x_{14} \\ x_3 \end{pmatrix}_{3_2} \otimes \begin{pmatrix} y_2 \\ y_{14} \\ y_3 \end{pmatrix}_{3_2} = \begin{pmatrix} x_2 y_2 \\ x_{14} y_{14} \\ x_3 y_3 \end{pmatrix}_{3_3} \oplus \begin{pmatrix} x_3 y_{14} \\ x_2 y_3 \\ x_{14} y_2 \end{pmatrix}_{\bar{3}_2} \oplus \begin{pmatrix} x_{14} y_3 \\ x_3 y_2 \\ x_2 y_{14} \end{pmatrix}_{\bar{3}_2}, \tag{11.60}$$

$$\begin{pmatrix} \bar{x}_{17} \\ \bar{x}_5 \\ \bar{x}_{16} \end{pmatrix}_{\bar{3}_2} \otimes \begin{pmatrix} \bar{y}_{17} \\ \bar{y}_5 \\ \bar{y}_{16} \end{pmatrix}_{\bar{3}_2} = \begin{pmatrix} \bar{x}_{17} \bar{y}_{17} \\ \bar{x}_5 \bar{y}_5 \\ \bar{x}_{16} \bar{y}_{16} \end{pmatrix}_{\bar{3}_3} \oplus \begin{pmatrix} \bar{x}_5 \bar{y}_{16} \\ \bar{x}_{16} \bar{y}_{17} \\ \bar{x}_{17} \bar{y}_5 \end{pmatrix}_{3_2} \oplus \begin{pmatrix} \bar{x}_{16} \bar{y}_5 \\ \bar{x}_{17} \bar{y}_{16} \\ \bar{x}_5 \bar{y}_{17} \end{pmatrix}_{3_2}, \tag{11.61}$$

$$\begin{pmatrix} x_2 \\ x_{14} \\ x_3 \end{pmatrix}_{3_2} \otimes \begin{pmatrix} \bar{y}_{17} \\ \bar{y}_5 \\ \bar{y}_{16} \end{pmatrix}_{\bar{3}_2} = \sum_{k=0,1,2} (x_2 \bar{y}_{17} + \omega^{2k} x_{14} \bar{y}_5 + \omega^k x_3 \bar{y}_{16})_{1_k} \oplus \begin{pmatrix} x_3 \bar{y}_{17} \\ x_2 \bar{y}_5 \\ x_{14} \bar{y}_{16} \end{pmatrix}_{3_1} \oplus \begin{pmatrix} x_2 \bar{y}_{16} \\ x_{14} \bar{y}_{17} \\ x_3 \bar{y}_5 \end{pmatrix}_{\bar{3}_1}, $$

$$\tag{11.62}$$

$$\begin{pmatrix} x_4 \\ x_9 \\ x_6 \end{pmatrix}_{3_3} \otimes \begin{pmatrix} y_4 \\ y_9 \\ y_6 \end{pmatrix}_{3_3} = \begin{pmatrix} x_9 y_9 \\ x_6 y_6 \\ x_4 y_4 \end{pmatrix}_{\bar{3}_1} \oplus \begin{pmatrix} x_6 y_9 \\ x_4 y_6 \\ x_9 y_4 \end{pmatrix}_{3_3} \oplus \begin{pmatrix} x_9 y_6 \\ x_6 y_4 \\ x_4 y_9 \end{pmatrix}_{\bar{3}_3} , \quad (11.63)$$

$$\begin{pmatrix} \bar{x}_{15} \\ \bar{x}_{10} \\ \bar{x}_{13} \end{pmatrix}_{\bar{3}_3} \otimes \begin{pmatrix} \bar{y}_{15} \\ \bar{y}_{10} \\ \bar{y}_{13} \end{pmatrix}_{\bar{3}_3} = \begin{pmatrix} \bar{x}_{10} \bar{y}_{10} \\ \bar{x}_{13} \bar{y}_{13} \\ \bar{x}_{15} \bar{y}_{15} \end{pmatrix}_{3_1} \oplus \begin{pmatrix} \bar{x}_{10} \bar{y}_{13} \\ \bar{x}_{13} \bar{y}_{15} \\ \bar{x}_{15} \bar{y}_{10} \end{pmatrix}_{3_3} \oplus \begin{pmatrix} \bar{x}_{13} \bar{y}_{10} \\ \bar{x}_{15} \bar{y}_{13} \\ \bar{x}_{10} \bar{y}_{15} \end{pmatrix}_{3_3} , \quad (11.64)$$

$$\begin{pmatrix} x_4 \\ x_9 \\ x_6 \end{pmatrix}_{3_3} \otimes \begin{pmatrix} \bar{y}_{15} \\ \bar{y}_{10} \\ \bar{y}_{13} \end{pmatrix}_{\bar{3}_3} = \sum_{k=0,1,2} (x_4 \bar{y}_{15} + \omega^{2k} x_9 \bar{y}_{10} + \omega^{k} x_6 \bar{y}_{13})_{1_k} \oplus \begin{pmatrix} x_6 \bar{y}_{15} \\ x_4 \bar{y}_{10} \\ x_9 \bar{y}_{13} \end{pmatrix}_{3_2} \oplus \begin{pmatrix} x_4 \bar{y}_{13} \\ x_9 \bar{y}_{15} \\ x_6 \bar{y}_{10} \end{pmatrix}_{\bar{3}_2} ,$$

$$(11.65)$$

$$\begin{pmatrix} x_1 \\ x_7 \\ x_{11} \end{pmatrix}_{3_1} \otimes \begin{pmatrix} y_2 \\ y_{14} \\ y_3 \end{pmatrix}_{3_2} = \begin{pmatrix} x_7 y_{14} \\ x_{11} y_3 \\ x_1 y_2 \end{pmatrix}_{3_2} \oplus \begin{pmatrix} x_1 y_3 \\ x_7 y_2 \\ x_{11} y_{14} \end{pmatrix}_{3_3} \oplus \begin{pmatrix} x_1 y_{14} \\ x_7 y_3 \\ x_{11} y_2 \end{pmatrix}_{\bar{3}_3} , \quad (11.66)$$

$$\begin{pmatrix} x_1 \\ x_7 \\ x_{11} \end{pmatrix}_{3_1} \otimes \begin{pmatrix} \bar{y}_{17} \\ \bar{y}_5 \\ \bar{y}_{16} \end{pmatrix}_{\bar{3}_2} = \begin{pmatrix} x_1 \bar{y}_{17} \\ x_7 \bar{y}_5 \\ x_{11} \bar{y}_{16} \end{pmatrix}_{\bar{3}_1} \oplus \begin{pmatrix} x_7 \bar{y}_{16} \\ x_{11} \bar{y}_{17} \\ x_1 \bar{y}_5 \end{pmatrix}_{3_3} \oplus \begin{pmatrix} x_1 \bar{y}_{16} \\ x_7 \bar{y}_{17} \\ x_{11} \bar{y}_5 \end{pmatrix}_{\bar{3}_2} , \quad (11.67)$$

$$\begin{pmatrix} x_1 \\ x_7 \\ x_{11} \end{pmatrix}_{3_1} \otimes \begin{pmatrix} y_4 \\ y_9 \\ y_6 \end{pmatrix}_{3_3} = \begin{pmatrix} x_{11} y_6 \\ x_1 y_4 \\ x_7 y_9 \end{pmatrix}_{\bar{3}_2} \oplus \begin{pmatrix} x_{11} y_4 \\ x_1 y_9 \\ x_7 y_6 \end{pmatrix}_{\bar{3}_3} \oplus \begin{pmatrix} x_{11} y_9 \\ x_1 y_6 \\ x_7 y_4 \end{pmatrix}_{3_1} , \quad (11.68)$$

$$\begin{pmatrix} x_1 \\ x_7 \\ x_{11} \end{pmatrix}_{3_1} \otimes \begin{pmatrix} \bar{y}_{15} \\ \bar{y}_{10} \\ \bar{y}_{13} \end{pmatrix}_{\bar{3}_3} = \begin{pmatrix} x_7 \bar{y}_{10} \\ x_{11} \bar{y}_{13} \\ x_1 \bar{y}_{15} \end{pmatrix}_{\bar{3}_2} \oplus \begin{pmatrix} x_7 \bar{y}_{13} \\ x_{11} \bar{y}_{15} \\ x_1 \bar{y}_{10} \end{pmatrix}_{3_1} \oplus \begin{pmatrix} x_{11} \bar{y}_{10} \\ x_1 \bar{y}_{13} \\ x_7 \bar{y}_{15} \end{pmatrix}_{3_2} , \quad (11.69)$$

$$\begin{pmatrix} x_2 \\ x_{14} \\ x_3 \end{pmatrix}_{3_2} \otimes \begin{pmatrix} \bar{y}_{18} \\ \bar{y}_{12} \\ \bar{y}_8 \end{pmatrix}_{\bar{3}_1} = \begin{pmatrix} x_2 \bar{y}_{18} \\ x_{14} \bar{y}_{12} \\ x_3 \bar{y}_8 \end{pmatrix}_{3_1} \oplus \begin{pmatrix} x_3 \bar{y}_{18} \\ x_2 \bar{y}_{12} \\ x_{14} \bar{y}_8 \end{pmatrix}_{3_2} \oplus \begin{pmatrix} x_3 \bar{y}_{12} \\ x_2 \bar{y}_8 \\ x_{14} \bar{y}_{18} \end{pmatrix}_{\bar{3}_3} , \quad (11.70)$$

$$\begin{pmatrix} \bar{x}_{18} \\ \bar{x}_{12} \\ \bar{x}_8 \end{pmatrix}_{\bar{3}_1} \otimes \begin{pmatrix} \bar{y}_{17} \\ \bar{y}_5 \\ \bar{y}_{16} \end{pmatrix}_{\bar{3}_2} = \begin{pmatrix} \bar{x}_{12} \bar{y}_5 \\ \bar{x}_8 \bar{y}_{16} \\ \bar{x}_{18} \bar{y}_{17} \end{pmatrix}_{\bar{3}_2} \oplus \begin{pmatrix} \bar{x}_{18} \bar{y}_5 \\ \bar{x}_{12} \bar{y}_{16} \\ \bar{x}_8 \bar{y}_{17} \end{pmatrix}_{3_3} \oplus \begin{pmatrix} \bar{x}_{18} \bar{y}_{16} \\ \bar{x}_{12} \bar{y}_{17} \\ \bar{x}_8 \bar{y}_5 \end{pmatrix}_{\bar{3}_3} , \quad (11.71)$$

$$\begin{pmatrix} x_4 \\ x_9 \\ x_6 \end{pmatrix}_{3_3} \otimes \begin{pmatrix} \bar{y}_{18} \\ \bar{y}_{12} \\ \bar{y}_8 \end{pmatrix}_{\bar{3}_1} = \begin{pmatrix} x_9 \bar{y}_{12} \\ x_6 \bar{y}_8 \\ x_4 \bar{y}_{18} \end{pmatrix}_{3_2} \oplus \begin{pmatrix} x_9 \bar{y}_8 \\ x_6 \bar{y}_{18} \\ x_4 \bar{y}_{12} \end{pmatrix}_{\bar{3}_2} \oplus \begin{pmatrix} x_6 \bar{y}_{12} \\ x_4 \bar{y}_8 \\ x_9 \bar{y}_{18} \end{pmatrix}_{\bar{3}_1} , \quad (11.72)$$

$$\begin{pmatrix} \bar{x}_{18} \\ \bar{x}_{12} \\ \bar{x}_8 \end{pmatrix}_{\bar{3}_1} \otimes \begin{pmatrix} \bar{y}_{15} \\ \bar{y}_{10} \\ \bar{y}_{13} \end{pmatrix}_{\bar{3}_3} = \begin{pmatrix} \bar{x}_8 \bar{y}_{13} \\ \bar{x}_{18} \bar{y}_{15} \\ \bar{x}_{12} \bar{y}_{10} \end{pmatrix}_{3_2} \oplus \begin{pmatrix} \bar{x}_8 \bar{y}_{15} \\ \bar{x}_{18} \bar{y}_{10} \\ \bar{x}_{12} \bar{y}_{13} \end{pmatrix}_{3_3} \oplus \begin{pmatrix} \bar{x}_8 \bar{y}_{10} \\ \bar{x}_{18} \bar{y}_{13} \\ \bar{x}_{12} \bar{y}_{15} \end{pmatrix}_{3_1} , \quad (11.73)$$

$$\begin{pmatrix} x_2 \\ x_{14} \\ x_3 \end{pmatrix}_{3_2} \otimes \begin{pmatrix} y_4 \\ y_9 \\ y_6 \end{pmatrix}_{3_3} = \begin{pmatrix} x_{14} y_9 \\ x_3 y_6 \\ x_2 y_4 \end{pmatrix}_{3_3} \oplus \begin{pmatrix} x_{14} y_6 \\ x_3 y_4 \\ x_2 y_9 \end{pmatrix}_{3_1} \oplus \begin{pmatrix} x_{14} y_4 \\ x_3 y_9 \\ x_2 y_6 \end{pmatrix}_{\bar{3}_1} , \quad (11.74)$$

$$\begin{pmatrix} x_2 \\ x_{14} \\ x_3 \end{pmatrix}_{3_2} \otimes \begin{pmatrix} \bar{y}_{15} \\ \bar{y}_{10} \\ \bar{y}_{13} \end{pmatrix}_{\bar{3}_3} = \begin{pmatrix} x_2 \bar{y}_{15} \\ x_{14} \bar{y}_{10} \\ x_3 \bar{y}_{13} \end{pmatrix}_{3_2} \oplus \begin{pmatrix} x_3 \bar{y}_{15} \\ x_2 \bar{y}_{10} \\ x_{14} \bar{y}_{13} \end{pmatrix}_{\bar{3}_1} \oplus \begin{pmatrix} x_2 \bar{y}_{13} \\ x_{14} \bar{y}_{15} \\ x_3 \bar{y}_{10} \end{pmatrix}_{\bar{3}_3} , \quad (11.75)$$

$$\begin{pmatrix} x_4 \\ x_9 \\ x_6 \end{pmatrix}_{3_3} \otimes \begin{pmatrix} \bar{y}_{17} \\ \bar{y}_5 \\ \bar{y}_{16} \end{pmatrix}_{\bar{3}_2} = \begin{pmatrix} x_4\bar{y}_{17} \\ x_9\bar{y}_5 \\ x_6\bar{y}_{16} \end{pmatrix}_{3_2} \oplus \begin{pmatrix} x_6\bar{y}_{17} \\ x_4\bar{y}_5 \\ x_9\bar{y}_{16} \end{pmatrix}_{3_3} \oplus \begin{pmatrix} x_4\bar{y}_{16} \\ x_9\bar{y}_{17} \\ x_6\bar{y}_5 \end{pmatrix}_{3_1} , \quad (11.76)$$

$$\begin{pmatrix} \bar{x}_{17} \\ \bar{x}_5 \\ \bar{x}_{16} \end{pmatrix}_{\bar{3}_2} \otimes \begin{pmatrix} \bar{y}_{15} \\ \bar{y}_{10} \\ \bar{y}_{13} \end{pmatrix}_{\bar{3}_3} = \begin{pmatrix} \bar{x}_5\bar{y}_{10} \\ \bar{x}_{16}\bar{y}_{13} \\ \bar{x}_{17}\bar{y}_{15} \end{pmatrix}_{\bar{3}_3} \oplus \begin{pmatrix} \bar{x}_5\bar{y}_{13} \\ \bar{x}_{16}\bar{y}_{15} \\ \bar{x}_{17}\bar{y}_{10} \end{pmatrix}_{\bar{3}_1} \oplus \begin{pmatrix} \bar{x}_5\bar{y}_{15} \\ \bar{x}_{16}\bar{y}_{10} \\ \bar{x}_{17}\bar{y}_{13} \end{pmatrix}_{3_1} . \quad (11.77)$$

The tensor products between singlets are obtained as

$$(x)_{1_0}(y)_{1_0} = (x)_{1_1}(y)_{1_2} = (x)_{1_2}(y)_{1_1} = (xy)_{1_0},$$
$$(x)_{1_1}(y)_{1_1} = (xy)_{1_2}, \ (x)_{1_2}(y)_{1_2} = (xy)_{1_1}. \quad (11.78)$$

The tensor products between triplets and singlets are found as

$$1_{k'} \otimes \begin{pmatrix} x_k \\ x_{km} \\ x_{km2} \end{pmatrix}_{3_{[k]}} = \begin{pmatrix} x_k \\ \omega^{k'} x_{km} \\ \omega^{2k'} x_{km2} \end{pmatrix}_{3_{[k]}} , \ 1_{k'} \otimes \begin{pmatrix} \bar{x}_{19-k} \\ \bar{x}_{(19-k)m} \\ \bar{x}_{(19-k)m2} \end{pmatrix}_{\bar{3}_{[19-k]}} = \begin{pmatrix} \bar{x}_{19-k} \\ \omega^{k'} \bar{x}_{(19-k)m} \\ \omega^{2k'} \bar{x}_{(19-k)m2} \end{pmatrix}_{\bar{3}_{[19-k]}} .$$
$$(11.79)$$

where $k' = 0, 1, 2$ and $k = 1, 2, 4$.

References

1. Bovier, A., Luling, M., Wyler, D.: J. Math. Phys. **22**, 1536 (1981)
2. Bovier, A., Luling, M., Wyler, D.: J. Math. Phys. **22**, 1543 (1981)
3. Fairbairn, W.M., Fulton, T.: J. Math. Phys. **23**, 1747 (1982)
4. King, S.F., Luhn, C.: JHEP **0910**, 093 (2009). arXiv:0908.1897 [hep-ph]

$\Sigma(3N^3)$ **12**

12.1 Generic Aspects

In this section, we study $\Sigma(3N^3)$ [1]. This discrete group is defined as a closed algebra of three Abelian symmetries, Z_N, Z'_N and Z''_N, which commute each other, and their Z_3 permutations. Let us denote the generators of Z_N, Z'_N and Z''_N by a, a' and a'', respectively, and the Z_3 generator by b. All of $\Sigma(3N^3)$ elements are given as

$$g = b^k a^m a'^n a''^\ell, \tag{12.1}$$

with $k = 0, 1, 2$, and $m, n, \ell = 0, \dots, N-1$, where a, a', a'' and b satisfy

$$a^N = a'^N = a''^N = 1, \quad aa' = a'a, \quad aa'' = a''a, \quad a''a' = a'a'', \quad b^3 = 1,$$
$$b^2ab = a'', \quad b^2a'b = a, \quad b^2a''b = a'. \tag{12.2}$$

These generators, a, a', a'' and b, are represented, e.g., as

$$b = \begin{pmatrix} 0 & 1 & 0 \\ 0 & 0 & 1 \\ 1 & 0 & 0 \end{pmatrix}, \quad a = \begin{pmatrix} 1 & 0 & 0 \\ 0 & 1 & 0 \\ 0 & 0 & \rho \end{pmatrix}, \quad a' = \begin{pmatrix} 1 & 0 & 0 \\ 0 & \rho & 0 \\ 0 & 0 & 1 \end{pmatrix}, \quad a'' = \begin{pmatrix} \rho & 0 & 0 \\ 0 & 1 & 0 \\ 0 & 0 & 1 \end{pmatrix}, \tag{12.3}$$

where $\rho = e^{2i\pi/N}$. Then, all of $\Sigma(3N^3)$ elements are given as

$$\begin{pmatrix} 0 & \rho^n & 0 \\ 0 & 0 & \rho^m \\ \rho^\ell & 0 & 0 \end{pmatrix}, \quad \begin{pmatrix} \rho^\ell & 0 & 0 \\ 0 & \rho^m & 0 \\ 0 & 0 & \rho^n \end{pmatrix}, \quad \begin{pmatrix} 0 & 0 & \rho^m \\ \rho^\ell & 0 & 0 \\ 0 & \rho^n & 0 \end{pmatrix}. \tag{12.4}$$

For the case of $N = 2$, the element $aa'a''$ commutes with all of the elements. In addition, when we define $\tilde{a} = aa''$ and $\tilde{a}' = a'a''$, we find the closed algebra among

© The Author(s), under exclusive license to Springer-Verlag GmbH, DE, part of Springer Nature 2022
T. Kobayashi et al., *An Introduction to Non-Abelian Discrete Symmetries for Particle Physicists*, Lecture Notes in Physics 995, https://doi.org/10.1007/978-3-662-64679-3_12

$\tilde{a}$, $\tilde{a}'$ and b, which corresponds to $\Delta(12)$. Since the element $aa'a''$ is not included in this closed algebra, this group is isomorphic to $Z_2 \times \Delta(12)$.

The situation for $N = 3$ is different. It is the same as the fact that the element $aa'a''$ commutes with all of the elements. When we define $\tilde{a} = a^2 a''$ and $\tilde{a}' = a'a''^2$, the closed algebra among $\tilde{a}$, $\tilde{a}'$ and b corresponds to $\Delta(27)$. On the other hand, the element $aa'a''$ is written by $aa'a'' = \tilde{a}^2\tilde{a}'$ in this case. Thus, $aa'a''$ is one of elements of $\Delta(27)$. That is, the group $\Sigma(81)$ is not $Z_3 \times \Delta(27)$, but isomorphic to $(Z_3 \times Z_3' \times Z_3'') \rtimes Z_3$.

Similarly, for generic value of N, the element $aa'a''$ commutes with all of the elements. When we define $\tilde{a} = a^{N-1} a''$ and $\tilde{a}' = a'a''^{N-1}$, the closed algebra among $\tilde{a}$, $\tilde{a}'$ and b corresponds to $\Delta(3N^2)$. For the case of $N/3 \neq$ integer, the element $aa'a''$ is not included in $\Delta(3N^2)$. Thus, we find that this group is isomorphic to $Z_N \times \Delta(3N^2)$. On the other hand, when $N/3 =$ integer, the element $aa'a''$ is included in $\Delta(3N^2)$. That is, the group $\Sigma(3N^3)$ cannot be $Z_N \times \Delta(3N^2)$.

• Conjugacy classes

Here we summarize the conjugacy classes of $\Sigma(3N^3)$,

$$
\begin{aligned}
C_1 : \quad & \{e\}, & h = 1, \\
C_1^{(\ell)} : \quad & \{a^\ell a'^\ell a''^\ell\}, & h = N/gcd(N, \ell), \\
C_3^{(\ell,m,n)} : \quad & \{a^\ell a'^m a''^n, a^m a'^n a''^\ell, a^n a'^\ell a''^m \,|\, \ell \neq m \neq n\}, & h = N/gcd(N, \ell, m, n), \\
C_{N^2}^{(p)} : \quad & \{ba^\ell a'^m a''^{p-\ell-m}\}, \quad \text{for } p = 0, \ldots, N-1. & h = 3N/gcd(N, p), \\
C_{N^2}'^{(p)} : \quad & \{b^2 a^\ell a'^m a''^{p-\ell-m}\}, \quad \text{for } p = 0, \ldots, N-1. & h = 3N/gcd(N, p),
\end{aligned}
$$
$$(12.5)$$

where ℓ, m, n run $0, \ldots, N-1$. For example, the conjugacy classes $C_{N^2}^{(p)}$ and $C_{N^2}'^{(p)}$ are obtained by using the following relations:

$$
a^p a'^q a''^r (ba^\ell a'^m a''^n) a^{-p} a'^{-q} a''^{-r} = (ba^\ell a'^m a''^n) a^{q-p} a'^{r-q} a''^{p-r},
$$

$$
ba^p a'^q a''^r (ba^\ell a'^m a''^n) a^{-p} a'^{-q} a''^{-r} b^{-1} = ba^{n+p-r} a'^{\ell+q-p} a''^{m+r-q}, \quad (12.6)
$$

$$
b^2 a^p a'^q a''^r (ba^\ell a'^m a''^n) a^{-p} a'^{-q} a''^{-r} b^{-2} = ba^{m+r-q} a'^{n+p-r} a''^{\ell+q-p},
$$

where we used $a^p a'^q a''^r b = ba^q a'^r a''^p$, $a^p a'^q a''^r b^2 = b^2 a^r a'^p a''^q$. Note that the sum of the factors of a, a', and a'' is the same. The numbers of the classes $C_1^{(\ell)}$, $C_3^{(\ell,m,n)}$, $C_{N^2}^{(p)}$ and $C_{N^2}'^{(p)}$ are $(N-1)$, $(N^3 - N)/3$, N and N, respectively. Thus the number of conjugacy classes is

$$
1 + (N-1) + (N^3 - N)/3 + N + N = \frac{1}{3} N(N^2 + 8). \qquad (12.7)
$$

The number of irreducible representations can be determined by

$$
m_1 + 4m_2 + 9m_3 + \cdots = 3N^3, \quad m_1 + m_2 + m_3 + \cdots = \frac{1}{3} N(N^2 + 8). \quad (12.8)
$$

Their solutions are $(m_1, m_2, m_3) = (3N, 0, N(N^2 - 1)/3)$. Hence the group $\Sigma(3N^3)$ has $N(N^2 + 8)/3$ conjugacy classes, $3N$ singlets, and $N(N^2 - 1)/3$ triplets.

• Characters and representations

The number of singlets is $3N$. For the representations of singlets, any operator a, a', a'', and b can commute each other. Then, by the algebraic relations it is found that the characters for a, a' and a'' must be the same. Thus we can represent the $3N$ singlets by $\chi_{1_{k,\ell}}(a) = \chi_{1_{k,\ell}}(a') = \chi_{1_{k,\ell}}(a'') = \rho^\ell$ and $\chi_{1_{k,\ell}}(b) = \omega^k$, as shown in Table 12.1.

The number of triplets is $N(N^2 - 1)/3$. We write a, a', a'' and b as

$$a = \begin{pmatrix} \rho^\ell & 0 & 0 \\ 0 & \rho^m & 0 \\ 0 & 0 & \rho^n \end{pmatrix}, \quad a' = \begin{pmatrix} \rho^m & 0 & 0 \\ 0 & \rho^n & 0 \\ 0 & 0 & \rho^\ell \end{pmatrix}, \quad a'' = \begin{pmatrix} \rho^n & 0 & 0 \\ 0 & \rho^\ell & 0 \\ 0 & 0 & \rho^m \end{pmatrix}, \quad b = \begin{pmatrix} 0 & 1 & 0 \\ 0 & 0 & 1 \\ 1 & 0 & 0 \end{pmatrix},$$

(12.9)

on the triplet $3_{[\ell][m][n]}$,

$$3_{[\ell][m][n]} = \begin{pmatrix} x_\ell \\ x_m \\ x_n \end{pmatrix},$$

(12.10)

with $[\ell][m][n] = (\ell, m, n)$, (m, n, ℓ), or (n, ℓ, m). These characters are shown in Table 12.1. Subscripts of the components describe Z_N charge.

• Tensor products

The tensor products between two triplets are given by

$$\begin{pmatrix} x_\ell \\ x_m \\ x_n \end{pmatrix}_{3_{[\ell][m][n]}} \otimes \begin{pmatrix} y_{\ell'} \\ y_{m'} \\ y_{n'} \end{pmatrix}_{3_{[\ell'][m'][n']}}$$

$$= \begin{pmatrix} x_\ell y_{\ell'} \\ x_m y_{m'} \\ x_n y_{n'} \end{pmatrix}_{3_{[\ell+\ell'][m+m'][n+n']}} \oplus \begin{pmatrix} x_m y_{n'} \\ x_n y_{\ell'} \\ x_\ell y_{m'} \end{pmatrix}_{3_{[m+n'][n+\ell'][\ell+m']}} \oplus \begin{pmatrix} x_n y_{m'} \\ x_\ell y_{n'} \\ x_m y_{\ell'} \end{pmatrix}_{3_{[n+m'][\ell+n'][m+\ell']}}.$$

(12.11)

If all the subscripts become equal, the triplet can be decomposed into singlets as $(x_a, x_b, x_c)_{3_{[k][k][k]}} = (x_a + x_b + x_c)_{1_{0,k}} + (x_a + \omega^2 x_b + \omega x_c)_{1_{1,k}} + (x_a + \omega x_b + \omega^2 x_c)_{1_{2,k}}$. The tensor products between singlet and triplet are

$$\begin{pmatrix} x_\ell \\ x_m \\ x_n \end{pmatrix}_{3_{[\ell][m][n]}} \otimes (y)_{1_{k',k}} = \begin{pmatrix} x_\ell y \\ \omega^{k'} x_m y \\ \omega^{2k'} x_n y \end{pmatrix}_{3_{[\ell+k][m+k][n+k]}}.$$

(12.12)

Table 12.1 Characters of $\Sigma(3N^3)$

	h	$\chi_{1_{\ell,m}}$	$\chi_{3_{[\ell][m][n]}}$
C_1	1	1	3
$C_1^{(p)}$	$\frac{N}{\gcd(N,p)}$	ρ^{pm}	$\rho^{p\ell} + \rho^{pm} + \rho^{pn}$
$C_3^{(p,q,r)}$	$\frac{N}{\gcd(N,p,q,r)}$	$\rho^{(p+q+r)m}$	$\rho^{p\ell+qn+mr} +$ $\rho^{pm+q\ell+rn} +$ $\rho^{pn+qm+r\ell}$
$C_{N^2}^{(p)}$	$\frac{3N}{\gcd(N,p)}$	$\omega^\ell \rho^{pm}$	0
$C_{N^2}^{\prime(p)}$	$\frac{3N}{\gcd(N,p)}$	$\omega^{2\ell} \rho^{pm}$	0

The multiplication between singlets is

$$(x)_{1_{k,\ell}} \otimes (y)_{1_{k',\ell'}} = (xy)_{1_{k+k',\ell+\ell'}}. \tag{12.13}$$

12.2 $\Sigma(81)$

We show details for the case of $N = 3$, that is $\Sigma(81)$. It has eighty-one elements and those are written as $b^k a^m a'^n a''^\ell$ for $k = 0, 1, 2$ and $m, n, \ell = 0, 1, 2$, where a, a', a'', and b satisfy $a^3 = a'^3 = a''^3 = 1$, $aa' = a'a$, $aa'' = a''a$, $a''a' = a'a''$, $b^3 = 1$, $b^2ab = a''$, $b^2a'b = a$ and $b^2a''b = a'$. These elements are classified into seventeen conjugacy classes as follows:

$$
\begin{array}{lll}
C_1 : & \{e\}, & h = 1, \\
C_1^{(1)} : & \{aa'a''\}, & h = 3, \\
C_1^{(2)} : & \{(aa'a'')^2\}, & h = 3, \\
C_3^{(0)} : & \{a'a''^2, a''aa'^2, a'a''a^2\}, & h = 3, \\
C_3^{\prime(0)} : & \{aa''^2, a''a'^2, a'a^2\}, & h = 3, \\
C_3^{(1)} : & \{a'', a''a', a\}, & h = 3, \\
C_3^{\prime(1)} : & \{a'^2a''^2, a^2a'^2, a''^2a^2\}, & h = 3, \\
C_3^{\prime\prime(1)} : & \{aa'a''^2, a''aa'^2, a'a''a^2\}, & h = 3, \\
C_3^{(2)} : & \{a''^2, a'^2, a^2\}, & h = 3, \\
C_3^{\prime(2)} : & \{a'a'', a''aa', a'a''a\}, & h = 3, \\
C_3^{\prime\prime(2)} : & \{aa'^2a''^2, a''a^2a'^2, a'a''^2a^2\}, & h = 3, \\
C_9^{(0)} : & \{ba^a a'^b a''^{-a-b}\} & h = 3, \\
C_9^{(1)} : & \{ba^a a'^b a''^{1-a-b}\} & h = 9, \\
C_9^{(2)} : & \{ba^a a'^b a''^{2-a-b}\} & h = 9, \\
C_9^{\prime(0)} : & \{b^2 a^a a'^b a''^{-a-b}\}, & h = 3, \\
C_9^{\prime(1)} : & \{b^2 a^a a'^b a''^{1-a-b}\}, & h = 9, \\
C_9^{\prime(2)} : & \{b^2 a^a a'^b a''^{2-a-b}\}, & h = 9,
\end{array}
\tag{12.14}
$$

Table 12.2 Characters of $\Sigma(81)$ for the 9 one-dimensional representations

	h	$\chi1_{0,0}$	$\chi1_{1,0}$	$\chi1_{2,0}$	$\chi1_{0,1}$	$\chi1_{1,1}$	$\chi1_{2,1}$	$\chi1_{0,2}$	$\chi1_{1,2}$	$\chi1_{2,2}$
C_1	1	1	1	1	1	1	1	1	1	1
$C_1^{(1)}$	1	1	1	1	1	1	1	1	1	1
$C_1^{(2)}$	1	1	1	1	1	1	1	1	1	1
$C_3^{(0)}$	3	1	1	1	1	1	1	1	1	1
$C_3'^{(0)}$	3	1	1	1	1	1	1	1	1	1
$C_3^{(1)}$	3	1	1	1	ω	ω	ω	ω^2	ω^2	ω^2
$C_3'^{(1)}$	3	1	1	1	ω	ω	ω	ω^2	ω^2	ω^2
$C_3''^{(1)}$	3	1	1	1	ω	ω	ω	ω^2	ω^2	ω^2
$C_3^{(2)}$	3	1	1	1	ω^2	ω^2	ω^2	ω	ω	ω
$C_3'^{(2)}$	3	1	1	1	ω^2	ω^2	ω^2	ω	ω	ω
$C_3''^{(2)}$	3	1	1	1	ω^2	ω^2	ω^2	ω	ω	ω
$C_9^{(0)}$	3	1	ω	ω^2	1	ω	ω^2	1	ω	ω^2
$C_9^{(1)}$	9	1	ω	ω^2	ω	ω^2	1	ω^2	1	ω
$C_9^{(2)}$	9	1	ω	ω^2	ω^2	1	ω	ω	ω^2	1
$C_9'^{(0)}$	3	1	ω^2	ω	1	ω^2	ω	1	ω^2	ω
$C_9'^{(1)}$	9	1	ω^2	ω	ω^2	ω	1	ω	1	ω^2
$C_9'^{(2)}$	9	1	ω^2	ω	ω	1	ω^2	ω^2	ω	1

where h denotes the order of each element in the conjugacy class.

The relations (2.18) and (2.19) for $\Sigma(81)$ gives

$$m_1 + 2^2 m_2 + 3^2 m_3 + \cdots = 81, \tag{12.15}$$

$$m_1 + m_2 + m_3 + \cdots = 17. \tag{12.16}$$

Then, we find $(m_1, m_3) = (9, 8)$. Thus, there are nine singlets $\mathbf{1}_{k,\ell}$ with $k, \ell = 0, 1, 2$ and eight triplets, $\mathbf{3}_A, \mathbf{3}_B, \mathbf{3}_C, \mathbf{3}_D, \bar{\mathbf{3}}_A, \bar{\mathbf{3}}_B, \bar{\mathbf{3}}_C$ and $\bar{\mathbf{3}}_D$. The character tables are presented by Tables 12.2 and 12.3.

On all of the triplets, the generator b is represented as

$$b = \begin{pmatrix} 0 & 1 & 0 \\ 0 & 0 & 1 \\ 1 & 0 & 0 \end{pmatrix}. \tag{12.17}$$

The generators, a, a' and a'', are represented on each triplet as

$$a = \begin{pmatrix} \omega & 0 & 0 \\ 0 & 1 & 0 \\ 0 & 0 & 1 \end{pmatrix}, \quad a' = \begin{pmatrix} 1 & 0 & 0 \\ 0 & 1 & 0 \\ 0 & 0 & \omega \end{pmatrix}, \quad a'' = \begin{pmatrix} 1 & 0 & 0 \\ 0 & \omega & 0 \\ 0 & 0 & 1 \end{pmatrix}, \tag{12.18}$$

Table 12.3 Characters of $\Sigma(81)$ for the 8 three-dimensional representations

Class	n	h	χ_{3_A}	$\chi_{\bar{3}_A}$	χ_{3_B}	$\chi_{\bar{3}_B}$	χ_{3_C}	$\chi_{\bar{3}_C}$	χ_{3_D}	$\chi_{\bar{3}_D}$
C_1	1	1	3	3	3	3	3	3	3	3
$C_1^{(1)}$	1	3	3ω	$3\omega^2$	3ω	$3\omega^2$	3ω	$3\omega^2$	3	3
$C_1^{(2)}$	1	3	$3\omega^2$	3ω	$3\omega^2$	3ω	$3\omega^2$	3ω	3	3
$C_3^{(0)}$	3	3	0	0	0	0	0	0	$3\omega^2$	3ω
$C_3^{'(0)}$	3	3	0	0	0	0	0	0	3ω	$3\omega^2$
$C_3^{(1)}$	3	3	$-i\sqrt{3}\omega$	$i\sqrt{3}\omega^2$	$-i\sqrt{3}$	$i\sqrt{3}$	$-i\sqrt{3}\omega^2$	$i\sqrt{3}\omega$	0	0
$C_3^{'(1)}$	3	3	$-i\sqrt{3}$	$i\sqrt{3}$	$-i\sqrt{3}\omega^2$	$i\sqrt{3}\omega$	$-i\sqrt{3}\omega$	$i\sqrt{3}\omega^2$	0	0
$C_3^{''(1)}$	3	3	$-i\sqrt{3}\omega^2$	$i\sqrt{3}\omega$	$-i\sqrt{3}\omega$	$i\sqrt{3}\omega^2$	$-i\sqrt{3}$	$i\sqrt{3}$	0	0
$C_3^{(2)}$	3	3	$i\sqrt{3}\omega^2$	$-i\sqrt{3}\omega$	$i\sqrt{3}$	$-i\sqrt{3}$	$i\sqrt{3}\omega$	$-i\sqrt{3}\omega^2$	0	0
$C_3^{'(2)}$	3	3	$i\sqrt{3}$	$-i\sqrt{3}$	$i\sqrt{3}\omega$	$-i\sqrt{3}\omega^2$	$i\sqrt{3}\omega^2$	$-i\sqrt{3}\omega$	0	0
$C_3^{''(2)}$	3	3	$i\sqrt{3}\omega$	$-i\sqrt{3}\omega^2$	$i\sqrt{3}\omega^2$	$-i\sqrt{3}\omega$	$i\sqrt{3}$	$-i\sqrt{3}$	0	0
$C_9^{(0)}$	9	3	0	0	0	0	0	0	0	0
$C_9^{(1)}$	9	9	0	0	0	0	0	0	0	0
$C_9^{(2)}$	9	9	0	0	0	0	0	0	0	0
$C_9^{'(0)}$	9	3	0	0	0	0	0	0	0	0
$C_9^{'(1)}$	9	9	0	0	0	0	0	0	0	0
$C_9^{'(2)}$	9	9	0	0	0	0	0	0	0	0

on 3_A,

$$a = \begin{pmatrix} 1 & 0 & 0 \\ 0 & \omega^2 & 0 \\ 0 & 0 & \omega^2 \end{pmatrix}, \quad a' = \begin{pmatrix} \omega^2 & 0 & 0 \\ 0 & \omega^2 & 0 \\ 0 & 0 & 1 \end{pmatrix}, \quad a'' = \begin{pmatrix} \omega^2 & 0 & 0 \\ 0 & 1 & 0 \\ 0 & 0 & \omega^2 \end{pmatrix}, \quad (12.19)$$

on 3_B,

$$a = \begin{pmatrix} \omega^2 & 0 & 0 \\ 0 & \omega & 0 \\ 0 & 0 & \omega \end{pmatrix}, \quad a' = \begin{pmatrix} \omega & 0 & 0 \\ 0 & \omega & 0 \\ 0 & 0 & \omega^2 \end{pmatrix}, \quad a'' = \begin{pmatrix} \omega & 0 & 0 \\ 0 & \omega^2 & 0 \\ 0 & 0 & \omega \end{pmatrix}, \quad (12.20)$$

on 3_C,

$$a = \begin{pmatrix} \omega^2 & 0 & 0 \\ 0 & 1 & 0 \\ 0 & 0 & \omega \end{pmatrix}, \quad a' = \begin{pmatrix} 1 & 0 & 0 \\ 0 & \omega & 0 \\ 0 & 0 & \omega^2 \end{pmatrix}, \quad a'' = \begin{pmatrix} \omega & 0 & 0 \\ 0 & \omega^2 & 0 \\ 0 & 0 & 1 \end{pmatrix}, \quad (12.21)$$

on 3_D. The representations of a, a' and a'' on $3_{\bar{A}}$, $3_{\bar{B}}$, $3_{\bar{C}}$ and $3_{\bar{D}}$ are obtained as complex conjugates of the representations on 3_A, 3_B, 3_C and 3_D, respectively.

On the other hand, for the singlet $1_{k,\ell}$, these generators are represented as $b = \omega^k$ and $a = a' = a'' = \omega^\ell$.

The tensor products between triplets are obtained as

$$\begin{pmatrix} x_1 \\ x_2 \\ x_3 \end{pmatrix}_{3_A} \otimes \begin{pmatrix} y_1 \\ y_2 \\ y_3 \end{pmatrix}_{3_A} = \begin{pmatrix} x_1 y_1 \\ x_2 y_2 \\ x_3 y_3 \end{pmatrix}_{\bar{3}_A} \oplus \begin{pmatrix} x_2 y_3 \\ x_3 y_1 \\ x_1 y_2 \end{pmatrix}_{\bar{3}_B} \oplus \begin{pmatrix} x_3 y_2 \\ x_1 y_3 \\ x_2 y_1 \end{pmatrix}_{\bar{3}_B} , \quad (12.22)$$

$$\begin{pmatrix} x_1 \\ x_2 \\ x_3 \end{pmatrix}_{3_A} \otimes \begin{pmatrix} y_1 \\ y_2 \\ y_3 \end{pmatrix}_{3_A} = \left(\sum_{k=0,1,2} (x_1 y_1 + \omega^{2k} x_2 y_2 + \omega^k x_3 y_3) \mathbf{1}_{k,0} \right)$$
$$\oplus \begin{pmatrix} x_2 y_1 \\ x_3 y_2 \\ x_1 y_3 \end{pmatrix}_{3_D} \oplus \begin{pmatrix} x_1 y_2 \\ x_2 y_3 \\ x_3 y_1 \end{pmatrix}_{3_D} , \quad (12.23)$$

$$\begin{pmatrix} x_1 \\ x_2 \\ x_3 \end{pmatrix}_{3_A} \otimes \begin{pmatrix} y_1 \\ y_2 \\ y_3 \end{pmatrix}_{3_B} = \begin{pmatrix} x_1 y_1 \\ x_2 y_2 \\ x_3 y_3 \end{pmatrix}_{\bar{3}_C} \oplus \begin{pmatrix} x_3 y_2 \\ x_1 y_3 \\ x_2 y_1 \end{pmatrix}_{\bar{3}_A} \oplus \begin{pmatrix} x_2 y_3 \\ x_3 y_1 \\ x_1 y_2 \end{pmatrix}_{\bar{3}_A} , \quad (12.24)$$

$$\begin{pmatrix} x_1 \\ x_2 \\ x_3 \end{pmatrix}_{3_A} \otimes \begin{pmatrix} y_1 \\ y_2 \\ y_3 \end{pmatrix}_{\bar{3}_B} = \left(\sum_{k=0,1,2} (x_1 y_1 + \omega^{2k} x_2 y_2 + \omega^k x_3 y_3) \mathbf{1}_{k,1} \right)$$
$$\oplus \begin{pmatrix} x_2 y_3 \\ x_3 y_1 \\ x_1 y_2 \end{pmatrix}_{\bar{3}_D} \oplus \begin{pmatrix} x_1 y_3 \\ x_2 y_1 \\ x_3 y_2 \end{pmatrix}_{3_D} , \quad (12.25)$$

$$\begin{pmatrix} x_1 \\ x_2 \\ x_3 \end{pmatrix}_{3_A} \otimes \begin{pmatrix} y_1 \\ y_2 \\ y_3 \end{pmatrix}_{3_C} = \begin{pmatrix} x_1 y_1 \\ x_2 y_2 \\ x_3 y_3 \end{pmatrix}_{\bar{3}_B} \oplus \begin{pmatrix} x_2 y_3 \\ x_3 y_1 \\ x_1 y_2 \end{pmatrix}_{\bar{3}_C} \oplus \begin{pmatrix} x_3 y_2 \\ x_1 y_3 \\ x_2 y_1 \end{pmatrix}_{\bar{3}_C} , \quad (12.26)$$

$$\begin{pmatrix} x_1 \\ x_2 \\ x_3 \end{pmatrix}_{3_A} \otimes \begin{pmatrix} y_1 \\ y_2 \\ y_3 \end{pmatrix}_{\bar{3}_C} = \left(\sum_{k=0,1,2} (x_1 y_1 + \omega^{2k} x_2 y_2 + \omega^k x_3 y_3) \mathbf{1}_{k,2} \right)$$
$$\oplus \begin{pmatrix} x_3 y_1 \\ x_1 y_2 \\ x_2 y_3 \end{pmatrix}_{\bar{3}_D} \oplus \begin{pmatrix} x_3 y_2 \\ x_1 y_3 \\ x_2 y_1 \end{pmatrix}_{3_D} , \quad (12.27)$$

$$\begin{pmatrix} x_1 \\ x_2 \\ x_3 \end{pmatrix}_{3_A} \otimes \begin{pmatrix} y_1 \\ y_2 \\ y_3 \end{pmatrix}_{3_D} = \begin{pmatrix} x_3 y_3 \\ x_1 y_1 \\ x_2 y_2 \end{pmatrix}_{3_A} \oplus \begin{pmatrix} x_3 y_2 \\ x_1 y_3 \\ x_2 y_1 \end{pmatrix}_{3_B} \oplus \begin{pmatrix} x_3 y_1 \\ x_1 y_2 \\ x_2 y_3 \end{pmatrix}_{3_C} , \quad (12.28)$$

$$\begin{pmatrix} x_1 \\ x_2 \\ x_3 \end{pmatrix}_{3_A} \otimes \begin{pmatrix} y_1 \\ y_2 \\ y_3 \end{pmatrix}_{\bar{3}_D} = \begin{pmatrix} x_2 y_1 \\ x_3 y_2 \\ x_1 y_3 \end{pmatrix}_{3_A} \oplus \begin{pmatrix} x_2 y_2 \\ x_3 y_3 \\ x_1 y_1 \end{pmatrix}_{3_B} \oplus \begin{pmatrix} x_2 y_3 \\ x_3 y_1 \\ x_1 y_2 \end{pmatrix}_{3_C} , \quad (12.29)$$

$$\begin{pmatrix} x_1 \\ x_2 \\ x_3 \end{pmatrix}_{\bar{3}_A} \otimes \begin{pmatrix} y_1 \\ y_2 \\ y_3 \end{pmatrix}_{\bar{3}_B} = \begin{pmatrix} x_1 y_1 \\ x_2 y_2 \\ x_3 y_3 \end{pmatrix}_{3_C} \oplus \begin{pmatrix} x_3 y_2 \\ x_1 y_3 \\ x_2 y_1 \end{pmatrix}_{3_A} \oplus \begin{pmatrix} x_2 y_3 \\ x_3 y_1 \\ x_1 y_2 \end{pmatrix}_{3_A} , \quad (12.30)$$

$$\begin{pmatrix} x_1 \\ x_2 \\ x_3 \end{pmatrix}_{\bar{3}_A} \otimes \begin{pmatrix} y_1 \\ y_2 \\ y_3 \end{pmatrix}_{\bar{3}_C} = \begin{pmatrix} x_1 y_1 \\ x_2 y_2 \\ x_3 y_3 \end{pmatrix}_{3_B} \oplus \begin{pmatrix} x_3 y_2 \\ x_1 y_3 \\ x_2 y_1 \end{pmatrix}_{3_C} \oplus \begin{pmatrix} x_2 y_3 \\ x_3 y_1 \\ x_1 y_2 \end{pmatrix}_{3_C} , \quad (12.31)$$

$$\begin{pmatrix} x_1 \\ x_2 \\ x_3 \end{pmatrix}_{3_B} \otimes \begin{pmatrix} y_1 \\ y_2 \\ y_3 \end{pmatrix}_{3_B} = \begin{pmatrix} x_1 y_1 \\ x_2 y_2 \\ x_3 y_3 \end{pmatrix}_{\bar{3}_B} \oplus \begin{pmatrix} x_3 y_2 \\ x_1 y_3 \\ x_2 y_1 \end{pmatrix}_{\bar{3}_C} \oplus \begin{pmatrix} x_2 y_3 \\ x_3 y_1 \\ x_1 y_2 \end{pmatrix}_{\bar{3}_C} , \quad (12.32)$$

$$\begin{pmatrix} x_1 \\ x_2 \\ x_3 \end{pmatrix}_{3_B} \otimes \begin{pmatrix} y_1 \\ y_2 \\ y_3 \end{pmatrix}_{3_C} = \begin{pmatrix} x_1 y_1 \\ x_2 y_2 \\ x_3 y_3 \end{pmatrix}_{\bar{3}_A} \oplus \begin{pmatrix} x_3 y_2 \\ x_1 y_3 \\ x_2 y_1 \end{pmatrix}_{\bar{3}_B} \oplus \begin{pmatrix} x_2 y_3 \\ x_3 y_1 \\ x_1 y_2 \end{pmatrix}_{\bar{3}_B} , \quad (12.33)$$

$$\begin{pmatrix} x_1 \\ x_2 \\ x_3 \end{pmatrix}_{3_B} \otimes \begin{pmatrix} y_1 \\ y_2 \\ y_3 \end{pmatrix}_{3_D} = \begin{pmatrix} x_3 y_3 \\ x_1 y_1 \\ x_2 y_2 \end{pmatrix}_{3_B} \oplus \begin{pmatrix} x_3 y_2 \\ x_1 y_3 \\ x_2 y_1 \end{pmatrix}_{3_C} \oplus \begin{pmatrix} x_3 y_1 \\ x_1 y_2 \\ x_2 y_3 \end{pmatrix}_{3_A} , \quad (12.34)$$

$$\begin{pmatrix} x_1 \\ x_2 \\ x_3 \end{pmatrix}_{\bar{3}_B} \otimes \begin{pmatrix} y_1 \\ y_2 \\ y_3 \end{pmatrix}_{3_B} = \sum_{k=0,1,2} [(x_1 y_1 + \omega^{2k} x_2 y_2 + \omega^k x_3 y_3)\mathbf{1}_{k,0}$$
$$\oplus \begin{pmatrix} x_1 y_2 \\ x_2 y_3 \\ x_3 y_1 \end{pmatrix}_{3_D} \oplus \begin{pmatrix} x_2 y_1 \\ x_3 y_2 \\ x_1 y_3 \end{pmatrix}_{\bar{3}_D} , \quad (12.35)$$

$$\begin{pmatrix} x_1 \\ x_2 \\ x_3 \end{pmatrix}_{\bar{3}_B} \otimes \begin{pmatrix} y_1 \\ y_2 \\ y_3 \end{pmatrix}_{3_C} = \sum_{k=0,1,2} [(x_1 y_1 + \omega^{2k} x_2 y_2 + \omega^k x_3 y_3)\mathbf{1}_{k,2}$$
$$\oplus \begin{pmatrix} x_2 y_3 \\ x_3 y_1 \\ x_1 y_2 \end{pmatrix}_{3_D} \oplus \begin{pmatrix} x_1 y_3 \\ x_2 y_1 \\ x_3 y_2 \end{pmatrix}_{\bar{3}_D} , \quad (12.36)$$

$$\begin{pmatrix} x_1 \\ x_2 \\ x_3 \end{pmatrix}_{\bar{3}_B} \otimes \begin{pmatrix} y_1 \\ y_2 \\ y_3 \end{pmatrix}_{3_D} = \begin{pmatrix} x_2 y_2 \\ x_3 y_3 \\ x_1 y_1 \end{pmatrix}_{\bar{3}_C} \oplus \begin{pmatrix} x_2 y_1 \\ x_3 y_2 \\ x_1 y_3 \end{pmatrix}_{\bar{3}_B} \oplus \begin{pmatrix} x_2 y_3 \\ x_3 y_1 \\ x_1 y_2 \end{pmatrix}_{\bar{3}_A} , \quad (12.37)$$

$$\begin{pmatrix} x_1 \\ x_2 \\ x_3 \end{pmatrix}_{3_C} \otimes \begin{pmatrix} y_1 \\ y_2 \\ y_3 \end{pmatrix}_{3_C} = \begin{pmatrix} x_1 y_1 \\ x_2 y_2 \\ x_3 y_3 \end{pmatrix}_{\bar{3}_C} \oplus \begin{pmatrix} x_3 y_2 \\ x_1 y_3 \\ x_2 y_1 \end{pmatrix}_{\bar{3}_A} \oplus \begin{pmatrix} x_2 y_3 \\ x_3 y_1 \\ x_1 y_2 \end{pmatrix}_{\bar{3}_A} , \quad (12.38)$$

$$\begin{pmatrix} x_1 \\ x_2 \\ x_3 \end{pmatrix}_{\bar{3}_C} \otimes \begin{pmatrix} y_1 \\ y_2 \\ y_3 \end{pmatrix}_{3_C} = \sum_{k=0,1,2} [(x_1 y_1 + \omega^{2k} x_2 y_2 + \omega^k x_3 y_3)_{1_{k,0}}$$

$$\oplus \begin{pmatrix} x_1 y_2 \\ x_2 y_3 \\ x_3 y_1 \end{pmatrix}_{3_D} \oplus \begin{pmatrix} x_2 y_1 \\ x_3 y_2 \\ x_1 y_3 \end{pmatrix}_{3_D} , \quad (12.39)$$

$$\begin{pmatrix} x_1 \\ x_2 \\ x_3 \end{pmatrix}_{\bar{3}_C} \otimes \begin{pmatrix} y_1 \\ y_2 \\ y_3 \end{pmatrix}_{3_D} = \begin{pmatrix} x_2 y_2 \\ x_3 y_3 \\ x_1 y_1 \end{pmatrix}_{\bar{3}_A} \oplus \begin{pmatrix} x_2 y_1 \\ x_3 y_2 \\ x_1 y_3 \end{pmatrix}_{\bar{3}_C} \oplus \begin{pmatrix} x_2 y_3 \\ x_3 y_1 \\ x_1 y_2 \end{pmatrix}_{\bar{3}_B} , \quad (12.40)$$

$$\begin{pmatrix} x_1 \\ x_2 \\ x_3 \end{pmatrix}_{3_D} \otimes \begin{pmatrix} y_1 \\ y_2 \\ y_3 \end{pmatrix}_{3_D} = \begin{pmatrix} x_1 y_1 \\ x_2 y_2 \\ x_3 y_3 \end{pmatrix}_{\bar{3}_D} \oplus \begin{pmatrix} x_2 y_3 \\ x_3 y_1 \\ x_1 y_2 \end{pmatrix}_{\bar{3}_D} \oplus \begin{pmatrix} x_3 y_2 \\ x_1 y_3 \\ x_2 y_1 \end{pmatrix}_{\bar{3}_D} , \quad (12.41)$$

$$\begin{pmatrix} x_1 \\ x_2 \\ x_3 \end{pmatrix}_{3_D} \otimes \begin{pmatrix} y_1 \\ y_2 \\ y_3 \end{pmatrix}_{\bar{3}_D} = \sum_{k=0,1,2} [(x_1 y_1 + \omega^{2k} x_2 y_2 + \omega^k x_3 y_3)_{1_{k,0}}$$

$$\oplus (x_2 y_3 + \omega^{2k} x_3 y_1 + \omega^k x_1 y_2)_{1_{k,2}}$$

$$\oplus (x_3 y_2 + \omega^{2k} x_1 y_3 + \omega^k x_2 y_1)_{1_{k,1}}]. \quad (12.42)$$

The tensor products between singlets are given as

$$\mathbf{1}_{k,\ell} \otimes \mathbf{1}_{k',\ell'} = \mathbf{1}_{k+k' \text{ (mod 3)}, \ell+\ell' \text{ (mod 3)}}. \quad (12.43)$$

The tensor products between singlets and triplets are given as follows:

$$(x)_{1_{k,0}} \otimes \begin{pmatrix} y_1 \\ y_2 \\ y_3 \end{pmatrix}_{3(\bar{3})_A} = \begin{pmatrix} x y_2 \\ \omega^k x y_3 \\ \omega^{2k} x y_1 \end{pmatrix}_{3(\bar{3})_A} , \quad (12.44)$$

$$(x)_{1_{k,1}} \otimes \begin{pmatrix} y_1 \\ y_2 \\ y_3 \end{pmatrix}_{3(\bar{3})_A} = \begin{pmatrix} x y_3 \\ \omega^k x y_1 \\ \omega^{2k} x y_2 \end{pmatrix}_{3_C(\bar{3}_B)} , \quad (12.45)$$

$$(x)_{1_{k,2}} \otimes \begin{pmatrix} y_1 \\ y_2 \\ y_3 \end{pmatrix}_{3(\bar{3})_A} = \begin{pmatrix} x y_1 \\ \omega^k x y_2 \\ \omega^{2k} x y_3 \end{pmatrix}_{3_B, (\bar{3}_C)} , \quad (12.46)$$

$$(x)_{\mathbf{1}_{k,0}} \otimes \begin{pmatrix} y_1 \\ y_2 \\ y_3 \end{pmatrix}_{\mathbf{3}(\bar{\mathbf{3}})_B} = \begin{pmatrix} xy_1 \\ \omega^k xy_2 \\ \omega^{2k} xy_3 \end{pmatrix}_{\mathbf{3}(\bar{\mathbf{3}})_B} , \qquad (12.47)$$

$$(x)_{\mathbf{1}_{k,1}} \otimes \begin{pmatrix} y_1 \\ y_2 \\ y_3 \end{pmatrix}_{\mathbf{3}(\bar{\mathbf{3}})_B} = \begin{pmatrix} xy_1 \\ \omega^k xy_2 \\ \omega^{2k} xy_3 \end{pmatrix}_{\mathbf{3}_A,(\bar{\mathbf{3}}_C)} , \qquad (12.48)$$

$$(x)_{\mathbf{1}_{0,2}} \otimes \begin{pmatrix} y_1 \\ y_2 \\ y_3 \end{pmatrix}_{\mathbf{3}(\bar{\mathbf{3}})_B} = \begin{pmatrix} xy_1 \\ \omega^k xy_2 \\ \omega^{2k} xy_3 \end{pmatrix}_{\mathbf{3}_C,(\bar{\mathbf{3}}_A)} , \qquad (12.49)$$

$$(x)_{\mathbf{1}_{k,0}} \otimes \begin{pmatrix} y_1 \\ y_2 \\ y_3 \end{pmatrix}_{\mathbf{3}(\bar{\mathbf{3}})_C} = \begin{pmatrix} xy_1 \\ \omega^k xy_2 \\ \omega^{2k} xy_3 \end{pmatrix}_{\mathbf{3}(\bar{\mathbf{3}})_C} , \qquad (12.50)$$

$$(x)_{\mathbf{1}_{k,1}} \otimes \begin{pmatrix} y_1 \\ y_2 \\ y_3 \end{pmatrix}_{\mathbf{3}(\bar{\mathbf{3}})_C} = \begin{pmatrix} xy_1 \\ \omega^k xy_2 \\ \omega^{2k} xy_3 \end{pmatrix}_{\mathbf{3}_B,(\bar{\mathbf{3}}_A)} , \qquad (12.51)$$

$$(x)_{\mathbf{1}_{k,2}} \otimes \begin{pmatrix} y_1 \\ y_2 \\ y_3 \end{pmatrix}_{\mathbf{3}(\bar{\mathbf{3}})_C} = \begin{pmatrix} xy_1 \\ \omega^k xy_2 \\ \omega^{2k} xy_3 \end{pmatrix}_{\mathbf{3}_A,(\bar{\mathbf{3}}_B)} , \qquad (12.52)$$

$$(x)_{\mathbf{1}_{k,0}} \otimes \begin{pmatrix} y_1 \\ y_2 \\ y_3 \end{pmatrix}_{\mathbf{3}(\bar{\mathbf{3}})_D} = \begin{pmatrix} xy_1 \\ \omega^k xy_2 \\ \omega^{2k} xy_3 \end{pmatrix}_{\mathbf{3}(\bar{\mathbf{3}})_D}, \tag{12.53}$$

$$(x)_{\mathbf{1}_{k,1}} \otimes \begin{pmatrix} y_1 \\ y_2 \\ y_3 \end{pmatrix}_{\mathbf{3}(\bar{\mathbf{3}})_D} = \begin{pmatrix} xy_3 \\ \omega^k xy_1 \\ \omega^{2k} xy_2 \end{pmatrix}_{\mathbf{3}(\bar{\mathbf{3}})_D}, \tag{12.54}$$

$$(x)_{\mathbf{1}_{k,2}} \otimes \begin{pmatrix} y_1 \\ y_2 \\ y_3 \end{pmatrix}_{\mathbf{3}(\bar{\mathbf{3}})_D} = \begin{pmatrix} xy_2 \\ \omega^k xy_3 \\ \omega^{2k} xy_1 \end{pmatrix}_{\mathbf{3}(\bar{\mathbf{3}})_D}, \tag{12.55}$$

where $k = 0, 1, 2$.

Reference

1. Ishimori, H., Kobayashi, T.: Phys. Rev. D **85**, 125004 (2012). [arXiv:1201.3429 [hep-ph]]

In this section, we discuss the discrete group $\Delta(6N^2)$, which is isomorphic to $(Z_N \times Z_N') \rtimes S_3$. (See also Ref. [1].) Let us denote the generators of Z_N and Z_N' by a and a', respectively. We denote S_3 generators by b and c, where b and c are Z_3 and Z_2 generators of S_3, respectively. These generators satisfy

$$
\begin{aligned}
a^N = a'^N = b^3 = c^2 = (bc)^2 = e, \quad aa' = a'a, \\
bab^{-1} = a^{-1}(a')^{-1}, \quad ba'b^{-1} = a, \\
cac^{-1} = (a')^{-1}, \quad ca'c^{-1} = a^{-1}.
\end{aligned}
\tag{13.1}
$$

Using them, all of $\Delta(6N^2)$ elements are written as

$$
g = b^k c^\ell a^m a'^n,
\tag{13.2}
$$

for $k = 0, 1, 2, \ell = 0, 1$ and $m, n = 0, 1, 2, \ldots, N - 1$.

It is noticed that the $\Delta(6N^2)$ group includes the subgroup, $\Delta(3N^2)$, whose elements are written by $b^k a^m a'^n$. Thus, some group-theoretical aspects of $\Delta(6N^2)$ can be derived from those of $\Delta(3N^2)$.

13.1 $\Delta(6N^2)$ with $N/3 \neq$ Integer

• **Conjugacy classes**

Now, let us study the conjugacy classes. It is found that

$$
\begin{aligned}
aba^{-1} &= ba^{-1}a', & a'ba'^{-1} &= ba^{-1}a'^{-2}, \\
aca^{-1} &= ca^{-1}a'^{-1}, & a'ca'^{-1} &= ca^{-1}a'^{-1}, \\
cbc^{-1} &= b^2, & bcb^{-1} &= b^2c.
\end{aligned}
\tag{13.3}
$$

© The Author(s), under exclusive license to Springer-Verlag GmbH, DE,
part of Springer Nature 2022
T. Kobayashi et al., *An Introduction to Non-Abelian Discrete Symmetries
for Particle Physicists*, Lecture Notes in Physics 995,
https://doi.org/10.1007/978-3-662-64679-3_13

By using these relations, one can obtain conjugacy classes of the $\Delta(6N^2)$. Indeed, these relations are nothing but those in the $\Delta(3N^2)$ except the relations including c. Hence, the conjugacy classes of $\Delta(3N^2)$ are useful to obtain those of $\Delta(6N^2)$.

First, we consider the elements $a^\ell a'^m$. As shown in Chap. 10, the element $a^\ell a'^m$ is conjugate to $a^{-\ell+m}a'^{-\ell}$ and $a^{-m}a'^{\ell-m}$ for the group $\Delta(3N^2)$ with $N/3 \neq$ integer. These elements must be conjugate to each other in $\Delta(6N^2)$, too. In addition, it is found that

$$ca^\ell a'^m c^{-1} = a^{-m}a'^{-\ell}, \quad ca^{-\ell+m}a'^{-\ell}c^{-1} = a^\ell a'^{\ell-m}, \quad ca^{-m}a'^{\ell-m}c^{-1} = a^{-\ell+m}a'^m. \tag{13.4}$$

Thus, the following elements,

$$a^\ell a'^m, \quad a^{-\ell+m}a'^{-\ell}, \quad a^{-m}a'^{\ell-m}, \quad a^{-m}a'^{-\ell}, \quad a^\ell a'^{\ell-m}, \quad a^{-\ell+m}a'^m, \tag{13.5}$$

are conjugate to each other in $\Delta(6N^2)$. However, the elements, $a^{-m}a'^{-\ell}$, $a^\ell a'^{\ell-m}$ and $a^{-\ell+m}a'^m$, are the same as $a^\ell a'^m$ when ℓ and m satisfy the conditions,

$$\ell + m = 0 \bmod(N), \quad 2\ell - m = 0 \bmod(N), \quad \ell - 2m = 0 \bmod(N), \tag{13.6}$$

respectively. Under these conditions, the above conjugate elements in Eq. (13.5) reduce to the three elements, $a^\ell a'^{-\ell}$, $a^{-2\ell}a'^{-\ell}$ and $a^\ell a'^{2\ell}$.

As a result, the elements $a^\ell a'^m$ are classified into the following conjugacy classes,

$$C_3^{(k)} = \left\{ a^k a'^{-k}, a^{-2k}a'^{-k}, a^k a'^{2k} \right\}, \quad k = 1, 2, \ldots, N-1, \tag{13.7}$$

$$C_6^{(\ell,m)} = \left\{ a^\ell a'^m, a^{m-\ell}a'^{-\ell}, a^{-m}a'^{\ell-m}, a^{-m}a'^{-\ell}, a^{m-\ell}a'^m, a^\ell a'^{\ell-m} \right\}, \tag{13.8}$$

for $N/3 \neq$ integer. Parameters ℓ and m of $C_6^{(\ell,m)}$ run from 0 to $N-1$ but they do not satisfy the conditions (13.6). The numbers of the conjugacy classes, $C_3^{(\ell)}$ and $C_6^{(\ell,m)}$, are equal to $(N-1)$ and $(N^2-3N+2)/6$, respectively.

Similarly, we can obtain conjugacy classes including $ba^\ell a'^m$ and $b^2 a^\ell a'^m$. As shown in Chap. 10, all of the elements $ba^\ell a'^m$ with $\ell, m = 0, \ldots, N-1$ belong to the same conjugacy classes in the $\Delta(3N^2)$ group. In addition, all of the elements $b^2 a^\ell a'^m$ with $\ell, m = 0, \ldots, N-1$ also belong to the same conjugacy classes in the $\Delta(3N^2)$ group. Furthermore, we obtain

$$cba^\ell a'^m c^{-1} = b^2 a^{-m}a'^{-\ell}. \tag{13.9}$$

Thus, all of the elements, $b^k a^\ell a'^m$, for $k = 1, 2$ and $\ell, m = 0, \ldots, N-1$ belong to the same conjugacy class in $\Delta(6N^2)$.

Finally, we consider the conjugate elements including c, which are not included in $\Delta(3N^2)$. It is found

$$(a^p a'^q)ca^\ell a'^m (a^p a'^q)^{-1} = ca^{\ell-p-q}a'^{m-p-q} = ca^{k+n}a'^n, \tag{13.10}$$

$$(ba^p a'^q)ca^\ell a'^m (ba^p a'^q)^{-1} = b^2 ca^{-\ell+m}a'^{-\ell+p+q} = b^2 ca^{-k}a'^{-k-n}, \tag{13.11}$$

$$(b^2 a^p a'^q)ca^\ell a'^m (b^2 a^p a'^q)^{-1} = bca^{-m+p+q}a'^{\ell-m} = bca^{-n}a'^k, \tag{13.12}$$

where $k = \ell - m$ and $n = m - p - q$. As a result, all of the elements,

$$ca^{k+n}a'^n, \quad b^2ca^{-k}a'^{-k-n}, \quad bca^{-n}a'^k, \tag{13.13}$$

with $n = 0, 1, \ldots, N - 1$ belong to the same conjugacy class.

Here, we summarize the conjugacy classes of $\Delta(6N^2)$. For $N/3 \neq$ integer, the $\Delta(6N^2)$ has the following conjugacy classes,

$$
\begin{aligned}
C_1 : & \quad \{e\}, \\
C_3^{(k)} : & \quad \{a^k a'^{-k}, a^{-2k} a'^{-k}, a^k a'^{2k} \mid k = 1, 2, \ldots, N - 1\}, \\
C_6^{(\ell,m)} : & \quad \{a^\ell a'^m, a^{m-\ell} a'^{-\ell}, a^{-m} a'^{\ell-m}, a^{-m} a'^{-\ell}, a^{m-\ell} a'^m, a^\ell a'^{\ell-m} \mid \\
& \quad \ell, m = 0, 1, \ldots, N - 1\}, \\
C_{2N^2} : & \quad \{b^k a^\ell a'^m \mid k = 1, 2, \ \ell, m = 0, 1, \ldots, N - 1\}, \\
C_{3N}^{(k)} : & \quad \{ca^{k+n}a'^n, b^2ca^{-k}a'^{-k-n}, bca^{-n}a'^k \mid n = 0, 1, \ldots, N - 1\},
\end{aligned}
\tag{13.14}
$$

where ℓ and m in $C_6^{(\ell,m)}$ do not satisfy the conditions, (13.6). The order, h, of each element in the conjugacy classes is found as

$$
\begin{aligned}
C_1 : & \quad h = 1, \\
C_3^{(k)} : & \quad h = N / \gcd(N, \ell), \\
C_6^{(\ell,m)} : & \quad h = N / \gcd(N, \ell, m), \\
C_{2N^2} : & \quad h = 3, \\
C_{3N}^{(k)} : & \quad h = 2N / \gcd(N, k).
\end{aligned}
\tag{13.15}
$$

The numbers of the conjugacy classes $C_3^{(\ell)}$, $C_6^{(\ell,m)}$, and $C_{3N}^{(k)}$ are equal to $N - 1$, $\frac{N^2-3N+2}{6}$, and N, respectively. Then, the total number of conjugacy classes is equal to $\frac{N^2+9N+8}{6}$. The relations (2.18) and (2.19) for $\Delta(6n^2)$ with $N/3 \neq$ integer give

$$m_1 + 2^2 m_2 + 3^2 m_3 + \cdots = 6N^2, \tag{13.16}$$

$$m_1 + m_2 + m_3 + \cdots = \frac{N^2 + 9N + 8}{6}. \tag{13.17}$$

We have the solution $(m_1, m_2, m_3, m_6) = (2, 1, 2(N - 1), \frac{N^2-3N+2}{6})$.

• Characters and representations

Now, we study characters and representations. The $\Delta(6N^2)$ group with $N/3 \neq$ integer has two singlets and one doublet. These are nothing but irreducible representations of S_3. Thus, on these representations, the Z_N and Z'_N generators, a and a', are identity matrices. Because $c^2 = e$, characters of two singlets have two possible values $\chi_{1_k}(c) = (-1)^k$ with $k = 0, 1$ and they correspond to two singlets, $\mathbf{1}_k$. Note that $\chi_{1_k}(a) = \chi_{1_k}(a') = \chi_{1_k}(b) = 1$. Similarly to S_3, the characters of the doublet $\mathbf{2}$ are obtained as

$$\chi_2(C_{2N^2}) = -1, \quad \chi_2(C_{3N}^{(k)}) = 0, \tag{13.18}$$

as well as $\chi_2(a) = \chi_2(a') = 2$. Then, the two-dimensional representations are obtained, e.g.

$$b(2) = \begin{pmatrix} \omega & 0 \\ 0 & \omega^2 \end{pmatrix}, \quad c(2) = \begin{pmatrix} 0 & 1 \\ 1 & 0 \end{pmatrix}, \quad a(2) = a'(2) = \begin{pmatrix} 1 & 0 \\ 0 & 1 \end{pmatrix}. \tag{13.19}$$

Next, let us consider sextet representations. We can obtain (6×6) matrix representations for generic sextet. The $\Delta(6N^2)$ group is represented as

$$b = \begin{pmatrix} b_1 & 0 \\ 0 & b_2 \end{pmatrix}, \quad c = \begin{pmatrix} 0 & 1 \\ 1 & 0 \end{pmatrix}, \quad a = \begin{pmatrix} a_1 & 0 \\ 0 & a_2 \end{pmatrix}, \quad a' = \begin{pmatrix} a'_1 & 0 \\ 0 & a'_2 \end{pmatrix}, \tag{13.20}$$

where

$$b_1 = \begin{pmatrix} 0 & 1 & 0 \\ 0 & 0 & 1 \\ 1 & 0 & 0 \end{pmatrix}, \qquad b_2 = \begin{pmatrix} 0 & 0 & 1 \\ 1 & 0 & 0 \\ 0 & 1 & 0 \end{pmatrix}, \tag{13.21}$$

$$a_1 = a'^{-1}_2 = \begin{pmatrix} \rho^l & 0 & 0 \\ 0 & \rho^k & 0 \\ 0 & 0 & \rho^{-l-k} \end{pmatrix}, \quad a_2 = a'^{-1}_1 = \begin{pmatrix} \rho^{l+k} & 0 & 0 \\ 0 & \rho^{-l} & 0 \\ 0 & 0 & \rho^{-k} \end{pmatrix}, \tag{13.22}$$

on the sextet $\mathbf{6}_{[[k],[\ell]]}$ with $(k, \ell) \neq (0, 0)$, where $[[k], [\ell]]$ denotes[1]

$$[[k], [\ell]] = (k, \ell), \quad (-k - \ell, k), \quad (\ell, -k - \ell),$$
$$(-\ell, -k), \quad (k + \ell, -\ell), \text{ or } (-k, k + \ell). \tag{13.23}$$

We also denote the vector of $\mathbf{6}_{[[k],[\ell]]}$ as

$$\mathbf{6}_{[[k],[\ell]]} = \begin{pmatrix} x_{\ell, -k-\ell} \\ x_{k, \ell} \\ x_{-k-\ell, k} \\ x_{k+\ell, -\ell} \\ x_{-\ell, -k} \\ x_{-k, k+\ell} \end{pmatrix}, \tag{13.24}$$

for $k, \ell = 0, 1, \ldots, N - 1$, where k and ℓ correspond to Z_N and Z'_N charges, respectively.

---

[1] The notation $[[k], [\ell]]$ corresponds to $\widetilde{(k, \ell)}$ in Ref. [1].

Table 13.1 Characters of $\Delta(6N^2)$ for $N/3 \neq$ integer

	h	χ_{1_r}	χ_2	$\chi_{3_{1k}}$	$\chi_{3_{2k}}$	$\chi_{6_{[[k],[\ell]]}}$
C_1	1	1	2	3	3	6
$C_3^{(m)}$	$\dfrac{N}{\gcd(N,m)}$	1	2	$\rho^{-2mk} + 2\rho^{mk}$	$\rho^{-2mk} + 2\rho^{mk}$	$2\rho^{m(k-\ell)} + 2\rho^{-m(2k+\ell)} + 2\rho^{m(k+2\ell)}$
$C_6^{(m,n)}$	$\dfrac{N}{\gcd(N,m,n)}$	1	2	$\rho^{mk} + \rho^{-nk}$ $+\rho^{(-m+n)k}$	$\rho^{mk} + \rho^{-nk}$ $+\rho^{(-m+n)k}$	$\rho^{mk+n\ell} +$ $\rho^{(-m+n)k-m\ell} +$ $\rho^{-nk+(m-n)\ell}$ $+\rho^{-nk-m\ell} +$ $\rho^{mk+(-m+n)\ell} +$ $\rho^{(-m+n)k+n\ell}$
C_{2N^2}	3	1	-1	0	0	0
$C_{3N}^{(m)}$	$\dfrac{2N}{\gcd(N,m)}$	$(-1)^r$	0	ρ^{-mk}	$-\rho^{-mk}$	0

In a certain case, the above representation becomes reducible. For $N/3 \neq$ integer, $6_{[[k],[\ell]]}$ is reducible if $k + \ell = 0 \pmod{N}$, $k = 0, \ell \neq 0$, and $\ell = 0, k \neq 0$, then the number of irreducible sextet representations is $(N^2 - N - 2(N - 1))/6$.

On the other hand, we diagonalize the above reducible six-dimensional representations so as to obtain $2(N - 1)$ irreducible triplets, 3_{1k} and 3_{2k}, with $k = 1, \ldots, N - 1$, where the generators are represented by

$$b(3_{1k}) = \begin{pmatrix} 0 & 1 & 0 \\ 0 & 0 & 1 \\ 1 & 0 & 0 \end{pmatrix}, \ c(3_{1k}) = \begin{pmatrix} 0 & 0 & 1 \\ 0 & 1 & 0 \\ 1 & 0 & 0 \end{pmatrix}, \ a(3_{1k}) = \begin{pmatrix} \rho^k & 0 & 0 \\ 0 & \rho^{-k} & 0 \\ 0 & 0 & 1 \end{pmatrix}, \ a'(3_{1k}) = \begin{pmatrix} 1 & 0 & 0 \\ 0 & \rho^k & 0 \\ 0 & 0 & \rho^{-k} \end{pmatrix},$$

$$(13.25)$$

$$b(3_{2k}) = b(3_{1k}), \quad c(3_{2k}) = -c(3_{1k}), \quad a(3_{2k}) = a(3_{1k}), \quad a'(3_{2k}) = a'(3_{1k}). \quad (13.26)$$

We also denote the vector of 3_{1k} and 3_{2k} as

$$3_{1k} = \begin{pmatrix} x_{k,0} \\ x_{-k,k} \\ x_{0,-k} \end{pmatrix}, \quad 3_{2k} = \begin{pmatrix} x_{k,0} \\ x_{-k,k} \\ x_{0,-k} \end{pmatrix}, \qquad (13.27)$$

for $k = 1, \ldots, N - 1$.

The characters are shown in Table 13.1.

● **Tensor products**

Now, we study tensor products. Because of their $Z_N \times Z'_N$ charges, tensor products of sextets **6** can be obtained as

$$
\begin{pmatrix} x_{\ell,-k-\ell} \\ x_{k,\ell} \\ x_{-k-\ell,k} \\ x_{k+\ell,-\ell} \\ x_{-\ell,-k} \\ x_{-k,k+\ell} \end{pmatrix}_{\mathbf{6}_{[[k],[\ell]]}} \otimes \begin{pmatrix} y_{\ell',-k'-\ell'} \\ y_{k',\ell'} \\ y_{-k'-\ell',k'} \\ y_{k'+\ell',-\ell'} \\ y_{-\ell',-k'} \\ y_{-k',k'+\ell'} \end{pmatrix}_{\mathbf{6}_{[[k'],[\ell']]}}
= \begin{pmatrix} x_{\ell,-k-\ell}\, y_{\ell',-k'-\ell'} \\ x_{k,\ell}\, y_{k',\ell'} \\ x_{-k-\ell,k}\, y_{-k'-\ell',k'} \\ x_{k+\ell,-\ell}\, y_{k'+\ell',-\ell'} \\ x_{-\ell,-k}\, y_{-\ell',-k'} \\ x_{-k,k+\ell}\, y_{-k',k'+\ell'} \end{pmatrix}_{\mathbf{6}_{[[k+k'],[\ell+\ell']]}}
\oplus \begin{pmatrix} x_{\ell,-k-\ell}\, y_{k',\ell'} \\ x_{k,\ell}\, y_{-k'-\ell',k'} \\ x_{-k-\ell,k}\, y_{\ell',-k'-\ell'} \\ x_{k+\ell,-\ell}\, y_{-\ell',-k'} \\ x_{-\ell,-k}\, y_{-k',k'+\ell'} \\ x_{-k,k+\ell}\, y_{k'+\ell',-\ell'} \end{pmatrix}_{\mathbf{6}_{[[k-k'-\ell'],[\ell+k'']]}}
$$

$$
\oplus \begin{pmatrix} x_{\ell,-k-\ell}\, y_{-k'-\ell',k'} \\ x_{k,\ell}\, y_{\ell',-k'-\ell'} \\ x_{-k-\ell,k}\, y_{k',\ell'} \\ x_{k+\ell,-\ell}\, y_{-k',k'+\ell'} \\ x_{-\ell,-k}\, y_{k'+\ell',-\ell'} \\ x_{-k,k+\ell}\, y_{-\ell',-k'} \end{pmatrix}_{\mathbf{6}_{[[k+\ell'],[\ell-k'-\ell']]}}
\oplus \begin{pmatrix} x_{\ell,-k-\ell}\, y_{-\ell',-k'} \\ x_{k,\ell}\, y_{-\ell',-k'} \\ x_{-k-\ell,k}\, y_{k'+\ell',-\ell'} \\ x_{k+\ell,-\ell}\, y_{-k'-\ell',k'} \\ x_{-\ell,-k}\, y_{k',\ell'} \\ x_{-k,k+\ell}\, y_{\ell',-k'-\ell'} \end{pmatrix}_{\mathbf{6}_{[[k-\ell'],[\ell-k']]}}
$$

$$
\oplus \begin{pmatrix} x_{\ell,-k-\ell}\, y_{-\ell',-k'} \\ x_{k,\ell}\, y_{k'+\ell',-\ell'} \\ x_{-k-\ell,k}\, y_{-k',k'+\ell'} \\ x_{k+\ell,-\ell}\, y_{\ell',\ell'} \\ x_{-\ell,-k}\, y_{\ell',-k'-\ell'} \\ x_{-k,k+\ell}\, y_{-k'-\ell',k'} \end{pmatrix}_{\mathbf{6}_{[[k+k'+\ell'],[\ell-\ell']]}}
\oplus \begin{pmatrix} x_{\ell,-k-\ell}\, y_{k'+\ell',-\ell'} \\ x_{k,\ell}\, y_{-k',k'+\ell'} \\ x_{-k-\ell,k}\, y_{-\ell',-k'} \\ x_{k+\ell,-\ell}\, y_{\ell',-k'-\ell'} \\ x_{-\ell,-k}\, y_{-k'-\ell',k'} \\ x_{-k,k+\ell}\, y_{k',\ell'} \end{pmatrix}_{\mathbf{6}_{[[k-k'],[\ell+k'+\ell']]}}
$$

$$\tag{13.28}$$

Similarly, products of sextets $\mathbf{6}$ and triplets $\mathbf{3}_{1k}$, $\mathbf{3}_{2k}$ become

$$
\begin{pmatrix} x_{\ell,-k-\ell} \\ x_{k,\ell} \\ x_{-k-\ell,k} \\ x_{k+\ell,-\ell} \\ x_{-\ell,-k} \\ x_{-k,k+\ell} \end{pmatrix}_{\mathbf{6}_{[[k],[\ell]]}} \otimes \begin{pmatrix} y_{k',0} \\ y_{-k',k'} \\ y_{0,-k'} \end{pmatrix}_{\mathbf{3}_{1k'}}
= \begin{pmatrix} x_{\ell,-k-\ell}\, y_{-k',k'} \\ x_{k,\ell}\, y_{0,-k'} \\ x_{-k-\ell,k}\, y_{k',0} \\ x_{k+\ell,-\ell}\, y_{-k',k'} \\ x_{-\ell,-k}\, y_{k',0} \\ x_{-k,k+\ell}\, y_{0,-k'} \end{pmatrix}_{\mathbf{6}_{[[k],[\ell-k']]}}
\oplus \begin{pmatrix} x_{\ell,-k-\ell}\, y_{k',0} \\ x_{k,\ell}\, y_{-k',k'} \\ x_{-k-\ell,k}\, y_{0,-k'} \\ x_{k+\ell,-\ell}\, y_{0,-k'} \\ x_{-\ell,-k}\, y_{-k',k'} \\ x_{-k,k+\ell}\, y_{k',0} \end{pmatrix}_{\mathbf{6}_{[[k-k'],[\ell+k']]}}
$$

$$
\oplus \begin{pmatrix} x_{\ell,-k-\ell}\, y_{0,-k'} \\ x_{k,\ell}\, y_{k',0} \\ x_{-k-\ell,k}\, y_{-k',k'} \\ x_{k+\ell,-\ell}\, y_{k',0} \\ x_{-\ell,-k}\, y_{0,-k'} \\ x_{-k,k+\ell}\, y_{-k',k'} \end{pmatrix}_{\mathbf{6}_{[[k+k'],[\ell]]}}
$$

$$\tag{13.29}$$

$$
\begin{pmatrix} x_{\ell,-k-\ell} \\ x_{k,\ell} \\ x_{-k-\ell,k} \\ x_{k+\ell,-\ell} \\ x_{-\ell,-k} \\ x_{-k,k+\ell} \end{pmatrix}_{\mathbf{6}_{[[k],[\ell]]}} \otimes \begin{pmatrix} y_{k',0} \\ y_{-k',k'} \\ y_{0,-k'} \end{pmatrix}_{\mathbf{3}_{2k'}}
= \begin{pmatrix} x_{\ell,-k-\ell}\, y_{-k',k'} \\ x_{k,\ell}\, y_{0,-k'} \\ x_{-k-\ell,k}\, y_{k',0} \\ -x_{k+\ell,-\ell}\, y_{-k',k'} \\ -x_{-\ell,-k}\, y_{k',0} \\ -x_{-k,k+\ell}\, y_{0,-k'} \end{pmatrix}_{\mathbf{6}_{[[k],[\ell-k']]}}
\oplus \begin{pmatrix} x_{\ell,-k-\ell}\, y_{k',0} \\ x_{k,\ell}\, y_{-k',k'} \\ x_{-k-\ell,k}\, y_{0,-k'} \\ -x_{k+\ell,-\ell}\, y_{0,-k'} \\ -x_{-\ell,-k}\, y_{-k',k'} \\ -x_{-k,k+\ell}\, y_{k',0} \end{pmatrix}_{\mathbf{6}_{[[k-k'],[\ell+k']]}}
$$

$$
\oplus \begin{pmatrix} x_{\ell,-k-\ell}\, y_{0,-k'} \\ x_{k,\ell}\, y_{k',0} \\ x_{-k-\ell,k}\, y_{-k',k'} \\ -x_{k+\ell,-\ell}\, y_{k',0} \\ -x_{-\ell,-k}\, y_{0,-k'} \\ -x_{-k,k+\ell}\, y_{-k',k'} \end{pmatrix}_{\mathbf{6}_{[[k+k'],[\ell]]}}
$$

$$\tag{13.30}$$

$$\begin{pmatrix} x_{k,0} \\ x_{-k,k} \\ x_{0,-k} \end{pmatrix}_{3_{1k}} \otimes \begin{pmatrix} y_{k',0} \\ y_{-k',k'} \\ y_{0,-k'} \end{pmatrix}_{3_{1k'}} = \begin{pmatrix} x_{k,0}y_{k',0} \\ x_{-k,k}y_{-k',k'} \\ x_{0,-k}y_{0,-k'} \end{pmatrix}_{3_{1(k+k')}} \oplus \begin{pmatrix} x_{0,-k}y_{-k',k'} \\ x_{k,0}y_{0,-k'} \\ x_{-k,k}y_{k',0} \\ x_{k,0}y_{-k',k'} \\ x_{0,-k}y_{k',0} \\ x_{-k,k}y_{0,-k'} \end{pmatrix}_{6_{[[k],[-k']]}} ,$$

$$\tag{13.31}$$

$$\begin{pmatrix} x_{k,0} \\ x_{-k,k} \\ x_{0,-k} \end{pmatrix}_{3_{1k}} \otimes \begin{pmatrix} y_{k',0} \\ y_{-k',k'} \\ y_{0,-k'} \end{pmatrix}_{3_{2k'}} = \begin{pmatrix} x_{k,0}y_{k',0} \\ x_{-k,k}y_{-k',k'} \\ x_{0,-k}y_{0,-k'} \end{pmatrix}_{3_{2(k+k')}} \oplus \begin{pmatrix} x_{0,-k}y_{-k',k'} \\ x_{k,0}y_{0,-k'} \\ x_{-k,k}y_{k',0} \\ -x_{k,0}y_{-k',k'} \\ -x_{0,-k}y_{k',0} \\ -x_{-k,k}y_{0,-k'} \end{pmatrix}_{6_{[[k],[-k']]}} ,$$

$$\tag{13.32}$$

$$\begin{pmatrix} x_{k,0} \\ x_{-k,k} \\ x_{0,-k} \end{pmatrix}_{3_{2k}} \otimes \begin{pmatrix} y_{k',0} \\ y_{-k',k'} \\ y_{0,-k'} \end{pmatrix}_{3_{2k'}} = \begin{pmatrix} x_{k,0}y_{k',0} \\ x_{-k,k}y_{-k',k'} \\ x_{0,-k}y_{0,-k'} \end{pmatrix}_{3_{1(k+k')}} \oplus \begin{pmatrix} x_{0,-k}y_{-k',k'} \\ x_{k,0}y_{0,-k'} \\ x_{-k,k}y_{k',0} \\ x_{k,0}y_{-k',k'} \\ x_{0,-k}y_{k',0} \\ x_{-k,k}y_{0,-k'} \end{pmatrix}_{6_{[[k],[-k']]}} ,$$

$$\tag{13.33}$$

$$\begin{pmatrix} x_{\ell,-k-\ell} \\ x_{k,\ell} \\ x_{-k-\ell,k} \\ x_{k+\ell,-\ell} \\ x_{-\ell,-k} \\ x_{-k,k+\ell} \end{pmatrix}_{6_{[[k],[\ell]]}} \otimes \begin{pmatrix} y_1 \\ y_2 \end{pmatrix}_{2} = \begin{pmatrix} x_{\ell,-k-\ell}y_1 \\ \omega x_{k,\ell}y_1 \\ \omega^2 x_{-k-\ell,k}y_1 \\ x_{k+\ell,-\ell}y_2 \\ \omega x_{-\ell,-k}y_2 \\ \omega^2 x_{-k,k+\ell}y_2 \end{pmatrix}_{6_{[[k],[\ell]]}} \oplus \begin{pmatrix} x_{\ell,-k-\ell}y_2 \\ \omega^2 x_{k,\ell}y_2 \\ \omega x_{-k-\ell,k}y_2 \\ x_{k+\ell,-\ell}y_1 \\ \omega^2 x_{-\ell,-k}y_1 \\ \omega x_{-k,k+\ell}y_1 \end{pmatrix}_{6_{[[k],[\ell]]}} ,$$

$$\tag{13.34}$$

$$\begin{pmatrix} x_{k,0} \\ x_{-k,k} \\ x_{0,-k} \end{pmatrix}_{3_{1k}} \otimes \begin{pmatrix} y_1 \\ y_2 \end{pmatrix}_{2} = \begin{pmatrix} x_{k,0}(y_1 + \omega^2 y_2) \\ \omega x_{-k,k}(y_1 + y_2) \\ x_{0,-k}(\omega^2 y_1 + y_2) \end{pmatrix}_{3_{1k}} \oplus \begin{pmatrix} x_{k,0}(y_1 - \omega^2 y_2) \\ \omega x_{-k,k}(y_1 - y_2) \\ x_{0,-k}(\omega^2 y_1 - y_2) \end{pmatrix}_{3_{2k}} ,$$

$$\tag{13.35}$$

$$\begin{pmatrix} x_{k,0} \\ x_{-k,k} \\ x_{0,-k} \end{pmatrix}_{3_{2k}} \otimes \begin{pmatrix} y_1 \\ y_2 \end{pmatrix}_2 = \begin{pmatrix} x_{k,0}(y_1 + \omega^2 y_2) \\ \omega x_{-k,k}(y_1 + y_2) \\ x_{0,-k}(\omega^2 y_1 + y_2) \end{pmatrix}_{3_{2k}} \oplus \begin{pmatrix} x_{k,0}(y_1 - \omega^2 y_2) \\ \omega x_{-k,k}(y_1 - y_2) \\ x_{0,-k}(\omega^2 y_1 - y_2) \end{pmatrix}_{3_{1k}},$$
$$(13.36)$$

$$\begin{pmatrix} x_1 \\ x_2 \end{pmatrix}_2 \otimes \begin{pmatrix} y_1 \\ y_2 \end{pmatrix}_2 = (x_1 y_2 + x_2 y_1)_{1_0} \oplus (x_1 y_2 - x_2 y_1)_{1_1} \oplus \begin{pmatrix} x_2 y_2 \\ x_1 y_1 \end{pmatrix}_2. \quad (13.37)$$

Tensor products with trivial singlet remain the same representation while products with non-trivial singlets are

$$\begin{pmatrix} x_1 \\ x_2 \\ x_3 \\ x_4 \\ x_5 \\ x_6 \end{pmatrix}_{6_{[[k],[\ell]]}} \otimes (y)_{1_1} = \begin{pmatrix} x_1 y \\ x_2 y \\ x_3 y \\ -x_4 y \\ -x_5 y \\ -x_6 y \end{pmatrix}_{6_{[[k],[\ell]]}}, \quad (13.38)$$

$$\begin{pmatrix} x_1 \\ x_2 \\ x_3 \end{pmatrix}_{3_{1k}} \otimes (y)_{1_1} = \begin{pmatrix} x_1 y \\ x_2 y \\ x_3 y \end{pmatrix}_{3_{2k}}, \quad (13.39)$$

$$\begin{pmatrix} x_1 \\ x_2 \\ x_3 \end{pmatrix}_{3_{2k}} \otimes (y)_{1_1} = \begin{pmatrix} x_1 y \\ x_2 y \\ x_3 y \end{pmatrix}_{3_{1k}}, \quad (13.40)$$

$$\begin{pmatrix} x_1 \\ x_2 \end{pmatrix}_2 \otimes (y)_{1_1} = \begin{pmatrix} x_1 y \\ -x_2 y \end{pmatrix}_2. \quad (13.41)$$

13.2 $\Delta(6N^2)$ with $N/3$ = Integer

• Conjugacy classes

Let us study the conjugacy classes. The conjugacy classes of $\Delta(3N^2)$ are useful to obtain those of $\Delta(6N^2)$ except the elements including c.

Conjugacy classes with the elements $a^\ell a'^m$ are almost the same to those of $\Delta(6N^2)$ with $N/3 \neq$ integer. A difference is that, when $N/3 =$ integer, there are classes with one element. Since

$$b a^\ell a'^{-\ell} b^{-1} = a^{-2\ell} a'^{-\ell}, \quad c a^\ell a'^{-\ell} c^{-1} = a^\ell a'^{-\ell}, \quad (13.42)$$

and $\ell = -2\ell$ if $\ell = N/3$ or $2N/3$, there are classes $\{a^{N/3} a'^{2N/3}\}$ and $\{a^{2N/3} a'^{N/3}\}$.

Next, we consider the elements $b a^\ell a'^m$ and $b^2 a^\ell a'^m$. In $\Delta(3N^2)$ with $N/3 =$ integer, the elements $b a^{p-n-3m} a'^n$ with $m = 0, 1, \ldots, (N-3)/3$ and $n = 0, 1, \ldots,$

$N - 1$ belong to the same conjugacy class. Also the elements $b^2 a^{p-n-3m} a'^n$ with $m = 0, 1, \ldots, (N-3)/3$ and $n = 0, 1, \ldots, N - 1$ belong to the same conjugacy class. Since

$$cba^{p-n-3m} a'^n c^{-1} = b^2 a^{-n} a'^{-p+n+3m}, \tag{13.43}$$

these elements belong to the same conjugacy class for $\Delta(6N^2)$.

Similarly to $\Delta(6N^2)$ with $N/3 \neq$ integer, we can study the conjugacy class of the elements including c. As a result, the conjugacy class of the elements including c are the same between the $\Delta(6N^2)$ group with $N/3 \neq$ integer and with $N/3 =$ integer.

Here, we summarize the conjugacy classes of the $\Delta(6N^2)$ group with $N/3 =$ integer,

$$
\begin{aligned}
&C_1: && \{e\}, \\
&C_1^{(p)}: && \{a^p a'^{-p}\}, \quad p = \tfrac{N}{3}, \tfrac{2N}{3}, \\
&C_3^{(k)}: && \{a^k a'^{-k}, a^{-2k} a'^{-k}, a^k a'^{2k}\}, k = 1, \ldots, N-1, \; k \neq \tfrac{N}{3}, \tfrac{2N}{3}, \\
&C_6^{(\ell,m)}: && \{a^\ell a'^m, a^{m-\ell} a'^{-\ell}, a^{-m} a'^{\ell-m}, a^{-m} a'^{-\ell}, a^{m-\ell} a'^m, a^\ell a'^{\ell-m}\}, \\
& && \qquad \ell, m = 0, \ldots, N-1 \\
&C_{2N^2/3}^{(q)}: && \{ba^{q-n-3m} a'^n, \, b^2 a^{-n} a'^{n+3m-q} \mid n = 0, 1, \ldots, N-1, \; m = 0, 1, \ldots, \tfrac{N-3}{3}\}, q = 0, 1, 2, \\
&C_{3N}^{(\ell)}: && \{ca^{\ell+n} a'^n, \, b^2 ca^{-\ell} a'^{-\ell-n}, \, bca^{-n} a'^\ell \mid n = 0, 1, \ldots, N-1\}, \\
& && \qquad \ell = 0, \ldots, N-1
\end{aligned}
\tag{13.44}
$$

where ℓ, m in $C_6^{(\ell,m)}$ do not satisfy (13.6). The order, h, of each element in the conjugacy classes is found as The order, h, of each element in the conjugacy classes is found as

$$
\begin{aligned}
&C_1: && h = 1, \\
&C_1^{(p)}: && h = 3, \\
&C_3^{(k)}: && h = N/\gcd(N, k), \\
&C_6^{(\ell,m)}: && h = N/\gcd(N, \ell, m), \\
&C_{2N^2/3}^{(q)}: && h = 3, \\
&C_{3N}^{(\ell)}: && h = 2N/\gcd(N, \ell).
\end{aligned}
\tag{13.45}
$$

The numbers of the conjugacy classes $C_1^{(\ell)}$, $C_3^{(\ell)}$, $C_6^{(\ell,m)}$, $C_{2N^2/3}^{(p)}$, and $C_{3N}^{(k)}$ are equal to 2, $N - 3$, $\frac{N^2 - 3N + 6}{6}$, 3, and N, respectively. The total number of conjugacy classes is equal to $\frac{N^2 + 9N + 24}{6}$. The relations (2.18) and (2.19) for $\Delta(6N^2)$ with $N/3 =$ integer lead to

$$m_1 + 2^2 m_2 + 3^2 m_3 + \cdots = 6N^2, \tag{13.46}$$

$$m_1 + m_2 + m_3 + \cdots = \frac{N^2 + 9N + 24}{6}. \tag{13.47}$$

The solution is found as $(m_1, m_2, m_3, m_6) = (2, 4, 2(N-1), \frac{N(N-3)}{6})$.

• Characters and representations

Now, we study characters. There are two singlets, $\mathbf{1}_k$. Their characters are the same as those of the $\Delta(6N^2)$ group with $N/3 = $ integer. That is, because $c^2 = e$, characters of two singlets have two possible values $\chi_{1_k}(c) = (-1)^k$ with $k = 0, 1$ and they correspond to two singlets, $\mathbf{1}_k$. Note that $\chi_{1_k}(a) = \chi_{1_k}(a') = \chi_{1_k}(b) = 1$. The $\Delta(6N^2)$ group with $N/3 = $ integer has four doublets $\mathbf{2}_k$ with $k = 1, 2, 3, 4$. One of them, $\mathbf{2}_1$, is the same as the doublet (13.119) of $\Delta(6N^2)$ with $N/3 \neq $ integer. The other three doublet representations will be shown after the discussion on the sextet representations.

Next, let us consider sextet representations, which are the same as those of $\Delta(6N^2)$ with $N/3 = $ integer, (13.120) and (13.121), i.e.

$$b = \begin{pmatrix} b_1 & 0 \\ 0 & b_2 \end{pmatrix}, \quad c = \begin{pmatrix} 0 & 1 \\ 1 & 0 \end{pmatrix}, \quad a = \begin{pmatrix} a_1 & 0 \\ 0 & a_2 \end{pmatrix}, \quad a' = \begin{pmatrix} a'_1 & 0 \\ 0 & a'_2 \end{pmatrix}, \quad (13.48)$$

where

$$b_1 = \begin{pmatrix} 0 & 1 & 0 \\ 0 & 0 & 1 \\ 1 & 0 & 0 \end{pmatrix}, \qquad b_2 = \begin{pmatrix} 0 & 0 & 1 \\ 1 & 0 & 0 \\ 0 & 1 & 0 \end{pmatrix}, \qquad (13.49)$$

$$a_1 = a'^{-1}_2 = \begin{pmatrix} \rho^l & 0 & 0 \\ 0 & \rho^k & 0 \\ 0 & 0 & \rho^{-l-k} \end{pmatrix}, \qquad a_2 = a'^{-1}_1 = \begin{pmatrix} \rho^{l+k} & 0 & 0 \\ 0 & \rho^{-l} & 0 \\ 0 & 0 & \rho^{-k} \end{pmatrix}. \quad (13.50)$$

In a certain case, the above representation becomes reducible. For $N/3 = $ integer, $\mathbf{6}_{[[k],[\ell]]}$ is reducible if $(k, \ell) = (N/3, N/3), (2N/3, 2N/3)$, $k + \ell = 0 \pmod{N}$, $k = 0, \ell \neq 0$, and $\ell = 0, k \neq 0$. Thus, the number of irreducible sextet representations is $(N^2 - 2 - N - 2(N - 1))/6$.

Among the above reducible sextet representations, the irreducible triplets are obtained in the same way as (13.125) in $\Delta(6N^2)$ with $N/3 \neq $ integer, i.e.

$$b(\mathbf{3}_{1k}) = \begin{pmatrix} 0 & 1 & 0 \\ 0 & 0 & 1 \\ 1 & 0 & 0 \end{pmatrix}, \quad c(\mathbf{3}_{1k}) = \begin{pmatrix} 0 & 0 & 1 \\ 0 & 1 & 0 \\ 1 & 0 & 0 \end{pmatrix}, \quad a(\mathbf{3}_{1k}) = \begin{pmatrix} \rho^k & 0 & 0 \\ 0 & \rho^{-k} & 0 \\ 0 & 0 & 1 \end{pmatrix}, \quad a'(\mathbf{3}_{1k}) = \begin{pmatrix} 1 & 0 & 0 \\ 0 & \rho^k & 0 \\ 0 & 0 & \rho^{-k} \end{pmatrix},$$
$$(13.51)$$

$$b(\mathbf{3}_{2k}) = b(\mathbf{3}_{1k}), \quad c(\mathbf{3}_{2k}) = -c(\mathbf{3}_{1k}), \quad a(\mathbf{3}_{2k}) = a(\mathbf{3}_{1k}), \quad a'(\mathbf{3}_{2k}) = a'(\mathbf{3}_{1k}). \quad (13.52)$$

In addition, one can obtain the three doublets by diagonalizing the above reducible sextet representations as

$$b(\mathbf{2}_2) = b(\mathbf{2}_1), \quad c(\mathbf{2}_2) = c(\mathbf{2}_1), \quad a(\mathbf{2}_2) = a'(\mathbf{2}_2) = \begin{pmatrix} \omega^2 & 0 \\ 0 & \omega \end{pmatrix}, \quad (13.53)$$

$$b(\mathbf{2}_3) = b(\mathbf{2}_1), \quad c(\mathbf{2}_3) = c(\mathbf{2}_1), \quad a(\mathbf{2}_3) = a'(\mathbf{2}_3) = \begin{pmatrix} \omega & 0 \\ 0 & \omega^2 \end{pmatrix}, \quad (13.54)$$

$$b(\mathbf{2}_4) = 1, \quad c(\mathbf{2}_4) = c(\mathbf{2}_1), \quad a(\mathbf{2}_4) = a'(\mathbf{2}_4) = \begin{pmatrix} \omega & 0 \\ 0 & \omega^2 \end{pmatrix}. \quad (13.55)$$

The characters are shown in Table 13.2.

• Tensor products

For $3N =$ integer, the number of doublets is increased. Since multiplications with other representations are the same as those for $3N \neq$ integer, it is enough to see multiplications with additional doublets:

$$
\begin{pmatrix} x_{\ell,-k-\ell} \\ x_{k,\ell} \\ x_{-k-\ell,k} \\ x_{k+\ell,-\ell} \\ x_{-\ell,-k} \\ x_{-k,k+\ell} \end{pmatrix}_{6_{[[k],[\ell]]}} \otimes \begin{pmatrix} y_1 \\ y_2 \end{pmatrix}_{2_2} = \begin{pmatrix} x_{\ell,-k-\ell}y_1 \\ \omega x_{k,\ell}y_1 \\ \omega^2 x_{-k-\ell,k}y_1 \\ x_{k+\ell,-\ell}y_2 \\ \omega x_{-\ell,-k}y_2 \\ \omega^2 x_{-k,k+\ell}y_2 \end{pmatrix}_{6_{[[k+2N/3],[\ell+2N/3]]}} \oplus \begin{pmatrix} x_{\ell,-k-\ell}y_2 \\ \omega^2 x_{k,\ell}y_2 \\ \omega x_{-k-\ell,k}y_2 \\ x_{k+\ell,-\ell}y_1 \\ \omega^2 x_{-\ell,-k}y_1 \\ \omega x_{-k,k+\ell}y_1 \end{pmatrix}_{6_{[[k+N/3],[\ell+N/3]]}} ,
$$
(13.56)

$$
\begin{pmatrix} x_{\ell,-k-\ell} \\ x_{k,\ell} \\ x_{-k-\ell,k} \\ x_{k+\ell,-\ell} \\ x_{-\ell,-k} \\ x_{-k,k+\ell} \end{pmatrix}_{6_{[[k],[\ell]]}} \otimes \begin{pmatrix} y_1 \\ y_2 \end{pmatrix}_{2_3} = \begin{pmatrix} x_{\ell,-k-\ell}y_1 \\ \omega x_{k,\ell}y_1 \\ \omega^2 x_{-k-\ell,k}y_1 \\ x_{k+\ell,-\ell}y_2 \\ \omega x_{-\ell,-k}y_2 \\ \omega^2 x_{-k,k+\ell}y_2 \end{pmatrix}_{6_{[[k+N/3],[\ell+N/3]]}} \oplus \begin{pmatrix} x_{\ell,-k-\ell}y_2 \\ \omega^2 x_{k,\ell}y_2 \\ \omega x_{-k-\ell,k}y_2 \\ x_{k+\ell,-\ell}y_1 \\ \omega^2 x_{-\ell,-k}y_1 \\ \omega x_{-k,k+\ell}y_1 \end{pmatrix}_{6_{[[k+2N/3],[\ell+2N/3]]}} ,
$$
(13.57)

$$
\begin{pmatrix} x_{\ell,-k-\ell} \\ x_{k,\ell} \\ x_{-k-\ell,k} \\ x_{k+\ell,-\ell} \\ x_{-\ell,-k} \\ x_{-k,k+\ell} \end{pmatrix}_{6_{[[k],[\ell]]}} \otimes \begin{pmatrix} y_1 \\ y_2 \end{pmatrix}_{2_4} = \begin{pmatrix} x_{\ell,-k-\ell}y_1 \\ x_{k,\ell}y_1 \\ x_{-k-\ell,k}y_1 \\ x_{k+\ell,-\ell}y_2 \\ x_{-\ell,-k}y_2 \\ x_{-k,k+\ell}y_2 \end{pmatrix}_{6_{[[k+N/3],[\ell+N/3]]}} \oplus \begin{pmatrix} x_{\ell,-k-\ell}y_2 \\ x_{k,\ell}y_2 \\ x_{-k-\ell,k}y_2 \\ x_{k+\ell,-\ell}y_1 \\ x_{-\ell,-k}y_1 \\ x_{-k,k+\ell}y_1 \end{pmatrix}_{6_{[[k+2N/3],[\ell+2N/3]]}} ,
$$
(13.58)

$$
\begin{pmatrix} x_{k,0} \\ x_{-k,k} \\ x_{0,-k} \end{pmatrix}_{3_{1k}} \otimes \begin{pmatrix} y_1 \\ y_2 \end{pmatrix}_{2_2} = \begin{pmatrix} x_{k,0}y_1 \\ \omega x_{-k,k}y_1 \\ \omega^2 x_{0,-k}y_1 \\ x_{0,-k}y_2 \\ \omega x_{-k,k}y_2 \\ \omega^2 x_{k,0}y_2 \end{pmatrix}_{6_{[[-k+2N/3],[k+2N/3]]}} ,
$$
(13.59)

Table 13.2 Characters of $\Delta(6N^2)$ for $N/3 =$ integer

	h	χ_{1_r}	χ_{2_1}	χ_{2_2}	χ_{2_3}	χ_{2_4}	$\chi_{3_{1k}}$	$\chi_{3_{2k}}$	$\chi_{6_{[[k],[\ell]]}}$
C_1	1	1	2	2	2	2	3	3	6
$C_1^{(m)}$	3	1	2	2	2	2	$\rho^{-2mk}+2\rho^{mk}$	$\rho^{-2mk}+2\rho^{mk}$	$2\rho^{m(k-\ell)}+$ $2\rho^{-m(2k+\ell)}+$ $2\rho^{m(k+2\ell)}$
$C_3^{(m)}$	$\dfrac{N}{\gcd(N,m)}$	1	2	2	2	2	$\rho^{-2mk}+2\rho^{mk}$	$\rho^{-2mk}+2\rho^{mk}$	$2\rho^{m(k-\ell)}+$ $2\rho^{-m(2k+\ell)}+$ $2\rho^{m(k+2\ell)}$
$C_6^{(m,n)}$	$\dfrac{N}{\gcd(N,m,n)}$	1	2	ω^{m+n} $+\omega^{2m+2n}$	ω^{m+n} $+\omega^{2m+2n}$	ω^{m+n} $+\omega^{2m+2n}$	$\rho^{mk}+\rho^{-nk}$ $+\rho^{(-m+n)k}$	$\rho^{mk}+\rho^{-nk}$ $+\rho^{(-m+n)k}$	$\rho^{mk+n\ell}+$ $\rho^{(-m+n)k-m\ell}+$ $\rho^{-nk+(m-n)\ell}$ $+\rho^{-nk-m\ell}+$ $\rho^{mk+(m-n)\ell}+$ $\rho^{(-m+n)k+n\ell}$
$C_{2N^2/3}^{(\tau)}$	3	1	-1	$\omega^{2+\tau}$ $+\omega^{(4+2\tau)}$	$\omega^{1+\tau}$ $+\omega^{2+2\tau}$	ω^{τ} $+\omega^{2\tau}$	0	0	0
$C_{3N}^{(m)}$	$\dfrac{2N}{\gcd(N,m)}$	$(-1)^r$	0	0	0	0	ρ^{-mk}	$-\rho^{-mk}$	0

$$
\begin{pmatrix} x_{k,0} \\ x_{-k,k} \\ x_{0,-k} \end{pmatrix}_{3_{2k}} \otimes \begin{pmatrix} y_1 \\ y_2 \end{pmatrix}_{2_2} = \begin{pmatrix} x_{k,0}y_1 \\ \omega x_{-k,k}y_1 \\ \omega^2 x_{0,-k}y_1 \\ -x_{0,-k}y_2 \\ -\omega x_{-k,k}y_2 \\ -\omega^2 x_{k,0}y_2 \end{pmatrix}_{6_{[[-k+2N/3],[k+2N/3]]}} , \qquad (13.60)
$$

$$
\begin{pmatrix} x_{k,0} \\ x_{-k,k} \\ x_{0,-k} \end{pmatrix}_{3_{1k}} \otimes \begin{pmatrix} y_1 \\ y_2 \end{pmatrix}_{2_3} = \begin{pmatrix} x_{k,0}y_1 \\ \omega x_{-k,k}y_1 \\ \omega^2 x_{0,-k}y_1 \\ x_{0,-k}y_2 \\ \omega x_{-k,k}y_2 \\ \omega^2 x_{k,0}y_2 \end{pmatrix}_{6_{[[-k+N/3],[k+N/3]]}} , \qquad (13.61)
$$

$$
\begin{pmatrix} x_{k,0} \\ x_{-k,k} \\ x_{0,-k} \end{pmatrix}_{3_{2k}} \otimes \begin{pmatrix} y_1 \\ y_2 \end{pmatrix}_{2_3} = \begin{pmatrix} x_{k,0}y_1 \\ \omega x_{-k,k}y_1 \\ \omega^2 x_{0,-k}y_1 \\ -x_{0,-k}y_2 \\ -\omega x_{-k,k}y_2 \\ -\omega^2 x_{k,0}y_2 \end{pmatrix}_{6_{[[-k+N/3],[k+N/3]]}} , \qquad (13.62)
$$

$$
\begin{pmatrix} x_{k,0} \\ x_{-k,k} \\ x_{0,-k} \end{pmatrix}_{3_{1k}} \otimes \begin{pmatrix} y_1 \\ y_2 \end{pmatrix}_{2_4} = \begin{pmatrix} x_{k,0}y_1 \\ x_{-k,k}y_1 \\ x_{0,-k}y_1 \\ x_{0,-k}y_2 \\ x_{-k,k}y_2 \\ x_{k,0}y_2 \end{pmatrix}_{6_{[[-k+N/3],[k+N/3]]}} , \qquad (13.63)
$$

$$
\begin{pmatrix} x_{k,0} \\ x_{-k,k} \\ x_{0,-k} \end{pmatrix}_{3_{2k}} \otimes \begin{pmatrix} y_1 \\ y_2 \end{pmatrix}_{2_4} = \begin{pmatrix} x_{k,0}y_1 \\ x_{-k,k}y_1 \\ x_{0,-k}y_1 \\ -x_{0,-k}y_2 \\ -x_{-k,k}y_2 \\ -x_{k,0}y_2 \end{pmatrix}_{6_{[[-k+N/3],[k+N/3]]}} . \qquad (13.64)
$$

The tensor products among doublets are obtained as

$$
\begin{pmatrix} x_1 \\ x_2 \end{pmatrix}_{2_k} \otimes \begin{pmatrix} y_1 \\ y_2 \end{pmatrix}_{2_k} = (x_1 y_2 + x_2 y_1)_{1_0} \oplus (x_1 y_2 - x_2 y_1)_{1_1} \oplus \begin{pmatrix} x_2 y_2 \\ x_1 y_1 \end{pmatrix}_{2_k} , \qquad (13.65)
$$

for $k = 1, 2, 3, 4$,

$$\begin{pmatrix} x_1 \\ x_2 \end{pmatrix}_{2_1} \otimes \begin{pmatrix} y_1 \\ y_2 \end{pmatrix}_{2_2} = \begin{pmatrix} x_2 y_2 \\ x_1 y_1 \end{pmatrix}_{2_3} \oplus \begin{pmatrix} x_1 y_2 \\ x_2 y_1 \end{pmatrix}_{2_4}, \tag{13.66}$$

$$\begin{pmatrix} x_1 \\ x_2 \end{pmatrix}_{2_1} \otimes \begin{pmatrix} y_1 \\ y_2 \end{pmatrix}_{2_3} = \begin{pmatrix} x_2 y_2 \\ x_1 y_1 \end{pmatrix}_{2_2} \oplus \begin{pmatrix} x_2 y_1 \\ x_1 y_2 \end{pmatrix}_{2_4}, \tag{13.67}$$

$$\begin{pmatrix} x_1 \\ x_2 \end{pmatrix}_{2_1} \otimes \begin{pmatrix} y_1 \\ y_2 \end{pmatrix}_{2_4} = \begin{pmatrix} x_1 y_2 \\ x_2 y_1 \end{pmatrix}_{2_2} \oplus \begin{pmatrix} x_1 y_1 \\ x_2 y_2 \end{pmatrix}_{2_3}, \tag{13.68}$$

$$\begin{pmatrix} x_1 \\ x_2 \end{pmatrix}_{2_2} \otimes \begin{pmatrix} y_1 \\ y_2 \end{pmatrix}_{2_3} = \begin{pmatrix} x_2 y_2 \\ x_1 y_1 \end{pmatrix}_{2_1} \oplus \begin{pmatrix} x_1 y_2 \\ x_2 y_1 \end{pmatrix}_{2_4}, \tag{13.69}$$

$$\begin{pmatrix} x_1 \\ x_2 \end{pmatrix}_{2_2} \otimes \begin{pmatrix} y_1 \\ y_2 \end{pmatrix}_{2_4} = \begin{pmatrix} x_1 y_1 \\ x_2 y_2 \end{pmatrix}_{2_1} \oplus \begin{pmatrix} x_1 y_2 \\ x_2 y_1 \end{pmatrix}_{2_3}, \tag{13.70}$$

$$\begin{pmatrix} x_1 \\ x_2 \end{pmatrix}_{2_3} \otimes \begin{pmatrix} y_1 \\ y_2 \end{pmatrix}_{2_4} = \begin{pmatrix} x_1 y_2 \\ x_2 y_1 \end{pmatrix}_{2_1} \oplus \begin{pmatrix} x_1 y_1 \\ x_2 y_2 \end{pmatrix}_{2_2}. \tag{13.71}$$

Tensor products with non-trivial singlets for additional doublets are the same to those of **2** with $N/3 \neq$ integer.

13.3 $\Delta(54)$

We present a simple example of $\Delta(6N^2)$. The $\Delta(6)$ is nothing but S_3 and the $\Delta(24)$ is isomorphic to the S_4 group. Thus, the simple and non-trivial example is $\Delta(54)$.

• Conjugacy classes

All of $\Delta(54)$ elements are written as $b^k c^\ell a^m a'^n$, where $k, m, n = 0, 1, 2$ and $\ell = 0, 1$. Half of them are the elements of $\Delta(27)$, whose conjugacy classes are shown in (10.37). Because of $cac^{-1} = a'^{-1}$ and $ca'c^{-1} = a^{-1}$, the conjugacy classes $C_1^{(1)}$ and $C_1^{(2)}$ of $\Delta(27)$ still correspond to the conjugacy classes of $\Delta(54)$. However, the conjugacy classes $C_3^{(0,1)}$ and $C_3^{(0,2)}$ of $\Delta(27)$ are combined into one class. Similarly, since $cba^k a'^\ell c^{-1} = b^2 ca^k a'^\ell c^{-1}$, the conjugacy classes $C_3^{(1,p)}$ and $C_3^{(2,p')}$ of $\Delta(27)$ for $p + p' = 0 \pmod 3$ are combined into one class of $\Delta(54)$.

Let us consider the conjugacy classes of elements including c. For example, we obtain

$$a^k a'^\ell (ca^m) a^{-k} a'^{-\ell} = ca^{m+p} a'^p, \tag{13.72}$$

where $p = -k - \ell$. Thus, the element ca^m is conjugate to $ca^{m+p} a'^p$ with $p = 0, 1, 2$. Furthermore, it is found that

$$b(ca^{m+p} a'^p)b^{-1} = b^2 ca^{-m} a'^{-m-p}, \tag{13.73}$$
$$b(ca^{-m} a'^{-m-p})b^{-1} = bca^{-p} a'^m. \tag{13.74}$$

Then, these elements belong to the same conjugacy class.

Using the above results, the $\Delta(54)$ elements are classified into the following conjugacy classes,

$$
\begin{aligned}
&C_1 : &&\{e\}, &&h = 1, \\
&C_1^{(1)} : &&\{aa'^2\}, &&h = 3, \\
&C_1^{(2)} : &&\{a^2 a'\}, &&h = 3, \\
&C_6^{(0,1)} : &&\{a', a, a^2 a'^2, a'^2, a^2, aa'\}, &&h = 3, \\
&C_6^{(0)} : &&\{b, ba^2 a', baa'^2, b^2, b^2 a^2 a', b^2 aa'^2\}, &&h = 3, \\
&C_6^{(1)} : &&\{ba, ba', ba^2 a'^2, b^2 a^2, b^2 a'^2, b^2 aa'\}, &&h = 3, \\
&C_6^{(2)} : &&\{ba^2, baa', ba'^2, b^2 a, b^2 a^2 a'^2, b^2 a'\}, &&h = 3, \\
&C_9^{(0)} : &&\{ca^p a'^p, b^2 ca'^{-p}, bca^{-p} \mid p = 0, 1, 2\}, &&h = 2, \\
&C_9^{(1)} : &&\{ca^{1+p} a'^p, b^2 ca^2 a'^{-1-p}, bca^{-p} a' \mid p = 0, 1, 2\}, &&h = 6, \\
&C_9^{(2)} : &&\{ca^{2+p} a'^p, b^2 caa'^{-2-p}, bca^{-p} a'^2 \mid p = 0, 1, 2\}, &&h = 6.
\end{aligned}
\tag{13.75}
$$

The total number of conjugacy classes is equal to ten. The relations (2.18) and (2.19) for $\Delta(54)$ lead to

$$m_1 + 2^2 m_2 + 3^2 m_3 + \cdots = 54, \tag{13.76}$$
$$m_1 + m_2 + m_3 + \cdots = 10. \tag{13.77}$$

The solution is given as $(m_1, m_2, m_3) = (2, 4, 4)$. Therefore, there are two singlets, four doublets and four triplets.

• Characters and representations

Let us study characters and representations. We start with discussing two singlets. It is straightforward to find $\chi_{1_k}(a) = \chi_{1_k}(a') = \chi_{1_k}(b) = 1$ for two singlets from the above structure of conjugacy classes. In addition, because of $c^2 = e$, the two values $\chi_{1_k}(c) = (-1)^k$ with $k = 0, 1$. They correspond to two singlets.

Next, we consider triplets. For example, the generators, a, a', b and c, are represented by

$$a = \begin{pmatrix} \omega^k & 0 & 0 \\ 0 & \omega^{2k} & 0 \\ 0 & 0 & 1 \end{pmatrix}, \quad a' = \begin{pmatrix} 1 & 0 & 0 \\ 0 & \omega^k & 0 \\ 0 & 0 & \omega^{2k} \end{pmatrix},$$

$$b = \begin{pmatrix} 0 & 1 & 0 \\ 0 & 0 & 1 \\ 1 & 0 & 0 \end{pmatrix}, \quad c = \begin{pmatrix} 0 & 0 & 1 \\ 0 & 1 & 0 \\ 1 & 0 & 0 \end{pmatrix}, \tag{13.78}$$

on $\mathbf{3}_{1k}$ for $k = 1, 2$. Obviously, the $\Delta(54)$ algebra is still satisfied when c is replaced by $-c$. That is, the generators, a, a', b and c, are represented by

$$a = \begin{pmatrix} \omega^k & 0 & 0 \\ 0 & \omega^{2k} & 0 \\ 0 & 0 & 1 \end{pmatrix}, \quad a' = \begin{pmatrix} 1 & 0 & 0 \\ 0 & \omega^k & 0 \\ 0 & 0 & \omega^{2k} \end{pmatrix},$$

$$b = \begin{pmatrix} 0 & 1 & 0 \\ 0 & 0 & 1 \\ 1 & 0 & 0 \end{pmatrix}, \quad c = \begin{pmatrix} 0 & 0 & -1 \\ 0 & -1 & 0 \\ -1 & 0 & 0 \end{pmatrix}, \tag{13.79}$$

on $\mathbf{3}_{2k}$ for $k = 1, 2$. Then, characters χ_3 for $\mathbf{3}_{1k}$ and $\mathbf{3}_{2k}$ are shown in Table 13.3.

Now, consider doublets. There are four doublets in $\Delta(54)$. The generators, a, a', b and c, are represented by

$$a = a' = \begin{pmatrix} 1 & 0 \\ 0 & 1 \end{pmatrix}, \quad b = \begin{pmatrix} \omega & 0 \\ 0 & \omega^2 \end{pmatrix}, \quad c = \begin{pmatrix} 0 & 1 \\ 1 & 0 \end{pmatrix}, \quad \text{on } \mathbf{2}_1, \tag{13.80}$$

$$a = a' = \begin{pmatrix} \omega^2 & 0 \\ 0 & \omega \end{pmatrix}, \quad b = \begin{pmatrix} \omega & 0 \\ 0 & \omega^2 \end{pmatrix}, \quad c = \begin{pmatrix} 0 & 1 \\ 1 & 0 \end{pmatrix}, \quad \text{on } \mathbf{2}_2, \tag{13.81}$$

$$a = a' = \begin{pmatrix} \omega & 0 \\ 0 & \omega^2 \end{pmatrix}, \quad b = \begin{pmatrix} \omega & 0 \\ 0 & \omega^2 \end{pmatrix}, \quad c = \begin{pmatrix} 0 & 1 \\ 1 & 0 \end{pmatrix}, \quad \text{on } \mathbf{2}_3, \tag{13.82}$$

$$a = a' = \begin{pmatrix} \omega & 0 \\ 0 & \omega^2 \end{pmatrix}, \quad b = \begin{pmatrix} 1 & 0 \\ 0 & 1 \end{pmatrix}, \quad c = \begin{pmatrix} 0 & 1 \\ 1 & 0 \end{pmatrix}, \quad \text{on } \mathbf{2}_4. \tag{13.83}$$

Then, characters χ_2 for $\mathbf{2}_{1,2,3,4}$ are shown in Table 13.3.

Table 13.3 Characters of $\Delta(54)$

	χ_{1_0}	χ_{1_1}	$\chi_{3_{1k}}$	$\chi_{3_{2k}}$	χ_{2_1}	χ_{2_2}	χ_{2_3}	χ_{2_4}
C_1	1	1	3	3	2	2	2	2
$C_1^{(1)}$	1	1	$3\omega^k$	$3\omega^{2k}$	2	2	2	2
$C_1^{(2)}$	1	1	$3\omega^{2k}$	$3\omega^k$	2	2	2	2
$C_6^{(0)}$	1	1	0	0	2	-1	-1	-1
$C_6^{(1,0)}$	1	1	0	0	-1	-1	-1	2
$C_6^{(1,1)}$	1	1	0	0	-1	2	-1	-1
$C_6^{(1,2)}$	1	1	0	0	-1	-1	2	-1
$C_9^{(0)}$	1	-1	1	-1	0	0	0	0
$C_9^{(1)}$	1	-1	ω^{2k}	$-\omega^{2k}$	0	0	0	0
$C_9^{(2)}$	1	-1	ω^k	$-\omega^k$	0	0	0	0

• Tensor products

The tensor products between triplets are given as follows:

$$
\begin{pmatrix} x_1 \\ x_2 \\ x_3 \end{pmatrix}_{3_{11}} \otimes \begin{pmatrix} y_1 \\ y_2 \\ y_3 \end{pmatrix}_{3_{11}} = \begin{pmatrix} x_1 y_1 \\ x_2 y_2 \\ x_3 y_3 \end{pmatrix}_{3_{12}} \oplus \begin{pmatrix} x_2 y_3 + x_3 y_2 \\ x_3 y_1 + x_1 y_3 \\ x_1 y_2 + x_2 y_1 \end{pmatrix}_{3_{12}} \oplus \begin{pmatrix} x_2 y_3 - x_3 y_2 \\ x_3 y_1 - x_1 y_3 \\ x_1 y_2 - x_2 y_1 \end{pmatrix}_{3_{22}} ,
$$
(13.84)

$$
\begin{pmatrix} x_1 \\ x_2 \\ x_3 \end{pmatrix}_{3_{12}} \otimes \begin{pmatrix} y_1 \\ y_2 \\ y_3 \end{pmatrix}_{3_{12}} = \begin{pmatrix} x_1 y_1 \\ x_2 y_2 \\ x_3 y_3 \end{pmatrix}_{3_{11}} \oplus \begin{pmatrix} x_2 y_3 + x_3 y_2 \\ x_3 y_1 + x_1 y_3 \\ x_1 y_2 + x_2 y_1 \end{pmatrix}_{3_{11}} \oplus \begin{pmatrix} x_2 y_3 - x_3 y_2 \\ x_3 y_1 - x_1 y_3 \\ x_1 y_2 - x_2 y_1 \end{pmatrix}_{3_{21}} ,
$$
(13.85)

$$
\begin{pmatrix} x_1 \\ x_2 \\ x_3 \end{pmatrix}_{3_{21}} \otimes \begin{pmatrix} y_1 \\ y_2 \\ y_3 \end{pmatrix}_{3_{21}} = \begin{pmatrix} x_1 y_1 \\ x_2 y_2 \\ x_3 y_3 \end{pmatrix}_{3_{12}} \oplus \begin{pmatrix} x_2 y_3 + x_3 y_2 \\ x_3 y_1 + x_1 y_3 \\ x_1 y_2 + x_2 y_1 \end{pmatrix}_{3_{12}} \oplus \begin{pmatrix} x_2 y_3 - x_3 y_2 \\ x_3 y_1 - x_1 y_3 \\ x_1 y_2 - x_2 y_1 \end{pmatrix}_{3_{22}} ,
$$
(13.86)

$$
\begin{pmatrix} x_1 \\ x_2 \\ x_3 \end{pmatrix}_{3_{22}} \otimes \begin{pmatrix} y_1 \\ y_2 \\ y_3 \end{pmatrix}_{3_{22}} = \begin{pmatrix} x_1 y_1 \\ x_2 y_2 \\ x_3 y_3 \end{pmatrix}_{3_{11}} \oplus \begin{pmatrix} x_2 y_3 + x_3 y_2 \\ x_3 y_1 + x_1 y_3 \\ x_1 y_2 + x_2 y_1 \end{pmatrix}_{3_{11}} \oplus \begin{pmatrix} x_2 y_3 - x_3 y_2 \\ x_3 y_1 - x_1 y_3 \\ x_1 y_2 - x_2 y_1 \end{pmatrix}_{3_{21}} ,
$$
(13.87)

$$\begin{pmatrix} x_1 \\ x_2 \\ x_3 \end{pmatrix}_{3_{11}} \otimes \begin{pmatrix} y_1 \\ y_2 \\ y_3 \end{pmatrix}_{3_{12}} = (x_1 y_1 + x_2 y_2 + x_3 y_3)_{1_0} \oplus \begin{pmatrix} x_1 y_1 + \omega^2 x_2 y_2 + \omega x_3 y_3 \\ \omega x_1 y_1 + \omega^2 x_2 y_2 + x_3 y_3 \end{pmatrix}_{2_1}$$

$$\oplus \begin{pmatrix} x_1 y_2 + \omega^2 x_2 y_3 + \omega x_3 y_1 \\ \omega x_1 y_3 + \omega^2 x_2 y_1 + x_3 y_2 \end{pmatrix}_{2_2} \oplus \begin{pmatrix} x_1 y_3 + \omega^2 x_2 y_1 + \omega x_3 y_2 \\ \omega x_1 y_2 + \omega^2 x_2 y_3 + x_3 y_1 \end{pmatrix}_{2_3}$$

$$\oplus \begin{pmatrix} x_1 y_3 + x_2 y_1 + x_3 y_2 \\ x_1 y_2 + x_2 y_3 + x_3 y_1 \end{pmatrix}_{2_4}, \tag{13.88}$$

$$\begin{pmatrix} x_1 \\ x_2 \\ x_3 \end{pmatrix}_{3_{11}} \otimes \begin{pmatrix} y_1 \\ y_2 \\ y_3 \end{pmatrix}_{3_{22}} = (x_1 y_1 + x_2 y_2 + x_3 y_3)_{1_1} \oplus \begin{pmatrix} x_1 y_1 + \omega^2 x_2 y_2 + \omega x_3 y_3 \\ -\omega x_1 y_1 - \omega^2 x_2 y_2 - x_3 y_3 \end{pmatrix}_{2_1}$$

$$\oplus \begin{pmatrix} x_1 y_2 + \omega^2 x_2 y_3 + \omega x_3 y_1 \\ -\omega x_1 y_3 - \omega^2 x_2 y_1 - x_3 y_2 \end{pmatrix}_{2_2} \oplus \begin{pmatrix} x_1 y_3 + \omega^2 x_2 y_1 + \omega x_3 y_2 \\ -\omega x_1 y_2 - \omega^2 x_2 y_3 - x_3 y_1 \end{pmatrix}_{2_3}$$

$$\oplus \begin{pmatrix} x_1 y_3 + x_2 y_1 + x_3 y_2 \\ -x_1 y_2 - x_2 y_3 - x_3 y_1 \end{pmatrix}_{2_4}, \tag{13.89}$$

$$\begin{pmatrix} x_1 \\ x_2 \\ x_3 \end{pmatrix}_{3_{12}} \otimes \begin{pmatrix} y_1 \\ y_2 \\ y_3 \end{pmatrix}_{3_{21}} = (x_1 y_1 + x_2 y_2 + x_3 y_3)_{1_1} \oplus \begin{pmatrix} x_1 y_1 + \omega^2 x_2 y_2 + \omega x_3 y_3 \\ -\omega x_1 y_1 - \omega^2 x_2 y_2 - x_3 y_3 \end{pmatrix}_{2_1}$$

$$\oplus \begin{pmatrix} x_1 y_3 + \omega^2 x_2 y_1 + \omega x_3 y_2 \\ -\omega x_1 y_2 - \omega^2 x_2 y_3 - x_3 y_1 \end{pmatrix}_{2_2} \oplus \begin{pmatrix} x_1 y_2 + \omega^2 x_2 y_3 + \omega x_3 y_1 \\ -\omega x_1 y_3 - \omega^2 x_2 y_1 - x_3 y_2 \end{pmatrix}_{2_3}$$

$$\oplus \begin{pmatrix} x_1 y_2 + x_2 y_3 + x_3 y_1 \\ -x_1 y_3 - x_2 y_1 - x_3 y_2 \end{pmatrix}_{2_4}. \tag{13.90}$$

The tensor products between doublets are given as follows:

$$\begin{pmatrix} x_1 \\ x_2 \end{pmatrix}_{2_k} \otimes \begin{pmatrix} y_1 \\ y_2 \end{pmatrix}_{2_k} = (x_1 y_2 + x_2 y_1)_{1_0} \oplus (x_1 y_2 - x_2 y_1)_{1_1} \oplus \begin{pmatrix} x_2 y_2 \\ x_1 y_1 \end{pmatrix}_{2_k}, \tag{13.91}$$

for $k = 1, 2, 3, 4$,

$$\begin{pmatrix} x_1 \\ x_2 \end{pmatrix}_{2_1} \otimes \begin{pmatrix} y_1 \\ y_2 \end{pmatrix}_{2_2} = \begin{pmatrix} x_2 y_2 \\ x_1 y_1 \end{pmatrix}_{2_3} \oplus \begin{pmatrix} x_1 y_2 \\ x_2 y_1 \end{pmatrix}_{2_4}, \tag{13.92}$$

$$\begin{pmatrix} x_1 \\ x_2 \end{pmatrix}_{2_1} \otimes \begin{pmatrix} y_1 \\ y_2 \end{pmatrix}_{2_3} = \begin{pmatrix} x_2 y_2 \\ x_1 y_1 \end{pmatrix}_{2_2} \oplus \begin{pmatrix} x_2 y_1 \\ x_1 y_2 \end{pmatrix}_{2_4}, \tag{13.93}$$

$$\begin{pmatrix} x_1 \\ x_2 \end{pmatrix}_{2_1} \otimes \begin{pmatrix} y_1 \\ y_2 \end{pmatrix}_{2_4} = \begin{pmatrix} x_1 y_2 \\ x_2 y_1 \end{pmatrix}_{2_2} \oplus \begin{pmatrix} x_1 y_1 \\ x_2 y_2 \end{pmatrix}_{2_3}, \tag{13.94}$$

$$\begin{pmatrix} x_1 \\ x_2 \end{pmatrix}_{2_2} \otimes \begin{pmatrix} y_1 \\ y_2 \end{pmatrix}_{2_3} = \begin{pmatrix} x_2 y_2 \\ x_1 y_1 \end{pmatrix}_{2_1} \oplus \begin{pmatrix} x_1 y_2 \\ x_2 y_1 \end{pmatrix}_{2_4}, \tag{13.95}$$

$$\begin{pmatrix} x_1 \\ x_2 \end{pmatrix}_{\mathbf{2_2}} \otimes \begin{pmatrix} y_1 \\ y_2 \end{pmatrix}_{\mathbf{2_4}} = \begin{pmatrix} x_1 y_1 \\ x_2 y_2 \end{pmatrix}_{\mathbf{2_1}} \oplus \begin{pmatrix} x_1 y_2 \\ x_2 y_1 \end{pmatrix}_{\mathbf{2_3}} , \tag{13.96}$$

$$\begin{pmatrix} x_1 \\ x_2 \end{pmatrix}_{\mathbf{2_3}} \otimes \begin{pmatrix} y_1 \\ y_2 \end{pmatrix}_{\mathbf{2_4}} = \begin{pmatrix} x_1 y_2 \\ x_2 y_1 \end{pmatrix}_{\mathbf{2_1}} \oplus \begin{pmatrix} x_1 y_1 \\ x_2 y_2 \end{pmatrix}_{\mathbf{2_2}} . \tag{13.97}$$

The tensor products between doublets and triplets are given as

$$\begin{pmatrix} x_1 \\ x_2 \end{pmatrix}_{\mathbf{2_1}} \otimes \begin{pmatrix} y_1 \\ y_2 \\ y_3 \end{pmatrix}_{\mathbf{3_{1k}}} = \begin{pmatrix} x_1 y_1 + \omega^2 x_2 y_1 \\ \omega x_1 y_2 + \omega x_2 y_2 \\ \omega^2 x_1 y_3 + x_2 y_3 \end{pmatrix}_{\mathbf{3_{1k}}} \oplus \begin{pmatrix} x_1 y_1 - \omega^2 x_2 y_1 \\ \omega x_1 y_2 - \omega x_2 y_2 \\ \omega^2 x_1 y_3 - x_2 y_3 \end{pmatrix}_{\mathbf{3_{2k}}} , \tag{13.98}$$

$$\begin{pmatrix} x_1 \\ x_2 \end{pmatrix}_{\mathbf{2_1}} \otimes \begin{pmatrix} y_1 \\ y_2 \\ y_3 \end{pmatrix}_{\mathbf{3_{2k}}} = \begin{pmatrix} x_1 y_1 + \omega^2 x_2 y_1 \\ \omega x_1 y_2 + \omega x_2 y_2 \\ \omega^2 x_1 y_3 + x_2 y_3 \end{pmatrix}_{\mathbf{3_{2k}}} \oplus \begin{pmatrix} x_1 y_1 - \omega^2 x_2 y_1 \\ \omega x_1 y_2 - \omega x_2 y_2 \\ \omega^2 x_1 y_3 - x_2 y_3 \end{pmatrix}_{\mathbf{3_{1k}}} , \tag{13.99}$$

$$\begin{pmatrix} x_1 \\ x_2 \end{pmatrix}_{\mathbf{2_2}} \otimes \begin{pmatrix} y_1 \\ y_2 \\ y_3 \end{pmatrix}_{\mathbf{3_{11}}} = \begin{pmatrix} \omega x_1 y_2 + x_2 y_3 \\ \omega^2 x_1 y_3 + \omega^2 x_2 y_1 \\ x_1 y_1 + \omega x_2 y_2 \end{pmatrix}_{\mathbf{3_{11}}} \oplus \begin{pmatrix} \omega x_1 y_2 - x_2 y_3 \\ \omega^2 x_1 y_3 - \omega^2 x_2 y_1 \\ x_1 y_1 - \omega x_2 y_2 \end{pmatrix}_{\mathbf{3_{21}}} , \tag{13.100}$$

$$\begin{pmatrix} x_1 \\ x_2 \end{pmatrix}_{\mathbf{2_2}} \otimes \begin{pmatrix} y_1 \\ y_2 \\ y_3 \end{pmatrix}_{\mathbf{3_{21}}} = \begin{pmatrix} \omega x_1 y_2 + x_2 y_3 \\ \omega^2 x_1 y_3 + \omega^2 x_2 y_1 \\ x_1 y_1 + \omega x_2 y_2 \end{pmatrix}_{\mathbf{3_{21}}} \oplus \begin{pmatrix} \omega x_1 y_2 - x_2 y_3 \\ \omega^2 x_1 y_3 - \omega^2 x_2 y_1 \\ x_1 y_1 - \omega x_2 y_2 \end{pmatrix}_{\mathbf{3_{11}}} , \tag{13.101}$$

$$\begin{pmatrix} x_1 \\ x_2 \end{pmatrix}_{\mathbf{2_2}} \otimes \begin{pmatrix} y_1 \\ y_2 \\ y_3 \end{pmatrix}_{\mathbf{3_{12}}} = \begin{pmatrix} \omega x_1 y_3 + x_2 y_2 \\ \omega^2 x_1 y_1 + \omega^2 x_2 y_3 \\ x_1 y_2 + \omega x_2 y_1 \end{pmatrix}_{\mathbf{3_{12}}} \oplus \begin{pmatrix} \omega x_1 y_3 - x_2 y_2 \\ \omega^2 x_1 y_1 - \omega^2 x_2 y_3 \\ x_1 y_2 - \omega x_2 y_1 \end{pmatrix}_{\mathbf{3_{22}}} , \tag{13.102}$$

$$\begin{pmatrix} x_1 \\ x_2 \end{pmatrix}_{2_2} \otimes \begin{pmatrix} y_1 \\ y_2 \\ y_3 \end{pmatrix}_{3_{22}} = \begin{pmatrix} \omega x_1 y_3 + x_2 y_2 \\ \omega^2 x_1 y_1 + \omega^2 x_2 y_3 \\ x_1 y_2 + \omega x_2 y_1 \end{pmatrix}_{3_{22}} \oplus \begin{pmatrix} \omega x_1 y_3 - x_2 y_2 \\ \omega^2 x_1 y_1 - \omega^2 x_2 y_3 \\ x_1 y_2 - \omega x_2 y_1 \end{pmatrix}_{3_{12}} ,$$

$$(13.103)$$

$$\begin{pmatrix} x_1 \\ x_2 \end{pmatrix}_{2_3} \otimes \begin{pmatrix} y_1 \\ y_2 \\ y_3 \end{pmatrix}_{3_{11}} = \begin{pmatrix} \omega x_1 y_3 + x_2 y_2 \\ \omega^2 x_1 y_1 + \omega^2 x_2 y_3 \\ x_1 y_2 + \omega x_2 y_1 \end{pmatrix}_{3_{11}} \oplus \begin{pmatrix} \omega x_1 y_3 - x_2 y_2 \\ \omega^2 x_1 y_1 - \omega^2 x_2 y_3 \\ x_1 y_2 - \omega x_2 y_1 \end{pmatrix}_{3_{21}} ,$$

$$(13.104)$$

$$\begin{pmatrix} x_1 \\ x_2 \end{pmatrix}_{2_3} \otimes \begin{pmatrix} y_1 \\ y_2 \\ y_3 \end{pmatrix}_{3_{21}} = \begin{pmatrix} \omega x_1 y_3 + x_2 y_2 \\ \omega^2 x_1 y_1 + \omega^2 x_2 y_3 \\ x_1 y_2 + \omega x_2 y_1 \end{pmatrix}_{3_{21}} \oplus \begin{pmatrix} \omega x_1 y_3 - x_2 y_2 \\ \omega^2 x_1 y_1 - \omega^2 x_2 y_3 \\ x_1 y_2 - \omega x_2 y_1 \end{pmatrix}_{3_{11}} ,$$

$$(13.105)$$

$$\begin{pmatrix} x_1 \\ x_2 \end{pmatrix}_{2_3} \otimes \begin{pmatrix} y_1 \\ y_2 \\ y_3 \end{pmatrix}_{3_{12}} = \begin{pmatrix} \omega x_1 y_2 + x_2 y_3 \\ \omega^2 x_1 y_3 + \omega^2 x_2 y_1 \\ x_1 y_1 + \omega x_2 y_2 \end{pmatrix}_{3_{12}} \oplus \begin{pmatrix} \omega x_1 y_2 - x_2 y_3 \\ \omega^2 x_1 y_3 - \omega^2 x_2 y_1 \\ x_1 y_1 - \omega x_2 y_2 \end{pmatrix}_{3_{22}} ,$$

$$(13.106)$$

$$\begin{pmatrix} x_1 \\ x_2 \end{pmatrix}_{2_3} \otimes \begin{pmatrix} y_1 \\ y_2 \\ y_3 \end{pmatrix}_{3_{22}} = \begin{pmatrix} \omega x_1 y_2 + x_2 y_3 \\ \omega^2 x_1 y_3 + \omega^2 x_2 y_1 \\ x_1 y_1 + \omega x_2 y_2 \end{pmatrix}_{3_{22}} \oplus \begin{pmatrix} \omega x_1 y_2 - x_2 y_3 \\ \omega^2 x_1 y_3 - \omega^2 x_2 y_1 \\ x_1 y_1 - \omega x_2 y_2 \end{pmatrix}_{3_{12}} ,$$

$$(13.107)$$

$$\begin{pmatrix} x_1 \\ x_2 \end{pmatrix}_{2_4} \otimes \begin{pmatrix} y_1 \\ y_2 \\ y_3 \end{pmatrix}_{3_{11}} = \begin{pmatrix} x_1 y_3 + x_2 y_2 \\ x_1 y_1 + x_2 y_3 \\ x_1 y_2 + x_2 y_1 \end{pmatrix}_{3_{11}} \oplus \begin{pmatrix} x_1 y_3 - x_2 y_2 \\ x_1 y_1 - x_2 y_3 \\ x_1 y_2 - x_2 y_1 \end{pmatrix}_{3_{21}} , \quad (13.108)$$

$$\begin{pmatrix} x_1 \\ x_2 \end{pmatrix}_{2_4} \otimes \begin{pmatrix} y_1 \\ y_2 \\ y_3 \end{pmatrix}_{3_{21}} = \begin{pmatrix} x_1 y_3 + x_2 y_2 \\ x_1 y_1 + x_2 y_3 \\ x_1 y_2 + x_2 y_1 \end{pmatrix}_{3_{21}} \oplus \begin{pmatrix} x_1 y_3 - x_2 y_2 \\ x_1 y_1 - x_2 y_3 \\ x_1 y_2 - x_2 y_1 \end{pmatrix}_{3_{11}} , \quad (13.109)$$

$$\begin{pmatrix} x_1 \\ x_2 \end{pmatrix}_{2_4} \otimes \begin{pmatrix} y_1 \\ y_2 \\ y_3 \end{pmatrix}_{3_{12}} = \begin{pmatrix} x_1 y_2 + x_2 y_3 \\ x_1 y_3 + x_2 y_1 \\ x_1 y_1 + x_2 y_2 \end{pmatrix}_{3_{12}} \oplus \begin{pmatrix} x_1 y_2 - x_2 y_3 \\ x_1 y_3 - x_2 y_1 \\ x_1 y_1 - x_2 y_2 \end{pmatrix}_{3_{22}} , \quad (13.110)$$

$$\begin{pmatrix} x_1 \\ x_2 \end{pmatrix}_{2_4} \otimes \begin{pmatrix} y_1 \\ y_2 \\ y_3 \end{pmatrix}_{3_{22}} = \begin{pmatrix} x_1 y_2 + x_2 y_3 \\ x_1 y_3 + x_2 y_1 \\ x_1 y_1 + x_2 y_2 \end{pmatrix}_{3_{22}} \oplus \begin{pmatrix} x_1 y_2 - x_2 y_3 \\ x_1 y_3 - x_2 y_1 \\ x_1 y_1 - x_2 y_2 \end{pmatrix}_{3_{12}} . \quad (13.111)$$

Furthermore, the tensor products of the non-trivial singlet $\mathbf{1}_1$ with other representations are given as

$$\mathbf{2}_k \otimes \mathbf{1}_1 = \mathbf{2}_k, \qquad \mathbf{3}_{1k} \otimes \mathbf{1}_1 = \mathbf{3}_{2k}, \qquad \mathbf{3}_{2k} \otimes \mathbf{1}_1 = \mathbf{3}_{1k}. \qquad (13.112)$$

13.4 $\Delta(96)$

We present a simple example of $\Delta(6N^2)$ in case of $N/3 \neq$ integer, that is $\Delta(96)$. It is isomorphic to $(Z_4 \times Z_4') \rtimes S_3$. (See also Refs. [2,3].) Let us denote the generators of Z_4 and Z_4' by a and a', respectively. We denote S_3 generators by b and c, where b and c are Z_3 and Z_2 generators of S_3, respectively. These generators satisfy

$$\begin{aligned} a^4 &= a'^4 = b^3 = c^2 = (bc)^2 = e, \quad aa' = a'a, \\ bab^{-1} &= a^{-1}(a')^{-1}, \quad ba'b^{-1} = a, \\ cac^{-1} &= (a')^{-1}, \quad ca'c^{-1} = a^{-1}. \end{aligned} \qquad (13.113)$$

Using them, all of $\Delta(96^2)$ elements are written as

$$g = b^k c^\ell a^m a'^n, \qquad (13.114)$$

for $k = 0, 1, 2, \ell = 0, 1$ and $m, n = 0, 1, 2, 3$.

It is noticed that the $\Delta(96)$ group includes the subgroup, $\Delta(48)$, whose elements are written by $b^k a^m a'^n$. Thus, some group-theoretical aspects of $\Delta(96)$ can be derived from those of $\Delta(48)$.

13.4.1 Conjugacy Classes

Here, we summarize the conjugacy classes of $\Delta(96)$ as follows: conjugacy classes,

$$
\begin{aligned}
C_1 : & \quad \{e\}, \\
C_3^{(1)} : & \quad \{aa'^3, a^2a'^3, aa'^2\}, \\
C_3^{(2)} : & \quad \{a^2a'^2, a'^2, a^2\}, \\
C_3^{(3)} : & \quad \{a^3a', a^2a', a^3a'^2\}, \\
C_6^{(1,1)} : & \quad \{aa', a'^3, a^3, a^3a'^3, a', a\}, \\
C_{32} : & \quad \{b^k a^\ell a'^m \mid k = 1, 2, \ \ell, m = 0, 1, 2, 3\}, \\
C_{12}^{(1)} : & \quad \{ca^{1+n}a'^n, \ b^2ca^{-1}a'^{-1-n}, \ bca^{-n}a' \mid n = 0, 1, 2, 3\}, \\
C_{12}^{(2)} : & \quad \{ca^{2+n}a'^n, \ b^2ca^2a'^{-2-n}, \ bca^{-n}a'^2 \mid n = 0, 1, 2, 3\}, \\
C_{12}^{(3)} : & \quad \{ca^{3+n}a'^n, \ b^2caa'^{-3-n}, \ bca^{-n}a'^3 \mid n = 0, 1, 2, 3\}, \\
C_{12}^{(0)} : & \quad \{ca^n a'^n, \ b^2ca'^{-n}, \ bca^{-n} \mid n = 0, 1, 2, 3\}.
\end{aligned}
$$

The order, h, of each element in the conjugacy classes is found as

$$
\begin{aligned}
C_1 : & \quad h = 1, \\
C_3^{(1)} : & \quad h = 4, \\
C_3^{(2)} : & \quad h = 2, \\
C_3^{(3)} : & \quad h = 4, \\
C_6^{(1,1)} : & \quad h = 4, \\
C_{32} : & \quad h = 3, \\
C_{12}^{(1)} : & \quad h = 8, \\
C_{12}^{(2)} : & \quad h = 4, \\
C_{12}^{(3)} : & \quad h = 8, \\
C_{12}^{(0)} : & \quad h = 2.
\end{aligned}
\qquad (13.115)
$$

The numbers of the conjugacy classes $C_3^{(k)}$, $C_6^{(1,1)}$, and $C_{12}^{(k)}$ are equal to 3, 1, and 4, respectively. Then, the total number of conjugacy classes is equal to 10. The relations (2.18) and (2.19) for $\Delta(96)$ give

$$
m_1 + 2^2 m_2 + 3^2 m_3 + 4^2 m_4 + 5^2 m_5 + 6^2 m_6 = 96, \qquad (13.116)
$$

$$
m_1 + m_2 + m_3 + m_4 + m_5 + m_6 = 10. \qquad (13.117)
$$

We have the solution $(m_1, m_2, m_3, m_6) = (2, 1, 6, 1)$.

13.4.2 Characters and Representations

Now, we study characters and representations. The $\Delta(96)$ group has two singlets and one doublet. These are nothing but irreducible representations of S_3. Thus, on these representations, the Z_4 and Z_4' generators, a and a', are identity matrices.

Because $c^2 = e$, characters of two singlets have two possible values $\chi_{1_k}(c) = (-1)^k$ with $k = 0, 1$ and they correspond to two singlets, $\mathbf{1}_k$. Note that $\chi_{1_k}(a) = \chi_{1_k}(a') = \chi_{1_k}(b) = 1$. Similarly to S_3, the characters of the doublet $\mathbf{2}$ are obtained as

$$\chi_2(C_{32}) = -1, \qquad \chi_2(C_{12}^{(k)}) = 0, \tag{13.118}$$

as well as $\chi_2(a) = \chi_2(a') = 2$. Then, the two-dimensional representations are obtained, e.g.

$$b(2) = \begin{pmatrix} \omega & 0 \\ 0 & \omega^2 \end{pmatrix}, \quad c(2) = \begin{pmatrix} 0 & 1 \\ 1 & 0 \end{pmatrix}, \quad a(2) = a'(2) = \begin{pmatrix} 1 & 0 \\ 0 & 1 \end{pmatrix}. \tag{13.119}$$

Next, let us consider sextet representations. We can obtain (6×6) matrix representations for generic sextet. The $\Delta(96)$ group is represented as

$$b = \begin{pmatrix} b_1 & 0 \\ 0 & b_2 \end{pmatrix}, \quad c = \begin{pmatrix} 0 & 1 \\ 1 & 0 \end{pmatrix}, \quad a = \begin{pmatrix} a_1 & 0 \\ 0 & a_2 \end{pmatrix}, \quad a' = \begin{pmatrix} a'_1 & 0 \\ 0 & a'_2 \end{pmatrix}, \tag{13.120}$$

where

$$b_1 = \begin{pmatrix} 0 & 1 & 0 \\ 0 & 0 & 1 \\ 1 & 0 & 0 \end{pmatrix}, \qquad b_2 = \begin{pmatrix} 0 & 0 & 1 \\ 1 & 0 & 0 \\ 0 & 1 & 0 \end{pmatrix}, \tag{13.121}$$

$$a_1 = a_2'^{-1} = \begin{pmatrix} \rho^l & 0 & 0 \\ 0 & \rho^k & 0 \\ 0 & 0 & \rho^{-l-k} \end{pmatrix}, \qquad a_2 = a_1'^{-1} = \begin{pmatrix} \rho^{l+k} & 0 & 0 \\ 0 & \rho^{-l} & 0 \\ 0 & 0 & \rho^{-k} \end{pmatrix}, \tag{13.122}$$

on the sextet $\mathbf{6}_{[[k],[\ell]]}$ with $(k, \ell) \neq (0, 0)$, where $\rho \equiv e^{2\pi i/4}$ and $[[k], [\ell]]$ denotes[2]

$$[[k], [\ell]] = (k, \ell), \quad (-k-\ell, k), \quad (\ell, -k-\ell),$$
$$(-\ell, -k), \quad (k+\ell, -\ell), \quad \text{or } (-k, k+\ell). \tag{13.123}$$

We also denote the vector of $\mathbf{6}_{[[k],[\ell]]}$ as

$$\mathbf{6}_{[[k],[\ell]]} = \begin{pmatrix} x_{\ell,-k-\ell} \\ x_{k,\ell} \\ x_{-k-\ell,k} \\ x_{k+\ell,-\ell} \\ x_{-\ell,-k} \\ x_{-k,k+\ell} \end{pmatrix}, \tag{13.124}$$

for $k, \ell = 0, 1, 2, 3$, where k and ℓ correspond to Z_4 and Z'_4 charges, respectively.

---

[2] The notation $[[k], [\ell]]$ corresponds to $\widetilde{(k, \ell)}$ in Ref. [1].

Table 13.4 Characters of $\Delta(96)$

	h	χ_{1_1}	χ_{1_2}	χ_2	$\chi_{3_{21}}$	$\chi_{3_{22}}$	$\chi_{3_{23}}$	$\chi_{3_{11}}$	$\chi_{3_{12}}$	$\chi_{3_{13}}$	$\chi_{6_{[[k],[\ell]]}}$
C_1	1	1	1	2	3	3	3	3	3	3	6
$C_3^{(1)}$	4	1	1	2	$-1+2i$	-1	$-1-2i$	$-1+2i$	-1	$-1-2i$	2
$C_3^{(2)}$	2	1	1	2	-1	3	-1	-1	3	-1	-2
$C_3^{(3)}$	4	1	1	2	$-1-2i$	-1	$-1+2i$	$-1-2i$	-1	$-1+2i$	2
$C_6^{(1,1)}$	4	1	1	2	1	-1	1	1	-1	1	-2
C_{32}	3	1	1	-1	0	0	0	0	0	0	0
$C_{12}^{(0)}$	2	1	-1	0	-1	-1	-1	1	1	1	0
$C_{12}^{(1)}$	8	1	-1	0	i	1	$-i$	$-i$	-1	i	0
$C_{12}^{(2)}$	4	1	-1	0	1	-1	1	-1	1	-1	0
$C_{12}^{(3)}$	8	1	-1	0	$-i$	1	i	i	-1	$-i$	0

In a certain case, the above representation becomes reducible. Notice here that $6_{[[k],[\ell]]}$ is reducible if $k + \ell = 0 \ (\mathrm{mod}\ 4)$, $k = 0$, $\ell \neq 0$, and $\ell = 0$, $k \neq 0$, then the number of irreducible sextet representations is 1.

On the other hand, we diagonalize the above reducible six-dimensional representations so as to obtain 6 irreducible triplets, 3_{1k} and 3_{2k}, with $k = 1, 2, 3$, where the generators are represented by

$$b(3_{1k}) = \begin{pmatrix} 0 & 1 & 0 \\ 0 & 0 & 1 \\ 1 & 0 & 0 \end{pmatrix}, \ c(3_{1k}) = \begin{pmatrix} 0 & 0 & 1 \\ 0 & 1 & 0 \\ 1 & 0 & 0 \end{pmatrix}, \ a(3_{1k}) = \begin{pmatrix} \rho^k & 0 & 0 \\ 0 & \rho^{-k} & 0 \\ 0 & 0 & 1 \end{pmatrix}, \ a'(3_{1k}) = \begin{pmatrix} 1 & 0 & 0 \\ 0 & \rho^k & 0 \\ 0 & 0 & \rho^{-k} \end{pmatrix},$$
$$(13.125)$$

$$b(3_{2k}) = b(3_{1k}), \quad c(3_{2k}) = -c(3_{1k}), \quad a(3_{2k}) = a(3_{1k}), \quad a'(3_{2k}) = a'(3_{1k}). \quad (13.126)$$

We also denote the vector of 3_{1k} and 3_{2k} as

$$3_{1k} = \begin{pmatrix} x_{k,0} \\ x_{-k,k} \\ x_{0,-k} \end{pmatrix}, \quad 3_{2k} = \begin{pmatrix} x_{k,0} \\ x_{-k,k} \\ x_{0,-k} \end{pmatrix}, \quad (13.127)$$

for $k = 1, 2, 3$.

The characters are shown in Table 13.4.

13.4.3 Tensor Products

The tensor products are given as follows:

$$(x)_{1'} \otimes \begin{pmatrix} y_1 \\ y_2 \end{pmatrix}_2 = \begin{pmatrix} xy_1 \\ -xy_2 \end{pmatrix}_2, \quad (x)_{1'} \otimes \begin{pmatrix} y_1 \\ y_2 \\ y_3 \\ y_4 \\ y_5 \\ y_6 \end{pmatrix}_6 = \begin{pmatrix} xy_1 \\ xy_2 \\ xy_3 \\ -xy_4 \\ -xy_5 \\ -xy_6 \end{pmatrix}_6 \quad (13.128)$$

$$(x)_{1'} \otimes \begin{pmatrix} y_1 \\ y_2 \\ y_3 \end{pmatrix}_{3_{21,22,23}} = \begin{pmatrix} xy_1 \\ xy_2 \\ xy_3 \end{pmatrix}_{3_{11,12,13}} \,, \quad (x)_{1'} \otimes \begin{pmatrix} y_1 \\ y_2 \\ y_3 \end{pmatrix}_{3_{11,12,13}} = \begin{pmatrix} xy_1 \\ xy_2 \\ xy_3 \end{pmatrix}_{3_{21,22,23}} \tag{13.129}$$

$$\begin{pmatrix} x_1 \\ x_2 \end{pmatrix}_2 \otimes \begin{pmatrix} y_1 \\ y_2 \end{pmatrix}_2 = (x_1 y_2 + x_2 y_1)_1 \oplus (x_1 y_2 - x_2 y_1)_{1'} + \oplus \begin{pmatrix} x_2 y_2 \\ x_1 y_1 \end{pmatrix}_2 \,, \tag{13.130}$$

$$\begin{pmatrix} x_1 \\ x_2 \end{pmatrix}_2 \otimes \begin{pmatrix} y_1 \\ y_2 \\ y_3 \\ y_4 \\ y_5 \\ y_6 \end{pmatrix}_6 = \begin{pmatrix} x_2 y_1 \\ \omega^2 x_2 y_2 \\ \omega x_2 y_3 \\ x_1 y_4 \\ \omega^2 x_1 y_5 \\ \omega x_1 y_6 \end{pmatrix}_6 \oplus \begin{pmatrix} x_1 y_1 \\ \omega x_1 y_2 \\ \omega^2 x_1 y_3 \\ x_2 y_4 \\ \omega x_2 y_5 \\ \omega^2 x_2 y_6 \end{pmatrix}_6 \,, \tag{13.131}$$

$$\begin{pmatrix} x_1 \\ x_2 \end{pmatrix}_2 \otimes \begin{pmatrix} y_1 \\ y_2 \\ y_3 \end{pmatrix}_{3_{21,22,23}} = \begin{pmatrix} (x_1 + \omega^2 x_2) y_1 \\ \omega(x_1 + x_2) y_2 \\ (\omega^2 x_1 + x_2) y_3 \end{pmatrix}_{3_{21,22,23}} \oplus \begin{pmatrix} (x_1 - \omega^2 x_2) y_1 \\ \omega(x_1 - x_2) y_2 \\ (\omega^2 x_1 - x_2) y_3 \end{pmatrix}_{3_{11,12,13}} \,, \tag{13.132}$$

$$\begin{pmatrix} x_1 \\ x_2 \end{pmatrix}_2 \otimes \begin{pmatrix} y_1 \\ y_2 \\ y_3 \end{pmatrix}_{3_{11,12,13}} = \begin{pmatrix} (x_1 - \omega^2 x_2) y_1 \\ \omega(x_1 - x_2) y_2 \\ (\omega^2 x_1 - x_2) y_3 \end{pmatrix}_{3_{11,22,23}} \oplus \begin{pmatrix} (x_1 + \omega^2 x_2) y_1 \\ \omega(x_1 + x_2) y_2 \\ (\omega^2 x_1 + x_2) y_3 \end{pmatrix}_{3_{21,12,13}} \,, \tag{13.133}$$

$$\begin{pmatrix} x_1 \\ x_2 \\ x_3 \end{pmatrix}_{3_{21(11)}} \otimes \begin{pmatrix} y_1 \\ y_2 \\ y_3 \end{pmatrix}_{3_{21(11)}} = \begin{pmatrix} x_1 y_1 \\ x_2 y_2 \\ x_3 y_3 \end{pmatrix}_{3_{12}} \oplus \begin{pmatrix} x_3 y_2 - x_2 y_3 \\ x_1 y_3 - x_3 y_1 \\ x_2 y_1 - x_1 y_2 \end{pmatrix}_{3_{23}} \oplus \begin{pmatrix} x_3 y_2 + x_2 y_3 \\ x_1 y_3 + x_3 y_1 \\ x_2 y_1 + x_1 y_2 \end{pmatrix}_{3_{13}} \,, \tag{13.134}$$

$$\begin{pmatrix} x_1 \\ x_2 \\ x_3 \end{pmatrix}_{3_{21}} \otimes \begin{pmatrix} y_1 \\ y_2 \\ y_3 \end{pmatrix}_{3_{11}} = \begin{pmatrix} x_1 y_1 \\ x_2 y_2 \\ x_3 y_3 \end{pmatrix}_{3_{22}} \oplus \begin{pmatrix} x_3 y_2 + x_2 y_3 \\ x_1 y_3 + x_3 y_1 \\ x_2 y_1 + x_1 y_2 \end{pmatrix}_{3_{13}} \oplus \begin{pmatrix} x_3 y_2 - x_2 y_3 \\ x_1 y_3 - x_3 y_1 \\ x_2 y_1 - x_1 y_2 \end{pmatrix}_{3_{23}} ,$$

$$(13.135)$$

$$\begin{pmatrix} x_1 \\ x_2 \\ x_3 \end{pmatrix}_{3_{21(11)}} \otimes \begin{pmatrix} y_1 \\ y_2 \\ y_3 \end{pmatrix}_{3_{22(12)}} = \begin{pmatrix} x_1 y_1 \\ x_2 y_2 \\ x_3 y_3 \end{pmatrix}_{3_{13}} \oplus \begin{pmatrix} x_1 y_3 \\ x_2 y_1 \\ x_3 y_2 \\ x_3 y_1 \\ x_2 y_3 \\ x_1 y_2 \end{pmatrix}_{6} ,$$

$$(13.136)$$

$$\begin{pmatrix} x_1 \\ x_2 \\ x_3 \end{pmatrix}_{3_{21(11)}} \otimes \begin{pmatrix} y_1 \\ y_2 \\ y_3 \end{pmatrix}_{3_{12(22)}} = \begin{pmatrix} x_1 y_1 \\ x_2 y_2 \\ x_3 y_3 \end{pmatrix}_{3_{23}} \oplus \begin{pmatrix} x_1 y_3 \\ x_2 y_1 \\ x_3 y_2 \\ -x_3 y_1 \\ -x_2 y_3 \\ -x_1 y_2 \end{pmatrix}_{6} ,$$

$$(13.137)$$

$$\begin{pmatrix} x_1 \\ x_2 \\ x_3 \end{pmatrix}_{3_{21(11)}} \otimes \begin{pmatrix} y_1 \\ y_2 \\ y_3 \end{pmatrix}_{3_{23(13)}} = (x_1 y_1 + x_2 y_2 + x_3 y_3)_{\mathbf{1}} \qquad (13.138)$$

$$\oplus \begin{pmatrix} x_1 y_1 + \omega^2 x_2 y_2 + \omega x_3 y_3 \\ \omega x_1 y_1 + \omega^2 x_2 y_2 + x_3 y_3 \end{pmatrix}_{\mathbf{2}} \oplus \begin{pmatrix} x_3 y_2 \\ x_1 y_3 \\ x_2 y_1 \\ x_1 y_2 \\ x_3 y_1 \\ x_2 y_3 \end{pmatrix}_{6} ,$$

$$
\begin{pmatrix} x_1 \\ x_2 \\ x_3 \end{pmatrix}_{3_{21(11)}} \otimes \begin{pmatrix} y_1 \\ y_2 \\ y_3 \end{pmatrix}_{3_{13(23)}} = (x_1 y_1 + x_2 y_2 + x_3 y_3)_{\mathbf{1'}} \tag{13.139}
$$

$$
\oplus \begin{pmatrix} x_1 y_1 + \omega^2 x_2 y_2 + \omega x_3 y_3 \\ -\omega x_1 y_1 - \omega^2 x_2 y_2 - x_3 y_3 \end{pmatrix}_{\mathbf{2}} \oplus \begin{pmatrix} x_3 y_2 \\ x_1 y_3 \\ x_2 y_1 \\ -x_1 y_2 \\ -x_3 y_1 \\ -x_2 y_3 \end{pmatrix}_{\mathbf{6}},
$$

$$
\begin{pmatrix} x_1 \\ x_2 \\ x_3 \end{pmatrix}_{3_{22(12)}} \otimes \begin{pmatrix} y_1 \\ y_2 \\ y_3 \end{pmatrix}_{3_{22(12)}} = (x_1 y_1 + x_2 y_2 + x_3 y_3)_{\mathbf{1}} \oplus \begin{pmatrix} x_1 y_1 + \omega^2 x_2 y_2 + \omega x_3 y_3 \\ \omega x_1 y_1 + \omega^2 x_2 y_2 + x_3 y_3 \end{pmatrix}_{\mathbf{2}}
$$

$$
\oplus \begin{pmatrix} x_3 y_2 - x_2 y_3 \\ x_1 y_3 - x_3 y_1 \\ x_2 y_1 - x_1 y_2 \end{pmatrix}_{3_{22}} \oplus \begin{pmatrix} x_3 y_2 + x_2 y_3 \\ x_1 y_3 + x_3 y_1 \\ x_2 y_1 + x_1 y_2 \end{pmatrix}_{3_{12}}, \tag{13.140}
$$

$$
\begin{pmatrix} x_1 \\ x_2 \\ x_3 \end{pmatrix}_{3_{12}} \otimes \begin{pmatrix} y_1 \\ y_2 \\ y_3 \end{pmatrix}_{3_{22}} = (x_1 y_1 + x_2 y_2 + x_3 y_3)_{\mathbf{1'}} \oplus \begin{pmatrix} x_1 y_1 + \omega^2 x_2 y_2 + \omega x_3 y_3 \\ -\omega x_1 y_1 - \omega^2 x_2 y_2 - x_3 y_3 \end{pmatrix}_{\mathbf{2}}
$$

$$
\oplus \begin{pmatrix} x_3 y_2 - x_2 y_3 \\ x_1 y_3 - x_3 y_1 \\ x_2 y_1 - x_1 y_2 \end{pmatrix}_{3_{12}} \oplus \begin{pmatrix} x_3 y_2 + x_2 y_3 \\ x_1 y_3 + x_3 y_1 \\ x_2 y_1 + x_1 y_2 \end{pmatrix}_{3_{22}}, \tag{13.141}
$$

$$
\begin{pmatrix} x_1 \\ x_2 \\ x_3 \end{pmatrix}_{3_{22(12)}} \otimes \begin{pmatrix} y_1 \\ y_2 \\ y_3 \end{pmatrix}_{3_{23(13)}} = \begin{pmatrix} x_1 y_1 \\ x_2 y_2 \\ x_3 y_3 \end{pmatrix}_{3_{11}} \oplus \begin{pmatrix} x_2 y_1 \\ x_3 y_2 \\ x_1 y_3 \\ x_2 y_3 \\ x_1 y_2 \\ x_3 y_1 \end{pmatrix}_{\mathbf{6}}, \tag{13.142}
$$

$$
\begin{pmatrix} x_1 \\ x_2 \\ x_3 \end{pmatrix}_{3_{22(12)}} \otimes \begin{pmatrix} y_1 \\ y_2 \\ y_3 \end{pmatrix}_{3_{13(23)}} = \begin{pmatrix} x_1 y_1 \\ x_2 y_2 \\ x_3 y_3 \end{pmatrix}_{3_{21}} \oplus \begin{pmatrix} x_2 y_1 \\ x_3 y_2 \\ x_1 y_3 \\ -x_2 y_3 \\ -x_1 y_2 \\ -x_3 y_1 \end{pmatrix}_{\mathbf{6}}, \tag{13.143}
$$

$$\begin{pmatrix} x_1 \\ x_2 \\ x_3 \end{pmatrix}_{\mathbf{3}_{23(13)}} \otimes \begin{pmatrix} y_1 \\ y_2 \\ y_3 \end{pmatrix}_{\mathbf{3}_{23(13)}} = \begin{pmatrix} x_3 y_2 - x_2 y_3 \\ x_1 y_3 - x_3 y_1 \\ x_2 y_1 - x_1 y_2 \end{pmatrix}_{\mathbf{3}_{21}} \oplus \begin{pmatrix} x_3 y_2 + x_2 y_3 \\ x_1 y_3 + x_3 y_1 \\ x_2 y_1 + x_1 y_2 \end{pmatrix}_{\mathbf{3}_{11}} \oplus \begin{pmatrix} x_1 y_1 \\ x_2 y_2 \\ x_3 y_3 \end{pmatrix}_{\mathbf{3}_{12}} . \quad (13.144)$$

$$\begin{pmatrix} x_1 \\ x_2 \\ x_3 \end{pmatrix}_{\mathbf{3}_{13}} \otimes \begin{pmatrix} y_1 \\ y_2 \\ y_3 \end{pmatrix}_{\mathbf{3}_{23}} = \begin{pmatrix} x_3 y_2 - x_2 y_3 \\ x_1 y_3 - x_3 y_1 \\ x_2 y_1 - x_1 y_2 \end{pmatrix}_{\mathbf{3}_{11}} \oplus \begin{pmatrix} x_3 y_2 + x_2 y_3 \\ x_1 y_3 + x_3 y_1 \\ x_2 y_1 + x_1 y_2 \end{pmatrix}_{\mathbf{3}_{21}} \oplus \begin{pmatrix} x_1 y_1 \\ x_2 y_2 \\ x_3 y_3 \end{pmatrix}_{\mathbf{3}_{22}} , \quad (13.145)$$

$$\begin{pmatrix} x_1 \\ x_2 \\ x_3 \end{pmatrix}_{\mathbf{3}_{21}} \otimes \begin{pmatrix} y_1 \\ y_2 \\ y_3 \\ y_4 \\ y_5 \\ y_6 \end{pmatrix}_{\mathbf{6}} = \begin{pmatrix} x_3 y_2 + x_2 y_4 \\ x_1 y_3 + x_3 y_6 \\ x_2 y_1 + x_1 y_5 \end{pmatrix}_{\mathbf{3}_{21}} \oplus \begin{pmatrix} x_3 y_3 + x_2 y_5 \\ x_1 y_1 + x_3 y_4 \\ x_2 y_2 + x_1 y_6 \end{pmatrix}_{\mathbf{3}_{22}} \quad (13.146)$$

$$\oplus \begin{pmatrix} x_3 y_2 - x_2 y_4 \\ x_1 y_3 - x_3 y_6 \\ x_2 y_1 - x_1 y_5 \end{pmatrix}_{\mathbf{3}_{11}} \oplus \begin{pmatrix} x_3 y_3 - x_2 y_5 \\ x_1 y_1 - x_3 y_4 \\ x_2 y_2 - x_1 y_6 \end{pmatrix}_{\mathbf{3}_{12}} \oplus \begin{pmatrix} x_2 y_3 \\ x_3 y_1 \\ x_1 y_2 \\ -x_2 y_6 \\ -x_1 y_4 \\ -x_3 y_5 \end{pmatrix}_{\mathbf{6}} ,$$

$$\begin{pmatrix} x_1 \\ x_2 \\ x_3 \end{pmatrix}_{\mathbf{3}_{11}} \otimes \begin{pmatrix} y_1 \\ y_2 \\ y_3 \\ y_4 \\ y_5 \\ y_6 \end{pmatrix}_{\mathbf{6}} = \begin{pmatrix} x_3 y_2 + x_2 y_4 \\ x_1 y_3 + x_3 y_6 \\ x_2 y_1 + x_1 y_5 \end{pmatrix}_{\mathbf{3}_{11}} \oplus \begin{pmatrix} x_3 y_3 + x_2 y_5 \\ x_1 y_1 + x_3 y_4 \\ x_2 y_2 + x_1 y_6 \end{pmatrix}_{\mathbf{3}_{12}} \quad (13.147)$$

$$\oplus \begin{pmatrix} x_3 y_2 - x_2 y_4 \\ x_1 y_3 - x_3 y_6 \\ x_2 y_1 - x_1 y_5 \end{pmatrix}_{\mathbf{3}_{21}} \oplus \begin{pmatrix} x_3 y_3 - x_2 y_5 \\ x_1 y_1 - x_3 y_4 \\ x_2 y_2 - x_1 y_6 \end{pmatrix}_{\mathbf{3}_{22}} \oplus \begin{pmatrix} x_2 y_3 \\ x_3 y_1 \\ x_1 y_2 \\ x_2 y_6 \\ x_1 y_4 \\ x_3 y_5 \end{pmatrix}_{\mathbf{6}} ,$$

$$\begin{pmatrix} x_1 \\ x_2 \\ x_3 \end{pmatrix}_{3_{22}} \otimes \begin{pmatrix} y_1 \\ y_2 \\ y_3 \\ y_4 \\ y_5 \\ y_6 \end{pmatrix}_6 = \begin{pmatrix} x_3 y_1 + x_2 y_6 \\ x_1 y_2 + x_3 y_5 \\ x_2 y_3 + x_1 y_4 \end{pmatrix}_{3_{21}} \oplus \begin{pmatrix} x_2 y_1 + x_3 y_6 \\ x_3 y_2 + x_1 y_5 \\ x_1 y_3 + x_2 y_4 \end{pmatrix}_{3_{23}} \tag{13.148}$$

$$\oplus \begin{pmatrix} x_3 y_1 - x_2 y_6 \\ x_1 y_2 - x_3 y_5 \\ x_2 y_3 - x_1 y_4 \end{pmatrix}_{3_{11}} \oplus \begin{pmatrix} x_2 y_1 - x_3 y_6 \\ x_3 y_2 - x_1 y_5 \\ x_1 y_3 - x_2 y_4 \end{pmatrix}_{3_{13}} \oplus \begin{pmatrix} x_1 y_6 \\ x_2 y_5 \\ x_3 y_4 \\ -x_3 y_3 \\ -x_2 y_2 \\ -x_1 y_1 \end{pmatrix}_6 ,$$

$$\begin{pmatrix} x_1 \\ x_2 \\ x_3 \end{pmatrix}_{3_{12}} \otimes \begin{pmatrix} y_1 \\ y_2 \\ y_3 \\ y_4 \\ y_5 \\ y_6 \end{pmatrix}_6 = \begin{pmatrix} x_3 y_1 + x_2 y_6 \\ x_1 y_2 + x_3 y_5 \\ x_2 y_3 + x_1 y_4 \end{pmatrix}_{3_{11}} \oplus \begin{pmatrix} x_2 y_1 + x_3 y_6 \\ x_3 y_2 + x_1 y_5 \\ x_1 y_3 + x_2 y_4 \end{pmatrix}_{3_{13}} \tag{13.149}$$

$$\oplus \begin{pmatrix} x_3 y_1 - x_2 y_6 \\ x_1 y_2 - x_3 y_5 \\ x_2 y_3 - x_1 y_4 \end{pmatrix}_{3_{21}} \oplus \begin{pmatrix} x_2 y_1 - x_3 y_6 \\ x_3 y_2 - x_1 y_5 \\ x_1 y_3 - x_2 y_4 \end{pmatrix}_{3_{23}} \oplus \begin{pmatrix} x_1 y_6 \\ x_2 y_5 \\ x_3 y_4 \\ x_3 y_3 \\ x_2 y_2 \\ x_1 y_1 \end{pmatrix}_6 ,$$

$$\begin{pmatrix} x_1 \\ x_2 \\ x_3 \end{pmatrix}_{3_{23}} \otimes \begin{pmatrix} y_1 \\ y_2 \\ y_3 \\ y_4 \\ y_5 \\ y_6 \end{pmatrix}_6 = \begin{pmatrix} x_2 y_2 + x_3 y_4 \\ x_3 y_3 + x_1 y_6 \\ x_1 y_1 + x_2 y_5 \end{pmatrix}_{3_{22}} \oplus \begin{pmatrix} x_2 y_3 + x_3 y_5 \\ x_3 y_1 + x_1 y_4 \\ x_1 y_2 + x_2 y_6 \end{pmatrix}_{3_{23}} \tag{13.150}$$

$$\oplus \begin{pmatrix} x_2 y_2 - x_3 y_4 \\ x_3 y_3 - x_1 y_6 \\ x_1 y_1 - x_2 y_5 \end{pmatrix}_{3_{12}} \oplus \begin{pmatrix} x_2 y_3 - x_3 y_5 \\ x_3 y_1 - x_1 y_4 \\ x_1 y_2 - x_2 y_6 \end{pmatrix}_{3_{13}} \oplus \begin{pmatrix} x_3 y_2 \\ x_1 y_3 \\ x_2 y_1 \\ -x_1 y_5 \\ -x_3 y_6 \\ -x_2 y_4 \end{pmatrix}_6 ,$$

$$
\begin{pmatrix} x_1 \\ x_2 \\ x_3 \end{pmatrix}_{\mathbf{3}_{13}} \otimes \begin{pmatrix} y_1 \\ y_2 \\ y_3 \\ y_4 \\ y_5 \\ y_6 \end{pmatrix}_{\mathbf{6}} = \begin{pmatrix} x_2 y_2 + x_3 y_4 \\ x_3 y_3 + x_1 y_6 \\ x_1 y_1 + x_2 y_5 \end{pmatrix}_{\mathbf{3}_{12}} \oplus \begin{pmatrix} x_2 y_3 + x_3 y_5 \\ x_3 y_1 + x_1 y_4 \\ x_1 y_2 + x_2 y_6 \end{pmatrix}_{\mathbf{3}_{13}} \tag{13.151}
$$

$$
\oplus \begin{pmatrix} x_2 y_2 - x_3 y_4 \\ x_3 y_3 - x_1 y_6 \\ x_1 y_1 - x_2 y_5 \end{pmatrix}_{\mathbf{3}_{22}} \oplus \begin{pmatrix} x_2 y_3 - x_3 y_5 \\ x_3 y_1 - x_1 y_4 \\ x_1 y_2 - x_2 y_6 \end{pmatrix}_{\mathbf{3}_{23}} \oplus \begin{pmatrix} x_3 y_2 \\ x_1 y_3 \\ x_2 y_1 \\ x_1 y_5 \\ x_3 y_6 \\ x_2 y_4 \end{pmatrix}_{\mathbf{6}},
$$

$$
\begin{pmatrix} x_1 \\ x_2 \\ x_3 \\ x_4 \\ x_5 \\ x_6 \end{pmatrix}_{\mathbf{6}} \otimes \begin{pmatrix} y_1 \\ y_2 \\ y_3 \\ y_4 \\ y_5 \\ y_6 \end{pmatrix}_{\mathbf{6}} = (x_6 y_1 + x_5 y_2 + x_4 y_3 + x_3 y_4 + x_2 y_5 + x_1 y_6)_{\mathbf{1}}
$$

$$
\oplus (-x_6 y_1 - x_5 y_2 - x_4 y_3 + x_3 y_4 + x_2 y_5 + x_1 y_6)_{\mathbf{1'}}
$$

$$
\oplus \begin{pmatrix} \omega x_6 y_1 + x_5 y_2 + \omega^2 x_4 y_3 + \omega^2 x_3 y_4 + x_2 y_5 + \omega x_1 y_6 \\ \omega^2 x_6 y_1 + x_5 y_2 + \omega x_4 y_3 + \omega x_3 y_4 + x_2 y_5 + \omega^2 x_1 y_6 \end{pmatrix}_{\mathbf{2}}
$$

$$
\oplus \begin{pmatrix} -\omega x_6 y_1 - x_5 y_2 - \omega^2 x_4 y_3 + \omega^2 x_3 y_4 + x_2 y_5 + \omega x_1 y_6 \\ \omega^2 x_6 y_1 + x_5 y_2 + \omega x_4 y_3 - \omega x_3 y_4 - x_2 y_5 - \omega^2 x_1 y_6 \end{pmatrix}_{\mathbf{2}}
$$

$$
\oplus \begin{pmatrix} x_5 y_3 - x_3 y_5 \\ x_4 y_1 - x_1 y_4 \\ x_6 y_2 - x_2 y_6 \end{pmatrix}_{\mathbf{3}_{21}} \oplus \begin{pmatrix} x_1 y_1 - x_6 y_6 \\ x_2 y_2 - x_5 y_5 \\ x_3 y_3 - x_4 y_4 \end{pmatrix}_{\mathbf{3}_{22}}
$$

$$
\oplus \begin{pmatrix} x_4 y_2 - x_2 y_4 \\ x_6 y_3 - x_3 y_6 \\ x_5 y_1 - x_1 y_5 \end{pmatrix}_{\mathbf{3}_{23}} \oplus \begin{pmatrix} x_5 y_3 + x_3 y_5 \\ x_4 y_1 + x_1 y_4 \\ x_6 y_2 + x_2 y_6 \end{pmatrix}_{\mathbf{3}_{11}}
$$

$$
\oplus \begin{pmatrix} x_1 y_1 + x_6 y_6 \\ x_2 y_2 + x_5 y_5 \\ x_3 y_3 + x_4 y_4 \end{pmatrix}_{\mathbf{3}_{12}} \oplus \begin{pmatrix} x_4 y_2 + x_2 y_4 \\ x_6 y_3 + x_3 y_6 \\ x_5 y_1 + x_1 y_5 \end{pmatrix}_{\mathbf{3}_{13}}
$$

$$
\oplus \begin{pmatrix} x_5 y_4 + x_4 y_5 \\ x_6 y_4 + x_4 y_6 \\ x_6 y_5 + x_5 y_6 \\ x_2 y_1 + x_1 y_2 \\ x_3 y_1 + x_1 y_3 \\ x_3 y_2 + x_2 y_3 \end{pmatrix}_{\mathbf{6}} \oplus \begin{pmatrix} x_5 y_4 - x_4 y_5 \\ -x_6 y_4 + x_4 y_6 \\ x_6 y_5 - x_5 y_6 \\ x_2 y_1 - x_1 y_2 \\ -x_3 y_1 + x_1 y_3 \\ x_3 y_2 - x_2 y_3 \end{pmatrix}_{\mathbf{6}}. \tag{13.152}
$$

References

1. Escobar, J.A., Luhn, C.: J. Math. Phys. **50**, 013524 (2009). arXiv:0809.0639 [hep-th]
2. King, S.F., Luhn, C., Stuart, A.J.: Nucl. Phys. B **867**, 203–235 (2013). arXiv:1207.5741 [hep-ph]
3. Ding, G.J., King, S.F.: Phys. Rev. D **89**(9), 093020 (2014). arXiv:1403.5846 [hep-ph]

Subgroups and Decompositions of Multiplets

14

In the particle physics, the symmetry is often broken to a subgroup to describe the low energy phenomena. Therefore, it is very important to study the breaking patterns of discrete groups and decompositions of multiplets. In this section, we discuss decompositions of multiplets for groups, which are studied in the previous sections. Suppose that a finite group G has the order N and M is a divisor of N. Then, Lagrange's theorem implies fine group H with the order M is a candidate for subgroups of G (See Appendix A.).

An irreducible representation r_G of G can be decomposed in terms of irreducible representations $r_{H,m}$ of its subgroup H as $r_G = \sum_m r_{H,m}$. If the trivial singlet of H is included in such a decomposition, $\sum_m r_{H,m}$, and a scalar field with such a trivial singlet develops its vacuum expectation value (VEV), the group G breaks to H. On the other hand, if a scalar field in a multiplet r_G develops its VEV and it does not correspond to the trivial singlet of H, the group G breaks not to H, but to another group.

In following subsections, we show decompositions of multiplets of G into multiplets of subgroups. For a finite group G, there are several chains of subgroups, $G \rightarrow G_1 \rightarrow \cdots \rightarrow G_k \rightarrow Z_N \rightarrow \{e\}, G \rightarrow G'_1 \rightarrow \cdots \rightarrow G'_m \rightarrow Z_M \rightarrow \{e\}$, etc. It would be obvious that the smallest non-trivial subgroup in those chains is an Abelian group such as Z_N or Z_M. Since we concentrate on subgroups, which are shown explicitly in the previous sections, we show the largest subgroup such as G_1 and G'_1 in each chain of subgroups.

© The Author(s), under exclusive license to Springer-Verlag GmbH, DE, part of Springer Nature 2022
T. Kobayashi et al., *An Introduction to Non-Abelian Discrete Symmetries for Particle Physicists*, Lecture Notes in Physics 995, https://doi.org/10.1007/978-3-662-64679-3_14

14.1 S_3

At first, we start with discussing S_3, because S_3 is the minimal non-Abelian discrete group, where the order is equal to $2 \times 3 = 6$. Thus, there are two candidates for subgroups. One is a group with the order two, and the other has the order three. The former corresponds to Z_2 and the latter corresponds to Z_3. As in Sect. 3.1, the S_3 consists of $\{e, a, b, ab, ba, bab\}$, where $a^2 = e$ and $(ab)^3 = e$. Indeed, the subgroup Z_2 consists of e.g. $\{e, a\}$, while the other combinations such as $\{e, b\}$ and $\{e, bab\}$ also correspond to Z_2. The subgroup Z_3 consists of $\{e, ab, ba = (ab)^2\}$. The S_3 has two singlets, **1** and **1'** and one doublet **2**. Both of subgroups, Z_2 and Z_3, are Abelian. Thus, decompositions of multiplets under Z_2 and Z_3 are rather simple. We show such decompositions in what follows. The breaking pattern of S_3 is summarized in Table 14.1.

- $S_3 \to Z_3$

The elements

$$\{e, ab, ba\},$$

of the S_3 group construct the Z_3 subgroup, which is the normal subgroup. There is no other choice to make a Z_3 subgroup. There are three singlet representations, $\mathbf{1}_k$ $k = 0, 1, 2$ for Z_3, that is, $ab = \omega^k$ on $\mathbf{1}_k$. Recall that $\chi_1(ab) = \chi_{1'}(ab) = 1$ for both **1** and **1'** of S_3. Thus, both **1** and **1'** of S_3 correspond to $\mathbf{1}_0$ of Z_3. On the other hand, the doublet **2** of S_3 decomposes into two singlets of Z_3. Since $\chi_2(ab) = -1$, the S_3 doublet **2** decomposes into $\mathbf{1}_1$ and $\mathbf{1}_2$ of Z_3.

In order to understand this decomposition explicitly, we take the two dimensional representations of the group element ab (2.28),

$$ab = \begin{pmatrix} -\frac{1}{2} & -\frac{\sqrt{3}}{2} \\ \frac{\sqrt{3}}{2} & -\frac{1}{2} \end{pmatrix}. \tag{14.1}$$

Then, the doublet (x_1, x_2) is decomposed into two non-trivial singlets as

$$\mathbf{1}_1 : x_1 - i x_2 \qquad \mathbf{1}_2 : x_1 + i x_2. \tag{14.2}$$

Table 14.1 Breaking pattern of S_3 is summarized

S_3	Z_3	S_3	Z_2
1	$\mathbf{1}_0$	**1**	$\mathbf{1}_0$
1'	$\mathbf{1}_0$	**1'**	$\mathbf{1}_1$
2	$\mathbf{1}_1 + \mathbf{1}_2$	**2**	$\mathbf{1}_0 + \mathbf{1}_1$

- $S_3 \to Z_2$

The subgroup Z_2 of S_3 consists of e.g.

$$\{e, a\}.$$

It has two singlet representations $\mathbf{1}_k$ $k = 0, 1$, for, that is, $a = (-1)^k$ on $\mathbf{1}_k$. Recall $\chi_1(a) = 1$ and $\chi_{1'}(a) = -1$ for $\mathbf{1}$ and $\mathbf{1}'$ of S_3. Thus, $\mathbf{1}$ and $\mathbf{1}'$ of S_3 correspond to $\mathbf{1}_0$ and $\mathbf{1}_1$ of Z_2, respectively. On the other hand, the doublet $\mathbf{2}$ of S_3 is decomposed into two singlets of Z_2. Since $\chi_2(a) = 0$, the S_3 doublet $\mathbf{2}$ is decomposed into $\mathbf{1}_0$ and $\mathbf{1}_1$ of Z_2. Indeed, the element a is represented on $\mathbf{2}$ in (2.28) as

$$a = \begin{pmatrix} 1 & 0 \\ 0 & -1 \end{pmatrix}. \tag{14.3}$$

Then, for the doublet (x_1, x_2), the elements x_1 and x_2 correspond to $x_1 = \mathbf{1}_0$ and $x_2 = \mathbf{1}_1$, respectively.

In addition to $\{e, a\}$, there are other Z_2 subgroups, $\{e, b\}$ and $\{e, aba\}$. In both cases, the same results are obtained when we choose a proper basis. This is an example of Abelian subgroups.

In non-Abelian subgroups, the same situation happens. That is, different elements of a finite group G construct the same subgroup. We consider a simple example, D_6. All of the D_6 elements are written by $a^m b^k$ for $m = 0, 1, \ldots, 5$ and $k = 0, 1$, where $a^6 = e$ and $bab = a^{-1}$. By denoting $\tilde{a} = a^2$, we find that the elements $\tilde{a}^m b^k$ for $m = 0, 1, 2$ and $k = 0, 1$ correspond to the subgroup $D_3 \simeq S_3$. On the other hand, by denoting $\tilde{b} = ab$, we find that the elements $\tilde{a}^m \tilde{b}^k$ for $m = 0, 1, 2$ and $k = 0, 1$ correspond to another D_3 subgroup. The decompositions of D_6 multiplets into D_3 multiplets are the same between both D_3 subgroups when we move to a proper basis.

14.2 S_4

As mentioned in Sect. 13, the S_4 group is isomorphic to $\Delta(24)$ and $(Z_2 \times Z_2) \rtimes S_3$. It would be convenient to use the terminology of $(Z_2 \times Z_2) \rtimes S_3$. That is, all of the elements are written as $b^k c^\ell a^m a'^n$ with $k = 0, 1, 2$ and $\ell, m, n = 0, 1$. (See Sect. 13.) The generators, a, a', b and c, are related with the notation in Sect. 3.2 as follows,

$$b = c_1, \qquad c = f_1, \qquad a = a_4, \qquad a' = a_2. \tag{14.4}$$

They satisfy the following algebraic relations

$$b^3 = c^2 = (bc)^2 = a^2 = a'^2 = e, \quad aa' = a'a,$$
$$bab^{-1} = a^{-1}a'^{-1}, \quad ba'b^{-1} = a, \quad cac^{-1} = a'^{-1}, \quad ca'a^{-1} = a^{-1}. \tag{14.5}$$

Furthermore, their representations on $\mathbf{1}$, $\mathbf{1}'$, $\mathbf{2}$, $\mathbf{3}$ and $\mathbf{3}'$ are shown in Table 14.2. As subgroups, the S_4 includes non-Abelian groups, S_3, A_4 and $\Sigma(8)$, which is $(Z_2 \times$

Table 14.2 Representations of S_4 elements

	1	1'	2	3	3'
b	1	1	$\begin{pmatrix} \omega & 0 \\ 0 & \omega^2 \end{pmatrix}$	$\begin{pmatrix} 0 & 1 & 0 \\ 0 & 0 & 1 \\ 1 & 0 & 0 \end{pmatrix}$	$\begin{pmatrix} 0 & 1 & 0 \\ 0 & 0 & 1 \\ 1 & 0 & 0 \end{pmatrix}$
c	1	-1	$\begin{pmatrix} 0 & 1 \\ 1 & 0 \end{pmatrix}$	$\begin{pmatrix} 0 & 0 & 1 \\ 0 & 1 & 0 \\ 1 & 0 & 0 \end{pmatrix}$	$\begin{pmatrix} 0 & 0 & -1 \\ 0 & -1 & 0 \\ -1 & 0 & 0 \end{pmatrix}$
a	1	1	$\begin{pmatrix} 1 & 0 \\ 0 & 1 \end{pmatrix}$	$\begin{pmatrix} -1 & 0 & 0 \\ 0 & -1 & 0 \\ 0 & 0 & 1 \end{pmatrix}$	$\begin{pmatrix} -1 & 0 & 0 \\ 0 & -1 & 0 \\ 0 & 0 & 1 \end{pmatrix}$
a'	1	1	$\begin{pmatrix} 1 & 0 \\ 0 & 1 \end{pmatrix}$	$\begin{pmatrix} 1 & 0 & 0 \\ 0 & -1 & 0 \\ 0 & 0 & -1 \end{pmatrix}$	$\begin{pmatrix} 1 & 0 & 0 \\ 0 & -1 & 0 \\ 0 & 0 & -1 \end{pmatrix}$

Table 14.3 Representations of S_3 elements

	1	1'	2
b	1	1	$\begin{pmatrix} \omega & 0 \\ 0 & \omega^2 \end{pmatrix}$
c	1	-1	$\begin{pmatrix} 0 & 1 \\ 1 & 0 \end{pmatrix}$

$Z_2) \rtimes Z_2$. Thus, the decompositions of S_4 are non-trivial compared with those of S_3. The breaking pattern of S_4 is summarized in Table 14.6.

- $S_4 \rightarrow S_3$

The subgroup S_3 elements are $\{a_1, b_1, c_1, d_1, e_1, f_1\}$. Alternatively, they are denoted by $b^k c^\ell$ with $k = 0, 1, 2$ and $\ell = 0, 1$, i.e., $\{e, b, b^2, c, bc, b^2c\}$. Among them, Table 14.3 shows the representations of the generators b and c on **1**, **1'** and **2** of S_3. Then singlets and doublet **1**, **1'**, and **2** remain the same representation of S_3, i.e. **1**, **1'**, and **2** for each one. Triplets **3** and **3'** are decomposed to **1 + 2** and **1' + 2**. The components of **3** (x_1, x_2, x_3) are decomposed to **1** and **2** as

$$\mathbf{1}: x_1 + x_2 + x_3, \quad \mathbf{2}: \left(x_1 + \omega^2 x_2 + \omega x_3, \omega x_1 + x_2 + \omega^2 x_3\right), \quad (14.6)$$

and components of **3'** are decomposed to **1'** and **2** as

$$\mathbf{1'}: x_1 + x_2 + x_3, \quad \mathbf{2}: \left(x_1 + \omega^2 x_2 + \omega x_3, -\omega x_1 - x_2 - \omega^2 x_3\right). \quad (14.7)$$

Table 14.4 Representations of A_4 elements

	1	1'	1''	3
b	1	ω	ω^2	$\begin{pmatrix} 0 & 1 & 0 \\ 0 & 0 & 1 \\ 1 & 0 & 0 \end{pmatrix}$
a	1	1	1	$\begin{pmatrix} -1 & 0 & 0 \\ 0 & -1 & 0 \\ 0 & 0 & 1 \end{pmatrix}$
a'	1	1	1	$\begin{pmatrix} 1 & 0 & 0 \\ 0 & -1 & 0 \\ 0 & 0 & -1 \end{pmatrix}$

Table 14.5 Representations of $\Sigma(8)$ elements

	1_{+0}	1_{+1}	1_{-0}	1_{-1}	$2_{1,0}$
c	1	1	-1	-1	$\begin{pmatrix} 0 & 1 \\ 1 & 0 \end{pmatrix}$
a	1	-1	1	-1	$\begin{pmatrix} 1 & 0 \\ 0 & -1 \end{pmatrix}$
a'	1	-1	1	-1	$\begin{pmatrix} -1 & 0 \\ 0 & 1 \end{pmatrix}$

- $S_4 \to A_4$

The A_4 subgroup consists of $b^k a^m a'^n$ with $k = 0, 1, 2$ and $m, n = 0, 1$. Recall that the A_4 is isomorphic to $\Delta(12)$. Table 14.4 shows the representations of the generators b, a and a' on **1**, **1'**, **1''** and **3** of A_4. Then each representation of S_4 **1**, **1'**, **2**, **3**, and **3'** is decomposed to **1**, **1**, **1'** + **1''**, **3**, and **3**, respectively.

- $S_4 \to \Sigma(8)$

The subgroup $\Sigma(8)$, i.e. $(Z_2 \times Z_2) \rtimes Z_2$, consists of $c^\ell a^m a'^n$ with $\ell, m, n = 0, 1$. Table 14.5 shows the representations of the generators c, a and a' on 1_{+0}, 1_{+1}, 1_{-0}, 1_{-1} and $2_{1,0}$ of $\Sigma(8)$. Then each representation of S_4 **1**, **1'**, **2**, **3**, and **3'** is decomposed to 1_{+0}, 1_{-0}, $1_{+0} + 1_{-0}$, $1_{+1} + 2$, and $1_{-1} + 2$, respectively. The components of **3** (x_1, x_2, x_3) are decomposed to 1_{+1} and **2** as

$$1_{+1} : x_2, \qquad \mathbf{2} : (x_3, x_1), \tag{14.8}$$

and the components of **3'** are decomposed to 1_{-1} and **2** as

$$1_{-1} : x_2, \qquad \mathbf{2} : (x_3, -x_1). \tag{14.9}$$

Table 14.6 Breaking pattern of S_4

S_4	S_3	S_4	A_4	S_4	$\Sigma(8)$
1	1	1	1	1	1_{+0}
$1'$	$1'$	$1'$	1	$1'$	1_{-0}
2	2	2	$1' + 1''$	2	$1_{+0} + 1_{-0}$
3	$1 + 2$	3	3	3	$1_{+1} + 2$
$3'$	$1' + 2$	$3'$	3	$3'$	$1_{-1} + 2$

14.3 A_4

The A_4 group is isomorphic to $\Delta(12)$. Here, we apply the generic results of $\Delta(3N^3)$ to the A_4 group. All of the $\Delta(12)$ elements are written by $b^k a^m a'^n$ with $k = 0, 1, 2$ and $m, n = 0, 1$. Table 14.7 shows the representations of generators a, a' and b on each representation. As subgroups, the A_4 includes Abelian groups, Z_3, $Z_2 \times Z_2$. The breaking pattern of A_4 is summarized in Table 14.8.

- $A_4 \rightarrow Z_3$

The Z_3 group consists of $\{e, b, b^2\}$. The representations of $\Delta(12)$ 1_k and 3 are decomposed to 1_k and $1_0 + 1_1 + 1_2$, respectively. Decomposition of triplet (x_1, x_2, x_3) is obtained by $1_0 : x_1 + x_2 + x_3$, $1_1 : x_1 + \omega^2 x_2 + \omega x_3$ and $1_2 : x_1 + \omega x_2 + \omega^2 x_3$.

Table 14.7 Representations of $\tilde{a}$, $\tilde{a}'$ and $\tilde{b}$ in $\Delta(12)$

	1_k	3
a	1	$\begin{pmatrix} -1 & 0 & 0 \\ 0 & 1 & 0 \\ 0 & 0 & -1 \end{pmatrix}$
a'	1	$\begin{pmatrix} -1 & 0 & 0 \\ 0 & -1 & 0 \\ 0 & 0 & 1 \end{pmatrix}$
b	ω^k	$\begin{pmatrix} 0 & 1 & 0 \\ 0 & 0 & 1 \\ 1 & 0 & 0 \end{pmatrix}$

Table 14.8 Breaking pattern of A_4

A_4	Z_3	A_4	$Z_2 \times Z_2$
1_k	1_k	1_k	$1_{0,0}$
3	$1_0 + 1_1 + 1_2$	3	$1_{1,1} + 1_{0,1} + 1_{1,0}$

• $A_4 \rightarrow Z_2 \times Z_2$

The subgroup $Z_2 \times Z_2$ consists of $\{e, a, a', aa'\}$. The representations of A_4 $\mathbf{1}_k$ and $\mathbf{3}$ are decomposed to $\mathbf{1}_{0,0}$ and $\mathbf{1}_{1,1} + \mathbf{1}_{0,1} + \mathbf{1}_{1,0}$, respectively.

14.4 A_5

All of the A_5 elements are written by products of $s = a$ and $t = bab$ as shown in Sect. 4.2. As subgroups, the A_5 includes non-Abelian groups, A_4, D_5 and S_3. The breaking pattern of A_5 is summarized in Table 14.10.

• $A_5 \rightarrow A_4$

The subgroup A_4 elements are $\{e, b, \tilde{a}, b\tilde{a}b^2, b^2\tilde{a}b, b\tilde{a}, \tilde{a}b, \tilde{a}b\tilde{a}, b^2\tilde{a}, b^2\tilde{a}b\tilde{a}b\}$ where $\tilde{a} = ab^2aba$. We denote $\tilde{t} = b$ and $\tilde{s} = \tilde{a}$. They satisfy the following relations

$$\tilde{s}^2 = \tilde{t}^3 = (\tilde{s}\tilde{t})^3 = e, \tag{14.10}$$

and correspond to the generators, s and t, of the A_4 group in Sect. 4.1. The representations of A_5 $\mathbf{1}$, $\mathbf{3}$, $\mathbf{3'}$, $\mathbf{4}$, and $\mathbf{5}$ are decomposed to $\mathbf{1}$, $\mathbf{3}$, $\mathbf{3}$, $\mathbf{1} + \mathbf{3}$, $\mathbf{1'} + \mathbf{1''} + \mathbf{3}$, respectively.

• $A_5 \rightarrow D_5$

The D_5 subgroup consists of $a^k \tilde{a}^m$ with $k = 0, 1$ and $m = 0, 1, 2, 3, 4$ where $\tilde{a} \equiv bab^2a$. They satisfy $a^2 = \tilde{a}^5 = e$ and $a\tilde{a}a = \tilde{a}^4$. In order to identify the D_5 basis used in Sect. 6, we define $\tilde{b} = abab^2a$. Table 14.9 shows the representations of these generators $\tilde{a}, \tilde{b}$ on $\mathbf{1}_+, \mathbf{1}_-, \mathbf{2}_1$ and $\mathbf{2}_2$ of D_5. Then, the representations of A_5 $\mathbf{1}$, $\mathbf{3}$, $\mathbf{3'}$, $\mathbf{4}$, and $\mathbf{5}$ are decomposed to $\mathbf{1}_+, \mathbf{1}_- + \mathbf{2}_1, \mathbf{1}_- + \mathbf{2}_2, \mathbf{2}_1 + \mathbf{2}_2, \mathbf{1}_+ + \mathbf{2}_1 + \mathbf{2}_2$, respectively.

• $A_5 \rightarrow S_3 \simeq D_3$

Recall that the S_3 group is isomorphic to the D_3 group. The subgroup D_3 consists of $b^k \tilde{a}^m$ with $k = 0, 1, 2$ and $m = 0, 1$ where we define $\tilde{a} = ab^2ab^2ab$. These generators satisfy $\tilde{a}^2 = e$ and $\tilde{a}b\tilde{a} = b^2$. Then, the representations of A_5 $\mathbf{1}$, $\mathbf{3}$, $\mathbf{3'}$, $\mathbf{4}$, and $\mathbf{5}$ are decomposed to $\mathbf{1}_+, \mathbf{1}_- + \mathbf{2}, \mathbf{1}_- + \mathbf{2}, \mathbf{1}_+ + \mathbf{1}_- + \mathbf{2}, \mathbf{1}_+ + \mathbf{2} + \mathbf{2}$, respectively.

Table 14.9 Representations of D_5 elements

	$\mathbf{1}_+$	$\mathbf{1}_-$	$\mathbf{2}_1$	$\mathbf{2}_2$
$\tilde{a}$	1	1	$\begin{pmatrix} \exp 2\pi i/5 & 0 \\ 0 & -\exp 2\pi i/5 \end{pmatrix}$	$\begin{pmatrix} \exp 4\pi i/5 & 0 \\ 0 & -\exp 4\pi i/5 \end{pmatrix}$
$\tilde{b}$	1	-1	$\begin{pmatrix} 0 & 1 \\ 1 & 0 \end{pmatrix}$	$\begin{pmatrix} 0 & 1 \\ 1 & 0 \end{pmatrix}$

Table 14.10 Breaking pattern of A_5

A_5	A_4	A_5	D_5	A_5	D_3
1	1	1	1_+	1	1_+
3	3	3	$1_- + 2_1$	3	$1_- + 2$
$3'$	3	$3'$	$1_- + 2_2$	$3'$	$1_- + 2$
4	$1 + 3$	4	$2_1 + 2_2$	4	$1_+ + 1_- + 2$
5	$1' + 1'' + 3$	5	$1_+ + 2_1 + 2_2$	5	$1_+ + 2 + 2$

14.5 T'

All of the T' elements are written in terms of the generators, s and t as well as r, which satisfy the algebraic relations, $s^2 = r, r^2 = t^3 = (st)^3 = e$ and $rt = tr$. Table 14.11 shows the representations of s, t and r on each representation. As subgroups, the T' includes Z_6, Z_4 and Q_4. The breaking pattern of T' is summarized in Table 14.12.

- $T' \to Z_6$

The subgroup Z_6 consists of a^m, with $m = 0, \ldots, 5$, where $a = rt$ and $a^6 = e$. The Z_6 group has six singlet representations, 1_n with $n = 0, \ldots, 5$. On the singlet 1_n, the generator a is represented as $a = e^{2\pi i n/6}$. Thus, the representations of T' 1, $1'$, $1''$, 2, $2'$, $2''$, 3 are decomposed to 1_0, 1_2, 1_4, $1_5 + 1_1$, $1_5 + 1_3$, $1_3 + 1_5$, $1_0 + 1_2 + 1_4$, respectively.

- $T' \to Z_4$

The subgroup Z_4 consists of $\{e, s, s^2, s^3\}$. The Z_4 group has four singlet representations, 1_m with $m = 0, 1, 2, 3$. On the singlet 1_m, the generator s is represented as $s = e^{\pi i m/2}$. All of the doublets of T', 2, $2'$, $2''$, are decomposed to two singlets of Z_4 1_1 and 1_3 as $1_1 : \frac{1+\sqrt{3}}{\sqrt{2}} x_1 + x_2$ and $1_3 : -\frac{-1+\sqrt{3}}{\sqrt{2}} i x_1 + x_2$, where (x_1, x_2) correspond to the doublets. In addition, the triplet 3 : (x_1, x_2, x_3) is decomposed to singlets, $1_0 + 1_2 + 1_2$ as $1_0 : (x_1 + x_2 + x_3)$, $1_2 : (-x_1 + x_3)$ and $1_2 : (-x_1 + x_2)$.

Table 14.11 Representations of T'

	1	1'	1"	2	2'	2"	3
s	1	1	1	$\frac{i}{\sqrt{3}}\begin{pmatrix} 1 & \sqrt{2} \\ \sqrt{2} & -1 \end{pmatrix}$	$\frac{i}{\sqrt{3}}\begin{pmatrix} 1 & \sqrt{2} \\ \sqrt{2} & -1 \end{pmatrix}$	$\frac{i}{\sqrt{3}}\begin{pmatrix} 1 & \sqrt{2} \\ \sqrt{2} & -1 \end{pmatrix}$	$\frac{1}{3}\begin{pmatrix} -1 & 2 & 2 \\ 2 & -1 & 2 \\ 2 & 2 & -1 \end{pmatrix}$
r	1	1	1	$\begin{pmatrix} -1 & 0 \\ 0 & -1 \end{pmatrix}$	$\begin{pmatrix} -1 & 0 \\ 0 & -1 \end{pmatrix}$	$\begin{pmatrix} -1 & 0 \\ 0 & -1 \end{pmatrix}$	$\begin{pmatrix} 1 & 0 & 0 \\ 0 & 1 & 0 \\ 0 & 0 & 1 \end{pmatrix}$
t	1	ω	ω^2	$\begin{pmatrix} \omega & 0 \\ 0 & \omega^2 \end{pmatrix}$	$\begin{pmatrix} \omega^2 & 0 \\ 0 & 1 \end{pmatrix}$	$\begin{pmatrix} 1 & 0 \\ 0 & \omega \end{pmatrix}$	$\frac{1}{3}\begin{pmatrix} 1 & 0 & 0 \\ 0 & \omega & 0 \\ 0 & 0 & \omega^2 \end{pmatrix}$

Table 14.12 Breaking pattern of T'

T'	Z_6	T'	Z_4	T'	Q_4
1	1_0	1	1_0	1	1_{++}
1'	1_2	1'	1_0	1'	1_{++}
1''	1_4	1''	1_0	1''	1_{++}
2	$1_1 + 1_5$	2	$1_1 + 1_3$	2	2
2'	$1_3 + 1_5$	2'	$1_1 + 1_3$	2'	2
2''	$1_3 + 1_5$	2''	$1_1 + 1_3$	2''	2
3	$1_0 + 1_2 + 1_4$	3	$1_0 + 1_2 + 1_2$	3	$1_{+-} + 1_{-+} +$ 1_{--}

- $T' \to Q_4$

We consider the subgroup Q_4, which consists of $s^m b^k$ with $m = 0, 1, 2, 3$ and $k = 0, 1$. The generator b is defined by $b = tst^2$. Then, the representations of T' **1, 1', 1''**, **2, 2', 2''**, and **3** are decomposed to 1_{++}, 1_{++}, 1_{++}, **2, 2, 2**, and $1_{+-} + 1_{-+} + 1_{--}$, respectively.

14.6 General D_N

Since the group D_N is isomorphic to $Z_N \rtimes Z_2$, D_M and Z_N as well as Z_2 appear as subgroups of D_N, where M is a divisor of N. Recall that all of D_N elements are written by $a^m b^k$ with $m = 0, \ldots, N - 1$ and $k = 0, 1$. There are singlets and doublets 2_k, where $k = 1, \ldots, N/2 - 1$ for $N =$ even and $k = 1, \ldots, (N - 1)/2$ for $N =$ odd. On the doublet 2_k, the generators a and b are represented as

$$a = \begin{pmatrix} \rho^k & 0 \\ 0 & \rho^{-k} \end{pmatrix}, \quad b = \begin{pmatrix} 0 & 1 \\ 1 & 0 \end{pmatrix}, \tag{14.11}$$

where $\rho = e^{2\pi i/N}$. For $N =$ even, there are four singlets $1_{\pm\pm}$. The generator b is represented as $b = 1$ on $1_{+\pm}$, while $b = -1$ on $1_{-\pm}$. The generator a is represented as $a = 1$ on 1_{++} and 1_{--}, while $a = -1$ on 1_{+-} and 1_{+-}. For $N =$ odd, there are two singlets $1_\pm$. The generator b is represented as $b = 1$ on 1_+ and $b = -1$ on 1_-, while $a = 1$ on both singlets. The general breaking patterns of D_N are summarized in Table 14.13 for even N and Table 14.14 for odd N.

- $D_N \to Z_2$

The two elements e and b construct the Z_2 subgroup. Obviously, there are two singlet representations $1_0, 1_1$, where the subscript denotes the Z_2 charge. That is, we have $b = 1$ on 1_0 and $b = -1$ on 1_1.

When N is even, the singlets 1_{++} and 1_{+-} of D_N become 1_0 of Z_2 and the singlets 1_{-+} and 1_{--} of D_N become 1_1 of Z_2. The doublets 2_k of D_N, (x_1, x_2), decompose two singlets as $1_0 : x_1 + x_2$ and $1_1 : x_1 - x_2$.

Table 14.13 Breaking pattern of D_N for even N

D_N	Z_2
1_{++}	1_0
1_{+-}	1_0
1_{-+}	1_1
1_{--}	1_1
2_k	$1_0 + 1_1$

D_N	Z_N
1_{++}	1_0
1_{+-}	$1_{N/2}$
1_{-+}	$1_{N/2}$
1_{--}	1_0
2_k	$1_k + 1_{N-k}$

D_N	D_M (M is even)
1_{++}	1_{++}
1_{+-}	1_{++} (M/N is even)
	1_{+-} (M/N is odd)
1_{-+}	1_{--} (M/N is even)
	1_{-+} (M/N is odd)
1_{--}	1_{--}
2_k	$2_{k'}(k = k' + Mn)$
	$\tilde{2}_{M/2-k'}(k = M/2 - k' + Mn)$
	$1_{+-} + 1_{-+}(k = M(2n + 1)/2)$
	$1_{++} + 1_{--}(k = Mn)$

D_N	D_M (M is odd)
1_{++}	1_+
1_{+-}	1_+
1_{-+}	1_-
1_{--}	1_-
2_k	$2_{k'}(k = k' + Mn)$
	$\tilde{2}_{M-k'}(k = M - k')$
	$1_+ + 1_-(k = Mn)$

Table 14.14 Breaking pattern of D_N for odd N

D_N	Z_2
1_+	1_0
1_-	1_0
2_k	$1_0 + 1_1$

D_N	Z_N
1_+	1_0
1_-	1_0
2_k	$1_k + 1_{N-k}$

D_N	D_M
1_+	1_+
1_-	1_-
2_k	$2_{k'}(N = k' + Mn)$
	$\tilde{2}_{M-k'}(k = Mn - k')$
	$1_+ + 1_-(k = Mn)$

When N is odd, the singlet 1_+ of D_N becomes 1_0 of Z_2 and the singlet 1_- of D_N becomes 1_1 of Z_2. The decompositions of doublets 2_k are the same as those for $N =$ even.

• $D_N \to Z_N$

The subgroup Z_N consists of the elements $\{e, a, \ldots, a^{N-1}\}$. Obviously it is the normal subgroups of D_N and there are N types of irreducible singlet representations $1_0, 1_1, \ldots, 1_{N-1}$. On the 1_k, the generator a is represented as $a = \rho^k$.

When N is even, the singlets $\mathbf{1}_{++}$ and $\mathbf{1}_{--}$ of D_N become $\mathbf{1}_0$ of Z_N and the singlets $\mathbf{1}_{+-}$ and $\mathbf{1}_{-+}$ of D_N become $\mathbf{1}_{N/2}$ of Z_N. The doublets $\mathbf{2}_k$, (x_1, x_2), decompose to two singlets as $\mathbf{1}_k : x_1$ and $\mathbf{1}_{N-k} : x_2$.

When N is odd, both $\mathbf{1}_+$ and $\mathbf{1}_-$ of D_N become $\mathbf{1}_0$ of Z_N. The decompositions of doublets $\mathbf{2}_k$ are the same as those for $N =$ even.

- $D_N \rightarrow D_M$

The above decompositions of D_N are rather straightforward, because subgroups are Abelian. Here we consider the D_M subgroup, where M is a divisor of N. The decompositions of D_N to D_M would be non-trivial. We denote $\tilde{a} = a^\ell$ with $\ell = N/M$, where ℓ is integer. The subgroup D_M consists of $\tilde{a}^m b^k$ with $m = 0, \ldots, M - 1$ and $k = 0, 1$. There are three combinations of (N, M), i.e. (N, M)=(even,even), (even,odd) and (odd,odd).

We start with the combination (N, M) =(even,even). Recall that (ab) of D_N is represented as $ab = 1$ on $\mathbf{1}_{\pm+}$ and $ab = -1$ on $\mathbf{1}_{\pm-}$. Thus, the representations of $(a^\ell b)$ depend on whether ℓ is even or odd. When ℓ is odd, (ab) and $(a^\ell b)$ are represented in the same way on each of the above singlets. On the other hand, when $\ell =$ even, we always have the singlet representations with $a^\ell = 1$. The doublets $\mathbf{2}_k$ of D_N correspond to the doublets $\mathbf{2}_{k'}$ of D_M when $k = k'$ (mod M). In addition, when $k = -k'$ (mod M), doublets $\mathbf{2}_k$ (x_1, x_2) of D_N correspond to the doublets $\mathbf{2}_{M-k'}$ (x_2, x_1) of D_M. That is, the components are exchanged each other and we denote it by $\tilde{\mathbf{2}}_{M-k'}$. Furthermore, the other doublets $\mathbf{2}_k$ of D_N decompose to two singlets of D_M as $\mathbf{1}_{+-} + \mathbf{1}_{-+}$ with $\mathbf{1}_{+-} : x_1 + x_2$ and $\mathbf{1}_{-+} : x_1 - x_2$ for $k = (M/2)$ (mod M) and $\mathbf{1}_{++} + \mathbf{1}_{--}$ with $\mathbf{1}_{++} : x_1 + x_2$ and $\mathbf{1}_{--} : x_1 - x_2$ for $k = 0$ (mod M).

Next we consider the case with $(N, M) =$ (even, odd). In this case, the singlet $\mathbf{1}_{++}, \mathbf{1}_{+-}, \mathbf{1}_{-+}$ and $\mathbf{1}_{--}$ of D_N become $\mathbf{1}_+, \mathbf{1}_+, \mathbf{1}_-$ and $\mathbf{1}_-$ of D_M, respectively. The doublets $\mathbf{2}_k$ of D_N correspond to the doublets $\mathbf{2}_{k'}$ of D_M when $k = k'$ (mod M). In addition, when $k = -k'$ (mod M), the doublets $\mathbf{2}_k$ (x_1, x_2) of D_N correspond to the doublets $\mathbf{2}_{M-k'}$ (x_2, x_1) of D_M. Furthermore, when $k = 0$ (mod M), other doublets $\mathbf{2}_k$ of D_N decompose to two singlets of D_M as $\mathbf{1}_+ + \mathbf{1}_-$, where $\mathbf{1}_+ : x_1 + x_2$ and $\mathbf{1}_- : x_1 - x_2$.

Now, let us consider the case with $(N, M) =$ (odd, odd). In this case, the singlets $\mathbf{1}_+$ and $\mathbf{1}_-$ of D_N become $\mathbf{1}_+$ and $\mathbf{1}_-$ of D_M. The doublets $\mathbf{2}_k$ of D_N correspond to the doublets $\mathbf{2}_{k'}$ of D_M when $k = k'$ (mod M). In addition, when $k = -k'$ (mod M), doublets $\mathbf{2}_k$ (x_1, x_2) of D_N correspond to the doublets $\mathbf{2}_{M-k'}$ (x_2, x_1) of D_M. Furthermore, when $k = 0$ (mod M), other doublets $\mathbf{2}_k$ of D_N decompose to two singlets of D_M as $\mathbf{1}_+ + \mathbf{1}_-$, where $\mathbf{1}_+ : x_1 + x_2$ and $\mathbf{1}_- : x_1 - x_2$.

14.7 D_4

Here, we study D_4, which is the second minimum discrete symmetry. All of the D_4 elements are written by $a^m b^k$ with $m = 0, 1, 2, 3$ and $k = 0, 1$. Since the order of D_4 is 8, it contains the order 2 and 4 subgroups. There are two types of the order

4 groups which are corresponding to $Z_2 \times Z_2$ and Z_4 groups. All of subgroups are Abelian. Thus, decompositions are rather simple.

• $D_4 \to Z_4$

The subgroup Z_4 is consist of the elements $\{e, a, a^2, a^3\}$. Obviously it is the normal subgroups of D_4 and there are four types of irreducible singlet representations $\mathbf{1}_m$ with $m = 0, 1, 2, 3$, where a is represented as $a = e^{\pi i m/2}$. From the characters of D_4 groups, it is found that $\mathbf{1}_{++}$ and $\mathbf{1}_{--}$ of D_4 correspond to $\mathbf{1}_0$ of Z_4 and $\mathbf{1}_{+-}$ and $\mathbf{1}_{-+}$ of D_4 correspond to $\mathbf{1}_2$ of D_4. For the D_4 doublet $\mathbf{2}$, it is convenient to use the diagonal base of matrix a as

$$a = \begin{pmatrix} i & 0 \\ 0 & -i \end{pmatrix}. \tag{14.12}$$

Then we can read the doublet $\mathbf{2} : (x_1, x_2)$ decomposes to two singlets as $\mathbf{1}_1 : x_1$ and $\mathbf{1}_3 : x_2$.

• $D_4 \to Z_2 \times Z_2$

We denote $\tilde{a} = a^2$. Then, the subgroup $Z_2 \times Z_2$ consists of $\{e, \tilde{a}, b, \tilde{a}b\}$, where $\tilde{a}b = b\tilde{a}$ and $\tilde{a}^2 = b^2 = e$. Obviously, their representations are quite simple, that is, $\mathbf{1}_{\pm\pm}$, whose $Z_2 \times Z_2$ charges are determined by $\tilde{a} = \pm 1$ and $b = \pm 1$. We use the notation that the first (second) subscript of $\mathbf{1}_{\pm\pm}$ denotes the Z_2 charge for $\tilde{a}$ (b). Then, the singlets $\mathbf{1}_{++}$ and $\mathbf{1}_{+-}$ of D_4 correspond to $\mathbf{1}_{++}$ of $Z_2 \times Z_2$ and $\mathbf{1}_{-+}$ and $\mathbf{1}_{--}$ of D_4 correspond to $\mathbf{1}_{+-}$ of $Z_2 \times Z_2$. The doublet $\mathbf{2}$ of D_4 decomposes to $\mathbf{1}_{-+}$ and $\mathbf{1}_{--}$ of Z_2.

In addition to the above, there is another choice of $Z_2 \times Z_2$ subgroup, which consists of $\{e, a^2, ab, a^3b\}$. In this case, we can obtain the same decomposition of D_4.

• $D_4 \to Z_2$

Furthermore, both Z_4 and $Z_2 \times Z_2$ include Z_2 subgroup. The decomposition of D_4 to Z_2 is rather straightforward.

14.8   General Q_N

Recall that all of the Q_N elements are written as $a^m b^k$ with $m = 0, \ldots, N-1$ and $k = 0, 1$, where $a^N = e$ and $b^2 = a^{N/2}$. Similarly to D_N with $N = $ even, there are four singlets $\mathbf{1}_{\pm\pm}$ and doublets $\mathbf{2}_k$ with $k = 1, \ldots, N/2 - 1$. Tables 14.15 and 14.16 show the representations of a and b on these representations for $N = 4n$ and $N = 4n + 2$. In general, the group Q_N includes Z_4, Z_N and Q_M as subgroups. The breaking patterns of Q_N are summarized in Table 14.17 for $N = 4n$ and Table 14.18 for $N = 4n + 2$.

Table 14.15 Representations of Q_N for $N = 4n$

$(N = 4n)$	1_{++}	1_{+-}	1_{-+}	1_-	$2_{k=\text{odd}}$	$2_{k=\text{even}}$
a	1	-1	-1	1	$\begin{pmatrix} \rho^k & 0 \\ 0 & \rho^{-k} \end{pmatrix}$	$\begin{pmatrix} \rho^k & 0 \\ 0 & \rho^{-k} \end{pmatrix}$
b	1	1	-1	-1	$\begin{pmatrix} 0 & i \\ i & 0 \end{pmatrix}$	$\begin{pmatrix} 0 & 1 \\ 1 & 0 \end{pmatrix}$

Table 14.16 Representations of Q_N for $N = 4n + 2$

$(N = 4n + 2)$	1_{++}	1_{+-}	1_{-+}	1_-	$2_{k=\text{odd}}$	$2_{k=\text{even}}$
a	1	-1	-1	1	$\begin{pmatrix} \rho^k & 0 \\ 0 & \rho^{-k} \end{pmatrix}$	$\begin{pmatrix} \rho^k & 0 \\ 0 & \rho^{-k} \end{pmatrix}$
b	1	i	$-i$	-1	$\begin{pmatrix} 0 & i \\ i & 0 \end{pmatrix}$	$\begin{pmatrix} 0 & 1 \\ 1 & 0 \end{pmatrix}$

Table 14.17 Breaking pattern of Q_N for $N = 4n$. All parameters n, m, k, k', n' are integers

Q_N	Z_4
1_{++}	1_0
1_{+-}	1_0
1_{-+}	1_2
1_{--}	1_2
2_k	$1_1 + 1_3$

Q_N	Z_N
1_{++}	1_0
1_{+-}	$1_{N/2}$
1_{-+}	1_2
1_{--}	$1_{N/2}$
2_k	$1_k + 1_{N-k}$

Q_N	$Q_M \ (M = 4m)$
1_{++}	1_{++}
1_{+-}	1_{++} (M/N is even)
	1_{+-} (M/N is odd)
1_{-+}	1_{--} (M/N is even)
	1_{-+} (M/N is odd)
1_{--}	1_{--}
2_k	$2_{k'} (k = k' + Mn')$
	$\tilde{2}_{M-k'} (k = Mn' - k')$
	$1_{+-} + 1_{-+} (k = M(2n' + 1)/2)$
	$1_{++} + 1_{--} (k = Mn')$

Q_N	$Q_M \ (M = 4m + 2)$
1_{++}	1_{++}
1_{+-}	1_{++}
1_{-+}	1_{--}
1_{--}	1_{--}
2_k	$2_{k'} (k = k' + Mn')$
	$\tilde{2}_{M-k'} (k = Mn' - k')$
	$1_{+-} + 1_{-+} (k = M(2n' + 1)/2)$
	$1_{++} + 1_{--} (k = Mn')$

Table 14.18 Breaking pattern of Q_N for $N = 4n + 2$ and $M = 4m + 2$. All parameters n, m, k, k', n' are integer

Q_N	Z_4
1_{++}	1_0
1_{+-}	1_1
1_{-+}	1_3
1_{--}	1_2
2_k	$1_1 + 1_3$

Q_N	Z_N
1_{++}	1_0
1_{+-}	$1_{N/2}$
1_{-+}	$1_{N/2}$
1_{--}	1_0
2_k	$1_k + 1_{N-k}$

Q_N	Q_M
1_{++}	1_{++}
1_{+-}	1_{++}
1_{-+}	1_{--}
1_{--}	1_{--}
2_k	$2_{k'} (N = k' + Mn')$
	$\tilde{2}_{M-k'} (N = Mn' - k')$
	$1_{+-} + 1_{-+} (N = M(2n' + 1)/2)$
	$1_{++} + 1_{--} (N = Mn')$

• $Q_N \to Z_4$

First, we consider the subgroup Z_4, which consists of the elements $\{e, b, b^2, b^3\}$. Obviously, there are four singlet representations 1_m for Z_4 and the generator b is represented as $b = e^{\pi i m/2}$ on 1_m.

When $N = 4n$, 1_{++} and 1_{+-} of Q_N correspond to 1_0 of Z_4 and 1_{-+} and 1_{--} of Q_N correspond to 1_2 of Z_4. The doublets 2_k of Q_N, (x_1, x_2) decompose to two singlets as $1_1 : (x_1 - ix_2)$ and $1_3 : (x_1 + ix_2)$.

When $N = 4n + 2$, $1_{++}, 1_{+-}, 1_{-+}$ and 1_{--} of Q_N correspond to $1_0, 1_1, 1_3$ and 1_2 of Z_4, respectively. The decompositions of doublets 2_k are the same as those for $N = 4n$.

• $Q_N \to Z_N$

Next, we consider the subgroup Z_N, which consist of the elements $\{e, a, \ldots, a^{N-1}\}$. Obviously it is the normal subgroups of Q_N and there are N types of irreducible singlet representations $1_0, 1_1, \ldots, 1_{N-1}$. On the singlet 1_m of Z_N, the generator a is represented as $a = \rho^m$. The singlets, 1_{++} and 1_{--} of Q_N correspond to 1_0 of Z_N and the singlets 1_{+-} and 1_{-+} of Q_N correspond to $1_{N/2}$ of Z_N. The doublets 2_k, (x_1, x_2), of Q_N decompose to two singlets as $1_k : x_1$ and $1_{N-k} : x_2$.

• $Q_N \to Q_M$

We consider the Q_M subgroup, where M is a divisor of N. We denote $\tilde{a} = a^{\ell}$ with $\ell = N/M$, where $\ell = $ integer. The subgroup Q_M consists of $\tilde{a}^m b^k$ with $m = 0, \ldots, M - 1$ and $k = 0, 1$. There are three combinations of (N, M), i.e. $(N, M) = (4n, 4m), (4n, 4m + 2)$ and $(4n + 2, 4m + 2)$.

We start with the combination $(N, M) = (4n, 4m)$, where $\ell = N/M$ can be even or odd. Recall that (ab) of Q_N is represented as $ab = 1$ on 1_{++} and $ab = -1$ on $1_{\pm-}$. Thus, the representations of $(a^{\ell} b)$ depend on whether ℓ is even or odd. When ℓ is odd, (ab) and $(a^{\ell} b)$ are represented in the same way as on each of the above singlets. On the other hand, when $\ell =$ even, we always have the singlet representations with $a^\ell = 1$. The doublets 2_k of Q_N correspond to the doublets $2_{k'}$ of Q_M when $k = k' \pmod M$. In addition, when $k = -k' \pmod M$, doublets 2_k (x_1, x_2) of Q_N correspond to the doublets $2_{M-k'}$ (x_2, x_1) of Q_M. Furthermore, other doublets 2_k of Q_N decompose to two singlets of Q_M as $1_{+-} + 1_{-+}$ with $1_{+-} : x_1 + x_2$ and $1_{-+} : x_1 - x_2$ for $k = (M/2) \pmod M$ and $1_{++} + 1_{--}$ with $1_{++} : x_1 + x_2$ and $1_{--} : x_1 - x_2$ for $k = 0 \pmod M$.

Next we consider the case with $(N, M) = (4n, 4m + 2)$, where ℓ must be even. Similarly to the above case with $\ell =$ even, the singlets 1_{++}, 1_{+-}, 1_{-+} and 1_{--} of Q_N correspond to 1_{++}, 1_{++}, 1_{--} and 1_{--} of Q_M. The results on decompositions of doublets are also the same as the above case with $(N, M) = (4n, 4m)$ and $\ell = N/M =$ even.

Next, we consider the case with $(N, M) = (4n + 2, 4m + 2)$, where ℓ must be odd. In this case, the results on decompositions are the same as the case with $(N, M) = (4n, 4m)$ and $\ell = N/M =$ odd.

14.9 Q_4

All elements Q_4 can be expressed in the form of $a^m b^k$ with $m = 0, 1, 2, 3$ and $k = 0, 1$. Since the order of Q_4 is equal to 8, it contains the order 2 and 4 subgroups. There are a few types of the order 4 groups which correspond to Z_4 groups.

- $Q_4 \rightarrow Z_4$

For example, the elements $\{e, a, a^2, a^3\}$ construct the Z_4 subgroup. Obviously it is the normal subgroups of Q_4 and there are four types of irreducible singlet representations 1_m with $m = 0, 1, 2, 3$, where a is represented as $a = e^{\pi i m/2}$. From the characters of Q_4 groups, it is found that 1_{++} and 1_{--} of Q_4 correspond to 1_0 of Z_4 and 1_{-+} and 1_{+-} of Q_4 correspond to 1_2 of Z_4. For the doublets of Q_4, it is convenient to use the diagonal base of matrix a as

$$a = \begin{pmatrix} i & 0 \\ 0 & -i \end{pmatrix}. \tag{14.13}$$

Then we find that the doublet 2 (x_1, x_2) decomposes to two singlets as $1_1 : x_1$ and $1_3 : x_2$.

In addition, other Z_4 subgroups consist of $\{e, b, b^2, b^3\}$ and $\{e, ab, (ab)^2, (ab)^3\}$. For those Z_4 subgroups, we obtain the same results when we choose proper basis. Furthermore, subgroups of Z_2 can appear from the above Z_4 groups. The decomposition of Z_4 to Z_2 is rather straightforward.

14.10 QD_{2N}

Since the group QD_{2N} is isomorphic to $Z_N \rtimes Z_2$, D_M and Z_N as well as Z_2 appear as subgroups of D_N, where M is a divisor of N. Recall that all of QD_{2N} elements are written by $a^m b^k$ with $m = 0, \ldots, N-1$ and $k = 0, 1$. There are four singlets and $(N/2 - 1)$ doublets $\mathbf{2}_k$, where $k = 1, \ldots, N/2 - 1$. On the doublet $\mathbf{2}_k$, the generators a and b are represented as

$$a = \begin{pmatrix} \rho^k & 0 \\ 0 & \rho^{k(N/2-1)} \end{pmatrix}, \quad b = \begin{pmatrix} 0 & 1 \\ 1 & 0 \end{pmatrix}, \tag{14.14}$$

where $\rho = e^{2\pi i/N}$. There are four singlets $\mathbf{1}_{ss'}$ with $s, s' = \pm$. The generator a is represented as $a = 1$ on $\mathbf{1}_{+s'}$, while $a = -1$ on $\mathbf{1}_{-s'}$. The generator b is represented as $b = 1$ on $\mathbf{1}_{s+}$, while $b = -1$ on $\mathbf{1}_{s-}$. The general breaking pattern of QD_{2N} is summarized in Table 14.19.

• $QD_{2N} \to Z_2$
The two elements e and b construct the Z_2 subgroup. Obviously, there are two singlet representations $\mathbf{1}_0, \mathbf{1}_1$, where the subscript denotes the Z_2 charge. That is, we have $b = 1$ on $\mathbf{1}_0$ and $b = -1$ on $\mathbf{1}_1$.

The singlets $\mathbf{1}_{++}$ and $\mathbf{1}_{-+}$ of QD_{2N} become $\mathbf{1}_0$ of Z_2 and the singlets $\mathbf{1}_{+-}$ and $\mathbf{1}_{--}$ of QD_{2N} become $\mathbf{1}_1$ of Z_2. The doublets $\mathbf{2}_k$ of QD_{2N}, (x_1, x_2), decompose to two singlets as $\mathbf{1}_0 : x_1 + x_2$ and $\mathbf{1}_1 : x_1 - x_2$.

• $QD_{2N} \to Z_N$
The subgroup Z_N consists of the elements $\{e, a, \ldots, a^{N-1}\}$. Obviously it is the normal subgroups of D_N and there are N types of irreducible singlet representations $\mathbf{1}_0, \mathbf{1}_1, \ldots, \mathbf{1}_{N-1}$. On the $\mathbf{1}_k$, the generator a is represented as $a = \rho^k$.

The singlets $\mathbf{1}_{++}$ and $\mathbf{1}_{+-}$ of QD_{2N} become $\mathbf{1}_0$ of Z_N and the singlets $\mathbf{1}_{-+}$ and $\mathbf{1}_{--}$ of QD_{2N} become $\mathbf{1}_{N/2}$ of Z_N. The doublets $\mathbf{2}_k$, (x_1, x_2), decompose to two singlets as $\mathbf{1}_k : x_1$ and $\mathbf{1}_{k(N/2-1)} : x_2$.

Table 14.19 Breaking pattern of QD_{2N}

QD_{2N}	Z_2
$\mathbf{1}_{++}$	$\mathbf{1}_0$
$\mathbf{1}_{-+}$	$\mathbf{1}_0$
$\mathbf{1}_{+-}$	$\mathbf{1}_1$
$\mathbf{1}_{--}$	$\mathbf{1}_1$
$\mathbf{2}_k$	$\mathbf{1}_0 + \mathbf{1}_1$

QD_{2N}	Z_N
$\mathbf{1}_{++}$	$\mathbf{1}_0$
$\mathbf{1}_{-+}$	$\mathbf{1}_{N/2}$
$\mathbf{1}_{+-}$	$\mathbf{1}_0$
$\mathbf{1}_{--}$	$\mathbf{1}_{N/2}$
$\mathbf{2}_k$	$\mathbf{1}_k + \mathbf{1}_{k(N/2-1)}$

QD_{2N}	$D_{N/2}$
$\mathbf{1}_{++}$	$\mathbf{1}_{++}$
$\mathbf{1}_{-+}$	$\mathbf{1}_{+-}$
$\mathbf{1}_{+-}$	$\mathbf{1}_{--}$
$\mathbf{1}_{--}$	$\mathbf{1}_{-+}$
$\mathbf{2}_k$	$\mathbf{2}_{k'}$ $(k = k' + N/4)$
	$\mathbf{1}_{+-} + \mathbf{1}_{-+}$ $(k = N/4)$

• $QD_{2N} \to D_{N/2}$

The above decompositions of QD_{2N} are rather straightforward, because subgroups are Abelian. The decompositions of QD_{2N} to $D_{N/2}$ would be non-trivial. We denote $\tilde{a} = a^2$. The subgroup $D_{N/2}$ consists of $\tilde{a}^m b^k$ with $m = 0, \ldots, N/2 - 1$ and $k = 0, 1$.

The singlet representations 1_{st} are decomposed to 1_{tu} of $D_{N/2}$ with $u = st$. The doublets 2_k of QD_{2N} correspond to the doublets $2_{k'}$ of $D_{N/2}$ when $k = k'$ (mod $N/2$). In addition, the doublet $2_{N/4}$ of D_N decompose to two singlets of $D_{N/2}$ as $1_{+-} + 1_{-+}$ with $1_{+-} : x_1 + x_2$ and $1_{-+} : x_1 - x_2$.

14.11 General $\Sigma(2N^2)$

Recall that all of the $\Sigma(2N^2)$ elements are written by $b^k a^m a'^n$ with $k = 0, 1$ and $m, n = 0, 1, \ldots, N - 1$. The generators, a, a' and b, satisfy $a^N = a'^N = b^2 = e$, $aa' = a'a$ and $bab = a'$, that is, a, a' and b correspond to Z_N, Z'_N and Z_2 of $(Z_N \times Z'_N) \rtimes Z_2$, respectively. Table 14.20 shows the representations of these generators on each representation. The number of doublets $2_{p,q}$ is equal to $N(N - 1)/2$ with the relation $p > q$.

In general, the group $\Sigma(2N^2)$ includes Z_{2N}, $Z_N \times Z_N$, D_N, Q_N and $\Sigma(2M^2)$ as subgroups. The breaking pattern of $\Sigma(2N^2)$ is summarized in Table 14.24.

• $\Sigma(2N^2) \to Z_{2N}$

The group $\Sigma(2N^2)$ generally includes a subgroup Z_{2N}. We consider the elements of Z_{2N} as $(ba)^m$ with $m = 0, \ldots, 2N - 1$. There are $2N$ singlet representations 1_m for Z_{2N} and the generator ba is represented as $b = \rho^m$ on 1_m where $\rho = e^{\pi i/N}$. Then, the representations of $\Sigma(2N^2)$ 1_{+n}, 1_{-n}, and $2_{\ell,m}$ are decomposed to 1_{2n}, 1_{2n+N}, and $1_{\ell+m} + 1_{\ell+m+N}$, respectively. The components of doublets (x_ℓ, x_m) correspond to $1_{\ell+m} : (\rho^\ell x_\ell + \rho^m x_m)$ and $1_{\ell+m+N} : (\rho^\ell x_\ell - \rho^m x_m)$.

• $\Sigma(2N^2) \to Z_N \times Z_N$

The subgroup $Z_N \times Z_N$ consists of the elements $a^m a'^n$ with $m, n = 0, \ldots, N - 1$. Obviously it is the normal subgroups of $\Sigma(2N^2)$. There are N^2 singlet representations $1_{m,n}$ and the generators a and a' are represented as $a = \rho^m$ and $a' = \rho^n$ on $1_{m,n}$.

Table 14.20 Representations of $\Sigma(2N^2)$

	1_{+n}	1_{-n}	$2_{p,q}$
a	ρ^n	ρ^n	$\begin{pmatrix} \rho^q & 0 \\ 0 & \rho^p \end{pmatrix}$
a'	ρ^n	ρ^n	$\begin{pmatrix} \rho^p & 0 \\ 0 & \rho^q \end{pmatrix}$
b	1	-1	$\begin{pmatrix} 0 & 1 \\ 1 & 0 \end{pmatrix}$

Table 14.21 Representations of $\tilde{a}$ and b in $\Sigma(2N^2)$

	$\mathbf{1}_{+n}$	$\mathbf{1}_{-n}$	$\mathbf{2}_{p,q}$
$\tilde{a}$	1	1	$\begin{pmatrix} \rho^{p-q} & 0 \\ 0 & \rho^{-(p-q)} \end{pmatrix}$
b	1	-1	$\begin{pmatrix} 0 & 1 \\ 1 & 0 \end{pmatrix}$

Table 14.22 Representations of $\tilde{a}$ and $\tilde{b}$ in $\Sigma(2N^2)$

	$\mathbf{1}_{+n}$	$\mathbf{1}_{-n}$	$\mathbf{2}_{p,q}$
$\tilde{a}$	1	1	$\begin{pmatrix} \rho^{p-q} & 0 \\ 0 & \rho^{-(p-q)} \end{pmatrix}$
$\tilde{b}$	1	-1	$\begin{pmatrix} 0 & (-1)^q \\ (-1)^p & 0 \end{pmatrix}$

Then, the representations of $\Sigma(2N^2)$ $\mathbf{1}_{+n}$, $\mathbf{1}_{-n}$, and $\mathbf{2}_{\ell,m}$ are decomposed to $\mathbf{1}_{n,n}$, $\mathbf{1}_{n,n}$, and $\mathbf{1}_{\ell,m} + \mathbf{1}_{m,\ell}$, respectively.

- $\Sigma(2N^2) \to D_N$
We consider D_N as a subgroup of $\Sigma(2N^2)$. We denote $\tilde{a} = a^{-1}a'$. Then, the subgroup D_N consists of the elements $\tilde{a}^m b^k$ with $k = 0, 1$ and $m = 0, \ldots, N-1$. Table 14.21 shows the representations of the generators, $\tilde{a}$ and b, on each representation of $\Sigma(2N^2)$.

At first, we consider the case that N is even. The doublets $\mathbf{2}_{p,q}$ of $\Sigma(2N^2)$ are still doublets of D_N except $p - q = \frac{N}{2}$. On the other hand, when $p - q = \frac{N}{2}$, the doublets decompose to two singlets of D_N. Then, the representations of $\Sigma(2N^2)$ $\mathbf{1}_{+n}$, $\mathbf{1}_{-n}$, $\mathbf{2}_{q+k',q}$, $\mathbf{2}_{q-k',q}$, and $\mathbf{2}_{q+\frac{N}{2},q}$ are decomposed to $\mathbf{1}_{++}$, $\mathbf{1}_{--}$, $\mathbf{2}_{k'}$, $\tilde{\mathbf{2}}_{k'}$, and $\mathbf{1}_{+-} + \mathbf{1}_{-+}$, respectively.

Next, we consider the case that N is odd. In this case, the representations of $\Sigma(2N^2)$ $\mathbf{1}_{+n}$, $\mathbf{1}_{-n}$, $\mathbf{2}_{q+k',q}$, and $\mathbf{2}_{q-k',q}$ are decomposed to $\mathbf{1}_+$, $\mathbf{1}_-$, $\mathbf{2}_{k'}$, and $\tilde{\mathbf{2}}_{N-k'}$, respectively.

- $\Sigma(2N^2) \to Q_N$
We consider Q_N as a subgroup of $\Sigma(2N^2)$ with $N =$ even. We denote $\tilde{a} = a^{-1}a'$ and $\tilde{b} = ba'^{N/2}$. Then, the subgroup Q_N consists of $\tilde{a}^m \tilde{b}^k$ with $m = 0, \ldots, N-1$ and $k = 0, 1$. Table 14.22 shows the representations of these generators $\tilde{a}$ and $\tilde{b}$ on each representation of $\Sigma(2N^2)$. Then the singlets of $\Sigma(2N^2)$ $\mathbf{1}_{+n}$ and $\mathbf{1}_{-n}$ become singlets of Q_N as $\mathbf{1}_{++}$, $\mathbf{1}_{--}$ for even n and $\mathbf{1}_{--}$, $\mathbf{1}_{++}$ for odd n. The decompositions of doublets are obtained in a way similar to the decomposition, $\Sigma(2N^2) \to D_N$: $\mathbf{2}_{q+k',q}$ and $\mathbf{2}_{q+\frac{N}{2}}$ are to $\mathbf{2}_{k'}$, $\mathbf{1}_{+-} + \mathbf{1}_{-+}$.

Table 14.23 Representations of $\tilde{a}$, $\tilde{a}'$ and $\tilde{b}$ in $\Sigma(2N^2)$

	$\mathbf{1}_{+n}$	$\mathbf{1}_{-n}$	$\mathbf{2}_{p,q}$
$\tilde{a}$	$\rho^{n\ell}$	$\rho^{n\ell}$	$\begin{pmatrix} \rho^{q\ell} & 0 \\ 0 & \rho^{p\ell} \end{pmatrix}$
$\tilde{a}'$	$\rho^{n\ell}$	$\rho^{n\ell}$	$\begin{pmatrix} \rho^{p\ell} & 0 \\ 0 & \rho^{q\ell} \end{pmatrix}$
b	1	-1	$\begin{pmatrix} 0 & 1 \\ 1 & 0 \end{pmatrix}$

Table 14.24 Breaking pattern of $\Sigma(2N^2)$. All parameters n, m, k, k', n' are integers

$\Sigma(2N^2)$	Z_{2N}
$\mathbf{1}_{+n}$	$\mathbf{1}_{2n}$
$\mathbf{1}_{-n}$	$\mathbf{1}_{2n+N}$
$\mathbf{2}_{\ell,m}$	$\mathbf{1}_{\ell+m} + \mathbf{1}_{\ell+m+N}$

$\Sigma(2N^2)$	$Z_N \times Z_N$
$\mathbf{1}_{+n}$	$\mathbf{1}_{n,n}$
$\mathbf{1}_{-n}$	$\mathbf{1}_{n,n}$
$\mathbf{2}_{\ell,m}$	$\mathbf{1}_{\ell,m} + \mathbf{1}_{m,\ell}$

$\Sigma(2N^2)$	D_N (N is even)
$\mathbf{1}_{+n}$	$\mathbf{1}_{++}$
$\mathbf{1}_{-n}$	$\mathbf{1}_{--}$
$\mathbf{2}_{\ell,m}$	$\mathbf{2}_{k'}(\ell = m + k')$
	$\tilde{\mathbf{2}}_{k'}(\ell = m - k')$
	$\mathbf{1}_{+-} + \mathbf{1}_{-+}(\ell = m + N/2)$

$\Sigma(2N^2)$	D_N (N is odd)
$\mathbf{1}_{+n}$	$\mathbf{1}_{+}$
$\mathbf{1}_{-n}$	$\mathbf{1}_{-}$
$\mathbf{2}_{\ell,m}$	$\mathbf{2}_{k'}(\ell = m + k')$
	$\tilde{\mathbf{2}}_{N-k'}(\ell = m - k')$

$\Sigma(2N^2)$	Q_N
$\mathbf{1}_{+n}$	$\mathbf{1}_{++}$ (n is even)
	$\mathbf{1}_{--}$ (n is odd)
$\mathbf{1}_{-n}$	$\mathbf{1}_{--}$ (n is even)
	$\mathbf{1}_{++}$ (n is odd)
$\mathbf{2}_{\ell,m}$	$\mathbf{2}_{k'}(\ell = m + k')$
	$\mathbf{1}_{+-} + \mathbf{1}_{-+}(\ell = m + N/2)$

$\Sigma(2N^2)$	$\Sigma(2M^2)$
$\mathbf{1}_{+n}$	$\mathbf{1}_{+n}$
$\mathbf{1}_{-n}$	$\mathbf{1}_{-n}$
$\mathbf{2}_{\ell,m}$	$\mathbf{2}_{\ell',m'}(\ell = \ell' + Mn, m = m' + Mn')$

- $\Sigma(2N^2) \to \Sigma(2M^2)$

We consider the subgroup $\Sigma(2M^2)$, where M is a divisor of N. We denote $\tilde{a} = a^\ell$ and $\tilde{a}' = a'^\ell$ with $\ell = N/M$, where $\ell =$ integer. The subgroup $\Sigma(2M^2)$ consists of $b^k \tilde{a}^m \tilde{a}'^n$ with $k = 0, 1$ and $m, n = 0, \ldots, M - 1$. Table 14.23 shows the representations of $\tilde{a}$, $\tilde{a}'$ and $\tilde{b}$ on each representation of $\Sigma(2N^2)$. Then, representations of $\Sigma(2N^2)$ $\mathbf{1}_{+n}$, $\mathbf{1}_{-n}$, and $\mathbf{2}_{p+Mn,q+Mn'}$ correspond to representations of $\Sigma(2M^2)$ as $\mathbf{1}_{+n}$, $\mathbf{1}_{-n}$, and $\mathbf{2}_{p,q}$, where n, n' are integers.

Table 14.25 Conjugacy classes and characters of $(Z_4 \times Z_2) \rtimes Z_2$

		h	$\chi_{\pm 0}$	$\chi_{\pm 1}$	$\chi_{\pm 2}$	$\chi_{\pm 3}$	χ_{2_1}	χ_{2_2}
C_1:	$\{e\}$,	1	1	1	1	1	2	2
$C_1^{(1)}$:	$\{\tilde{a}\tilde{a}'\}$,	4	1	-1	1	-1	$2i$	$-2i$
$C_1^{(2)}$:	$\{\tilde{a}^2\tilde{a}'^2\}$,	2	1	1	1	1	-2	-2
$C_1^{(3)}$:	$\{\tilde{a}^3\tilde{a}'^3\}$,	4	1	-1	1	-1	$-2i$	$2i$
$C_2^{(0)}$:	$\{b, b\tilde{a}^2\tilde{a}'^2\}$,	2	± 1	± 1	± 1	± 1	0	0
$C_2^{(0)}$:	$\{b\tilde{a}\tilde{a}', b\tilde{a}^3\tilde{a}'^3\}$,	4	± 1	∓ 1	± 1	∓ 1	0	0
$C_2^{(0)}$:	$\{b\tilde{a}^2, b\tilde{a}'^2\}$,	4	± 1	∓ 1	∓ 1	± 1	0	0
$C_2^{(0)}$:	$\{b\tilde{a}\tilde{a}'^3, b\tilde{a}^3\tilde{a}'\}$,	2	± 1	± 1	∓ 1	∓ 1	0	0
$C_2^{(2,0)}$:	$\{\tilde{a}^2, \tilde{a}'^2\}$,	2	1	-1	-1	1	0	0
$C_2^{(3,1)}$:	$\{\tilde{a}\tilde{a}'^3, \tilde{a}^3\tilde{a}'\}$,	2	1	1	-1	-1	0	0

14.12 $\Sigma(32)$

The $\Sigma(32)$ group includes subgroups, D_4, Q_4 and $\Sigma(8)$ as well as Abelian groups. In addition, it is useful to construct a discrete group as a subgroup of known groups. Here, we show one example $(Z_4 \times Z_2) \rtimes Z_2$ as a subgroup of $\Sigma(32) \simeq (Z_4 \times Z_4) \rtimes Z_2$.

All of the $\Sigma(32)$ elements are written by $b^k a^m a'^n$ with $k = 0, 1$ and $m, n = 0, 1, 2, 3$. The generators, a, a' and b, satisfy $a^4 = a'^4 = b^2 = e$, $aa' = a'a$ and $bab = a'$. Here we define $\tilde{a} = aa'$ and $\tilde{a}' = a^2$, where $\tilde{a}^4 = e$ and $\tilde{a}'^2 = e$. Then, the elements $b^k \tilde{a}^m \tilde{a}'^n$ with $k, n = 0, 1$ and $m = 0, 1, 2, 3$ construct a closed subalgebra, i.e. $(Z_4 \times Z_2) \rtimes Z_2$. It has ten conjugacy classes and eight singlets, $\mathbf{1}_{\pm 0}, \mathbf{1}_{\pm 1}, \mathbf{1}_{\pm 2}$ and $\mathbf{1}_{\pm 3}$, and two doublets, $\mathbf{2}_1$ and $\mathbf{2}_2$. These conjugacy classes and characters are shown in Table 14.25. From this table, we can find decompositions of $\Sigma(32)$ representations $\mathbf{1}_{\pm 0, \pm 1, \pm 2, \pm 3}, \mathbf{2}_{1,0}, \mathbf{2}_{3,2}, \mathbf{2}_{3,0}, \mathbf{2}_{2,1}, \mathbf{2}_{2,0}$, and $\mathbf{2}_{3,1}$ to representation of $(Z_4 \times Z_2) \rtimes Z_2$ as $\mathbf{1}_{\pm 0, \pm 1, \pm 0, \pm 1}, \mathbf{2}_1, \mathbf{2}_2, \mathbf{1}_{+3} + \mathbf{1}_{-3}$, and $\mathbf{1}_{+2} + \mathbf{1}_{-2}$, respectively.

14.13 General $\Delta(3N^2)$

All of the $\Delta(3N^2)$ elements are written by $b^k a^m a'^n$ with $k = 0, 1, 2$ and $m, n = 0, \ldots, N - 1$, where the generators, b, a and a', correspond to Z_3, Z_N and Z'_N of $(Z_N \times Z'_N) \rtimes Z_3$, respectively. Table 14.26 shows the representations of generators, b, a and a' on each representation of $\Delta(3N^2)$ for $N/3 \neq$ integer. Also Table 14.27 shows the same for $N/3 =$ integer.

In general, the $\Delta(3N^2)$ group includes subgroups, Z_3, $Z_N \times Z_N$ and $\Delta(3M^2)$ where M is a divisor of N. In addition, when N is a special value, it includes the T_N subgroup which is shown in Sect. 11. Here, we show above decompositions in what follows. The summary of the breaking patterns are shown in Table 14.30 for $N/3 =$ integer and Table 14.31 for $N/3 \neq$ integer.

Table 14.26 Representations of a, a' and b in $\Delta(3N^2)$ for $N/3 \neq$ integer

	$\mathbf{1}_k$	$\mathbf{3}_{[k][\ell]}$
a	1	$\begin{pmatrix} \rho^\ell & 0 & 0 \\ 0 & \rho^k & 0 \\ 0 & 0 & \rho^{-k-\ell} \end{pmatrix}$
a'	1	$\begin{pmatrix} \rho^{-k-\ell} & 0 & 0 \\ 0 & \rho^\ell & 0 \\ 0 & 0 & \rho^k \end{pmatrix}$
b	ω^k	$\begin{pmatrix} 0 & 1 & 0 \\ 0 & 0 & 1 \\ 1 & 0 & 0 \end{pmatrix}$

Table 14.27 Representations of a, a' and b in $\Delta(3N^2)$ for $N/3 =$ integer

	$\mathbf{1}_{k,\ell}$	$\mathbf{3}_{[k][\ell]}$
a	ω^ℓ	$\begin{pmatrix} \rho^\ell & 0 & 0 \\ 0 & \rho^k & 0 \\ 0 & 0 & \rho^{-k-\ell} \end{pmatrix}$
a'	ω^ℓ	$\begin{pmatrix} \rho^{-k-\ell} & 0 & 0 \\ 0 & \rho^\ell & 0 \\ 0 & 0 & \rho^k \end{pmatrix}$
b	ω^k	$\begin{pmatrix} 0 & 1 & 0 \\ 0 & 0 & 1 \\ 1 & 0 & 0 \end{pmatrix}$

- $\Delta(3N^2) \rightarrow Z_3$

The subgroup Z_3 consists of $\{e, b, b^2\}$. There are three singlet representations $\mathbf{1}_m$ with $m = 0, 1, 2$ for Z_3 and the generator b is represented as $b = \omega^m$ on $\mathbf{1}_m$. When $N/3 \neq$ integer, the representations of $\Delta(3N^2)$ $\mathbf{1}_k$ and $\mathbf{3}_{[k][\ell]}$ are decomposed to $\mathbf{1}_k$ and $\mathbf{1}_0 + \mathbf{1}_1 + \mathbf{1}_2$, respectively. On the other hand, when $N/3 =$ integer the representations of $\Delta(3N^2)$ $\mathbf{1}_{k,\ell}$ and $\mathbf{3}_{[k][\ell]}$ are decomposed to $\mathbf{1}_k$ and $\mathbf{1}_0 + \mathbf{1}_1 + \mathbf{1}_2$. In both cases, the triplet components (x_1, x_2, x_3) of $\Delta(3N^2)$ are decomposed to singlets of Z_3 as $\mathbf{1}_0 : x_1 + x_2 + x_3$, $\mathbf{1}_1 : x_1 + \omega^2 x_2 + \omega x_3$ and $\mathbf{1}_2 : x_1 + \omega x_2 + \omega^2 x_3$.

- $\Delta(3N^2) \rightarrow Z_N \times Z_N$

The subgroup $Z_N \times Z_N$ consists of $\{a^m a'^n\}$ with $m, n = 0, 1, \ldots N - 1$. There are N^2 singlet representations $\mathbf{1}_{m,n}$ and the generators a and a' are represented as $a = \rho^m$ and $a' = \rho^n$ on $\mathbf{1}_{m,n}$. When $N/3 \neq$ integer, the representations of $\Delta(3N^2)$ $\mathbf{1}_k$ and $\mathbf{3}_{[k][\ell]}$ are decomposed to $\mathbf{1}_{0,0}$ and $\mathbf{1}_{\ell,-k-\ell} + \mathbf{1}_{k,\ell} + \mathbf{1}_{-k-\ell,k}$, respectively. In addition, when $N/3 =$ integer, the representations $\mathbf{1}_{k,\ell}$ and $\mathbf{3}_{[k][\ell]}$ are decomposed to $\mathbf{1}_{N\ell/3,N\ell/3}$ and $\mathbf{1}_{\ell,-k-\ell} + \mathbf{1}_{k,\ell} + \mathbf{1}_{-k-\ell,k}$, respectively.

• $\Delta(3N^2) \to T_N$

When N is a prime number or its power except 3, $\Delta(3N^2)$ has a subgroup of T_N. In the basis of $\Delta(3N^2)$ and T_N, used in the text, the subgroup T_N consists of $\{\tilde{a}^n b^k\}$ where $\tilde{a} = a^{-m^2} a'^m$, $n = 0, 1, \ldots N - 1$, $k = 0, 1, 2$, and m is defined in the section of T_N. For each representation $\mathbf{3}_{[k][\ell]}$, we have

$$\tilde{a} = \begin{pmatrix} \rho^{\ell-mk} & 0 & 0 \\ 0 & \rho^{m(\ell-mk)} & 0 \\ 0 & 0 & \rho^{m^2(\ell-mk)} \end{pmatrix}. \tag{14.15}$$

This can be compared with representations of $\mathbf{3}_{[k]}$, i.e. the triplets of T_N:

$$\tilde{a} = \begin{pmatrix} \rho^k & 0 & 0 \\ 0 & \rho^{km} & 0 \\ 0 & 0 & \rho^{km^2} \end{pmatrix}. \tag{14.16}$$

Thus the triplets $\mathbf{3}_{[k][\ell]}$ of $\Delta(3N^2)$ are decomposed to $\mathbf{3}_{[\ell-mk]}$ of T_N. On the other hand, if $\ell - mk = 0 \pmod{N}$, the triplets are decomposed to singlets $\mathbf{1}_0 + \mathbf{1}_1 + \mathbf{1}_2$ of T_N and their components (x_1, x_2, x_3) correspond to

$$\mathbf{1}_0 : x_1 + x_2 + x_3, \qquad \mathbf{1}_1 : x_1 + \omega^2 x_2 + \omega x_3, \qquad \mathbf{1}_2 : x_1 + \omega x_2 + \omega^2 x_3. \tag{14.17}$$

For instance, when $N = 7$, there are 16 triplets in $\Delta(147)$ and two triplets in T_7. Triplets $\mathbf{3}_{[0][1]}$, $\mathbf{3}_{[0][2]}$, $\mathbf{3}_{[0][4]}$, $\mathbf{3}_{[1][3]}$, $\mathbf{3}_{[1][4]}$, $\mathbf{3}_{[2][6]}$, $\mathbf{3}_{[4][5]}$ are decomposed to $\mathbf{3}_1$ of T_7. In the same way, triplets $\mathbf{3}_{[0][3]}$, $\mathbf{3}_{[0][5]}$, $\mathbf{3}_{[0][6]}$, $\mathbf{3}_{[1][1]}$, $\mathbf{3}_{[2][2]}$, $\mathbf{3}_{[3][5]}$, $\mathbf{3}_{[4][4]}$ are decomposed to $\mathbf{3}_2$ of T_7. Remaining ones, i.e. $\mathbf{3}_{[1][2]}$ and $\mathbf{3}_{[3][6]}$, are decomposed to $\mathbf{1}_0 + \mathbf{1}_1 + \mathbf{1}_2$ of T_7.

For singlets, because representations are $a = 1$, $a' = 1$, $\tilde{a} = 1$, and $b = \omega^n$ in both of $\Delta(3N^2)$ and T_N, the ones of them are correspondent.

• $\Delta(3N^2) \to \Delta(3M^2)$

We consider the subgroup $\Delta(3M^2)$, where M is a divisor of N. We denote $\tilde{a} = a^p$ and $\tilde{a}' = a'^p$ with $p = N/M$, where $p =$ integer. The subgroup $\Delta(3M^2)$ consists of $b^k \tilde{a}^m \tilde{a}'^n$ with $k = 0, 1, 2$ and $m, n = 0, \ldots, M - 1$. Table 14.28 shows the representations of $\tilde{a}$, $\tilde{a}'$ and $\tilde{b}$ on each representation of $\Delta(3N^2)$ for $N/3 =$ integer. In addition, Table 14.29 shows the representations of $\tilde{a}$, $\tilde{a}'$ and $\tilde{b}$ on each representation of $\Delta(3N^2)$ for $N/3 \neq$ integer. There are three types of combinations (N, M), i.e. (1) both $N/3$ and $M/3$ are integers, (2) $N/3$ is integer, but $M/3$ is not integer, (3) neither $N/3$ nor $M/3$ is integer.

When both $N/3$ and $M/3$ are integers, the representations of $\Delta(3N^2)$ $\mathbf{1}_{k,\ell}$ and $\mathbf{3}_{[k+Mn][\ell+Mn']}$ are decomposed to representations of $\Delta(3M^2)$ as $\mathbf{1}_{k,p\ell}$ and $\mathbf{3}_{[k][\ell]}$, where n and n' are integers.

Table 14.28 Representations of $\tilde{a}$, $\tilde{a}'$ and $\tilde{b}$ in $\Delta(3N^2)$ for $N/3 = $ integer

	$\mathbf{1}_{k,\ell}$	$\mathbf{3}_{[k][\ell]}$
$\tilde{a}$	$\omega^{p\ell}$	$\begin{pmatrix} \rho^{p\ell} & 0 & 0 \\ 0 & \rho^{pk} & 0 \\ 0 & 0 & \rho^{-p(k+\ell)} \end{pmatrix}$
$\tilde{a}'$	$\omega^{p\ell}$	$\begin{pmatrix} \rho^{-p(k+\ell)} & 0 & 0 \\ 0 & \rho^{p\ell} & 0 \\ 0 & 0 & \rho^{pk} \end{pmatrix}$
b	ω^{k}	$\begin{pmatrix} 0 & 1 & 0 \\ 0 & 0 & 1 \\ 1 & 0 & 0 \end{pmatrix}$

Table 14.29 Representations of $\tilde{a}$, $\tilde{a}'$ and $\tilde{b}$ in $\Delta(3N^2)$ for $N/3 \neq$ integer

	$\mathbf{1}_{k}$	$\mathbf{3}_{[k][\ell]}$
$\tilde{a}$	1	$\begin{pmatrix} \rho^{p\ell} & 0 & 0 \\ 0 & \rho^{pk} & 0 \\ 0 & 0 & \rho^{-p(k+\ell)} \end{pmatrix}$
$\tilde{a}'$	1	$\begin{pmatrix} \rho^{-p(k+\ell)} & 0 & 0 \\ 0 & \rho^{p\ell} & 0 \\ 0 & 0 & \rho^{pk} \end{pmatrix}$
b	ω^{k}	$\begin{pmatrix} 0 & 1 & 0 \\ 0 & 0 & 1 \\ 1 & 0 & 0 \end{pmatrix}$

Next we consider the case that $N/3 = $ integer and $M/3 \neq$ integer, where $p = N/M$ must be $3n$. In this case, the representations of $\Delta(3N^2)$ $\mathbf{1}_{k,\ell}$ and $\mathbf{3}_{[k+Mn][\ell+Mn']}$ are decomposed to representations of $\Delta(3M^2)$ as $\mathbf{1}_k$ and $\mathbf{3}_{[k][\ell]}$, respectively.

The last case is that neither $N/3$ nor $M/3$ is integer. In this case, the representations of $\Delta(3N^2)$ $\mathbf{1}_k$ and $\mathbf{3}_{[k+Mn][\ell+Mn']}$ are decomposed to representations of $\Delta(3M^2)$ as $\mathbf{1}_k$ and $\mathbf{3}_{[k][\ell]}$.

14.14 $\Delta(27)$

$\Delta(27)$ elements are written by $b^k a^m a'^n$ with $k = 0, 1, 2$ and $m, n = 0, 1, 2$, where the generators, b, a and a', correspond to Z_3 groups of $(Z_3 \times Z_3) \rtimes Z_3$. Table 14.32 shows the representations of generators, b, a and a' on each representation of $\Delta(27)$. As subgroups, the $\Delta(27)$ group includes Abelian subgroups Z_3 and $Z_3 \times Z_3$. The breaking pattern of $\Delta(27)$ is summarized in Table 14.33.

Table 14.30 Breaking pattern of $\Delta(3N^2)$ for $N/3 =$ integer

$\Delta(3N^2)$	Z_3	$\Delta(3N^2)$	$Z_N \times Z_N$
$\mathbf{1}_{k,\ell}$	$\mathbf{1}_k$	$\mathbf{1}_{k,\ell}$	$\mathbf{1}_{N\ell/3,N\ell/3}$
$\mathbf{3}_{[k][\ell]}$	$\mathbf{1}_0 + \mathbf{1}_1 + \mathbf{1}_2$	$\mathbf{3}_{[k][\ell]}$	$\mathbf{1}_{\ell,-k-\ell} + \mathbf{1}_{k,\ell} + \mathbf{1}_{-k-\ell,k}$

$\Delta(3N^2)$	$\Delta(3M^2)$ ($M/3 =$ integer)
$\mathbf{1}_{k,\ell}$	$\mathbf{1}_{k,N\ell/M}$
$\mathbf{3}_{[k][\ell]}$	$\mathbf{3}_{[k'][\ell']}$ ($k = k' + Mn, \ell = \ell' + Mn'$)

$\Delta(3N^2)$	$\Delta(3M^2)$ ($M/3 \neq$ integer)
$\mathbf{1}_{k,\ell}$	$\mathbf{1}_k$
$\mathbf{3}_{[k][\ell]}$	$\mathbf{3}_{[k'][\ell']}$ ($k = k' + Mn, \ell = \ell' + Mn'$)

Table 14.31 Breaking pattern of $\Delta(3N^2)$ for $N/3 \neq$ integer

$\Delta(3N^2)$	Z_3	$\Delta(3N^2)$	$Z_N \times Z_N$
$\mathbf{1}_k$	$\mathbf{1}_k$	$\mathbf{1}_k$	$\mathbf{1}_{0,0}$
$\mathbf{3}_{[k][\ell]}$	$\mathbf{1}_0 + \mathbf{1}_1 + \mathbf{1}_2$	$\mathbf{3}_{[k][\ell]}$	$\mathbf{1}_{\ell,-k-\ell} + \mathbf{1}_{k,\ell} + \mathbf{1}_{-k-\ell,k}$

$\Delta(3N^2)$	T_N
$\mathbf{1}_k$	$\mathbf{1}_k$
$\mathbf{3}_{[k][\ell]}$	$\mathbf{3}_{[l-mk]}$

$\Delta(3N^2)$	$\Delta(3M^2)$
$\mathbf{1}_k$	$\mathbf{1}_k$
$\mathbf{3}_{[k][\ell]}$	$\mathbf{3}_{[k'][\ell']}$ ($k = k' + Mn, \ell = \ell' + Mn'$)

- $\Delta(27) \rightarrow Z_3$

The subgroup Z_3 consists of $\{e, b, b^2\}$. There are three singlet representations $\mathbf{1}_m$ with $m = 0, 1, 2$ for Z_3 and the generator b is represented as $b = \omega^m$ on $\mathbf{1}_m$. When $N/3 \neq$ integer, the representations of $\Delta(27)$ $\mathbf{1}_{k,\ell}$ and $\mathbf{3}_{[k][\ell]}$ are decomposed to $\mathbf{1}_k$ and $\mathbf{1}_0 + \mathbf{1}_1 + \mathbf{1}_2$, respectively. The triplet components (x_1, x_2, x_3) are decomposed to singlets of Z_3 as $\mathbf{1}_0 : x_1 + x_2 + x_3$, $\mathbf{1}_1 : x_1 + \omega^2 x_2 + \omega x_3$ and $\mathbf{1}_2 : x_1 + \omega x_2 + \omega^2 x_3$.

Table 14.32 Representations of a, a' and b in $\Delta(27)$

	$\mathbf{1}_{k,\ell}$	$\mathbf{3}_{[k][\ell]}$
a	ω^ℓ	$\begin{pmatrix} \omega^\ell & 0 & 0 \\ 0 & \omega^k & 0 \\ 0 & 0 & \omega^{-k-\ell} \end{pmatrix}$
a'	ω^ℓ	$\begin{pmatrix} \omega^{-k-\ell} & 0 & 0 \\ 0 & \omega^\ell & 0 \\ 0 & 0 & \omega^k \end{pmatrix}$
b	ω^k	$\begin{pmatrix} 0 & 1 & 0 \\ 0 & 0 & 1 \\ 1 & 0 & 0 \end{pmatrix}$

Table 14.33 Breaking pattern of $\Delta(27)$

$\Delta(27)$	Z_3	$\Delta(27)$	$Z_3 \times Z_3$
$\mathbf{1}_{k,\ell}$	$\mathbf{1}_k$	$\mathbf{1}_{k,\ell}$	$\mathbf{1}_{\ell,\ell}$
$\mathbf{3}_{[k][\ell]}$	$\mathbf{1}_0 + \mathbf{1}_1 + \mathbf{1}_2$	$\mathbf{3}_{[k][\ell]}$	$\mathbf{1}_{\ell,-k-\ell} + \mathbf{1}_{k,\ell} + \mathbf{1}_{-k-\ell,k}$

- $\Delta(27) \to Z_3 \times Z_3$

The subgroup $Z_3 \times Z_3$ consists of $\{a^m a'^n\}$ with $m, n = 0, 1, 2$. There are 9 singlet representations $\mathbf{1}_{m,n}$ and the generators a and a' are represented as $a = \omega^m$ and $a' = \omega^n$ on $\mathbf{1}_{m,n}$. The representations of $\Delta(27)$ $\mathbf{1}_{k,\ell}$ and $\mathbf{3}_{[k][\ell]}$ are decomposed to $\mathbf{1}_{\ell,\ell}$ and $\mathbf{1}_{\ell,-k-\ell} + \mathbf{1}_{k,\ell} + \mathbf{1}_{-k-\ell,k}$, respectively.

14.15 General T_N

Since the group T_N is isomorphic to $Z_N \rtimes Z_3$, Z_N and Z_3 appear as subgroups. Recall that all of T_N elements are written by $a^m b^k$ with $m = 0, \ldots, N-1$ and $k = 0, 1, 2$. There are three singlets $\mathbf{1}_k$ and $2(N-1)/6$ triplets $\mathbf{3}_{[k']}$ and $\bar{\mathbf{3}}_{[k']}$ with $k' = 1, \ldots, (N-1)/6$. On the triplets $\mathbf{3}_{[k]}$, the generators a and b are represented as

$$b = \begin{pmatrix} 0 & 1 & 0 \\ 0 & 0 & 1 \\ 1 & 0 & 0 \end{pmatrix}, \quad a = \begin{pmatrix} \rho^k & 0 & 0 \\ 0 & \rho^{km} & 0 \\ 0 & 0 & \rho^{km^2} \end{pmatrix}, \tag{14.18}$$

where $\rho = e^{2\pi i/N}$. For each N, we have the value of m, see the section of T_N. The representations of $\bar{\mathbf{3}}_{[k]}$ can be obtained by changing ρ to ρ^{-1}. In the followings, we show the general breaking patterns of $T_N \to Z_3$ and $T_N \to Z_N$ in detail. The summary of the breaking patterns is shown in Table 14.34.

Table 14.34 Breaking pattern of T_N

T_N	Z_3	T_N	Z_N
1_k	1_k	1_k	1_0
$3_{[k]}$	$1_0 + 1_1 + 1_2$	$3_{[k]}$	$1_k + 1_{km} + 1_{km^2}$
$\bar{3}_{[k]}$	$1_0 + 1_1 + 1_2$	$\bar{3}_{[k]}$	$1_{-k} + 1_{-km} + 1_{-km^2}$

- $T_N \twoheadrightarrow Z_3$

The two elements e and b construct the Z_3 subgroup. Obviously, there are three singlet representations $1_0, 1_1, 1_2$. That is, we have $b = 1$ on 1_0, $b = \omega$ on 1_1 and $b = \omega^2$ on 1_2. The singlets 1_k of T_N become 1_k of Z_3. The triplets $3_{[k]}$ of T_N, (x_k, x_{km}, x_{km^2}), are decomposed to three singlets as $1_0 : x_k + x_{km} + x_{km^2}, 1_1 : x_k + \omega^2 x_{km} + \omega x_{km^2}$, and $1_2 : x_k + \omega x_{km} + \omega^2 x_{km^2}$. Decompositions of triplets $\bar{3}_{[k]}$ are the same.

- $T_N \twoheadrightarrow Z_N$

The subgroup Z_N consists of the elements $\{e, a, \ldots, a^{N-1}\}$. This is the normal subgroups of T_N and there are N types of irreducible singlet representations $1_0, 1_1, \ldots, 1_{N-1}$. On the 1_k of T_N, the generator a is represented as $a = 1$. Therefore the singlets 1_k of T_N become 1_0 of Z_N. The triplets $3_{[k]} : (x_k, x_{km}, x_{km^2})$ are decomposed to three singlets $1_k, 1_{km}$, and 1_{km^2}. The triplets $\bar{3}_{[k]}$ are decomposed to $1_{-k}, 1_{-km}$, and 1_{-km^2}.

14.16 T_7

All of the T_7 elements are written as $b^m a^n$ with $m = 0, 1, 2$ and $n = 0, \ldots, 6$, where $b^3 = e$ and $a^7 = e$. Table 14.35 shows the representation of generators a and b on each representation of T_7. As subgroups, the T_7 group includes Abelian subgroups Z_3 and Z_7. The summary of the breaking pattern is shown in Table 14.36.

Table 14.35 Representations of a and b in T_7

	1_0	1_1	1_2	3	$\bar{3}$
a	1	1	1	$\begin{pmatrix} \rho & 0 & 0 \\ 0 & \rho^2 & 0 \\ 0 & 0 & \rho^4 \end{pmatrix}$	$\begin{pmatrix} \rho^{-1} & 0 & 0 \\ 0 & \rho^{-2} & 0 \\ 0 & 0 & \rho^{-4} \end{pmatrix}$
b	1	ω	ω^2	$\begin{pmatrix} 0 & 1 & 0 \\ 0 & 0 & 1 \\ 1 & 0 & 0 \end{pmatrix}$	$\begin{pmatrix} 0 & 1 & 0 \\ 0 & 0 & 1 \\ 1 & 0 & 0 \end{pmatrix}$

Table 14.36 Breaking pattern of T_7

T_7	Z_3	T_7	Z_7
1_k	1_k	1_k	1_0
3	$1_0 + 1_1 + 1_2$	3	$1_1 + 1_2 + 1_4$
$\bar{3}$	$1_0 + 1_1 + 1_2$	$\bar{3}$	$1_3 + 1_5 + 1_6$

- $T_7 \to Z_3$

The subgroup Z_3 consists of $\{e, b, b^2\}$. The three singlet representations 1_m of Z_3 with $m = 0, 1, 2$ are specified such that $b = \omega^m$ on 1_m. Then, the representations of T_7 $1_0, 1_1, 1_2, 3$, and $\bar{3}$ are decomposed to $1_0, 1_1, 1_2, 1_0 + 1_1 + 1_2$, and $1_0 + 1_1 + 1_2$, respectively. Here the T_7 triplet $3 : (x_1, x_2, x_4)$ and triplet $\bar{3} : (x_6, x_5, x_3)$ decompose to three singlets, $1_0 + 1_1 + 1_2$, and their components correspond to

$$1_0 : x_1 + x_2 + x_4, \quad 1_1 : x_1 + \omega^2 x_2 + \omega x_4, \quad 1_2 : x_1 + \omega x_2 + \omega^2 x_4,$$
$$1_0 : x_6 + x_5 + x_3, \quad 1_1 : x_6 + \omega^2 x_5 + \omega x_3, \quad 1_2 : x_6 + \omega x_5 + \omega^2 x_3.$$

$$(14.19)$$

- $T_7 \to Z_7$

The subgroup Z_7 consists of a^n with $n = 0, \ldots, 6$. The seven singlets 1_m of Z_7 with $m = 0, \ldots, 6$ are specified such that $b = \rho^m$ on 1_m where $\rho = e^{2\pi i/7}$. Then, the representations of T_7 $1_0, 1_1, 1_2, 3$, and $\bar{3}$ are decomposed to $1_0, 1_0, 1_0, 1_1 + 1_2 + 1_4$, and $1_3 + 1_5 + 1_6$, respectively.

14.17 General $\Sigma(3N^3)$

Recall that all of the $\Sigma(3N^3)$ elements are written by $b^k a^\ell a'^m a''^n$ with $k = 0, 1, 2$ and $\ell, m, n = 0, 1, \ldots, N - 1$. As subgroups, the group $\Sigma(3N^3)$ includes $Z_3, Z_N \times Z_N \times Z_N, \Delta(3N^2)$ and $\Sigma(3M^3)$ where M is a divisor of N. The summary of the breaking pattern of $\Sigma(3N^3)$ is shown in Table 14.37.

- $\Sigma(3N^2) \to Z_N \times Z_N \times Z_N$

The subgroup $Z_N \times Z_N \times Z_N$ consists of the elements $a^k a'^\ell a''^m$ with $k, \ell, m = 0, \ldots, N - 1$. Obviously it is the normal subgroups of $\Sigma(3N^3)$. There are N^3 singlet representations $1_{k,\ell,m}$ and the generators a, a', and a'' are represented as $a = \rho^k$, $a' = \rho^\ell$, and $a'' = \rho^m$ on $1_{k,\ell,m}$. Then, the representations of $\Sigma(3N^3)$ $1_{k,\ell}$ and $3_{[k][\ell][m]}$ are decomposed to $1_{\ell,\ell,\ell}$ and $1_{k,\ell,m} + 1_{m,k,\ell} + 1_{\ell,m,k}$, respectively.

Table 14.37 Breaking pattern of $\Sigma(3N^3)$

$\Sigma(3N^3)$	Z_3
$\mathbf{1}_{k,\ell}$	$\mathbf{1}_k$
$\mathbf{3}_{[k][\ell][m]}$	$\mathbf{1}_0 + \mathbf{1}_1 + \mathbf{1}_2$

$\Sigma(3N^3)$	$Z_N \times Z_N \times Z_N$
$\mathbf{1}_{k,\ell}$	$\mathbf{1}_{\ell,\ell,\ell}$
$\mathbf{3}_{[k][\ell][m]}$	$\mathbf{1}_{k,\ell,m} + \mathbf{1}_{m,k,\ell} + \mathbf{1}_{\ell,m,k}$

$\Sigma(3N^3)$	$\Delta(3N^2)$ ($N/3 \neq$ integer)
$\mathbf{1}_{k,\ell}$	$\mathbf{1}_k$
$\mathbf{3}_{[k][\ell][m]}$	$\mathbf{3}_{[\ell-k][k-m]}$

$\Sigma(3N^3)$	$\Delta(3N^2)$ ($N/3 =$ integer)
$\mathbf{1}_{k,\ell}$	$\mathbf{1}_{k,\ell}$
$\mathbf{3}_{[k][\ell][m]}$	$\mathbf{3}_{[\ell-k][k-m]}$

$\Sigma(3N^3)$	$\Sigma(3M^2)$
$\mathbf{1}_{k,\ell+Mn}$	$\mathbf{1}_{k,\ell}$
$\mathbf{3}_{[k+Mn][\ell+Mn'][m+Mn'']}$	$\mathbf{3}_{[k][\ell][m]}$

- $\Sigma(3N^3) \to \Delta(3N^2)$

The subgroup $\Delta(3N^2)$ consists of the elements $\tilde{a}^k \tilde{a}'^\ell b^m$ with $k, \ell = 0, \ldots, N-1$ and $m = 0, 1, 2$. There are 3 singlet representations $\mathbf{1}_k$ and $(N^2-1)/3$ triplet representations $\mathbf{3}_{[k][\ell]}$ for $N/3 \neq$ integer and 9 singlet representations $\mathbf{1}_{k,\ell}$ and $(N^2-3)/3$ triplet representations $\mathbf{3}_{[k][\ell]}$ for $N/3 =$ integer. The generators $\tilde{a}, \tilde{a}'$ are represented as $\tilde{a} = aa'^{-1}$ and $\tilde{a} = a'a''^{-1}$. Then, the representations of $\Sigma(3N^3)$ $\mathbf{1}_{k,\ell}$ and $\mathbf{3}_{[k][\ell][m]}$ are decomposed to $\mathbf{1}_k$ and $\mathbf{3}_{[\ell-k][k-m]}$ for $N/3 \neq$ integer and $\mathbf{1}_{k,\ell}$ and $\mathbf{3}_{[\ell-k][k-m]}$ for $N/3 =$ integer.

- $\Sigma(3N^3) \to \Sigma(3M^3)$

We consider the subgroup $\Sigma(3M^3)$, where M is a divisor of N. We denote $\tilde{a} = a^p$, $\tilde{a}' = a'^p$, and $\tilde{a}'' = a''^p$ with $p = N/M$, where $p =$ integer. The subgroup $\Sigma(3M^3)$ consists of $b^k \tilde{a}^\ell \tilde{a}'^m \tilde{a}''^n$ with $k = 0, 1, 2$ and $\ell, m, n = 0, \ldots, M-1$. The singlets $\mathbf{1}_{k,\ell+Mn}$ and triplets $\mathbf{3}_{[k+Mn][\ell+Mn'][m+Mn'']}$ of $\Sigma(3N^3)$ correspond to singlets $\mathbf{1}_{k,\ell}$ and triplets $\mathbf{3}_{[k][\ell][m]}$ of $\Sigma(3M^3)$ where n, n', n'' are integers.

14.18 $\Sigma(81)$

All of the $\Sigma(81)$ elements are written as $b^k a^\ell a'^m a''^n$ with $k, \ell, m, n = 0, 1, 2$, where these generators satisfy $a^3 = a'^3 = a''^3 = 1$, $aa' = a'a$, $aa'' = a''a$, $a''a' = a'a''$, $b^3 = 1$, $b^2 ab = a''$, $b^2 a'b = a$ and $b^2 a''b = a'$. Table 14.38 shows the representations of generators, b, a, a' and a'' on each representation of $\Sigma(81)$. As subgroups, the group $\Sigma(81)$ includes $Z_3 \times Z_3 \times Z_3$ and $\Delta(27)$. The breaking pattern of $\Sigma(81)$ is summarized in Table 14.40.

Table 14.38 Representations of a, a', a'' and b in $\Sigma(81)$

	$1_{k,\ell}$	3_A	3_B	3_C	3_D
a	ω^ℓ	$\begin{pmatrix} \omega & 0 & 0 \\ 0 & 1 & 0 \\ 0 & 0 & 1 \end{pmatrix}$	$\begin{pmatrix} \omega^2 & 0 & 0 \\ 0 & \omega & 0 \\ 0 & 0 & \omega \end{pmatrix}$	$\begin{pmatrix} 1 & 0 & 0 \\ 0 & \omega^2 & 0 \\ 0 & 0 & \omega^2 \end{pmatrix}$	$\begin{pmatrix} \omega^2 & 0 & 0 \\ 0 & 1 & 0 \\ 0 & 0 & \omega \end{pmatrix}$
a'	ω^ℓ	$\begin{pmatrix} 1 & 0 & 0 \\ 0 & \omega & 0 \\ 0 & 0 & 1 \end{pmatrix}$	$\begin{pmatrix} \omega & 0 & 0 \\ 0 & \omega^2 & 0 \\ 0 & 0 & \omega \end{pmatrix}$	$\begin{pmatrix} \omega^2 & 0 & 0 \\ 0 & 1 & 0 \\ 0 & 0 & \omega^2 \end{pmatrix}$	$\begin{pmatrix} \omega & 0 & 0 \\ 0 & \omega^2 & 0 \\ 0 & 0 & 1 \end{pmatrix}$
a''	ω^ℓ	$\begin{pmatrix} 1 & 0 & 0 \\ 0 & 1 & 0 \\ 0 & 0 & \omega \end{pmatrix}$	$\begin{pmatrix} \omega & 0 & 0 \\ 0 & \omega & 0 \\ 0 & 0 & \omega^2 \end{pmatrix}$	$\begin{pmatrix} \omega^2 & 0 & 0 \\ 0 & \omega^2 & 0 \\ 0 & 0 & 1 \end{pmatrix}$	$\begin{pmatrix} 1 & 0 & 0 \\ 0 & \omega & 0 \\ 0 & 0 & \omega^2 \end{pmatrix}$
b	ω^k	$\begin{pmatrix} 0 & 1 & 0 \\ 0 & 0 & 1 \\ 1 & 0 & 0 \end{pmatrix}$	$\begin{pmatrix} 0 & 1 & 0 \\ 0 & 0 & 1 \\ 1 & 0 & 0 \end{pmatrix}$	$\begin{pmatrix} 0 & 1 & 0 \\ 0 & 0 & 1 \\ 1 & 0 & 0 \end{pmatrix}$	$\begin{pmatrix} 0 & 1 & 0 \\ 0 & 0 & 1 \\ 1 & 0 & 0 \end{pmatrix}$

Table 14.39 Representations of b, $\tilde{a}$ and $\tilde{a}'$ of $\Delta(27)$ in $\Sigma(81)$

	$1_{k,\ell}$	3_A	3_B	3_C	3_D
$\tilde{a}$	1	$\begin{pmatrix} \omega & 0 & 0 \\ 0 & 1 & 0 \\ 0 & 0 & \omega^2 \end{pmatrix}$	$\begin{pmatrix} \omega & 0 & 0 \\ 0 & 1 & 0 \\ 0 & 0 & \omega^2 \end{pmatrix}$	$\begin{pmatrix} \omega & 0 & 0 \\ 0 & 1 & 0 \\ 0 & 0 & \omega^2 \end{pmatrix}$	$\begin{pmatrix} \omega^2 & 0 & 0 \\ 0 & \omega^2 & 0 \\ 0 & 0 & \omega^2 \end{pmatrix}$
$\tilde{a}'$	1	$\begin{pmatrix} \omega^2 & 0 & 0 \\ 0 & \omega & 0 \\ 0 & 0 & 1 \end{pmatrix}$	$\begin{pmatrix} \omega^2 & 0 & 0 \\ 0 & \omega & 0 \\ 0 & 0 & 1 \end{pmatrix}$	$\begin{pmatrix} \omega^2 & 0 & 0 \\ 0 & \omega & 0 \\ 0 & 0 & 1 \end{pmatrix}$	$\begin{pmatrix} \omega^2 & 0 & 0 \\ 0 & \omega^2 & 0 \\ 0 & 0 & \omega^2 \end{pmatrix}$
b	ω^k	$\begin{pmatrix} 0 & 1 & 0 \\ 0 & 0 & 1 \\ 1 & 0 & 0 \end{pmatrix}$	$\begin{pmatrix} 0 & 1 & 0 \\ 0 & 0 & 1 \\ 1 & 0 & 0 \end{pmatrix}$	$\begin{pmatrix} 0 & 1 & 0 \\ 0 & 0 & 1 \\ 1 & 0 & 0 \end{pmatrix}$	$\begin{pmatrix} 0 & 1 & 0 \\ 0 & 0 & 1 \\ 1 & 0 & 0 \end{pmatrix}$

- $\Sigma(81) \to Z_3 \times Z_3 \times Z_3$

The subgroup $Z_3 \times Z_3 \times Z_3$ consists of $\{e, a, a^2, a', a'^2, a'', a''^2, \ldots\}$. There are 3^3 singlets $1_{k,\ell,m}$ of $Z_3 \times Z_3 \times Z_3$ and the generators, a, a' and a'', are represented on $1_{k,\ell,m}$ as $a = \omega^k$, $a' = \omega^\ell$ and $a'' = \omega^m$. Then, the singlet of $\Sigma(81)$ $1_{k,\ell}$ is decomposed to $1_{\ell,\ell,\ell}$. For triplets, they become three different singlets of $Z_3 \times Z_3 \times Z_3$, see Table 14.18 for the detail.

- $\Sigma(81) \to \Delta(27)$

The subgroup $\Delta(27)$ consists of $b^k \tilde{a}^m \tilde{a}'^n$, where $\tilde{a} = a^2 a''$ and $\tilde{a}' = a' a''^2$. Table 14.39 shows the representations of the generators, b, $\tilde{a}$ and $\tilde{a}'$ on each representation of $\Sigma(81)$. Then, the singlet $1_{k,\ell}$ is decomposed to $1_{k,0}$ of $\Delta(27)$. Triplets $3_{A,B,C}$ and $\bar{3}_{A,B,C}$ become triplets $3_{[0][1]}$ and $3_{[0][2]}$, respectively, while 3_D and $\bar{3}_D$ correspond to three singlets $1_{0,2} + 1_{1,2} + 1_{2,2}$ and $1_{0,1} + 1_{2,1} + 1_{1,1}$.

Table 14.40 Breaking pattern of $\Sigma(81)$

$\Sigma(81)$	$Z_3 \times Z_3 \times Z_3$	$\Sigma(81)$	$\Delta(27)$
$1_{k,\ell}$	$1_{\ell,\ell,\ell}$	$1_{k,\ell}$	$1_{k,0}$
3_A	$1_{0,0,0} + 1_{0,1,0} + 1_{0,0,1}$	3_A	$3_{[0][1]}$
3_B	$1_{2,1,1} + 1_{1,2,1} + 1_{1,1,2}$	3_B	$3_{[0][1]}$
3_C	$1_{0,2,2} + 1_{2,0,2} + 1_{2,2,0}$	3_C	$3_{[0][1]}$
3_D	$1_{2,1,0} + 1_{0,2,1} + 1_{1,0,2}$	3_D	$1_{0,2} + 1_{1,2} + 1_{2,2}$
$\bar{3}_A$	$1_{2,0,0} + 1_{0,2,0} + 1_{0,0,2}$	$\bar{3}_A$	$3_{[0][2]}$
$\bar{3}_B$	$1_{1,2,2} + 1_{2,1,2} + 1_{2,2,1}$	$\bar{3}_B$	$3_{[0][2]}$
$\bar{3}_C$	$1_{0,1,1} + 1_{1,0,1} + 1_{1,1,0}$	$\bar{3}_C$	$3_{[0][2]}$
$\bar{3}_D$	$1_{1,2,0} + 1_{0,1,2} + 1_{2,0,1}$	$\bar{3}_D$	$1_{0,1} + 1_{1,1} + 1_{2,1}$

14.19 General $\Delta(6N^2)$

All of the $\Delta(6N^2)$ elements are written by $b^k c^\ell a^m a'^n$ with $k = 0, 1, 2, \ell = 0, 1,$ and $m, n = 0, \ldots, N - 1$, where the generators, b, c, a and a', correspond to Z_3, Z_2, Z_N and Z_N' of $(Z_N \times Z_N') \rtimes S_3$, respectively. We show the possible different breaking patterns, precisely, $\Delta(6N^2)$ can break into $\Sigma(2N^2)$ and $\Delta(3N^2)$. The breaking pattern of $\Delta(6N^2)$ is summarized in Table 14.43 for $N/3 \neq$ integer and Table 14.44 for $N/3 =$ integer. For further breaking of these two groups, see Sect. 14.11 for $\Sigma(2N^2)$ and Sect. 14.13 for $\Delta(3N^2)$.

- $\Delta(6N^2) \to \Sigma(2N^2)$
The subgroup $\Sigma(2N^2)$ consists of $\{c^\ell a^{-m} a'^n\}$. There are $2N$ singlets and $N(N - 1)/2$ doublets for $\Sigma(2N^2)$. The generator is summarized in Table 14.41. When $N/3 \neq$ integer, singlets 1_0 and 1_1 correspond to 1_{+0} and 1_{-0}. The doublet 2 becomes $1_{+0} + 1_{-0}$. Triplets 3_{1k} and 3_{2k} are decomposed to $1_{+k} + 2_{0,-k}$ and $1_{-k} + 2_{0,-k}$. The sextet $6_{[[k],[\ell]]}$ is decomposed to $2_{-k-\ell,-\ell} + 2_{\ell,-k} + 2_{k,k+\ell}$. The decomposition of 3_{2k}, written by (x_1, x_2, x_3), is non-trivial, and it is decomposed to $(x_2)1_{-k} + (x_1, -x_3)2_{0,-k}$. Similarly when $N/3 =$ integer, singlets 1_0 and 1_1 correspond to 1_{+0} and 1_{-0}. Doublets $2_1, 2_2, 2_3,$ and 2_4 become $1_{+0} + 1_{-0}$, $2_{2N/3,N/3}$, $2_{N/3,2N/3}$, and $2_{N/3,2N/3}$, respectively. Triplets 3_{1k} and 3_{2k} are decomposed to $1_{+k} + 2_{0,-k}$ and $1_{-k} + 2_{0,-k}$ The sextet $6_{[[k],[\ell]]}$ is decomposed to $2_{-k-\ell,-\ell} + 2_{\ell,-k} + 2_{k,k+\ell}$.

- $\Delta(6N^2) \to \Delta(3N^2)$
The subgroup $\Delta(3N^2)$ consists of $\{b^k a^m a'^n\}$. The generator is summarized in Table 14.42. When $N/3 \neq$ integer, singlets 1_0 and 1_1 correspond to 1_0 and 1_0. The doublet 2 becomes $1_1 + 1_2$. Triplets 3_{1k} and 3_{2k} correspond to $3_{[-k][k]}$ and $3_{[-k][k]}$. The sextet $6_{[[k],[\ell]]}$ is decomposed to $3_{[k][\ell]} + 3_{[\ell][k+\ell]}$. When $N/3 =$ integer, singlets 1_0 and 1_1 correspond to $1_{0,0}$ and $1_{0,0}$. The doublet $2_1, 2_2, 2_3,$ and 2_4 become $1_{1,0} + 1_{2,0}$, $1_{1,2} + 1_{2,1}$ $1_{1,1} + 1_{2,2}$, and $1_{0,1} + 1_{0,2}$, respectively. Triplets

Table 14.41 Representations of $\Sigma(2N^2)$

	1_{+n}	1_{-n}	$2_{p,q}$
a^{-1}	ρ^n	ρ^n	$\begin{pmatrix} \rho^q & 0 \\ 0 & \rho^p \end{pmatrix}$
a'	ρ^n	ρ^n	$\begin{pmatrix} \rho^p & 0 \\ 0 & \rho^q \end{pmatrix}$
b	1	-1	$\begin{pmatrix} 0 & 1 \\ 1 & 0 \end{pmatrix}$

Table 14.42 Representations of $\Delta(3N^2)$ for $3N \neq$ integer (left) and $3N =$ integer (right)

	1_k	$3_{[k][\ell]}$		$1_{k,\ell}$	$3_{[k][\ell]}$
a	1	$\begin{pmatrix} \rho^\ell & 0 & 0 \\ 0 & \rho^k & 0 \\ 0 & 0 & \rho^{-k-\ell} \end{pmatrix}$	a	ω^ℓ	$\begin{pmatrix} \rho^\ell & 0 & 0 \\ 0 & \rho^k & 0 \\ 0 & 0 & \rho^{-k-\ell} \end{pmatrix}$
a'	1	$\begin{pmatrix} \rho^{-k-\ell} & 0 & 0 \\ 0 & \rho^\ell & 0 \\ 0 & 0 & \rho^k \end{pmatrix}$	a'	ω^ℓ	$\begin{pmatrix} \rho^{-k-\ell} & 0 & 0 \\ 0 & \rho^\ell & 0 \\ 0 & 0 & \rho^k \end{pmatrix}$
b	ω^k	$\begin{pmatrix} 0 & 1 & 0 \\ 0 & 0 & 1 \\ 1 & 0 & 0 \end{pmatrix}$	b	ω^k	$\begin{pmatrix} 0 & 1 & 0 \\ 0 & 0 & 1 \\ 1 & 0 & 0 \end{pmatrix}$

3_{1k} and 3_{2k} correspond to $3_{[-k][k]}$ and $3_{[-k][k]}$. The sextet $6_{[[k],[\ell]]}$ is decomposed to $3_{[k][\ell]} + 3_{[\ell][k+\ell]}$.

- $\Delta(6N^2) \rightarrow \Delta(6M^2)$

We consider the subgroup $\Delta(6M^2)$, where M is a divisor of N. We denote $\tilde{a} = a^p$ and $\tilde{a}' = a'^p$ with $p = N/M$, where $p =$ integer. The subgroup $\Delta(6M^2)$ consists of $b^k \tilde{c}^\ell a^m \tilde{a}'^n$ with $k = 0, 1, 2$, $\ell = 0, 1$, and $m, n = 0, \ldots, M - 1$. There are three types of combinations (N, M), i.e. (1) both $N/3$ and $M/3$ are integers, (2) $N/3$ is integer, but $M/3$ is not integer, (3) either $N/3$ or $M/3$ is not integer.

When both $N/3$ and $M/3$ are integers, each representation of $\Delta(6N^2)$ is decomposed to representations of $\Delta(6M^2)$ as follows: singlets and doublet $1_0, 1_1, 2_k$ remain the same representations, triplets $3_{1(k+Mn)}$ and $3_{2(k+Mn)}$ are decomposed to 3_{1k} and 3_{2k}, and sextet $6_{[[k+Mn],[\ell+Mn']]}$ corresponds to $6_{[[k],[\ell]]}$, where n and n' are integers.

Next we consider the case that $N/3 =$ integer and $M/3 \neq$ integer, where $p = N/M$ must be $3n$. In this case, singlets 1_0 and 1_1 remain the same representations while doublets $2_1, 2_2$, and 2_3 correspond to 2 and 2_4 is decomposed to $1_0 + 1_1$ because p has a factor of three. Triplets $3_{1(k+Mn)}$ and $3_{2(k+Mn)}$ correspond to 3_{1k} and 3_{2k}, and sextet $6_{[[k+Mn],[\ell+Mn']]}$ corresponds to $6_{[[k],[\ell]]}$.

Table 14.43 Breaking pattern of $\Delta(6N^2)$ for $N/3 \neq$ integer

$\Delta(6N^2)$	$\Sigma(2N^2)$	$\Delta(6N^2)$	$\Delta(3N^2)$
1_0	1_{+0}	1_0	1_0
1_1	1_{-0}	1_1	1_0
2	$1_{+0} + 1_{-0}$	2	$1_1 + 1_2$
3_{1k}	$1_{+k} + 2_{0,-k}$	3_{1k}	$3_{[-k][k]}$
3_{2k}	$1_{-k} + 2_{0,-k}$	3_{2k}	$3_{[-k][k]}$
$6_{[[k],[\ell]]}$	$2_{-k-\ell,-\ell} + 2_{\ell,-k} + 2_{k,k+\ell}$	$6_{[[k],[\ell]]}$	$3_{[k][\ell]} + 3_{[-\ell][k+\ell]}$

$\Delta(6N^2)$	$\Delta(6M^2)\ (M/3 \neq$integer$)$
1_0	1_0
1_1	1_1
2	2
$3_{1(k+Mn)}$	3_{1k}
$3_{2(k+Mn)}$	3_{2k}
$6_{[[k+Mn],[\ell+Mn']]}$	$6_{[[k],[\ell]]}$

The last case is that either $N/3$ or $M/3$ is not integer. Singlets and doublet $1_0, 1_1$, 2 remain the same representations. Triplets $3_{1(k+Mn)}$ and $3_{2(k+Mn)}$ are decomposed to 3_{1k} and 3_{2k} while sextet $6_{[[k+Mn],[\ell+Mn']]}$ is to $6_{[[k],[\ell]]}$.

14.20 $\Delta(54)$

All of the $\Delta(54)$ elements are written as $b^k c^\ell a^m a'^n$ with $k, m, n = 0, 1, 2$ and $\ell = 0, 1$. Here, the generators a and a' correspond to Z_3 and Z_3' of $(Z_3 \times Z_3') \rtimes S_3$, respectively, while b and c correspond to Z_3 and Z_2 in S_3 of $(Z_3 \times Z_3') \rtimes S_3$, respectively. Table 14.45 shows the representations of generators, b, c, a and a' on each representation of $\Delta(54)$. As subgroups, the group $\Delta(54)$ includes $S_3 \times Z_3$, $\Sigma(18)$ and $\Delta(27)$. The breaking pattern of $\Delta(54)$ is summarized in Table 14.48.

- $\Delta(54) \to S_3 \times Z_3$
The $\Delta(54)$ group includes $S_3 \times Z_3$ as a subgroup. The subgroup S_3 consists of $\{e, b, c, b^2, bc, b^2c\}$. The Z_3 part of $S_3 \times Z_3$ consists of $\{e, aa'^2, a^2a'\}$, where $(aa'^2)^3 = e$ and the element aa'^2 commutes with all of the S_3 elements. Representations, r_k, for $S_3 \times Z_3$ are specified by representations r of S_3 and the Z_3 charge k, where $r = 1, 1', 2$ and $k = 0, 1, 2$. That is, the element aa'^2 is represented as $aa'^2 = \omega^k$ on r_k for $k = 0, 1, 2$. For the decomposition of $\Delta(54)$ to $S_3 \times Z_3$, it would be convenient to use the basis for S_3 representations, 1, $1'$ and 2, which is shown in Table 14.46. Then, the representations of $\Delta(54)$ $1_0, 1_1, 2_1, 2_2, 2_3, 2_4$, $3_{1(k)}$, and $3_{2(k)}$ are decomposed to representations of $S_3 \times Z_3$ as $1_0, 1'_0, 2_0, 2_0, 2_0$,

Table 14.44 Breaking pattern of $\Delta(6N^2)$ for $N/3 =$ integer

$\Delta(6N^2)$	$\Sigma(2N^2)$
1_0	1_{+0}
1_1	1_{-0}
2_1	$1_{+0} + 1_{-0}$
2_2	$2_{2N/3,N/3}$
2_3	$2_{N/3,2N/3}$
2_4	$2_{N/3,2N/3}$
3_{1k}	$1_{+k} + 2_{0,-k}$
3_{2k}	$1_{-k} + 2_{0,-k}$
$6_{[[k],[\ell]]}$	$2_{-k-\ell,-\ell} + 2_{\ell,-k} + 2_{k,k+\ell}$

$\Delta(6N^2)$	$\Delta(3N^2)$
1_0	$1_{0,0}$
1_1	$1_{0,0}$
2_1	$1_{1,0} + 1_{2,0}$
2_2	$1_{1,2} + 1_{2,1}$
2_3	$1_{1,1} + 1_{2,2}$
2_4	$1_{0,1} + 1_{0,2}$
3_{1k}	$3_{[-k][k]}$
3_{2k}	$3_{[-k][k]}$
$6_{[[k],[\ell]]}$	$3_{[k][\ell]} + 3_{[-\ell][k+\ell]}$

$\Delta(6N^2)$	$\Delta(6M^2)$ ($M/3 \neq$ integer)
1_0	1_0
1_1	1_1
2_1	2
2_2	2
2_3	2
2_4	$1_0 + 1_1$
$3_{1(k+Mn)}$	3_{1k}
$3_{2(k+Mn)}$	3_{2k}
$6_{[[k+Mn],[\ell+Mn']]}$	$6_{[[k],[\ell]]}$

$\Delta(6N^2)$	$\Delta(6M^2)$ ($M/3 =$ integer)
1_0	1_0
1_1	1_1
2_1	2_1
2_2	2_2
2_3	2_3
2_4	2_4
$3_{1(k+Mn)}$	3_{1k}
$3_{2(k+Mn)}$	3_{2k}
$6_{[[k+Mn],[\ell+Mn']]}$	$6_{[[k],[\ell]]}$

Table 14.45 Representations of a, a', b and c in $\Delta(54)$

	1_0	1_1	2_1	2_2	2_3	2_4	3_{1k}	3_{2k}
a	1	1	$\begin{pmatrix} 1 & 0 \\ 0 & 1 \end{pmatrix}$	$\begin{pmatrix} \omega^2 & 0 \\ 0 & \omega \end{pmatrix}$	$\begin{pmatrix} \omega & 0 \\ 0 & \omega^2 \end{pmatrix}$	$\begin{pmatrix} \omega & 0 \\ 0 & \omega^2 \end{pmatrix}$	$\begin{pmatrix} \omega^k & 0 & 0 \\ 0 & \omega^{2k} & 0 \\ 0 & 0 & 1 \end{pmatrix}$	$\begin{pmatrix} \omega^k & 0 & 0 \\ 0 & \omega^{2k} & 0 \\ 0 & 0 & 1 \end{pmatrix}$
a'	1	1	$\begin{pmatrix} 1 & 0 \\ 0 & 1 \end{pmatrix}$	$\begin{pmatrix} \omega^2 & 0 \\ 0 & \omega \end{pmatrix}$	$\begin{pmatrix} \omega & 0 \\ 0 & \omega^2 \end{pmatrix}$	$\begin{pmatrix} \omega & 0 \\ 0 & \omega^2 \end{pmatrix}$	$\begin{pmatrix} 1 & 0 & 0 \\ 0 & \omega^k & 0 \\ 0 & 0 & \omega^{2k} \end{pmatrix}$	$\begin{pmatrix} 1 & 0 & 0 \\ 0 & \omega^k & 0 \\ 0 & 0 & \omega^{2k} \end{pmatrix}$
b	1	1	$\begin{pmatrix} \omega & 0 \\ 0 & \omega^2 \end{pmatrix}$	$\begin{pmatrix} \omega & 0 \\ 0 & \omega^2 \end{pmatrix}$	$\begin{pmatrix} \omega & 0 \\ 0 & \omega^2 \end{pmatrix}$	$\begin{pmatrix} 1 & 0 \\ 0 & 1 \end{pmatrix}$	$\begin{pmatrix} 0 & 1 & 0 \\ 0 & 0 & 1 \\ 1 & 0 & 0 \end{pmatrix}$	$\begin{pmatrix} 0 & 1 & 0 \\ 0 & 0 & 1 \\ 1 & 0 & 0 \end{pmatrix}$
c	1	-1	$\begin{pmatrix} 0 & 1 \\ 1 & 0 \end{pmatrix}$	$\begin{pmatrix} 0 & 1 \\ 1 & 0 \end{pmatrix}$	$\begin{pmatrix} 0 & 1 \\ 1 & 0 \end{pmatrix}$	$\begin{pmatrix} 0 & 1 \\ 1 & 0 \end{pmatrix}$	$\begin{pmatrix} 0 & 0 & 1 \\ 0 & 1 & 0 \\ 1 & 0 & 0 \end{pmatrix}$	$\begin{pmatrix} 0 & 0 & -1 \\ 0 & -1 & 0 \\ -1 & 0 & 0 \end{pmatrix}$

Table 14.46 Representations of b and c of S_3 in $\Delta(54)$

	1	1'	2
b	1	1	$\begin{pmatrix} \omega & 0 \\ 0 & \omega^2 \end{pmatrix}$
c	1	-1	$\begin{pmatrix} 0 & 1 \\ 1 & 0 \end{pmatrix}$

$1_0 + 1_0'$, $1_k + 2_k$, and $1_k' + 2_k$, for $k = 1, 2$. Components of S_3 doublets and singlets obtained from $\Delta(54)$ triplets are the same ones as considered in the decomposition for $S_4 \to S_3$.

● $\Delta(54) \to \Sigma(18)$

We consider the subgroup $\Sigma(18)$, which consists of $\tilde{b}^\ell \tilde{a}^m a'^n$ with $\ell = 0, 1$ and $m, n = 0, 1, 2$, where $\tilde{b} = c$ and $\tilde{a} = a^2$. Table 14.47 shows the representations of the generators, $\tilde{a}, a'$ and $\tilde{b}$ on each representation of $\Delta(54)$. Then, the representations of $\Delta(54)$ 1_0, 1_1, 2_1, 2_2, 2_3, 2_4, 3_{1k}, and 3_{2k} are decomposed to representations of $\Sigma(18)$ as 1_{+0}, 1_{-0}, $2_{0,0}$, $2_{2,1}$, $2_{1,2}$, $2_{1,2}$, $1_{+k} + 2_{0,2k}$, and $1_{-k} + 2_{2k,0}$, respectively. The decomposition of triplet components is obtained as follows,

$$\begin{pmatrix} x_1 \\ x_2 \\ x_3 \end{pmatrix}_{3_{1k}} \to (x_2)1_{+k} \oplus \begin{pmatrix} x_1 \\ x_3 \end{pmatrix}_{2_{0,2k}}, \quad \begin{pmatrix} x_1 \\ x_2 \\ x_3 \end{pmatrix}_{3_{2k}} \to (x_2)1_{-k} \oplus \begin{pmatrix} x_3 \\ -x_1 \end{pmatrix}_{2_{2k,0}} \quad (14.20)$$

● $\Delta(54) \to \Delta(27)$

We consider the subgroup $\Delta(27)$, which consists of $b^k a^m a'^n$ with $k, m, n = 0, 1, 2$. By use of Table 14.45, it is found that the representations of $\Delta(54)$ 1_0, 1_1, 2_1, 2_2, 2_3,

Table 14.47 Representations of $\tilde{b}$, $\tilde{a}$ and a' of $\Sigma(18)$ in $\Delta(54)$

	1_0	1_1	2_1	2_2	2_3	2_4	3_{1k}	3_{2k}
$\tilde{a}$	1	1	$\begin{pmatrix} 1 & 0 \\ 0 & 1 \end{pmatrix}$	$\begin{pmatrix} \omega & 0 \\ 0 & \omega^2 \end{pmatrix}$	$\begin{pmatrix} \omega^2 & 0 \\ 0 & \omega \end{pmatrix}$	$\begin{pmatrix} \omega^2 & 0 \\ 0 & \omega \end{pmatrix}$	$\begin{pmatrix} \omega^{2k} & 0 & 0 \\ 0 & \omega^k & 0 \\ 0 & 0 & 1 \end{pmatrix}$	$\begin{pmatrix} \omega^{2k} & 0 & 0 \\ 0 & \omega^k & 0 \\ 0 & 0 & 1 \end{pmatrix}$
a'	1	1	$\begin{pmatrix} 1 & 0 \\ 0 & 1 \end{pmatrix}$	$\begin{pmatrix} \omega^2 & 0 \\ 0 & \omega \end{pmatrix}$	$\begin{pmatrix} \omega & 0 \\ 0 & \omega^2 \end{pmatrix}$	$\begin{pmatrix} \omega & 0 \\ 0 & \omega^2 \end{pmatrix}$	$\begin{pmatrix} 1 & 0 & 0 \\ 0 & \omega^k & 0 \\ 0 & 0 & \omega^{2k} \end{pmatrix}$	$\begin{pmatrix} 1 & 0 & 0 \\ 0 & \omega^k & 0 \\ 0 & 0 & \omega^{2k} \end{pmatrix}$
$\tilde{b}$	1	-1	$\begin{pmatrix} 0 & 1 \\ 1 & 0 \end{pmatrix}$	$\begin{pmatrix} 0 & 1 \\ 1 & 0 \end{pmatrix}$	$\begin{pmatrix} 0 & 1 \\ 1 & 0 \end{pmatrix}$	$\begin{pmatrix} 0 & 1 \\ 1 & 0 \end{pmatrix}$	$\begin{pmatrix} 0 & 0 & 1 \\ 0 & 1 & 0 \\ 1 & 0 & 0 \end{pmatrix}$	$\begin{pmatrix} 0 & 0 & -1 \\ 0 & -1 & 0 \\ -1 & 0 & 0 \end{pmatrix}$

Table 14.48 Breaking pattern of $\Delta(54)$

$\Delta(54)$	$S_3 \times Z_3$	$\Delta(54)$	$\Sigma(18)$	$\Delta(54)$	$\Delta(27)$
1_0	1_0	1_0	1_{+0}	1_0	$1_{0,0}$
1_1	$1'_0$	1_1	1_{-0}	1_1	$1_{0,0}$
2_1	2_0	2_1	$2_{0,0}$	2_1	$1_{1,0} + 1_{2,0}$
2_2	2_0	2_2	$2_{2,1}$	2_2	$1_{1,1} + 1_{2,2}$
2_3	2_0	2_3	$2_{1,2}$	2_3	$1_{1,2} + 1_{2,1}$
2_4	$1_0 + 1'_0$	2_4	$2_{1,2}$	2_4	$1_{0,1} + 1_{0,2}$
3_{1k}	$1_k + 2_k$	3_{1k}	$1_{+k} + 2_{0,2k}$	3_{1k}	$3_{[0][k]}$
3_{2k}	$1'_k + 2_k$	3_{2k}	$1_{-k} + 2_{2k,0}$	3_{2k}	$3_{[0][k]}$

$2_4, 3_{1k}$, and 3_{2k} are decomposed to representations of $\Delta(27)$ as $1_{0,0}, 1_{0,0}, 1_{1,0} + 1_{2,0}$, $1_{1,1} + 1_{2,2}, 1_{1,2} + 1_{2,1}, 1_{0,2} + 1_{0,1}, 3_{[0][k]}$, and $3_{[0][k]}$, respectively.

Finite Subgroups of Continuous Groups

We study non-Abelian discrete groups and their representations as finite subgroups of continuous groups. There exist continuous symmetries in nature such as Lorentz and gauge symmetries. In particle physics, since there are three families in quarks and leptons, continuous flavor symmetries such as the $SU(3)$ group have long been considered. Although the non-Abelian discrete flavor symmetries has become one of the most attractive models to understand the origin of the observed lepton mixing matrix, there exists an important question how such a non-Abelian discrete flavor symmetry could arise. One desired solution is that the discrete flavor symmetry is a remnant symmetry of the continuous (gauge) symmetry. In the framework of the quantum field theory this could occur due to the spontaneous symmetry breaking of a non-zero VEV of a scalar field. This leads to a reduction of the original group to a subgroup of unbroken symmetry in the low-energy physics. In general, a different VEV alignment can lead to a different breaking pattern (different low-energy physics), so that a classification of the symmetry breaking patterns is important.

The structure of vacuum alignments (breaking patterns) due to the scalar VEV can be mathematically defined as isotropy (or stabilizer) subgroups. Let x be a scalar VEV, which is a vector in a finite-dimensional vector space (an irreducible representation) of the continuous group G. The isotropy subgroup of G with respect to x, G_x, is the subset of G defined as:

$$G_x = \{g \in G | g \cdot x = x\}, \tag{15.1}$$

which becomes unbroken symmetry subgroup. We note that the structure of the isotropy subgroups becomes complicated if a vector x (scalar VEV) has a higher-dimensional representation. We also note that the VEV alignment depends on a potential that is invariant under the continuous group G and involves a scalar with a certain representation. In order to obtain a desired discrete subgroup H, one requires

T. Kobayashi et al., *An Introduction to Non-Abelian Discrete Symmetries for Particle Physicists*, Lecture Notes in Physics 995, https://doi.org/10.1007/978-3-662-64679-3_15

a non-zero VEV for an irreducible representation of G that contains the trivial singlet of H. Therefore it is important to know which irreducible representation includes a trivial singlet of H. We should stress, however, that this is not a sufficient condition for obtaining a desired subgroup, since it is not guaranteed that the desired group H that leaves x invariant is the maximal invariant subgroup of G.

There has been a number of works on the study of the spontaneous breaking of the continuous groups to discrete subgroups. As for $SO(3)$, the structures of the isotropy subgroup and the clarification of the G-invariant potential have been investigated using lower dimensional irreducible representations in [1–4], where it has been shown that a phenomenologically popular group A_4 can be realized in a scalar potential that uses a single $\underline{7}$ representation of $SO(3)$. The applications to flavor models based on these works can be found in [5,6]. Extension to other continuous symmetries of $SU(3)$ has been given in [7], where it has turns out that a scalar VEV with a dimension up to 8 can not break $SU(3)$ to A_4. To obtain A_4, higher-dimensional representations with multiple scalar VEVs may be required [8].

In the following we summarize the finite subgroups of continuous symmetries $G = SO(3)$, $SU(2)$, and $SU(3)$, and branching rules of the irreducible representations for the non-Abelian discrete groups discussed in the text. Most of the results presented here are also given in [8–10]. In addition, we explicitly show some calculations for the branching rules with lower dimensional representations using the tensor products [8,10]. This is easier to understand and may be useful for model building, while it is not a systematic approach and is not sufficient to identify the isotropy subgroups. Since there is no general criteria to simply check if H is the maximal subgroup, i.e., a scalar VEV in a particular direction breaks G to H or a larger subgroup, one has to examine each model individually to identify the maximal unbroken symmetry by explicitly constructing unbroken generators. This is in general a non-trivial problem and requires detailed information on the scalar potential, so that we will not discuss the potential and the resulting isotropy subgroups in detail. For some systematic approaches using software to calculate the branching rules, see [9,11].

15.1 Finite Subgroups of $SO(3)$

The $SO(3)$ group is a three-dimensional rotation group widely used in physics and mathematics. Let D_J be the spin J irreducible representation of $SO(3)$. Since D_J has $2J + 1$ dimensions, we denote it as a $\underline{\mathbf{2J+1}}$ (In what follows we use an underline for any representation of continuous groups to distinguish with the one of finite subgroups).

The $SO(3)$ is the continues group of three-dimensional rotation, while its subgroup $SO(2)$ is the continues group of two-dimensional rotation. Discrete subgroups of $SO(2)$ must be symmetries of N-polygons, i.e. Z_N. When we include a reflection of N-polygons, the symmetry groups are enhanced to D_N, which are subgroups of $SO(3)$. The other discrete subgroups of $SO(3)$ are symmetries of polyhedrons, i.e., the Platonic solids. There are five Platonic solids, the tetrahedron, $\mathcal{T}$, the cube $\mathcal{O}$,

Table 15.1 The isotropy non-Abelian discrete subgroups in lower-dimensional irreducible representations of $SO(3)$

$SO(3)$	Isotropy non-Abelian discrete subgroups
3	–
5	–
7	D_3, A_4
9	D_3, D_4, S_4
11	D_3, D_4, D_5
13	$D_3, D_4, D_5, D_6, A_4, A_5, S_4$

the octahedron, the dodecahedron, and the icosahedron, $\mathcal{I}$. The cube and octahedron are dual each other under replacing vertices and faces, and have the same symmetry isomorphic to S_4. Also, the icosahedron and the dodecahedron are dual each other, and have the same symmetry isomorphic to A_5. The tetrahedron is self dual, and has the symmetry isomorphic to A_4. Thus, the finite subgroups of $SO(3)$ have been classified as the following three different types[1]:

- the cyclic group Z_N with order N
- the dihedral group D_N with order $2N$
- the three polyhedral groups of $\mathcal{T}, \mathcal{O}, \mathcal{I}$.

Since we are interested in non-Abelian groups, which have a multi-dimensional irreducible representation, we only consider the latter two types, i.e., D_N, A_4, S_4, A_5.

Before going to the branching rules for each non-Abelian discrete subgroup, we note that a classification for the isotropy subgroups with lower dimensional irreducible representations has been investigated up to $J = 4$ in the context of the Higgs potential [1]. A systematic analysis which describes isotropy subgroups of $SO(3)$ in an arbitrary irreducible representation has been developed in [2]. We show the results for some lower dimensions in Table 15.1 restricting ourselves on non-Abelian discrete groups. Its details will be shown in what follows.

15.1.1 $SO(3) \rightarrow D_N$

We first consider D_N as a finite subgroup of $SO(3)$. The representations D_J for $SO(3)$ should also be a representation of its subgroup D_N, which is called a subduced representation $D_J^{D_N}$ onto D_N. This representation should be decomposed into the direct sum of irreducible subrepresentations of D_N. In fact, we see from Eq. (6.8) that both two generators a and b in D_N can be embedded into elements of $SO(3)$ as follows,

---

[1] For a mathematical proof, see e.g., [12].

$$D_{\underline{3}}(a) = \begin{pmatrix} a & 0 \\ 0 & 1 \end{pmatrix}, \qquad D_{\underline{3}}(b) = \begin{pmatrix} b & 0 \\ 0 & -1 \end{pmatrix}, \tag{15.2}$$

from which, we have a multiplet decomposition of $\underline{3} = \mathbf{1}_{--} + \mathbf{2}_1$ for even N, and $\underline{3} = \mathbf{1}_- + \mathbf{2}_1$ for odd N. Once we obtain the multiplet decomposition for the fundamental representation ($J = 1$) of $SO(3)$, the branching rules for higher representations ($J > 1$) can also be obtained by comparing a direct product of two lower representations with those of D_N. In the case of $SO(3)$, products of two representations of J and J' can be decomposed into

$$\underline{(2J + 1)} \times \underline{(2J' + 1)} = \underline{(2|J - J'|+1)} + \underline{(2|J - J'|+3)}$$
$$+ \cdots + \underline{(2|J + J'|+1)}. \tag{15.3}$$

For example, we obtain

$$\underline{3} \times \underline{3} = \underline{1} + \underline{3} + \underline{5}. \tag{15.4}$$

This irreducible decomposition should be compared with the tensor decompositions of D_N (See Chap. 6),

$$(\mathbf{1}_{--} + \mathbf{2}_1) \times (\mathbf{1}_{--} + \mathbf{2}_1) = \mathbf{1}_{++} + \mathbf{1}_{++} + \mathbf{1}_{--} + \mathbf{2}_1 + \mathbf{2}_1 + \mathbf{2}_2, \tag{15.5}$$

for even N, and

$$(\mathbf{1}_- + \mathbf{2}_1) \times (\mathbf{1}_- + \mathbf{2}_1) = \mathbf{1}_+ + \mathbf{1}_+ + \mathbf{1}_- + \mathbf{2}_1 + \mathbf{2}_1 + \mathbf{2}_2, \tag{15.6}$$

for odd N. Since $\underline{1} = \mathbf{1}_{++} = \mathbf{1}_+$, we obtain

$$\underline{5} = \mathbf{1}_{++} + \mathbf{2}_1 + \mathbf{2}_2, \tag{15.7}$$

for even N, and

$$\underline{5} = \mathbf{1}_+ + \mathbf{2}_1 + \mathbf{2}_2, \tag{15.8}$$

for odd N. The decompositions for higher representations can be easily obtained by repeating the above procedure.

A summary of the branching rules for $SO(3) \to D_N$ for general N is shown in Table 15.2. In this table, we use the following notation,

$$\mathbf{2}_{nN+k} = \mathbf{2}_k, \qquad \mathbf{2}_{nN-k} = \mathbf{2}_k, \tag{15.9}$$

for both even and odd N, and

$$\mathbf{2}_{nN} = \mathbf{1}_+ + \mathbf{1}_- \tag{15.10}$$

for odd N,

$$\mathbf{2}_{nN} = \mathbf{1}_{++} + \mathbf{1}_{--}, \qquad \mathbf{2}_{nN+N/2} = \mathbf{1}_{+-} + \mathbf{1}_{-+}, \tag{15.11}$$

Table 15.2 Branching rules for $SO(3) \to D_N$ up to $J = 6$

$SO(3)$	D_N (even N)	D_N (odd N)
$\underline{3}$	$1_{--} + 2_1$	$1_- + 2_1$
$\underline{5}$	$1_{++} + 2_1 + 2_2$	$1_+ + 2_1 + 2_2$
$\underline{7}$	$1_{--} + 2_1 + 2_2 + 2_3$	$1_- + 2_1 + 2_2 + 2_3$
$\underline{9}$	$1_{++} + 2_1 + 2_2 + 2_3 + 2_4$	$1_+ + 2_1 + 2_2 + 2_3 + 2_4$
$\underline{11}$	$1_{--} + 2_1 + 2_2 + 2_3 + 2_4 + 2_5$	$1_- + 2_1 + 2_2 + 2_3 + 2_4 + 2_5$
$\underline{13}$	$1_{++} + 2_1 + 2_2 + 2_3 + 2_4 + 2_5 + 2_6$	$1_+ + 2_1 + 2_2 + 2_3 + 2_4 + 2_5 + 2_6$

for even N. Thus, 2_N for D_N is reducible and includes a trivial singlet, i.e. 1_{++} for $N = $ even and 1_+ for $N = $ odd. Such a singlet VEV can break $SO(3)$ to D_N as the isotropy subgroup. That is, the VEV of $\underline{(2N+1)}$ scalar field can break $SO(3)$ to D_N as the isotropy subgroup.

Here we show examples with some lower N in detail.

$SO(3) \to D_3 \simeq S_3$

$D_3 \simeq S_3$ has two singlets 1_+, 1_-, and a doublet 2_1. The doublet 2_2 in the above notation is identified with 2_1. The doublet 2_3 in the above notation is reducible and can be decomposed into 1_+ and 1_-. It is convenient to use the notations 2_2 and 2_3. Since $2_{1,2} \otimes 2_{1,2} = 1_+ + 1_- + 2_{2,1}$ and $2_1 \otimes 2_2 = 2_1 + 2_3$, we obtain the tensor product of $\underline{3} \times \underline{3}$ as,

$$(1_- + 2_1) \otimes (1_- + 2_1) = 1_+ + 1_+ + 1_- + 2_1 + 2_1 + 2_2, \qquad (15.12)$$

which should correspond to $\underline{1} + \underline{3} + \underline{5}$. Thus we see $\underline{5} = 1_+ + 2_1 + 2_2$. To obtain the decomposition of $\underline{7}$, we calculate the tensor product of $\underline{3} \otimes \underline{5}$, which is given as

$$(1_- + 2_1) \otimes (1_+ + 2_1 + 2_2) = 1_+ + 1_- + 1_-$$
$$+ 2_1 + 2_1 + 2_2 + 2_2 + 2_3. \qquad (15.13)$$

Thus we see $\underline{7} = 1_- + 2_1 + 2_2 + 2_3$. As mentioned above, the reducible representation 2_3 includes 1_+, whose scalar VEV can break $SO(3)$ to D_3 as the isotropy subgroup.

$SO(3) \to D_4$

Next, we study D_4. Since $2 \otimes 2 = 1_{++} + 1_{+-} + 1_{-+} + 1_{--}$, and $2 \times 1_{--} = 2$, we obtain the tensor product of $\underline{3} \times \underline{3}$ as,

$$(1_{--} + 2) \otimes (1_{--} + 2) = 1_{++} + 1_{++} + 1_{+-} + 1_{-+} + 1_{--} + 2 + 2, \qquad (15.14)$$

Table 15.3 Branching rules for $SO(3) \rightarrow D_3$ up to $J = 6$

$SO(3)$	D_3
$\underline{3}$	$1_- + 1_+$
$\underline{5}$	$1_+ + 2 \times 2$
$\underline{7}$	$1_+ + 2 \times 1_- + 2 \times 2$
$\underline{9}$	$1_+ + 2 \times 1_- + 3 \times 2$
$\underline{11}$	$1_+ + 2 \times 1_- + 4 \times 2$
$\underline{13}$	$2 \times 1_+ + 3 \times 1_- + 4 \times 2$

Table 15.4 Branching rules for $SO(3) \rightarrow D_4$ up to $J = 6$

$SO(3)$	D_4
$\underline{3}$	$1_{--} + 2$
$\underline{5}$	$1_{++} + 1_{+-} + 1_{-+} + 2$
$\underline{7}$	$1_{+-} + 1_{-+} + 1_{--} + 2 + 2$
$\underline{9}$	$1_{++} + 1_{++} + 1_{+-} + 1_{-+} + 1_{--} + 2 \times 2$
$\underline{11}$	$1_{++} + 1_{+-} + 1_{-+} + 1_{--} + 1_{--} + 3 \times 2$
$\underline{13}$	$2 \times 1_{++} + 1_{--} + 1_{+-} + 1_{-+} + 3 \times 2$

which should correspond to $\underline{1} + \underline{3} + \underline{5}$. Thus we see $\underline{5} = 1_{++} + 1_{+-} + 1_{-+} + 2$. To obtain the decomposition of $\underline{7}$, we calculate the tensor product of $\underline{3} \otimes \underline{5}$, which is given as

$$(1_{--} + 2) \otimes (1_{+-} + 1_{-+} + 1_{--} + 2) = 1_{++} + 2 \times 1_{+-} + 2 \times 1_{-+} + 2 \times 1_{--}$$
$$+ 4 \times 2. \tag{15.15}$$

Thus we see $\underline{7} = 1_{+-} + 1_{-+} + 1_{--} + 2 \times 2$.

The above results for $SO(3) \rightarrow D_{3,4}$ are summarized in Tables 15.3 and 15.4. We note that these results are also consistent with the ones obtained for generic N in Table 15.2.

15.1.2 $SO(3) \rightarrow A_4$

We consider the tetrahedral group $\mathcal{T}$, which is isomorphic to A_4. Since A_4 has single triplet irreducible representation $\mathbf{3}$, we immediately see that this is a subduced representation $D_J^{A_4}$ onto A_4, namely $\underline{3} = \mathbf{3}$. Higher representations of $SO(3)$ can also be identified by comparing the direct products of the $J = 1$ representations with those of A_4. For example, from Eq. (4.10), we obtain

$$\mathbf{3} \times \mathbf{3} = \mathbf{1} + \mathbf{1}' + \mathbf{1}'' + \mathbf{3} + \mathbf{3}, \tag{15.16}$$

Table 15.5 Branching rules for $SO(3) \rightarrow A_4$ up to $J = 6$

$SO(3)$	A_4
$\underline{3}$	3
$\underline{5}$	$1' + 1'' + 3$
$\underline{7}$	$1 + 3 + 3$
$\underline{9}$	$1 + 1' + 1'' + 3 + 3$
$\underline{11}$	$1' + 1'' + 3 + 3 + 3$
$\underline{13}$	$1 + 1 + 1' + 1'' + 3 + 3 + 3$

which should be compared with $\underline{3} \times \underline{3} = \underline{1} + \underline{3} + \underline{5}$. Since $\underline{1} = 1$, we obtain $\underline{5} = 1' + 1'' + 3$. Repeating the procedure, we obtain the branching rules for embedding the discrete groups into continuous groups as shown in Table 15.5. From the table, we find that the trivial singlet 1 of A_4 first appears in the decomposition of 7 in $SO(3)$, which implies that in order to break $SO(3) \rightarrow A_4$, a scalar VEV in a higher representation of dimension 7 or more is required.

15.1.3 $SO(3) \rightarrow S_4$

Next, we consider the octahedral group, which is isomorphic to S_4. S_4 has two triplet representations, 3 and $3'$. This is isomorphic to the cubic group.[2] From Eqs. (3.23) and (3.31) we see that the representative matrices for the triplet 3 can not be group elements of $SU(3)$ (e.g. det $(d_1(3)) = -1$). We thus obtain $\underline{3} = 3'$. The tensor product of two $3'$ is obtained by

$$3' \times 3' = 1 + 2 + 3 + 3', \tag{15.17}$$

which should be compared with $\underline{3} \times \underline{3} = \underline{1} + \underline{3} + \underline{5}$. Since $\underline{1} = 1$, we obtain $\underline{5} = 2 + 3$.

The decompositions of higher dimensional representations are also obtained by comparing the tensor products. The branching rules for $SO(3) \rightarrow S_4$ is summarized in Table 15.6. The trivial singlet appears in $\underline{9}$ representation as well as $\underline{13}$ of $SO(3)$. A scalar VEV in either representation can break $SO(3)$ to S_4 (See Table 15.1).

15.1.3.1 $SO(3) \rightarrow A_5$
Icosahedral symmetry (A_5) has two triplets 3, and $3'$. We note that both two representative matrices for 3, and $3'$ can be group elements of $SO(3)$ (See Sect. 4.2), which means that there are two different ways of identifying both elements of $SO(3)$ and

---

[2] Five irreducible representations of the cubic group are given in terms of the point group as $A_{1,2}$ E, $T_{1,2}$ in the Schoenflies notation, which have dimensions 1, 1, 2, 3, 3. From the character table of S_4 these are identified as $A_1 = 1$, $A_2 = 1'$, $E = 2$, $T_1 = 3'$, $T_2 = 3$.

Table 15.6 Branching rules for $SO(3) \rightarrow S_4$ up to $J = 5$

$SO(3)$	S_4
$\underline{3}$	$3'$
$\underline{5}$	$2 + 3$
$\underline{7}$	$1' + 3 + 3'$
$\underline{9}$	$1 + 2 + 3 + 3$
$\underline{11}$	$2 + 3 + 3' + 3'$
$\underline{13}$	$1 + 1' + 2 + 3 + 3' + 3'$

Table 15.7 Branching rules for $SO(3) \rightarrow A_5$ up to $J = 5$

$SO(3)$	A_5
$\underline{3}$	$3 (3')$
$\underline{5}$	5
$\underline{7}$	$3'(3) + 4$
$\underline{9}$	$4 + 5$
$\underline{11}$	$3 + 3' + 5$
$\underline{13}$	$1 + 3(3') + 4 + 5$

A_5. In fact, the triplet representation $\underline{3}$ can be identified with either 3 or $3'$, where both representations are interchangeable. The result for branching rules in $SO(3) \rightarrow A_5$ (Icosahedral symmetry) are obtained as shown in Table 15.7. In either case, the trivial singlet appears in $\underline{13}$ representation of $SO(3)$. Its scalar VEV can break $SO(3)$ to A_5 (See Table 15.1).

15.2   Finite Subgroups of $SU(2)$

The continuous symmetry $SU(2)$ is the double covering group of the $SO(3)$. From this fact with the classification of $SO(3)$, the finite subgroups of $SU(2)$ have been classified as the following five different types:

- the cyclic group Z_N with order N,
- the binary dihedral group Q_N with order $2N$,
- the binary tetrahedral group, the double covering group of A_4 (T'),
- the binary octahedral group, the double covering group of S_4,
- the binary icosahedral group, the double covering group of A_5.

Here, we study Q_N and T'. Similar to the $SO(3)$, the d-dimensional irreducible representation is refereed to as the spin-S representation with $d = 2S + 1$, where the spin S is half-integer ($S = 0, 1/2, 1, 3/2, 2, ...$). Let D_S be the spin S irreducible representation, then a tensor product of two irreducible representations is decomposed

Table 15.8 Branching rules for $SU(2) \to Q_N$ up to $S = 5$

$SU(2)$	Q_N
$\underline{2}$	2_1
$\underline{3}$	$1_{--} + 2_2$
$\underline{4}$	$2_1 + 2_3$
$\underline{5}$	$1_{++} + 2_2 + 2_4$
$\underline{6}$	$2_1 + 2_3 + 2_5$
$\underline{7}$	$1_{--} + 2_2 + 2_4 + 2_6$
$\underline{8}$	$2_1 + 2_3 + 2_5 + 2_7$
$\underline{9}$	$1_{++} + 2_2 + 2_4 + 2_6 + 2_8$
$\underline{10}$	$2_1 + 2_3 + 2_5 + 2_7 + 2_9$
$\underline{11}$	$1_{--} + 2_2 + 2_4 + 2_6 + 2_8 + 2_{10}$

into

$$D_S \times D_{S'} = D_{S+S'} + D_{S+S'-1} + \cdots + D_{|S-S'|}, \qquad (15.18)$$

which is used to calculate the branching rules.

15.2.1 $SU(2) \to Q_N$

We notice that the generators a, b for 2_k, ($k =$odd) in Eq. (7.3) are identical to elements of $SU(2)$, thus the fundamental representation of $SU(2)$ should be identified as $\underline{2} = 2_1$ for both $N = 4n$ and $N = 4n + 2$. We note that the multiplication rules for $N = 4n + 2$ are the same as those for $N = 4n$, i.e.,

$$2_1 \times 2_1 = 1_{++} + 1_{--} + 2_2, \qquad (15.19)$$

which should be compared with $\underline{2} \times \underline{2} = \underline{1} + \underline{3}$. This leads to $\underline{3} = 1_{--} + 2_2$. Thus the branching rules for both $N = 4n$ and $N = 4n + 2$ are the same as well. When $N = 4$, 2_2 is reducible and decomposed to $1_{+-} + 1_{-+}$. The decomposition of the higher representations are then obtained from tensor products. The general results for branching rules are summarized in Table 15.8.

In this table some of 2_k are reducible, i.e.,

$$2_{nN+k} = 2_k, \qquad 2_{nN-k} = 2_k,$$
$$2_{nN} = 1_{++} + 1_{--}, \qquad (15.20)$$
$$2_{(n+1/2)N} = 1_{+-} + 1_{-+}.$$

$SU(2) \to Q_6$

Table 15.9 Branching rules for $SU(2) \rightarrow Q_4$ up to $S = 5$

$SU(2)$	Q_4
$\underline{2}$	2
$\underline{3}$	$1_{--} + 1_{+-} + 1_{-+}$
$\underline{4}$	$2 + 2$
$\underline{5}$	$2 \times (1_{++}) + 1_{--} + 1_{+-} + 1_{-+}$
$\underline{6}$	$2 + 2 + 2$
$\underline{7}$	$1_{++} + 2 \times (1_{--} + 1_{+-} + 1_{-+})$
$\underline{8}$	$4 \times (2)$
$\underline{9}$	$3 \times (1_{++}) + 2 \times (1_{--} + 1_{+-} + 1_{-+})$
$\underline{10}$	$5 \times (2)$
$\underline{11}$	$2 \times (1_{++}) + 3 \times (1_{--} + 1_{+-} + 1_{-+})$

As an explicit example, we consider Q_6 in detail. It has four singlets $1_{++}, 1_{+-}$, $1_{-+}, 1_{--}$ and two doublets $2_1, 2_2$. With $\underline{3} = 1_{--} + 2_2$ we calculate the tensor product of $\underline{2} \times \underline{3} = \underline{2} + \underline{4}$, which yields

$$2_1 \times (1_{--} + 2_2) = 1_{+-} + 1_{-+} + 2_1 + 2_1, \tag{15.21}$$

from which we see $\underline{4} = 1_{+-} + 1_{-+} + 2_1$. To obtain the decomposition of $\underline{5}$, we again calculate a tensor product such as $\underline{2} \times \underline{4}$, which yields

$$2_1 \times (1_{+-} + 1_{-+} + 2_1) = 1_{++} + 1_{--} + 2_2 + 2_2 + 2_2. \tag{15.22}$$

We have used the following multiplication rules for the singlets and doublets,

$$1_{++} \times 2_{1,2} = 2_{1,2}, \quad 1_{--} \times 2_{1,2} = 2_{1,2}$$
$$1_{+-} \times 2_{1,2} = 2_{2,1}, \quad 1_{-+} \times 2_{1,2} = 2_{2,1}. \tag{15.23}$$

Thus we see $\underline{5} = 1_{++} + 2_2 + 2_2$.

We summarize explicitly the branching rules for some lower N, i.e., Q_4 and Q_6 in Tables 15.9 and 15.10.

15.2.2 $SU(2) \rightarrow T'$

T' is a double covering group of A_4, and has three singlets $\mathbf{1}, \mathbf{1}', \mathbf{1}''$, three doublets $\mathbf{2}, \mathbf{2}', \mathbf{2}''$, and one triplets $\mathbf{3}$. From Eqs. (5.20), (5.21), and (5.22), the generators of T' can be elements of $SU(2)$ in $\underline{2}$ representation, so that we uniquely identify $\underline{2} = 2$. We then obtain the branching rules for higher representations by a decomposition of tensor products, such as

$$2 \times 2 = 1 + 3, \tag{15.24}$$

Table 15.10 Branching rules for $SU(2) \to Q_6$ up to $S = 5$

$SU(2)$	Q_6
2	2_1
3	$1_{--} + 2_2$
4	$1_{+-} + 1_{-+} + 2_1$
5	$1_{++} + 2_2 + 2_2$
6	$1_{+-} + 1_{-+} + 2_1 + 2_1$
7	$1_{++} + 2 \times (1_{--}) + 2_1 + 2_1$
8	$1_{+-} + 1_{-+} + 3 \times 2_1$
9	$2 \times (1_{++}) + 1_{--} + 2 \times 2_2$
10	$2 \times (1_{+-} + 1_{-+}) + 3 \times 2_1$
11	$1_{++} + 2 \times (1_{--}) + 4 \times 2_2$

Table 15.11 Branching rules for $SU(2) \to T'$ up to $S = 5$

$SU(2)$	T'
2	**2**
3	**3**
4	$2' + 2''$
5	$1' + 1'' + 3$
6	$2 + 2' + 2''$
7	$1 + 2 \times 3$
8	$2 \times 2 + 2' + 2''$
9	$1 + 1' + 1'' + 2 \times 3$
10	$2 + 2 \times 2' + 2 \times 2'''$
11	$1' + 1'' + 3 \times 3$

thus we see $\underline{3} = 3$. The general results for branching rules are summarized in Table 15.11.

15.3   Finite Subgroups of $SU(3)$

The finite subgroups of $SU(3)$ have been classified as the following five different types [3] :

- Abelian finite subgroups,
- Non-Abelian finite subgroups of $SO(3)$ and $SU(2)$
- $\Sigma(n\phi)$, $n = 36, 72, 216, 360$,

---

[3] See e.g., [13] and references therein.

Table 15.12 $SU(3)$ irreducible representations up to 27

Dynkin label	Dimension
(10)	$\underline{3}$
(01)	$\underline{3}^*$
(20)	$\underline{6}$
(11)	$\underline{8}$
(30)	$\underline{10}$
(21)	$\underline{15}$
(40)	$\underline{15}$'
(05)	$\underline{21}$
(13)	$\underline{24}$
(22)	$\underline{27}$

- $\Sigma(60) \simeq A_5$, $\Sigma(168) \simeq PSL(2, 7)$
- $\Delta(3N^2)$,
- $\Delta(6N^2)$.

Here, we study $\Delta(3N^2)$ and $\Delta(6N^2)$, and non-Abelian finite subgroups of $SO(3)$ and $SU(2)$ such as A_4, A_5, S_4, T', Q_N. Note that $A_4 \simeq \Delta(12)$ and $S_4 \simeq \Delta(24)$. Before studying the subgroups, we briefly summarize the structure of $SU(3)$ group. As discussed, we use the method to compare the tensor products in both groups. It is useful to summarize the tensor product decompositions for some higher dimensional representations, in which we use the notation for irreducible representations of $SU(3)$ given in Ref. [14] as summarized in Table 15.12.

Starting with the fundamental representation $\underline{3}$ of $SU(3)$ and its complex conjugate $\underline{3}^*$, we obtain the following sets of the tensor decompositions,

$$
\begin{array}{ll}
\underline{3} \otimes \underline{3} = \underline{3}^* \oplus \underline{6}, & \underline{3}^* \otimes \underline{3} = \underline{1} \oplus \underline{8}, \\
\underline{6} \otimes \underline{3} = \underline{8} \oplus \underline{10}, & \underline{8} \otimes \underline{3} = \underline{3} \oplus \underline{6}^* \oplus \underline{15}, \\
\underline{10} \otimes \underline{3} = \underline{15} \oplus \underline{15}', & \underline{10}^* \otimes \underline{3} = \underline{6}^* \oplus \underline{24} \\
\underline{15}' \otimes \underline{3} = \underline{21}^* \oplus \underline{24}^* & \underline{15}^* \otimes \underline{3} = \underline{8} \oplus \underline{10}^* \oplus \underline{27},
\end{array} \tag{15.25}
$$

which are sufficient to obtain the branching rules for all representations up to $D = 27$.

15.3.1 $SU(3) \rightarrow A_4$

As in the case of $SO(3)$, all the elements for triplet representation of A_4 can be elements of $SU(3)$, so that $\underline{3} = 3$. We also notice that the triplet representations of A_4 is real, and $\underline{3}^*$ should correspond to 3. We use the multiplication rules for A_4,

$$
3 \otimes 3 = 1 + 1' + 1'' + 3 + 3, \tag{15.26}
$$

Table 15.13 Branching rules for $SU(3) \rightarrow A_4$

$SU(3)$	A_4
$\underline{3}, \underline{3}^*$	$\mathbf{3}$
$\underline{6}, \underline{6}^*$	$\mathbf{1} + \mathbf{1}' + \mathbf{1}'' + \mathbf{3}$
$\underline{8}$	$\mathbf{1}' + \mathbf{1}'' + 2 \times \mathbf{3}$
$\underline{10}, \underline{10}^*$	$\mathbf{1} + 3 \times \mathbf{3}$
$\underline{15}, \underline{15}^*$	$\mathbf{1} + \mathbf{1}' + \mathbf{1}'' + 4 \times \mathbf{3}$
$\underline{15}', \underline{15}'^*$	$2 \times (\mathbf{1} + \mathbf{1}' + \mathbf{1}'') + 3 \times \mathbf{3}$
$\underline{21}, \underline{21}^*$	$\mathbf{1} + \mathbf{1}' + \mathbf{1}'' + 6 \times \mathbf{3}$
$\underline{24}, \underline{24}^*$	$2 \times (\mathbf{1} + \mathbf{1}' + \mathbf{1}'') + 6 \times \mathbf{3}$
$\underline{27}$	$3 \times (\mathbf{1} + \mathbf{1}' + \mathbf{1}'') + 6 \times \mathbf{3}$

which should be compared with Eq. (15.25),

$$\underline{3} \otimes \underline{3} = \underline{3}^* \oplus \underline{6}, \quad \underline{3}^* \otimes \underline{3} = \underline{1} \oplus \underline{8}. \tag{15.27}$$

Since $\underline{3} = \underline{3}^*$ and $\underline{1} = \mathbf{1}$, we obtain the following branching rules

$$\underline{6} = \mathbf{1} + \mathbf{1}' + \mathbf{1}'' + \mathbf{3},$$
$$\underline{6}^* = \mathbf{1} + \mathbf{1}' + \mathbf{1}'' + \mathbf{3},$$
$$\underline{8} = \mathbf{1}' + \mathbf{1}'' + \mathbf{3} + \mathbf{3}. \tag{15.28}$$

We summarize the branching rules up to $D = 27$ in Table 15.13.

15.3.2 $SU(3) \rightarrow A_5$

As in the case of $SO(3) \rightarrow A_5$, both two representative matrices for the triplets $\underline{3}$ and $\underline{3}'$ can be group elements of $SU(3)$. The result for branching rules in $SU(3) \rightarrow A_5$ are shown in Table 15.14.

15.3.3 $SU(3) \rightarrow S_4$

As in the case of $SO(3)$, we notice that the fundamental representation $\underline{3}$ in $SU(3)$ is identified with $\mathbf{3}'$ in S_4. We then obtain $\underline{3} = \underline{3}^* = \mathbf{3}'$, since $\mathbf{3}'$ is a real representation in Eq. (3.23). The result for branching rules in $SU(3) \rightarrow S_4$ are shown in Table 15.15.

Table 15.14 Branching rules for $SU(3) \rightarrow A_5$

$SU(3)$	A_5
3, **3***	**3** (3)
6, **6***	**1** + **5**
8	**3**(**3**$'$) + **5**
10, **10***	**3** + **3**$'$ + **4**
15, **15***	**3** + **3**$'$ + **4** + **5**
15', **15'***	**1** + **4** + 2 × **5**
21, **21***	2 × (**3** + **3**$'$) + **4** + **5**
24, **24***	**3** + **3**$'$ + 2 × **4** + 2 × **5**
27	**1** + **3**$'$(3) + 2 × **4** + 3 × **5**

Table 15.15 Branching rules for $SU(3) \rightarrow S_4$

$SU(3)$	S_4
3, **3***	**3**$'$
6, **6***	**1** + **2** + **3**
8	**2** + **3** + **3**$'$
10, **10***	**1**$'$ + **3** + 2 × **3**$'$
15, **15***	**1**$'$ + **2** + 2 × **3** + 2 × **3**$'$
15', **15'***	2 × **1** + 2 × **2** + 2 × **3** + **3**$'$
21, **21***	**1**$'$ + **2** + 2 × **3** + 4 × **3**$'$
24, **24***	**1** + **1**$'$ + 2 × **2** + 3 × **3** + 3 × **3**$'$
27	2 × **1** + **1**$'$ + 3 × **2** + 4 × **3** + 2 × **3**$'$

15.3.4 $SU(3) \rightarrow D_N$

As in the case of $SO(3)$, both two generators a and b in Eq. (6.4) can be embedded into elements of $SU(3)$ as follows,

$$D_{\underline{3}}(a) = \begin{pmatrix} a & 0 \\ 0 & 1 \end{pmatrix}, \qquad D_{\underline{3}}(b) = \begin{pmatrix} b & 0 \\ 0 & -1 \end{pmatrix}. \tag{15.29}$$

We note that the doublet representation is real, i.e., $(2_k)^* = 2_{-k} = 2_k$, so that the both subduced representations are equivalent, $D_{\underline{3}^*}^{D_N} = D_{\underline{3}}^{D_N}$. The irreducible decompositions are given as

$$\underline{3} = \begin{cases} \mathbf{1}_{--} + \mathbf{2}_1 & N = 2m \\ \mathbf{1}_- + \mathbf{2}_1 & N = 2m+1 \end{cases}, \tag{15.30}$$

where $m \in Z$. The result for branching rules in $SU(3) \rightarrow D_N$ is shown in Tables 15.16 and 15.17.

Table 15.16 Branching rules for $SU(3) \to D_N$ with $N =$ even

$SU(3)$	D_N (even N)
$\underline{3}, \underline{3}^*$	$1_{--} + 2_1$
$\underline{6}, \underline{6}^*$	$2 \times 1_{++} + 2_1 + 2_2$
$\underline{8}$	$1_{++} + 1_{--} + 2 \times 2_1 + 2_2$
$\underline{10}, \underline{10}^*$	$2 \times 1_{--} + 2 \times 2_1 + 2_2 + 2_3$
$\underline{15}, \underline{15}^*$	$1_{++} + 2 \times 1_{--} + 3 \times 2_1 + 2 \times 2_2 + 2_3$
$\underline{15}', \underline{15}'^*$	$3 \times 1_{++} + 2 \times 2_1 + 2 \times 2_2 + 2_3 + 2_4$
$\underline{21}, \underline{21}^*$	$3 \times 1_{--} + 3 \times 2_1 + 2 \times 2_2 + 2 \times 2_3 + 2_4 + 2_5$
$\underline{24}, \underline{24}^*$	$2 \times (1_{++} + 1_{--}) + 4 \times 2_1 + 3 \times 2_2 + 2 \times 2_3 + 2_4$
$\underline{27}, \underline{27}^*$	$4 \times 1_{++} + 1_{--} + 4 \times 2_1 + 4 \times 2_2 + 2 \times 2_3 + 2_4$

Table 15.17 Branching rules for $SU(3) \to D_N$ with $N =$ odd

$SU(3)$	D_N (odd N)
$\underline{3}, \underline{3}^*$	$1_- + 2_1$
$\underline{6}, \underline{6}^*$	$2 \times 1_+ + 2_1 + 2_2$
$\underline{8}$	$1_+ + 1_- + 2 \times 2_1 + 2_2$
$\underline{10}, \underline{10}^*$	$2 \times 1_- + 2 \times 2_1 + 2_2 + 2_3$
$\underline{15}, \underline{15}^*$	$1_+ + 2 \times 1_- + 3 \times 2_1 + 2 \times 2_2 + 2_3$
$\underline{15}', \underline{15}'^*$	$3 \times 1_+ + 2 \times 2_1 + 2 \times 2_2 + 2_3 + 2_4$
$\underline{21}, \underline{21}^*$	$3 \times 1_- + 3 \times 2_1 + 2 \times 2_2 + 2 \times 2_3 + 2_4 + 2_5$
$\underline{24}, \underline{24}^*$	$2 \times (1_+ + 1_-) + 4 \times 2_1 + 3 \times 2_2 + 2 \times 2_3 + 2_4$
$\underline{27}, \underline{27}^*$	$4 \times 1_+ + 1_- + 4 \times 2_1 + 4 \times 2_2 + 2 \times 2_3 + 2_4$

We show some concrete examples of D_3 and D_4. Using the irreducible decompostions of D_N in Eqs. (15.10) and (15.11) in Tables 15.16 and 15.17, we obtain the result for $SU(3) \to D_{3,4}$ as shown in Tables 15.18 and 15.19.

15.3.5 $SU(3) \to T'$

Since all three generators s, t, and r in Eq. (5.23) are elements of $SU(3)$ in $\underline{3}$ representation, we see $\underline{3} = 3$. The complex conjugate representation of $\underline{3}$ is equivalent to $\underline{3}$, so that we also see $\underline{3}^* = 3$. The general results for branching rules are summarized in Table 15.20.

Table 15.18 Branching rules for $SU(3) \rightarrow D_3$

$SU(3)$	D_3
3, 3*	$1_- + 2$
6, 6*	$2 \times 1_+ + 2 \times 2$
8	$1_+ + 1_- + 3 \times 2$
10, 10*	$1_+ + 3 \times 1_- + 3 \times 2$
15, 15*	$2 \times 1_+ + 3 \times 1_- + 5 \times 2$
15', 15'*	$4 \times 1_+ + 1_- + 5 \times 2$
21, 21*	$2 \times 1_+ + 5 \times 1_- + 7 \times 2$
24, 24*	$4 \times 1_+ + 4 \times 1_- + 8 \times 2$
27, 27*	$6 \times 1_+ + 3 \times 1_- + 9 \times 2$

Table 15.19 Branching rules for $SU(3) \rightarrow D_4$

$SU(3)$	D_4
3, 3*	$1_{--} + 2$
6, 6*	$2 \times 1_{++} + 1_{+-} + 1_{-+} + 2$
8	$1_{++} + 1_{+-} + 1_{-+} + 1_{--} + 2 \times 2$
10, 10*	$1_{+-} + 1_{-+} + 2 \times 1_{--} + 3 \times 2$
15, 15*	$1_{++} + 2 \times 1_{+-} + 2 \times 1_{-+} + 2 \times 1_{--} + 4 \times 2$
15', 15'*	$4 \times 1_{++} + 2 \times 1_{+-} + 2 \times 1_{-+} + 1_{--} + 3 \times 2$
21, 21*	$1_{++} + 2 \times 1_{+-} + 2 \times 1_{-+} + 4 \times 1_{--} + 6 \times 2$
24, 24*	$3 \times 1_{++} + 3 \times 1_{+-} + 3 \times 1_{-+} + 3 \times 1_{--} + 6 \times 2$
27, 27*	$5 \times 1_{++} + 4 \times 1_{+-} + 4 \times 1_{-+} + 2 \times 1_{--} + 6 \times 2$

Table 15.20 Branching rules for $SU(3) \rightarrow T'$

$SU(3)$	T'
3, 3*	**3**
6, 6*	$1 + 1' + 1'' + 3$
8	$1' + 1'' + 2 \times 3$
10, 10*	$1 + 3 \times 3$
15, 15*	$1 + 1' + 1'' + 4 \times 3$
15', 15'*	$2 \times (1 + 1' + 1'') + 3 \times 3$
21, 21*	$1 + 1' + 1'' + 6 \times 3$
24, 24*	$2 \times (1 + 1' + 1'') + 6 \times 3$
27, 27*	$3 \times (1 + 1' + 1'') + 6 \times 3$

Table 15.21 Branching rules for $SU(3) \rightarrow Q_N$ for generic N

$SU(3)$	Q_N
$\underline{3}, \underline{3}^*$	$1_{++} + 2_1$
$\underline{6}, \underline{6}^*$	$1_{++} + 1_{--} + 2_1 + 2_2$
$\underline{8}$	$1_{++} + 1_{--} + 2 \times 2_1 + 2_2$
$\underline{10}, \underline{10}^*$	$1_{++} + 1_{--} + 2 \times 2_1 + 2_2 + 2_3$
$\underline{15}, \underline{15}^*$	$1_{++} + 21_{--} + 3 \times 2_1 + 2 \times 2_2 + 2_3$
$\underline{15}', \underline{15}'^*$	$2 \times 1_{++} + 1_{--} + 2 \times 2_1 + 2 \times 2_2 + 2_3 + 2_4$
$\underline{21}, \underline{21}^*$	$2 \times 1_{++} + 1_{--} + 3 \times 2_1 + 2 \times 2_2 + 2 \times 2_3 + 2_4 + 2_5$
$\underline{24}, \underline{24}^*$	$2 \times 1_{++} + 2 \times 1_{--} + 4 \times 2_1 + 3 \times 2_2 + 2 \times 2_3 + 2_4$
$\underline{27}, \underline{27}^*$	$2 \times 1_{++} + 3 \times 1_{--} + 4 \times 2_1 + 4 \times 2_2 + 2 \times 2_3 + 2_4$

15.3.5.1 $SU(3) \rightarrow Q_N$

Similar to the case of $SU(3) \rightarrow D_N$, both two generators a and b in Eq. (6.4) can be embedded into a direct sum of elements of $SU(3)$ as follows,

$$D_{\underline{3}}(a) = \begin{pmatrix} a & 0 \\ 0 & 1 \end{pmatrix}, \qquad D_{\underline{3}}(b) = \begin{pmatrix} b & 0 \\ 0 & 1 \end{pmatrix}. \tag{15.31}$$

Thus we see $\underline{3} = \underline{3}^* = 2_1 + 1_{++}$ for both $N = 4m$ and $N = 4m + 2$. We show the result for branching rules in $SU(3) \rightarrow Q_N$ in Table 15.21.

We show a concrete example of Q_4. Using the irreducible decompositions of 2_k for finite N in Eqs. (15.20) in Table 15.21, we obtain the result for $SU(3) \rightarrow Q_4$ as shown in Table 15.22.

15.3.6 $SU(3) \rightarrow \Delta(3N^2)$

$\Delta(3N^2)$ has three generators of a, a', and b. From Eq. (10.3) they are group elements of $SU(3)$, so that $\underline{3} = 3_{[0][1]}$ and $\underline{3}^* = 3_{[0][-1]}$. In general, a complex conjugate representation of $3_{[k][l]}$ is given as $3_{[-k][-l]}$. Similar to the previous cases, we obtain the irreducible decompositions of the higher dimensional representations. It is convenient to use general (reducible) representations to calculate a tensor product, for example,

$$3_{[k][\ell]} \times 3_{[k'][\ell']} = 3_{[k+k'][\ell+\ell']} + 3_{[-k-\ell+\ell'][k-k'-\ell']} + 3_{[\ell-k'-\ell'][-k-\ell+k']}. \tag{15.32}$$

It is noted that this result is applicable to either case with $N/3$ being integer or not. In fact, we obtain the tensor product of $\underline{3} \times \underline{3}^* = \underline{1} + \underline{8}$ as,

$$3_{[0][1]} \times 3_{[0][-1]} = 3_{[0][0]} + 3_{[-2][1]} + 3_{[2][-1]]}. \tag{15.33}$$

Table 15.22 Branching rules for $SU(3) \to Q_4$

$SU(3)$	Q_4
$\underline{3}, \underline{3}^*$	$1_{++} + 2$
$\underline{6}, \underline{6}^*$	$1_{++} + 1_{+-} + 1_{-+} + 1_{--} + 2$
$\underline{8}$	$1_{++} + 1_{+-} + 1_{-+} + 1_{--} + 2 \times 2$
$\underline{10}, \underline{10}^*$	$1_{++} + 1_{+-} + 1_{-+} + 1_{--} + 3 \times 2$
$\underline{15}, \underline{15}^*$	$1_{++} + 2 \times 1_{+-} + 2 \times 1_{-+} + 2 \times 1_{--} + 4 \times 2$
$\underline{15}', \underline{15}'^*$	$3 \times 1_{++} + 2 \times 1_{+-} + 2 \times 1_{-+} + 2 \times 1_{--} + 3 \times 2$
$\underline{21}, \underline{21}^*$	$3 \times 1_{++} + 2 \times 1_{+-} + 2 \times 1_{-+} + 2 \times 1_{--} + 6 \times 2$
$\underline{24}, \underline{24}^*$	$3 \times 1_{++} + 3 \times 1_{+-} + 3 \times 1_{-+} + 3 \times 1_{--} + 6 \times 2$
$\underline{27}, \underline{27}^*$	$3 \times 1_{++} + 4 \times 1_{+-} + 4 \times 1_{-+} + 4 \times 1_{--} + 6 \times 2$

Table 15.23 Branching rules for $SU(3) \to \Delta(3N^2)$

$SU(3)$	$\Delta(3N^2)$
$\underline{3}$	$3_{[0][1]}$
$\underline{6}$	$3_{[0][2]} + 3_{[0][-1]}$
$\underline{8}$	$1_1 + 1_2 + 3_{[1][1]} + 3_{[2][-1]}$
$\underline{10}$	$1_0 + 3_{[0][3]} + 3_{[1][1]} + 3_{[2][-1]}$
$\underline{15}$	$2 \times 3_{[0][1]} + 3_{[0][-2]} + 3_{[1][2]} + 3_{[3][-2]}$
$\underline{15}'$	$3_{[0][4]} + 3_{[0][1]} + 3_{[0][-2]} + 3_{[1][2]} + 3_{[3][-2]}$
$\underline{21}$	$3_{[0][1]} + 3_{[0][-2]} + 3_{[0][-5]} + 3_{[1][2]} + 3_{[1][-4]} + 3_{[3][-2]} + 3_{[4][-1]}$
$\underline{24}$	$2 \times 3_{[0][1]} + 2 \times 3_{[0][-2]} + 3_{[1][2]} + 3_{[1][-4]} + 3_{[3][-2]} + 3_{[4][-1]}$
$\underline{27}$	$1_0 + 1_1 + 1_2 + 3_{[0][-3]} + 3_{[0][3]} + 2 \times 3_{[1][-2]} + 2 \times 3_{[2][-1]} + 3_{[2][2]} + 3_{[4][-2]}$

It is understood that $3_{[0][0]}$ is a reducible representation, whose irreducible decomposition is given as

$$
3_{[0][0]} = \begin{cases} 1_0 + 1_1 + 1_2 & N \neq 3m, \\ 1_{00} + 1_{10} + 1_{20} & N = 3m, \end{cases} \tag{15.34}
$$

with $m \in Z$. We only show the results for branching rules for generic $N(\neq 3m)$ in Table 15.23. It is straightforward to extend the result to the case with $N = 3m$, which will be explained in a concrete example below.

Table 15.24 Branching rules for $SU(3) \to \Delta(27)$

$SU(3)$	$\Delta(27)$
$\underline{3}$	$3_{[0][1]}$
$\underline{6}$	$2 \times 3_{[0][2]}$
$\underline{8}$	$1_{01} + 1_{02} + 1_{10} + 1_{11} + 1_{12} + 1_{20} + 1_{21} + 1_{22}$
$\underline{10}$	$2 \times 1_{00} + 1_{01} + 1_{02} + 1_{10} + 1_{11} + 1_{12} + 1_{20} + 1_{21} + 1_{22}$
$\underline{15}$	$5 \times 3_{[0][1]}$
$\underline{15}'$	$5 \times 3_{[0][1]}$
$\underline{21}$	$7 \times 3_{[0][1]}$
$\underline{24}$	$8 \times 3_{[0][1]}$
$\underline{27}$	$3 \times (1_{00} + 1_{10} + 1_{20} + 1_{01} + 1_{11} + 1_{21} + 1_{02} + 1_{12} + 1_{22})$

$SU(3) \to \Delta(27)$

We show a concrete example of $N = 3$, namely $SU(3) \to \Delta(27)$. Since $N/3$ is integer, we use the irreducible decompositions for triplet representations

$$3_{[3m+k][3m'+k]} = 1_{0k} + 1_{1k} + 1_{2k},$$

where $m, m' \in Z$ and $k = 0, 1, 2$. The three singlet representations 1_0, 1_1, and 1_2 in $N \neq 3m$ should correspond to $1_{00}, 1_{10}, 1_{20}$ in $N = 3m$, respectively. The result for $SU(3) \to \Delta(27)$ is shown in Table 15.24.

15.3.7 $SU(3) \to \Delta(6N^2)$

$\Delta(6N^2)$ has $2(N - 1)$ triplet irreducible representations of 3_{1k} and 3_{2k} with $k = 1, \cdots, N - 1 \mod N$ in either case of $N/3 =$integer or $N/3 \neq$ integer, and has four generators of a, a', b, and c. Among them, a, a', and b are the same as those in $\Delta(3N^2)$, so that those in the triplet representations are group elements of $SU(3)$. On the other hand, since $\det(c(3_{1k})) = -1$ and $\det(c(3_{2k})) = 1$ in Eqs. (13.51) and (13.26), 3_{2k} can be a subduced representation of $SU(3)$. Thus we identify $\underline{3} = 3_{21}$. Similar to $\Delta(3N^2)$, the complex conjugate representations of $3_{1\ell}$ and $3_{2\ell}$ are given as $3_{1(-\ell)}$ and $3_{2(-\ell)}$, respectively. It is convenient to use reducible representations in general case with $N/3 \neq$integer to calculate a tensor product. We only show the results for branching rules for generic $N(\neq 3m)$ in Table 15.25. Similar to the case of $SU(3) \to \Delta(3N^2)$, it is straightforward to extend the result to the case with $N/3 =$ integer, where some representations become reducible as shown in a concrete example below.

$SU(3) \to \Delta(54)$

We show a concrete example of $N = 3$, namely $SU(3) \to \Delta(54)$. We use the following irreducible decompositions for sextet representations for the case with

Table 15.25 Branching rules for $SU(3) \rightarrow \Delta(6N^2)$

$SU(3)$	$\Delta(6N^2)$
3	$\mathbf{3}_{2(1)}$
6	$\mathbf{3}_{12} + \mathbf{3}_{1(-1)}$
8	$\mathbf{2} + \mathbf{6}_{[[1],[1]]}$
10	$\mathbf{1}_1 + \mathbf{3}_{23} + \mathbf{6}_{[[2],[-1]]}$
15	$\mathbf{3}_{11} + \mathbf{3}_{21} + \mathbf{3}_{2(-2)} + \mathbf{6}_{[[2],[1]]}$
15′	$\mathbf{3}_{11} + \mathbf{3}_{1(-2)} + \mathbf{3}_{14} + \mathbf{6}_{[[2],[1]]}$
21	$\mathbf{3}_{21} + \mathbf{3}_{2(-2)} + \mathbf{3}_{2(-5)} + \mathbf{6}_{[[1],[1]]} + \mathbf{6}_{[[1],[3]]}$
24	$\mathbf{3}_{11} + \mathbf{3}_{1(-2)} + \mathbf{3}_{21} + \mathbf{3}_{2(-2)} + \mathbf{6}_{[[1],[3]]} + \mathbf{6}_{[[1],[-3]]}$
27	$\mathbf{1}_0 + \mathbf{2} + \mathbf{3}_{1(-3)} + \mathbf{3}_{13} + 2 \times \mathbf{6}_{[[1],[1]]} + \mathbf{6}_{[[2],[2]]}$

Table 15.26 Branching rules for $SU(3) \rightarrow \Delta(54)$

$SU(3)$	$\Delta(54)$
3	$\mathbf{3}_{21}$
6	$2 \times \mathbf{3}_{12}$
8	$\mathbf{2}_1 + \mathbf{2}_2 + \mathbf{2}_3 + \mathbf{2}_4$
10	$2 \times \mathbf{1}_1 + \mathbf{2}_1 + \mathbf{2}_2 + \mathbf{2}_3 + \mathbf{2}_4$
15	$2 \times \mathbf{3}_{11} + 3 \times \mathbf{3}_{21}$
15′	$4 \times \mathbf{3}_{11} + \mathbf{3}_{21}$
21	$\mathbf{3}_{11} + 4 \times \mathbf{3}_{21} + \mathbf{2}_2 + \mathbf{2}_3 + \mathbf{2}_4$
24	$4 \times (\mathbf{3}_{11} + \mathbf{3}_{21})$
27	$3 \times (\mathbf{1}_0 + \mathbf{2}_1 + \mathbf{2}_2 + \mathbf{2}_3 + \mathbf{2}_4)$

$N/3 =$ integer,

$$\mathbf{6}_{[1][-1]} = \mathbf{3}_{1(-1)} + \mathbf{3}_{2(-1)},$$
$$\mathbf{6}_{[k][0]} = \mathbf{3}_{1k} + \mathbf{3}_{2k},$$
$$\mathbf{6}_{[0][\ell]} = \mathbf{3}_{1(-\ell)} + \mathbf{3}_{2(-\ell)},$$
$$\mathbf{6}_{[0][0]} = 3 \times \mathbf{2}_1,$$
$$\mathbf{6}_{[1][1]} = \mathbf{6}_{[2][2]} = \mathbf{2}_2 + \mathbf{2}_3 + \mathbf{2}_4,$$

where $k, \ell \in Z$. We note that the double **2** for $N \neq 3m$ in Table 15.25 should correspond to $\mathbf{2}_1$ in $N/3 =$ integer. The triplet representations $\mathbf{3}_{1k}$ and $\mathbf{3}_{2k}$ become reducible when $k = 3m$ ($m \in Z$),

$$\mathbf{3}_{1(3m)} = \mathbf{1}_0 + \mathbf{2}_1,$$
$$\mathbf{3}_{2(3m)} = \mathbf{1}_1 + \mathbf{2}_1.$$

The result for $SU(3) \rightarrow \Delta(54)$ is shown in Table 15.26.

References

1. Ovrut, B.A.: J. Math. Phys. **19**, 418 (1978). https://doi.org/10.1063/1.523660
2. Etesi, G.: J. Math. Phys. **37**, 1596–1602 (1996). https://doi.org/10.1063/1.531470. arXiv:hep-th/9706029 [hep-th]
3. Koca, M., Al-Barwani, M., Koc, R.: J. Phys. A **30**, 2109–2125 (1997). https://doi.org/10.1088/0305-4470/30/6/032
4. Koca, M., Koc, R., Tutunculer, H.: Int. J. Mod. Phys. A **18**, 4817–4827 (2003) https://doi.org/10.1142/S0217751X03015891. arXiv:hep-ph/0410270 [hep-ph]
5. Berger, J., Grossman, Y.: arXiv:0910.4392 [hep-ph]
6. King, S.F., Zhou, Y.L.: JHEP **11**, 173 (2018). https://doi.org/10.1007/JHEP11(2018)173. arXiv:1809.10292 [hep-ph]
7. Adulpravitchai, A., Blum, A., Lindner, M.: JHEP **0909**, 018 (2009). arXiv:0907.2332 [hep-ph]
8. Luhn, C.: JHEP **03**, 108 (2011). arXiv:1101.2417 [hep-ph]
9. Fallbacher, M.: Nucl. Phys. B **898**, 229–247 (2015). arXiv:1506.03677 [hep-th]
10. Rachlin, B.L., Kephart, T.W.: JHEP **08**, 110 (2017). https://doi.org/10.1007/JHEP08(2017)110. arXiv:1702.08073 [hep-ph]
11. Merle, A., Zwicky, R.: JHEP **02**, 128 (2012). https://doi.org/10.1007/JHEP02(2012)128. arXiv:1110.4891 [hep-ph]
12. Lamotke, K.: Regular Solids and Isolated Singularities. Vieweg Verlag (1986)
13. Ludl, P.O.: arXiv:0907.5587 [hep-ph]
14. Slansky, R.: Phys. Rept. **79**, 1–128 (1981)

Modular Symmetry

16

The *modular symmetry* is a geometrical symmetry of the two-dimensional torus, T^2. The two-dimensional torus is constructed as division of the two-dimensional Euclidean space R^2 by a lattice Λ, $T^2 = R^2/\Lambda$. Instead of R^2, one can use the one-dimensional complex plane. As shown in Fig. 16.1, the lattice is spanned by two basis vectors, e_1 and e_2 as $m_1 e_1 + m_2 e_2$, where m_1 and m_2 are integer. Their ratio,

$$\tau = \frac{e_2}{e_1}, \tag{16.1}$$

in the complex plane, represents the shape of T^2, and the parameter τ is called the *modulus*.

The same lattice can be spanned by other basis vectors such as

$$\begin{pmatrix} e'_2 \\ e'_1 \end{pmatrix} = \begin{pmatrix} a & b \\ c & d \end{pmatrix} \begin{pmatrix} e_2 \\ e_1 \end{pmatrix}, \tag{16.2}$$

where a, b, c, d are integer satisfying $ad - bc = 1$. That is the $SL(2, Z)$.

Under the above transformation, the modulus τ transforms as follows,

$$\tau \longrightarrow \tau' = \gamma\tau = \frac{a\tau + b}{c\tau + d}. \tag{16.3}$$

That is the modular symmetry [1–4]. For the element $-e$ in $SL(2, Z)$,

$$-e = \begin{pmatrix} -1 & 0 \\ 0 & -1 \end{pmatrix}, \tag{16.4}$$

the modulus τ is invariant, $\tau \longrightarrow \tau' = (-\tau)/(-1) = \tau$. Thus, the modular group is $\bar{\Gamma} = PSL(2, Z) = SL(2, Z)/\{e, -e\}$. It is sometimes called the *inhomogeneous*

© The Author(s), under exclusive license to Springer-Verlag GmbH, DE, part of Springer Nature 2022
T. Kobayashi et al., *An Introduction to Non-Abelian Discrete Symmetries for Particle Physicists*, Lecture Notes in Physics 995, https://doi.org/10.1007/978-3-662-64679-3_16

Fig. 16.1 Lattice Λ and two
basis vectors, e_1 and e_2

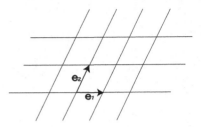

Fig. 16.2 Basis change with
$(a, b, c, d) = (1, 1, 0, 1)$
correspnds to T

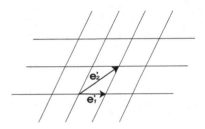

modular group. On the other hand, the group, $\Gamma = SL(2, Z)$ is called the *homogeneous modular group* or the *full modular group.*

The generators of $\Gamma \simeq SL(2, Z)$ are written by S and T,

$$S = \begin{pmatrix} 0 & 1 \\ -1 & 0 \end{pmatrix}, \qquad T = \begin{pmatrix} 1 & 1 \\ 0 & 1 \end{pmatrix}, \tag{16.5}$$

as shown in Appendix H. They satisfy the following algebraic relations,

$$S^4 = (ST)^3 = e. \tag{16.6}$$

Note that

$$S^2 = -e. \tag{16.7}$$

On $\bar{\Gamma} = PSL(2, Z)$, they satisfy

$$S^2 = (ST)^3 = e. \tag{16.8}$$

These relations are also confirmed explicitly by the following transformations:

$$S : \tau \longrightarrow -\frac{1}{\tau}, \qquad T : \tau \longrightarrow \tau + 1, \tag{16.9}$$

which are shown on the lattice Λ in Figs 16.2 and 16.3.

In addition to the above algebraic relations of $\bar{\Gamma} = PSL(2, Z)$, we can require $T^N = e$, i.e.

$$S^2 = (ST)^3 = T^N = e. \tag{16.10}$$

Fig. 16.3 Basis change with $(a, b, c, d) = (0, 1, -1, 0)$ corresponds to S

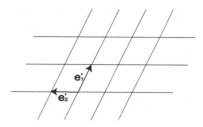

They can correspond to finite groups such as S_3, A_4, S_4, A_5 for $N = 2, 3, 4, 5$. In practice, we define the *principal congruence subgroup* $\Gamma(N)$ as

$$\Gamma(N) = \left\{ \begin{pmatrix} a & b \\ c & d \end{pmatrix} \in \Gamma \, \middle| \, \begin{pmatrix} a & b \\ c & d \end{pmatrix} = \begin{pmatrix} 1 & 0 \\ 0 & 1 \end{pmatrix} \ (\text{mod } N) \right\}. \tag{16.11}$$

It includes T^N, but not S or T. Then, we define the quotient $\Gamma_N = \bar{\Gamma}/\bar{\Gamma}(N)$, where the above algebraic relations are satisfied. It is found that Γ_N with $N = 2, 3, 4, 5$ are isomorphic to S_3, A_4, S_4, A_5, respectively [5]. See for S_3, where b and a in Eq. (3.6) correspond to S and T in the modular group, respectively. For A_4, s and t in Eq. (4.7) correspond to S and T in the modular group. For S_4, the elements $\tilde{s}$ and $\tilde{t}$ in the Basis IV of Appendix B.4 correspond to S and T. For A_5, the elements a and ab in Eq. (D.16) of the Basis II of Appendix D.2 correspond to S and T, respectively. We define $SL(2, Z_N)$ by

$$SL(2, Z_N) = \left\{ \begin{pmatrix} a & b \\ c & d \end{pmatrix} \middle| a, b, c, d \in Z_N, ad - bc = 1 \right\}, \tag{16.12}$$

where Z_N denotes integers modulo N. The group Γ_N is isomorphic to $PSL(2, Z_N) = SL(2, Z_N)/\{e, -e\}$ for $N > 2$, while Γ_2 is isomorphic to $SL(2, Z_2)$, because $e = -e$ in $SL(2, Z_2)$.

Similar to Γ_N, we can define $\Gamma'_N = SL(2, Z)/\Gamma(N)$, and it is the double cover of Γ_N. That is, the groups Γ'_N for $N = 3, 4, 5$ are isomorphic to the double covering groups of A_4, S_4, A_5, i.e. T', S'_4, A'_5, respectively, although Γ'_2 is isomorphic to S_3.

The upper half-plane of the modulus space τ is mapped onto itself. For example, Γ does not include the basis change. $(e_1, e_2) \longrightarrow (e_1, -e_2)$, i.e. $\tau \to -\tau$. In practice, we find

$$\text{Im}(\gamma\tau) = |c\tau + d|^{-2}\text{Im}(\tau). \tag{16.13}$$

Thus, the modular group is represented on the upper half-plane of τ. Obviously one can map any value of τ on the upper half-plane into the region, $-\frac{1}{2} \leq \text{Re}(\tau) \leq \frac{1}{2}$ by T^n. Furthermore, by the modular transformation one can map any value of τ on the upper half-plane into the following region:

$$-\frac{1}{2} \leq \text{Re}(\tau) \leq \frac{1}{2}, \quad |\tau| > 1, \tag{16.14}$$

which is called the *fundamental domain*. Suppose that $\text{Im}(\gamma\tau)$ is a maximum value among all of γ for a fixed value of τ. If $|\gamma\tau| < 1$, we map it by S, and we find

$$\text{Im}(S\gamma\tau) = \frac{\text{Im}(\gamma\tau)}{|\gamma\tau|^2} > \text{Im}(\gamma\tau). \tag{16.15}$$

That is inconsistent with the assumption that $\text{Im}(\gamma\tau)$ is a maximum value among all of γ. That is, we find $|\gamma\tau| > 1$. Thus, we can map τ on the upper half-plane into the fundamental region by the modular transformation. The point $\tau = i$ is the fixed point under S because $S : i \rightarrow -\frac{1}{i} = i$, where Z_2 symmetry remains. Similarly, the point $\tau = e^{2\pi i/3}$ is the fixed point under ST, where Z_3 symmetry remains.

References

1. Gunning, R.C.: Lectures on Modular Forms. Princeton University Press, Princeton, NJ (1962)
2. Schoeneberg, B.: Elliptic Modular Functions. Springer (1974)
3. Koblitz, N.: Introduction to Elliptic Curves and Modular Forms. Springer (1984)
4. Bruinier, J.H., Geer, G.V.D., Harder, G., Zagier, D.: The 1-2-3 of Modular Forms. Springer (2008)
5. de Adelhart Toorop, R., Feruglio, F., Hagedorn, C.: Nucl. Phys. B **858**, 437 (2012). arXiv:1112.1340
6. Fallbacher, M.: Nucl. Phys. B **898**, 229–247 (2015). [arXiv:1506.03677 [hep-th]]

Automorphism

17

As stated in (2.29) of Chap. 2, an automorphism f of a group of G is a *bijective* homomorphism $f: G \to G$. Therefore, it satisfies

$$f(g_1) \cdot f(g_2) = f(g_1 g_2) \tag{17.1}$$

for $^\forall g_1, ^\forall g_2$. These are nothing but nature of homomorphism. An identity mapping is an identity element of automorphism of G. The automorphism also forms a group G', where $G' \neq G$ in general. G', which is denoted by $\mathrm{Aut}(G)$, is the set of all automorphisms of G with composition as group multiplication. An inner automorphism f_h of a group G is an automorphism that is described by conjugation hgh^{-1} for $^\forall h \in G$. Consider a map $f_h(g) = hgh^{-1}$ for $^\forall g \in G$. Then, one finds

$$f_h(g) \cdot f_h(g') = hgh^{-1}hg'h^{-1} = h(gg')h^{-1} = f(gg'), \tag{17.2}$$

that satisfies Eq. (17.1). Otherwise, it is called an outer automorphism. Both the inner and outer automorphisms form groups. Inner automorphism is denoted by $\mathrm{Inn}(G)$, while outer automorphism is symbolized by $\mathrm{Out}(G)$. $\mathrm{Out}(G)$ is understood by transformations among different conjugacy classes. Here, there are two relations among them:

$$\mathrm{Out}(G) \equiv \mathrm{Aut}(G)/\mathrm{Inn}(G), \tag{17.3}$$

$$\mathrm{Inn}(G) \cong G/Z(G), \tag{17.4}$$

where $Z(G)$ is center of G that commutes with all the elements of G.

© The Author(s), under exclusive license to Springer-Verlag GmbH, DE,
part of Springer Nature 2022
T. Kobayashi et al., *An Introduction to Non-Abelian Discrete Symmetries for Particle Physicists*, Lecture Notes in Physics 995,
https://doi.org/10.1007/978-3-662-64679-3_17

17.1 Z_3

Let us consider an automorphism of Abelian discrete group Z_3. It has three elements $1, \omega, \omega^2$, where $\omega \equiv e^{2\pi i/3}$. Since Z_3 is an Abelian group, one straightforwardly finds $Z(Z_3) \cong Z_3$. Therefore, in Eq. (17.4), one finds

$$\text{Inn}(Z_3) \cong Z_3/Z_3 = Z_1. \tag{17.5}$$

Combining the above equation with Eq. (17.3), it suggests

$$\text{Out}(Z_3) \equiv \text{Aut}(Z_3). \tag{17.6}$$

In general, the inner automorphism is trivial for any Abelian groups. That leads to $\text{Out}(G) = \text{Aut}(G)$. Thus, all one should do is to find $\text{Out}(Z_3)$. The total number of mapping is equal to three, since all the elements can be written by ω only due to nature of homomorphism. Then, considering the bijective nature of automorphism, the possible mapping are as follows:

$$\begin{pmatrix} \omega \\ \omega^2 \end{pmatrix} \rightarrow \begin{pmatrix} \omega \\ \omega^2 \end{pmatrix} = \begin{pmatrix} 1 & 0 \\ 0 & 1 \end{pmatrix} \begin{pmatrix} \omega \\ \omega^2 \end{pmatrix}, \tag{17.7}$$

$$\begin{pmatrix} \omega \\ \omega^2 \end{pmatrix} \rightarrow \begin{pmatrix} \omega^2 \\ \omega \end{pmatrix} = \begin{pmatrix} 0 & 1 \\ 1 & 0 \end{pmatrix} \begin{pmatrix} \omega \\ \omega^2 \end{pmatrix}. \tag{17.8}$$

Thus, $\text{Out}(Z_3)$ is represented by the following two matrices:

$$\begin{pmatrix} 1 & 0 \\ 0 & 1 \end{pmatrix}, \quad \begin{pmatrix} 0 & 1 \\ 1 & 0 \end{pmatrix}, \tag{17.9}$$

which is nothing but Z_2 group. Therefore,

$$\text{Aut}(Z_3) = \text{Out}(Z_3) \cong Z_2. \tag{17.10}$$

17.2 $Z_2 \times Z_2'$

Let us consider an automorphism of another Abelian discrete group $Z_2 \times Z_2'$, which is called Klein four-group. Since $Z_2 \times Z_2'$ is an Abelian group, one finds $Z(Z_2 \times Z_2') \cong Z_2 \times Z_2'$. Therefore, in Eq. (17.4), one finds

$$\text{Inn}(Z_2 \times Z_2') \cong (Z_2 \times Z_2')/(Z_2 \times Z_2') = Z_1. \tag{17.11}$$

Similar to Z_3, one finds

$$\text{Out}(Z_2 \times Z_2') \equiv \text{Aut}(Z_2 \times Z_2'). \tag{17.12}$$

The four elements of this group are constructed by $1, S, T, ST$, where

$$S = \begin{pmatrix} 1 & 0 \\ 0 & -1 \end{pmatrix}, \quad T = \begin{pmatrix} -1 & 0 \\ 0 & 1 \end{pmatrix}. \tag{17.13}$$

Once S and T are fixed to be mapped, all the mapping are determined. Considering bijective nature, the total number of mapping is $4 \times 3 = 12$. However, imposing again the bijective one for the twelve mapping, they are reduced to be the following six mapping:

$$\begin{pmatrix} S \\ T \\ ST \end{pmatrix} \rightarrow \begin{pmatrix} S \\ T \\ ST \end{pmatrix} = \begin{pmatrix} 1 & 0 & 0 \\ 0 & 1 & 0 \\ 0 & 0 & 1 \end{pmatrix} \begin{pmatrix} S \\ T \\ ST \end{pmatrix},$$

$$\begin{pmatrix} S \\ T \\ ST \end{pmatrix} \rightarrow \begin{pmatrix} S \\ ST \\ T \end{pmatrix} = \begin{pmatrix} 1 & 0 & 0 \\ 0 & 0 & 1 \\ 0 & 1 & 0 \end{pmatrix} \begin{pmatrix} S \\ T \\ ST \end{pmatrix},$$

$$\begin{pmatrix} S \\ T \\ ST \end{pmatrix} \rightarrow \begin{pmatrix} T \\ S \\ ST \end{pmatrix} = \begin{pmatrix} 0 & 1 & 0 \\ 1 & 0 & 0 \\ 0 & 0 & 1 \end{pmatrix} \begin{pmatrix} S \\ T \\ ST \end{pmatrix},$$

$$\begin{pmatrix} S \\ T \\ ST \end{pmatrix} \rightarrow \begin{pmatrix} T \\ ST \\ S \end{pmatrix} = \begin{pmatrix} 0 & 1 & 0 \\ 0 & 0 & 1 \\ 1 & 0 & 0 \end{pmatrix} \begin{pmatrix} S \\ T \\ ST \end{pmatrix},$$

$$\begin{pmatrix} S \\ T \\ ST \end{pmatrix} \rightarrow \begin{pmatrix} ST \\ S \\ T \end{pmatrix} = \begin{pmatrix} 0 & 0 & 1 \\ 1 & 0 & 0 \\ 0 & 1 & 0 \end{pmatrix} \begin{pmatrix} S \\ T \\ ST \end{pmatrix},$$

$$\begin{pmatrix} S \\ T \\ ST \end{pmatrix} \rightarrow \begin{pmatrix} ST \\ T \\ S \end{pmatrix} = \begin{pmatrix} 0 & 0 & 1 \\ 0 & 1 & 0 \\ 1 & 0 & 0 \end{pmatrix} \begin{pmatrix} S \\ T \\ ST \end{pmatrix}. \tag{17.14}$$

Thus, one finds $\text{Aut}(Z_2 \times Z'_2) = \text{Out}(Z_2 \times Z'_2) \cong S_3$.

17.3 A_4

Let us consider an automorphism of a non-Abelian discrete group A_4, which is written by $(Z_2 \times Z'_2) \rtimes Z_3$. All the twelve elements are written by S and T, where they are represented by

$$S = \begin{pmatrix} 1 & 0 & 0 \\ 0 & -1 & 0 \\ 0 & 0 & -1 \end{pmatrix}, \quad T = \begin{pmatrix} 0 & 0 & 1 \\ 1 & 0 & 0 \\ 0 & 1 & 0 \end{pmatrix}, \tag{17.15}$$

satisfying $S^2 = T^3 = (ST)^3 = e$. Similar to the cases for Abelian groups, 12×11 mappings are reduced to be 24 mappings when the bijective nature is taken into consideration. Furthermore, the set of this twenty four mappings automatically satisfies $S'^2 = T'^3 = (S'T')^3 = e$, where S' and T' corresponds to S and T after mapping. One finds that Aut(A_4) has 24 elements. Because of $Z(A_4) \sim Z_1$ and Inn(A_4) $\cong A_4/Z_1 = A_4$, one finds Out(A_4) $\equiv$ Aut(A_4)/Inn(A_4) $= Z_2$. That is, the full automorphism of A_4 is Aut(A_4) $\cong A_4 \rtimes Z_2 \cong S_4$.

Here, let us think another way to derive the automorphism of A_4. Focusing on conjugacy classes that belong to s, t, one finds s belongs to C_3, and t belongs to C_4. Then, the inner transformations are give by P and Q, where P: $s \leftrightarrow tst^2$, and Q: $t \rightarrow ts$, since these transformations can derive all the elements of C_3 and C_4. One straightforwardly checks the relations $P^3 = Q^2 = (QP)^3 = e$ that is nothing but the generators of A_4. Thus, the inner automorphism is A_4. While the outer transformation is given by R, where R: $t \rightarrow t^2$, since R transfers C_4 to C_4' and vice varsa. Clearly, $R^2 = e$ is satisfied, and it is nothing but Z_2. Thus, the outer automorphism is Z_2. Finally, Aut(A_4) $\cong A_4 \rtimes Z_2$ is obtained, and that is isomorphic to S_4. Notice here that one also finds $QR \neq RQ$ and $PR \neq RP$, this is because the inner and outer automorphisms are connected via semi-direct product.

Notice here that the general feature of automorphism for alternative group A_N as well as symmetric group S_N is found [1].

17.4 $\Sigma(18)$

Let us consider an automorphism of a non-Abelian discrete group $\Sigma(18)$, which is written by $(Z_3 \times Z_3') \rtimes Z_2$. All the 18 elements are written by a, a' and b, where they are represented by

$$a = \begin{pmatrix} 1 & 0 \\ 0 & \omega \end{pmatrix}, \quad a' = \begin{pmatrix} \omega & 0 \\ 0 & 1 \end{pmatrix}, \quad b = \begin{pmatrix} 0 & 1 \\ 1 & 0 \end{pmatrix}, \tag{17.16}$$

satisfying $a^3 = a'^3 = b^2 = e$, $aa' = a'a$, and $bab = a'$. Similar to the cases for Abelian groups, 18×17 mappings are reduced only to be twelve mappings when the bijective nature and the above relations among a, a', b are taken into consideration. One finds that Aut($\Sigma(18)$) has 12 elements. Because of $Z(\Sigma(18)) \sim Z_3$, one obtains Inn($\Sigma(18)$) $\cong \Sigma(18)/Z_3 = Z_3 \rtimes Z_2$, and that is isomorphic to S_3. Thus, one finds Out($\Sigma(18)$) $\equiv$ Aut($\Sigma(18)$)/Inn($\Sigma(18)$) $= Z_2$ and Aut(A_4) $\cong (Z_3 \rtimes Z_2) \rtimes Z_2$.

Here, let us think another way to derive the automorphism of $\Sigma(18)$. Focusing on conjugacy classes that belong to b, a, a', one finds b belongs to $C_3'^{(0)}$, and a, a' belong to $C_2^{(1,0)}$. Then, the inner transformations are give by S and T, where S: $a \leftrightarrow a'$, and T: $b \rightarrow baa'^{-1}$, since these transformations can derive all the elements of $C_3'^{(0)}$ and $C_2^{(1,0)}$. One can straightforwardly check the relations $S^2 = T^3 = (ST)^2 = e$, that is written by $Z_3 \rtimes Z_2$. Thus, the inner automorphism is $(Z_3 \rtimes Z_2)$. While the outer transformation is given by A, where A: $(a, a') \rightarrow (a^2, a'^2)$, since A runs over $C_2^{(1,0)}$ and $C_2^{(2,0)}$. Clearly, $A^2 = e$ is satisfied, and it is nothing but Z_2. Thus, outer automorphism is Z_2. Finally, Aut(A_4) $\cong (Z_3 \rtimes Z_2) \rtimes Z_2$ is obtained. Notice here

Table 17.1 Automorphisms for groups. Here T' is double covering group of A_4

	Z	Aut	Inn	Out
S_3	Z_1	S_3	S_3	Z_1
S_4	Z_1	S_4	S_4	Z_1
A_4	Z_1	S_4	A_4	Z_2
A_5	Z_1	S_5	A_5	Z_2
T'	Z_2	S_4	A_4	Z_2
D_4	Z_2	D_4	$Z_2 \times Z_2$	Z_2
D_5	Z_1	$Z_5 \rtimes Z_4$	D_5	Z_2
Q_4	Z_2	S_4	$Z_2 \times Z_2$	S_3
Q_6	Z_2	D_6	S_3	Z_2
QD_{16}	Z_2	$Z_8 \rtimes Z_2$	Z_8	Z_2
$\Sigma(18)$	Z_3	$(Z_3 \rtimes Z_2) \rtimes Z_2$	$Z_3 \rtimes Z_2$	Z_2
$\Sigma(32)$	Z_4	$(Z_4 \rtimes Z_2) \rtimes (Z_2 \times Z_2)$	$Z_4 \rtimes Z_2$	$Z_2 \times Z_2$
$\Sigma(50)$	Z_5	$(Z_5 \rtimes Z_2) \rtimes (Z_4 \times Z_2)$	$Z_5 \rtimes Z_2$	$Z_4 \times Z_2$
$\Delta(27)$	Z_3	$(Z_3 \times Z_3') \rtimes GL(2,3)$	$Z_3 \times Z_3'$	$GL(2,3)$
$\Delta(48)$	Z_1	$(((Z_4 \times Z_4) \rtimes Z_3) \rtimes Z_4) \rtimes Z_2$	$\Delta(48)$	D_4
T_7	Z_1	$T_7 \rtimes Z_2$	T_7	Z_2
T_{13}	Z_1	$T_{13} \rtimes Z_4$	T_{13}	Z_4
T_{19}	Z_1	$T_{19} \rtimes Z_6$	T_{19}	Z_6
$\Sigma(81)$	Z_3	$Z_3 \times (((Z_3 \times Z_3) \rtimes Z_3) \rtimes (Z_2 \times Z_2))$	$(Z_3 \times Z_3) \rtimes Z_3$	$Z_6 \times Z_2$
$\Delta(54)$	Z_1	$\Delta(54) \rtimes Z_2$	$\Delta(54)$	Z_2
$\Delta(96)$	Z_1	$\Delta(96) \rtimes Z_2$	$\Delta(96)$	Z_2
$Z_9 \rtimes Z_3$	Z_3	$((Z_3 \times Z_3) \rtimes Z_3) \rtimes Z_2$	$Z_3 \times Z_3$	S_3

that one also finds $TA \neq AT$ and $AS = SA$, this is because the inner and outer automorphisms are connected via semi-direct product.

In general, the center of $\Sigma(2N^2)$ is straightforwardly found to be Z_N from the character table. Thus, the inner automorphism is $Z_N \rtimes Z_2$ that is isomorphic to D_N. This can also be derived by the same way as $\Sigma(18)$. In case of $\Sigma(2N^2)$, b belongs to $C_N'^{(0)}$, and a, a' belong to $C_2^{(1,0)}$. Then, the inner transformations are give by S and T, where $S: a \leftrightarrow a'$, and $T: b \to baa'^{-1}$, since these transformations can derive all the elements of $C_N'^{(0)}$ and $C_2^{(1,0)}$. One straightforwardly finds the relations $S^2 = T^N = (ST)^2 = e$ that is written by $Z_N \rtimes Z_2$. Thus, inner automorphism is $(Z_N \rtimes Z_2)$.

17.5 $\Delta(27)$

Let us consider a rather complicated automorphism of a non-Abelian discrete group $\Delta(27)$, which is written by $(Z_3 \times Z'_3) \rtimes Z''_3$. All the twenty seven elements are written by a, a' and b, where they are represented by

$$
a = \begin{pmatrix} \omega & 0 & 0 \\ 0 & 1 & 0 \\ 0 & 0 & \omega^2 \end{pmatrix}, \quad
a' = \begin{pmatrix} \omega^2 & 0 & 0 \\ 0 & \omega & 0 \\ 0 & 0 & 1 \end{pmatrix}, \quad
b = \begin{pmatrix} 0 & 1 & 0 \\ 0 & 0 & 1 \\ 1 & 0 & 0 \end{pmatrix},
\tag{17.17}
$$

satisfying $a^3 = a'^3 = b^3 = e$, $aa' = a'a$, $a'aba = b$, $ba' = ab$. Similar to the previous groups, $27 \times 26 \times 25$ mappings are reduced to be 6048 mappings when the bijective nature is taken into consideration. However, it does not lead to the correct number of element for Aut($\Delta(27)$). To get the correct one, we need to impose $a^3 = a'^3 = b^3 = e$, $aa' = a'a$, $a'aba = b$, $ba' = ab$ after mapping. Then, it is reduced to be 432. Since $Z(\Delta(27))$ is easily found as Z_3 whose elements are $\{e, aa'^2, a^2a'\}$, one finds that Inn($\Delta(27)$) $\cong \Delta(27)/Z(\Delta(27)) = Z_3 \times Z'_3$. Thus, the number of Out($\Delta(27)$) is found as $432/9 = 48$. To identify the group of Out($\Delta(27)$), we classify to which conjugacy classes the 432 elements are mapped. Since now the center is removed, we redefine the notation of conjugacy class as $C_i^{(j+k)}$, where i, j, k, which are written by $b^i a^j a'^k$ and run 1-3 with $j + k$(mod3). By the new notation, one finds that a, a' are mapped to the same conjugacy class, and 48 elements are characterized by $(C_i^{(j+k)}, C_{i'}^{(j'+k')})$ for $(a(a'), b)$. When the above is interpreted by the following two by two matrix

$$
\begin{pmatrix} i & j+k \\ i' & j'+k' \end{pmatrix},
\tag{17.18}
$$

all the elements are explicitly written by

$$
\begin{pmatrix} 1 & 0 \\ 0 & 1 \end{pmatrix}, \begin{pmatrix} 2 & 0 \\ 0 & 2 \end{pmatrix}, \begin{pmatrix} 0 & 1 \\ 1 & 0 \end{pmatrix}, \begin{pmatrix} 0 & 2 \\ 2 & 0 \end{pmatrix}, \begin{pmatrix} 0 & 2 \\ 1 & 2 \end{pmatrix}, \begin{pmatrix} 0 & 1 \\ 2 & 1 \end{pmatrix}, \begin{pmatrix} 0 & 1 \\ 2 & 2 \end{pmatrix}, \begin{pmatrix} 0 & 2 \\ 1 & 1 \end{pmatrix},
$$

$$
\begin{pmatrix} 1 & 0 \\ 0 & 2 \end{pmatrix}, \begin{pmatrix} 2 & 0 \\ 0 & 1 \end{pmatrix}, \begin{pmatrix} 0 & 2 \\ 1 & 0 \end{pmatrix}, \begin{pmatrix} 0 & 1 \\ 2 & 0 \end{pmatrix}, \begin{pmatrix} 2 & 2 \\ 1 & 0 \end{pmatrix}, \begin{pmatrix} 1 & 1 \\ 2 & 0 \end{pmatrix}, \begin{pmatrix} 2 & 1 \\ 2 & 0 \end{pmatrix}, \begin{pmatrix} 1 & 2 \\ 1 & 0 \end{pmatrix},
$$

$$
\begin{pmatrix} 2 & 2 \\ 2 & 0 \end{pmatrix}, \begin{pmatrix} 1 & 1 \\ 1 & 0 \end{pmatrix}, \begin{pmatrix} 2 & 0 \\ 1 & 2 \end{pmatrix}, \begin{pmatrix} 1 & 0 \\ 2 & 1 \end{pmatrix}, \begin{pmatrix} 2 & 0 \\ 2 & 2 \end{pmatrix}, \begin{pmatrix} 1 & 0 \\ 1 & 1 \end{pmatrix}, \begin{pmatrix} 2 & 1 \\ 0 & 2 \end{pmatrix}, \begin{pmatrix} 1 & 2 \\ 0 & 1 \end{pmatrix},
$$

$$
\begin{pmatrix} 1 & 1 \\ 0 & 2 \end{pmatrix}, \begin{pmatrix} 2 & 2 \\ 0 & 1 \end{pmatrix}, \begin{pmatrix} 2 & 0 \\ 2 & 1 \end{pmatrix}, \begin{pmatrix} 1 & 0 \\ 1 & 2 \end{pmatrix}, \begin{pmatrix} 2 & 0 \\ 1 & 1 \end{pmatrix}, \begin{pmatrix} 1 & 0 \\ 2 & 2 \end{pmatrix}, \begin{pmatrix} 1 & 2 \\ 0 & 2 \end{pmatrix}, \begin{pmatrix} 2 & 1 \\ 0 & 1 \end{pmatrix},
$$

$$
\begin{pmatrix} 1 & 2 \\ 2 & 2 \end{pmatrix}, \begin{pmatrix} 2 & 1 \\ 1 & 1 \end{pmatrix}, \begin{pmatrix} 1 & 1 \\ 1 & 2 \end{pmatrix}, \begin{pmatrix} 2 & 2 \\ 2 & 1 \end{pmatrix}, \begin{pmatrix} 2 & 1 \\ 2 & 2 \end{pmatrix}, \begin{pmatrix} 2 & 2 \\ 1 & 2 \end{pmatrix}, \begin{pmatrix} 1 & 2 \\ 1 & 1 \end{pmatrix}, \begin{pmatrix} 1 & 1 \\ 2 & 1 \end{pmatrix},
$$

$$
\begin{pmatrix} 0 & 1 \\ 1 & 1 \end{pmatrix}, \begin{pmatrix} 1 & 2 \\ 2 & 0 \end{pmatrix}, \begin{pmatrix} 0 & 2 \\ 2 & 1 \end{pmatrix}, \begin{pmatrix} 1 & 1 \\ 1 & 0 \end{pmatrix}, \begin{pmatrix} 2 & 2 \\ 2 & 0 \end{pmatrix}, \begin{pmatrix} 0 & 1 \\ 1 & 2 \end{pmatrix}, \begin{pmatrix} 2 & 1 \\ 1 & 0 \end{pmatrix}, \begin{pmatrix} 0 & 2 \\ 2 & 2 \end{pmatrix}. \tag{17.19}
$$

The set of matrices is corresponding to GL(2, 3) that is isomorphic to $Q_4 \rtimes S_3$. It is the general linear group of degree two: 2×2 invertible matrices over the field of three elements; $\{0,1,2\}$. GL(2, 3) is known as the automorphism group of $Z_3 \times Z_3'$. Finally, Aut($\Delta(27)$) is written by

$$(Z_3 \times Z_3') \rtimes GL(2, 3). \tag{17.20}$$

In the following, we summarize the results of automorphisms for non-Abelian groups.

References

1. Holthausen, M., Lindner, M., Schmidt, M.A.: JHEP **1304**, 122 (2013). arXiv:1211.6953 [hep-ph]
2. Fallbacher, M.: Nucl. Phys. B **898**, 229–247 (2015). arXiv:1506.03677 [hep-th]

Anomalies

18

18.1 Generic Aspects

Several interesting applications of Abelian and non-Abelian discrete symmetries have been studied in particle physics such as flavor physics (see Chap. 19). In general, symmetries at the tree-level can be broken by quantum effects, that is, anomalies. When symmetries are anomalous, symmetry breaking terms are induced. On the other hand, symmetries should be anomaly-free to be exact even including any quantum effects.

Anomalies of continuous symmetries, in particular gauge symmetries, have been studied well. However, anomalies of non-Abelian discrete symmetries as well as Abelian ones may not be well-known. Here we review about anomalies of Abelian and non-Abelian discrete symmetries.

We study the gauge theory with a (non-Abelian) gauge group G_g and a set of fermions $\Psi = [\psi^{(1)}, \ldots, \psi^{(M)}]$. Then, we assume that their Lagrangian is invariant under the following chiral transformation,

$$\Psi(x) \rightarrow U\Psi(x), \qquad (18.1)$$

with $U = \exp(i\alpha P_L)$ and $\alpha = \alpha^A T_A$, where T_A denote the generators of the transformation, α^A are the transformation parameters and P_L is the left-chiral projector. Here, the above transformation is not necessary a gauge transformation. The fermions $\Psi(x)$ are the (irreducible) M-plet representation $\boldsymbol{R}^M$. For the moment, we consider the Abelian flavor symmetry and we suppose that $\Psi(x)$ correspond to (non-trivial) singlets under the flavor symmetry while they correspond to the $\boldsymbol{R}^M$ representation under the gauge group G_g. Since the generator T_A as well as α is represented on $\boldsymbol{R}^M$ as a $(M \times M)$ matrix, we use the notation, $T_A(\boldsymbol{R}^M)$ and $\alpha(\boldsymbol{R}^M) = \alpha^A T_A(\boldsymbol{R}^M)$.

For our purpose, the path integral approach is convenient. Thus, we use Fujikawa's method [1,2] to derive anomalies of continuous and discrete symmetries (see e.g.

© The Author(s), under exclusive license to Springer-Verlag GmbH, DE,
part of Springer Nature 2022
T. Kobayashi et al., *An Introduction to Non-Abelian Discrete Symmetries for Particle Physicists*, Lecture Notes in Physics 995,
https://doi.org/10.1007/978-3-662-64679-3_18

Ref. [3]). We calculate the transformation of the path integral measure,

$$\mathcal{D}\Psi\mathcal{D}\bar{\Psi} \to \mathcal{D}\Psi\mathcal{D}\bar{\Psi} J(\alpha), \tag{18.2}$$

where the Jacobian, $J(\alpha)$, is written as

$$J(\alpha) = \exp\left(i \int d^4x \mathcal{A}(x; \alpha)\right). \tag{18.3}$$

The anomaly function $\mathcal{A}$ consists of a gauge part and a gravitational part [4–6]

$$\mathcal{A} = \mathcal{A}_{\text{gauge}} + \mathcal{A}_{\text{grav}}. \tag{18.4}$$

The gauge part is given by

$$\mathcal{A}_{\text{gauge}}(x; \alpha) = \frac{1}{32\pi^2}\text{Tr}\left[\alpha(\boldsymbol{R}^M) F^{\mu\nu}(x) \widetilde{F}_{\mu\nu}(x)\right], \tag{18.5}$$

where $F^{\mu\nu}$ denotes the field strength of the gauge fields, $F_{\mu\nu} = [D_\mu, D_\nu]$, and $\widetilde{F}_{\mu\nu}$ denotes its dual, $\widetilde{F}^{\mu\nu} = \varepsilon^{\mu\nu\rho\sigma} F_{\rho\sigma}$. The trace 'Tr' runs over all internal indices. When the transformation corresponds to a continuous symmetry, this anomaly can be calculated by the triangle diagram with external lines of two gauge bosons and one current corresponding to the symmetry for Eq. (18.1).

Similarly, the gravitation part is obtained as [4–6]

$$\mathcal{A}_{\text{grav}} = -\mathcal{A}_{\text{grav}}^{\text{Weyl fermion}}\text{tr}\left[\alpha(\boldsymbol{R}^{(M)})\right], \tag{18.6}$$

where 'tr' is the trace for the $(M \times M)$ matrix $T_A(\boldsymbol{R}^M)$. The contribution of a single Weyl fermion to the gravitational anomaly is given by [4–6]

$$\mathcal{A}_{\text{grav}}^{\text{Weyl fermion}} = \frac{1}{384\pi^2}\frac{1}{2}\varepsilon^{\mu\nu\rho\sigma} R_{\mu\nu}{}^{\lambda\gamma} R_{\rho\sigma\lambda\gamma}. \tag{18.7}$$

When other sets of M_i-plet fermions Ψ_{M_i} are included in a theory, the total gauge and gravity anomalies are obtained as their summations, $\sum_{\Psi_{M_i}} \mathcal{A}_{\text{gauge}}$ and $\sum_{\Psi_{M_i}} \mathcal{A}_{\text{grav}}$.

We evaluate these anomalies by using the following index theorems [4,5],

$$\int d^4x \frac{1}{32\pi^2} \varepsilon^{\mu\nu\rho\sigma} F_{\mu\nu}^a F_{\rho\sigma}^b \, \text{tr}\,[t_a t_b] \in \mathbb{Z}, \tag{18.8a}$$

$$\frac{1}{2}\int d^4x \frac{1}{384\pi^2}\frac{1}{2}\varepsilon^{\mu\nu\rho\sigma} R_{\mu\nu}{}^{\lambda\gamma} R_{\rho\sigma\lambda\gamma} \in \mathbb{Z}, \tag{18.8b}$$

where t_a are generators of G_g in the fundamental representation. We use the convention that $\text{tr}[t_a t_b] = \frac{1}{2}\delta_{ab}$. The factor $\frac{1}{2}$ in Eq. (18.8b) follows from Rohlin's theorem [7], as discussed in [8]. Of course, these indices are independent of each other. The path integral includes all possible configurations corresponding to different index numbers.

First of all, we study anomalies of the continuous $U(1)$ symmetry. We consider a theory with a (non-Abelian) gauge symmetry G_g as well as the continuous $U(1)$ symmetry, which may be gauged. This theory include fermions with $U(1)$ charges, $q^{(f)}$ and representations $\boldsymbol{R}^{(f)}$. Those anomalies vanish if and only if the Jacobian is trivial, i.e. $J(\alpha) = 1$ for an arbitrary value of α. Using the index theorems, one can find that the anomaly-free conditions require

$$A_{U(1)-G_g-G_g} \equiv \sum_{\boldsymbol{R}^{(f)}} q^{(f)} T_2(\boldsymbol{R}^{(f)}) = 0, \tag{18.9}$$

for the mixed $U(1) - G_g - G_g$ anomaly, and

$$A_{U(1)-\text{grav}-\text{grav}} \equiv \sum_f q^{(f)} = 0, \tag{18.10}$$

for the $U(1)$–gravity–gravity anomaly. Here, $T_2(\boldsymbol{R}^{(f)})$ is the Dynkin index of the $\boldsymbol{R}^f$ representation, i.e.

$$\text{tr}\left[t_a\left(\boldsymbol{R}^{(f)} \right) t_b\left(\boldsymbol{R}^{(f)} \right) \right] = \delta_{ab} T_2(\boldsymbol{R}^{(f)}) . \tag{18.11}$$

Next, let us study anomalies of the Abelian discrete symmetry, i.e. the Z_N symmetry. For the Z_N symmetry, we write $\alpha = 2\pi Q_N/N$, where Q_N is the Z_N charge operator and its eigenvalues are integers. Here we denote Z_N charges of fermions as $q_N^{(f)}$. Then we can evaluate the $Z_N - G_g - G_g$ and Z_N-gravity-gravity anomalies in a way similar to the above $U(1)$ anomalies. However, the important difference is that α is a discrete value, although α is a continuous parameter in the $U(1)$ symmetry. Then, the anomaly-free conditions, i.e., $J(\alpha) = 1$ for a discrete transformation, require

$$A_{Z_N-G_g-G_g} = \frac{1}{N} \sum_{\boldsymbol{R}^{(f)}} q_N^{(f)} \left(2\, T_2(\boldsymbol{R}^{(f)}) \right) \in \mathbb{Z} , \tag{18.12}$$

for the $Z_N - G_g - G_g$ anomaly, and

$$A_{Z_N-\text{grav}-\text{grav}} = \frac{2}{N} \sum_f q_N^{(f)} \dim \boldsymbol{R}^{(f)} \in \mathbb{Z} , \tag{18.13}$$

for the Z_N-gravity-gravity anomaly. These anomaly-free conditions reduce to

$$\sum_{\boldsymbol{R}^{(f)}} q_N^{(f)} T_2(\boldsymbol{R}^{(f)}) = 0 \mod N/2 , \tag{18.14a}$$

$$\sum_f q_N^{(f)} \dim \boldsymbol{R}^{(f)} = 0 \mod N/2 . \tag{18.14b}$$

Note that the Z_2 symmetry is always free from the Z_2-gravity-gravity anomaly.

Finally, we study anomalies of non-Abelian discrete symmetries G [3,9]. A discrete group G consists of the finite number of elements, g_i. Hence, the non-Abelian discrete symmetry is anomaly-free if and only if the Jacobian is vanishing for the transformation corresponding to each element g_i. Furthermore, recall that $(g_i)^{N_i} = 1$. That is, each element g_i in the non-Abelian discrete group generates a Z_{N_i} symmetry. Thus, the analysis on non-Abelian discrete anomalies reduces to one on Abelian discrete anomalies. One can take the field basis such that g_i is represented in a diagonal form. In such a basis, each field has a definite Z_{N_i} charge, $q_{N_i}^{(f)}$. The anomaly-free conditions for the g_i transformation are written as

$$\sum_{\boldsymbol{R}^{(f)}} q_{N_i}^{(f)}\, T_2(\boldsymbol{R}^{(f)}) \ = \ 0 \quad \mathrm{mod}\ N_i/2 \, , \tag{18.15a}$$

$$\sum_f q_{N_i}^{(f)}\, \dim \boldsymbol{R}^{(f)} \ = \ 0 \quad \mathrm{mod}\ N_i/2 \, . \tag{18.15b}$$

If these conditions are satisfied for all of $g_i \in G$, there are no anomalies of the full non-Abelian symmetry G. Otherwise, the non-Abelian symmetry is broken to its subgroup by quantum effects, where the subgroup does not include anomalous g_i elements. Furthermore, the non-Abelian symmetry is broken completely if all of elements $g_i \in G$ except the identity are anomalous.

In principle, we can investigate anomalies of non-Abelian discrete symmetries G following the above procedure. However, we give a practically simpler way to analyze those anomalies [3,9]. Here, we consider again the transformation similar to (18.1) for a set of fermions $\Psi = [\psi^{(1)}, \dots, \psi^{(Md_\alpha)}]$, which correspond to the $\boldsymbol{R}^M$ irreducible representation of the gauge group G_g and the $\boldsymbol{r}^\alpha$ irreducible representation of the non-Abelian discrete symmetry G with the dimension d_α. Let U correspond to one of group elements $g_i \in G$, which is represented by the matrix $D_\alpha(g_i)$ on $\boldsymbol{r}^\alpha$. Then, the Jacobian is proportional to its determinant, $\det D(g_i)$. Thus, the representations with $\det D_\alpha(g_i) = 1$ do not contribute to anomalies. Therefore, the non-trivial Jacobian, i.e. anomalies are originated from representations with $\det D_\alpha(g_i) \neq 1$. Note that $\det D_\alpha(g_i) = \det D_\alpha(gg_ig^{-1})$ for $g \in G$, that is, the determinant is constant in a conjugacy class. Thus, it would be useful to calculate the determinants of elements on each irreducible representation. Such a determinant for the conjugacy class C_i can be written by

$$\det(C_i)_\alpha = e^{2\pi i q_{\hat{N}_i}^\alpha / \hat{N}_i} \, , \tag{18.16}$$

on the irreducible representation $\boldsymbol{r}^\alpha$. Note that $\hat{N}_i$ is a divisor of N_i, where N_i is the order of g_i in the conjugacy class C_i, i.e. $g^{N_i} = e$, such that $q_{\hat{N}_i}^\alpha$ are normalized to be integers for all of the irreducible representations $\boldsymbol{r}^\alpha$. We consider the $Z_{\hat{N}_i}$ symmetries and their anomalies. Then, we obtain the anomaly-free conditions similar to (18.15).

That is, the anomaly-free conditions for the conjugacy classes C_i are written as

$$\sum_{r^{(\alpha)}, \mathbf{R}^{(f)}} q_{\hat{N}_i}^{\alpha(f)} T_2(\mathbf{R}^{(f)}) = 0 \mod \hat{N}_i/2 \,, \tag{18.17a}$$

$$\sum_{\alpha, f} q_{\hat{N}_i}^{\alpha(f)} \dim \mathbf{R}^{(f)} = 0 \mod \hat{N}_i/2 \,, \tag{18.17b}$$

for the theory including fermions with the $\mathbf{R}^{(f)}$ representations of the gauge group G_g and the $r^{\alpha(f)}$ representations of the flavor group G, which correspond to the $Z_{\hat{N}_i}$ charges, $q_{\hat{N}_i}^{\alpha(f)}$. Note that the fermion fields with the d_α-dimensional representation r^α contribute to these anomalies, $q_{\hat{N}_i}^{\alpha(f)} T_2(\mathbf{R}^{(f)})$ and $q_{\hat{N}_i}^{\alpha(f)} \dim \mathbf{R}^{(f)}$, but not $d_\alpha q_{\hat{N}_i}^{\alpha(f)} T_2(\mathbf{R}^{(f)})$ and $d_\alpha q_{\hat{N}_i}^{\alpha(f)} \dim \mathbf{R}^{(f)}$. If these conditions are satisfied for all of conjugacy classes of G, the full non-Abelian symmetry G is free from anomalies. Otherwise, the non-Abelian symmetry is broken by quantum effects. As we will see below, in concrete examples, the above anomaly-free conditions often lead to the same conditions between different conjugacy classes. Note, when $\hat{N}_i = 2$, the symmetry is always free from the mixed gravitational anomalies. We study explicitly more for concrete groups in what follows.

18.2 Explicit Calculations

Here, we apply the above studies on anomalies for concrete groups.

• S_3

We start with S_3. As shown in Sect. 3.1, the S_3 group has the three conjugacy classes, $C_1 = \{e\}$, $C_2 = \{ab, ba\}$ and $C_3 = \{a, b, bab\}$, and three irreducible representations, $\mathbf{1}$, $\mathbf{1}'$ and $\mathbf{2}$. Note that the determinants of elements are constant in a conjugacy class. The determinants of elements in singlet representations are equal to characters. Obviously, the determinants of elements in a trivial singlet representation $\mathbf{1}$ are always equal to 1. On the doublet representation $\mathbf{2}$, the determinants of representation matrices in C_1, C_2 and C_3 are obtained as 1, 1 and -1, respectively. These determinants are shown in Table 18.1.

From these results, it is found that only the conjugacy class C_3 is relevant to anomalies and the only Z_2 symmetry can be anomalous. Under such a Z_2 symmetry,

Table 18.1 Determinants on S_3 representations

	1	**1'**	**2**
$\det(C_1)$	1	1	1
$\det(C_2)$	1	1	1
$\det(C_3)$	1	-1	-1

Table 18.2 Determinants on S_4 representations

	1	1′	2	3	3′
$\det(C_1)$	1	1	1	1	1
$\det(C_3)$	1	1	1	1	1
$\det(C_6)$	1	−1	−1	−1	1
$\det(C_6')$	1	−1	−1	−1	1
$\det(C_8)$	1	1	1	1	1

the trivial singlet has vanishing Z_2 charge, while the other representations, **1′** and **2** have the Z_2 charges $q_2 = 1$, that is,

$$Z_2 \text{ even} : \mathbf{1},$$
$$Z_2 \text{ odd} : \mathbf{1'}, \quad \mathbf{2}. \tag{18.18}$$

Thus, the anomaly-free conditions for the $Z_2 - G_g - G_g$ mixed anomaly (18.17) are written as

$$\sum_{\mathbf{1'}} \sum_{\mathbf{R}^{(f)}} T_2(\mathbf{R}^{(f)}) + \sum_{\mathbf{2}} \sum_{\mathbf{R}^{(f)}} T_2(\mathbf{R}^{(f)}) = 0 \quad \text{mod } 1 . \tag{18.19}$$

Note that a doublet **2** contributes on the anomaly coefficient by not $2T_2(\mathbf{R}^{(f)})$ but $T_2(\mathbf{R}^{(f)})$, which is the same as **1′**. To show this explicitly, we have written the summations on **1′** and **2** separately.

• S_4

Similarly, we can study anomalies of S_4. As seen in Sect. 3.2, the S_4 group has five the conjugacy classes, C_1, C_3, C_6, C_6' and C_8 and the five irreducible representations, **1**, **1′**, **2**, **3** and **3′**. The determinants of group elements in each representation are shown in Table 18.2. These results imply that only the Z_2 symmetry can be anomalous. Under such a Z_2 symmetry, each representation has the following behaviors,

$$Z_2 \text{ even} : \mathbf{1}, \quad \mathbf{3'},$$
$$Z_2 \text{ odd} : \mathbf{1'}, \quad \mathbf{2}, \quad \mathbf{3}. \tag{18.20}$$

Then, the anomaly-free conditions for the $Z_2 - G_g - G_g$ mixed anomaly (18.15) are written as

$$\sum_{\mathbf{1'}} \sum_{\mathbf{R}^{(f)}} T_2(\mathbf{R}^{(f)}) + \sum_{\mathbf{2}} \sum_{\mathbf{R}^{(f)}} T_2(\mathbf{R}^{(f)}) + \sum_{\mathbf{3}} \sum_{\mathbf{R}^{(f)}} T_2(\mathbf{R}^{(f)}) = 0 \quad \text{mod } 1 . \tag{18.21}$$

• A_4

We study anomalies of A_4. As shown in Sect. 4.1, there are four conjugacy classes, C_1, C_3, C_4 and C_4', and four irreducible representations, **1**, **1′**, **1″** and **3**. The deter-

Table 18.3 Determinants on A_4 representations

	1	1′	1″	3
$\det(C_1)$	1	1	1	1
$\det(C_3)$	1	1	1	1
$\det(C_4)$	1	ω	ω^2	1
$\det(C_4')$	1	ω^2	ω	1

minants of group elements in each representation are shown in Table 18.3, where $\omega = e^{2\pi i/3}$. These results imply that only the Z_3 symmetry can be anomalous. Under such a Z_3 symmetry, each representation has the following Z_3 charge q_3,

$$
\begin{aligned}
q_3 &= 0 : \mathbf{1}, \quad \mathbf{3}, \\
q_3 &= 1 : \mathbf{1'}, \\
q_3 &= 2 : \mathbf{1''}.
\end{aligned}
\tag{18.22}
$$

This corresponds to the Z_3 symmetry for the conjugacy class C_4. There is another Z_3 symmetry for the conjugacy class C_4', but it is not independent of the former Z_3. Then, the anomaly-free conditions are written as

$$
\sum_{\mathbf{1'}} \sum_{R^{(f)}} T_2(R^{(f)}) + 2 \sum_{\mathbf{1''}} \sum_{R^{(f)}} T_2(R^{(f)}) = 0 \mod 3/2 ,
\tag{18.23}
$$

for the $Z_3 - G_g - G_g$ anomaly and

$$
\sum_{\mathbf{1'}} \sum_{R^{(f)}} \dim R^{(f)} + 2 \sum_{\mathbf{1''}} \sum_{R^{(f)}} \dim R^{(f)} = 0 \mod 3/2 ,
\tag{18.24}
$$

for the Z_3-gravity-gravity anomaly.

● **A$_5$**

We study anomalies of A_5. As shown in Sect. 4.2, there are five conjugacy classes, $C_1, C_{15}, C_{20}, C_{12}$ and C_{12}', and five irreducible representations, $\mathbf{1}, \mathbf{3}, \mathbf{3'}, \mathbf{4}$ and $\mathbf{5}$. The determinants of group elements in each representation are shown in Table 18.4. That is, the determinants of all the A_5 elements are equal to one on any representation. This result can be understood as follows. All of the A_5 elements are written by products of $s = a$ and $t = bab$. The generators, s and t, are written as real matrices on all of representations, $\mathbf{1}, \mathbf{3}, \mathbf{3'}, \mathbf{4}$ and $\mathbf{5}$. Thus, it is found $\det(t) = 1$, because $t^5 = e$. Similarly, since $s^2 = b^3 = e$, the possible values are obtained as $\det(s) = \pm 1$ and $\det(b) = \omega^k$ with $k = 0, 1, 2$. By imposing $\det(bab) = \det(t) = 1$, we find $\det(s) = \det(b) = 1$. Thus, it is found that $\det(g) = 1$ for all of the A_5 elements on any representation. Therefore, the A_5 symmetry is always anomaly-free (Table 18.4).

Table 18.4 Determinants on A_5 representations

	1	**3**	**3'**	**4**	**5**
$\det(C_1)$	1	1	1	1	1
$\det(C_{15})$	1	1	1	1	1
$\det(C_{20})$	1	1	1	1	1
$\det(C_{12})$	1	1	1	1	1
$\det(C'_{12})$	1	1	1	1	1

• **T'**

We study anomalies of T'. As shown in Chap. 5, the T' group has seven conjugacy classes, C_1, C'_1, C_4, C'_4, C''_4, C'''_4 and C_6, and seven irreducible representations, **1**, **1'**, **1''**, **2**, **2'**, **2''** and **3**. The determinants of group elements on each representation are shown in Table 18.5. These results imply that only the Z_3 symmetry can be anomalous. Under such a Z_3 symmetry, each representation has the following Z_3 charge q_3,

$$
\begin{aligned}
q_3 &= 0 : \mathbf{1}, \quad \mathbf{2}, \quad \mathbf{3}, \\
q_3 &= 1 : \mathbf{1'}, \quad \mathbf{2''}, \\
q_3 &= 2 : \mathbf{1''} \quad \mathbf{2'}.
\end{aligned}
\tag{18.25}
$$

This corresponds to the Z_3 symmetry for the conjugacy class C_4. There is other Z_3 symmetries for the conjugacy classes C'_4, C''_4 and C'''_4, but those are not independent of the former Z_3. Then, the anomaly-free conditions are written as

$$
\sum_{\mathbf{1'}} \sum_{\mathbf{R}^{(f)}} T_2(\mathbf{R}^{(f)}) + 2 \sum_{\mathbf{1''}} \sum_{\mathbf{R}^{(f)}} T_2(\mathbf{R}^{(f)}) + \sum_{\mathbf{2''}} \sum_{\mathbf{R}^{(f)}} T_2(\mathbf{R}^{(f)})
$$
$$
+ 2 \sum_{\mathbf{2'}} \sum_{\mathbf{R}^{(f)}} T_2(\mathbf{R}^{(f)}) = 0 \mod 3/2,
\tag{18.26}
$$

Table 18.5 Determinants on T' representations

	1	**1'**	**1''**	**2**	**2'**	**2''**	**3**
$\det(C_1)$	1	1	1	1	1	1	1
$\det(C'_1)$	1	1	1	1	1	1	1
$\det(C_4)$	1	ω	ω^2	1	ω^2	ω	1
$\det(C'_4)$	1	ω^2	ω	1	ω	ω^2	1
$\det(C''_4)$	1	ω	ω^2	1	ω^2	ω	1
$\det(C'''_4)$	1	ω^2	ω	1	ω	ω^2	1
$\det(C_6)$	1	1	1	1	1	1	1

Table 18.6 Determinants on D_N representations for $N = $ even

	1_{++}	1_{+-}	1_{-+}	1_{--}	2_k
$\det(b)$	1	1	-1	-1	-1
$\det(ab)$	1	-1	1	-1	-1

for the $Z_3 - G_g - G_g$ anomaly and

$$\sum_{1'}\sum_{R^{(f)}} \dim R^{(f)} + 2\sum_{1''}\sum_{R^{(f)}} \dim R^{(f)} + \sum_{2''}\sum_{R^{(f)}} \dim R^{(f)}$$
$$+ 2\sum_{2'}\sum_{R^{(f)}} \dim R^{(f)} = 0 \mod 3/2 , \qquad (18.27)$$

for the Z_3-gravity-gravity anomaly.

• **D_N (N = even)**

We study anomalies of D_N with $N = $ even. As shown in Chap. 6, the D_N group with $N = $ even has the four singlets $1_{\pm\pm}$ and $(N/2 - 1)$ doublets 2_k. All of the D_N elements can be written as products of two elements, a and b. Their determinants on 2_k are obtained as $\det(a) = 1$ and $\det(b) = -1$. Similarly, we can obtain determinants of a and b on four singlets, $1_{\pm\pm}$. Indeed, four singlets are classified by values of $\det(b)$ and $\det(ab)$, that is, $\det(b) = 1$ for $1_{+\pm}$, $\det(b) = -1$ for $1_{-\pm}$, $\det(ab) = 1$ for $1_{\pm+}$ and $\det(ab) = -1$ for $1_{\pm-}$. Thus, the determinants of b and ab are essential for anomalies. Those determinants are summarized in Table 18.6. This implies that two Z_2 symmetries can be anomalous. One Z_2 corresponds to b and the other Z_2' corresponds to ab. Under these $Z_2 \times Z_2'$ symmetry, each representation has the following behavior,

$$Z_2 \text{ even} : 1_{+\pm},$$
$$Z_2 \text{ odd} : 1_{-\pm}, \quad 2_k, \qquad (18.28)$$

$$Z_2' \text{ even} : 1_{\pm+},$$
$$Z_2' \text{ odd} : 1_{\pm-}, \quad 2_k. \qquad (18.29)$$

Then, the anomaly-free conditions are written as

$$\sum_{1_{-\pm}}\sum_{R^{(f)}} T_2(R^{(f)}) + \sum_{2_k}\sum_{R^{(f)}} T_2(R^{(f)}) = 0 \mod 1 , \qquad (18.30)$$

for the $Z_2 - G_g - G_g$ anomaly and

$$\sum_{1_{\pm-}}\sum_{R^{(f)}} T_2(R^{(f)}) + \sum_{2_k}\sum_{R^{(f)}} T_2(R^{(f)}) = 0 \mod 1 , \qquad (18.31)$$

Table 18.7 Determinants on D_N representations for $N = $ odd

	1_+	1_-	2_k
$\det(b)$	1	-1	-1
$\det(a)$	1	1	1

for the $Z_2' - G_g - G_g$ anomaly.

● **D_N (N = odd)**

Similarly, we study anomalies of D_N with $N = $ odd. As shown in Chap. 6, the D_N group with $N = $ odd has the two singlets $1_\pm$ and $(N-1)/2$ doublets 2_k. Similarly to D_N with $N = $ even, all elements of D_N with $N = $ odd are written by products of two elements, a and b. The determinants of a are obtained as $\det(a) = 1$ on all of representations, $1_\pm$ and 2_k. The determinants of b are obtained as $\det b = 1$ on 1_+ and $\det(b) = -1$ on 1_- and 2_k. These are shown in Table 18.7. Thus, only the Z_2 symmetry corresponding to b can be anomalous. Under such a Z_2 symmetry, each representation has the following behavior,

$$Z_2 \text{ even} : 1_+,$$
$$Z_2 \text{ odd} : 1_-, \quad 2_k. \tag{18.32}$$

Then, the anomaly-free condition is written as

$$\sum_{1_-} \sum_{R^{(f)}} T_2(R^{(f)}) + \sum_{2_k} \sum_{R^{(f)}} T_2(R^{(f)}) = 0 \mod 1, \tag{18.33}$$

for the $Z_2 - G_g - G_g$ anomaly.

● **Q_N (N = 4n)**

We study anomalies of Q_N with $N = 4n$. As shown in Chap. 7, the Q_N group with $N = 4n$ has four singlets $1_{\pm\pm}$ and $(N/2 - 1)$ doublets 2_k. All elements of Q_N are written by products of a and b. The determinant of a is obtained as $\det(a) = 1$ on all of doublets, 2_k. On the other hand, the determinant of b is obtained as $\det(b) = 1$ on the doublets 2_k with $k = $ odd and $\det(b) = -1$ on the doublets 2_k with $k = $ even. Similarly to D_N with $N = $ even, the four singlets $1_{\pm\pm}$ are classified by values of $\det(b)$ and $\det(ab)$, that is, $\det(b) = 1$ for $1_{+\pm}$, $\det(b) = -1$ for $1_{-\pm}$, $\det(b) = 1$ for $1_{\pm+}$ and $\det(b) = -1$ for $1_{\pm-}$. Thus, the determinants of b and ab are essential for anomalies. Those determinants are summarized in Table 18.8. Similarly to D_N with $N = $ even, two Z_2 symmetries can be anomalous. One Z_2 corresponds to b and the other Z_2' corresponds to ab. Under these $Z_2 \times Z_2'$ symmetry, each representation has the following behavior,

$$Z_2 \text{ even} : 1_{+\pm}, \quad 2_{k = \text{ odd}},$$
$$Z_2 \text{ odd} : 1_{-\pm}, \quad 2_{k = \text{ even}}. \tag{18.34}$$

Table 18.8 Determinants on Q_N representations for $N/2 =$ even

	1_{++}	1_{+-}	1_{-+}	1_{--}	$2_{k=\text{odd}}$	$2_{k=\text{even}}$
$\det(b)$	1	1	-1	-1	1	-1
$\det(ab)$	1	-1	1	-1	1	-1

$$Z_2' \text{ even} : 1_{\pm+}, \quad 2_{k=\text{odd}},$$
$$Z_2' \text{ odd} : 1_{\pm-}, \quad 2_{k=\text{even}}. \tag{18.35}$$

Then, the anomaly-free conditions are written as

$$\sum_{1_{-\pm}} \sum_{R^{(f)}} T_2(R^{(f)}) + \sum_{2_{k=\text{even}}} \sum_{R^{(f)}} T_2(R^{(f)}) = 0 \mod 1, \tag{18.36}$$

for the $Z_2 - G_g - G_g$ anomaly and

$$\sum_{1_{\pm-}} \sum_{R^{(f)}} T_2(R^{(f)}) + \sum_{2_{k=\text{even}}} \sum_{R^{(f)}} T_2(R^{(f)}) = 0 \mod 1, \tag{18.37}$$

for the $Z_2' - G_g - G_g$ anomaly.

• **Q_N ($N = 4n + 2$)**

Similarly, we study anomalies of Q_N with $N = 4n+2$. As shown in Chap. 7, the Q_N group with $N = 4n + 2$ has four singlets $1_{\pm\pm}$ and $(N/2 - 1)$ doublets 2_k. All elements of Q_N are written by products of a and b. The determinants of a are obtained as $\det(a) = 1$ on all of doublets, 2_k. On the other hand, the determinants of b are obtained as $\det(b) = 1$ on the doublets 2_k with $k =$ odd and $\det(b) = -1$ on the doublets 2_k with $k =$ even. For all of singlets, it is found that $\chi_\alpha(a) = \chi_\alpha(b^2)$, i.e. $\det(a) = \det(b^2)$. This implies that the determinants of b are more essential for anomalies than a. Indeed, the determinants of b are obtained as $\det(b) = 1$ on 1_{++}, $\det(b) = i$ on 1_{+-}, $\det(b) = -i$ on 1_{-+} and $\det(b) = -1$ on 1_{--}. Those determinants are summarized in Table 18.9. This result implies that only the Z_4 symmetry corresponding to b can be anomalous. Under such a Z_4 symmetry, each representation has the following Z_4 charge q_4,

$$q_4 = 0 : 1_{++}, \quad 2_{k=\text{odd}},$$
$$q_4 = 1 : 1_{+-},$$
$$q_4 = 2 : 1_{--}, \quad 2_{k=\text{even}},$$
$$q_4 = 3 : 1_{-+}. \tag{18.38}$$

That includes the Z_2 symmetry corresponding to a and the Z_2 charge q_2 for each representation is defined as $q_2 = q_4 \mod 2$. The anomaly-free conditions are written

Table 18.9 Determinants on Q_N representations for $N/2 = $ odd

	1_{++}	1_{+-}	1_{-+}	1_{--}	$2_{k=\text{ odd}}$	$2_{k=\text{ even}}$
$\det(b)$	1	i	$-i$	-1	1	-1
$\det(a)$	1	-1	-1	1	1	1

as

$$\sum_{1_{+-}}\sum_{\boldsymbol{R}^{(f)}} T_2(\boldsymbol{R}^{(f)}) + 2\sum_{1_{--}}\sum_{\boldsymbol{R}^{(f)}} T_2(\boldsymbol{R}^{(f)}) + 3\sum_{1_{-+}}\sum_{\boldsymbol{R}^{(f)}} T_2(\boldsymbol{R}^{(f)})$$
$$+2\sum_{2_{k=\text{ even}}}\sum_{\boldsymbol{R}^{(f)}} T_2(\boldsymbol{R}^{(f)}) = 0 \mod 2 \,, \qquad (18.39)$$

for the $Z_4 - G_g - G_g$ anomaly and

$$\sum_{1_{+-}}\sum_{\boldsymbol{R}^{(f)}} \dim \boldsymbol{R}^{(f)} + 2\sum_{1_{--}}\sum_{\boldsymbol{R}^{(f)}} \dim \boldsymbol{R}^{(f)} + 3\sum_{1_{-+}}\sum_{\boldsymbol{R}^{(f)}} \dim \boldsymbol{R}^{(f)}$$
$$+2\sum_{2_{k=\text{ even}}}\sum_{\boldsymbol{R}^{(f)}} \dim \boldsymbol{R}^{(f)} = 0 \mod 2 \,, \qquad (18.40)$$

for the Z_4-gravity-gravity anomaly. Similarly, we can obtain the anomaly-free condition on the Z_2 symmetry corresponding to a as

$$\sum_{1_{+-}}\sum_{\boldsymbol{R}^{(f)}} T_2(\boldsymbol{R}^{(f)}) + \sum_{1_{-+}}\sum_{\boldsymbol{R}^{(f)}} T_2(\boldsymbol{R}^{(f)}) = 0 \mod 1 \,, \qquad (18.41)$$

for the $Z_2 - G_g - G_g$ anomaly.

• **QD_{2N}**

We study anomalies of QD_{2N}. As shown in Chap. 8, the QD_{2N} group has four singlets $1_{\pm\pm}$ and $(N/2 - 1)$ doublets 2_k. All elements of QD_{2N} are written by products of a and b. The determinants of b are obtained as $\det(b) = -1$ on all of doublets, 2_k. On the other hand, the determinants of a are obtained as $\det(a) = -1$ on the doublets 2_k with $k = $ odd and $\det(a) = 1$ on the doublets 2_k with $k = $ even. For singlets, it is found that $\det(a) = 1$ for $1_{\pm+}$, $\det(a) = -1$ for $1_{\pm-}$, $\det(b) = 1$ for $1_{+\pm}$, and $\det(b) = -1$ for $1_{-\pm}$. Those determinants are summarized in Table 18.10. This implies that two Z_2 symmetries can be anomalous. One Z_2 corresponds to a and the other Z_2' corresponds to b. Under these $Z_2 \times Z_2'$ symmetry, each representation has the following behavior,

$$Z_2 \text{ even} : 1_{+\pm}, \quad 2_{k=\text{ even}}$$
$$Z_2 \text{ odd} : 1_{-\pm}, \quad 2_{k=\text{ odd}}, \qquad (18.42)$$

Table 18.10 Determinants on QD_{2N} representations

	1_{++}	1_{-+}	1_{+-}	1_{--}	$2_{k=\text{odd}}$	$2_{k=\text{even}}$
$\det(b)$	1	1	-1	-1	-1	-1
$\det(a)$	1	-1	1	-1	-1	1

Table 18.11 Determinants on $\Sigma(2N^2)$ representations

	1_{+n}	1_{-n}	2_k
$\det(b)$	1	-1	-1
$\det(a)$	ρ^n	ρ^n	ρ^{p+q}
$\det(a')$	ρ^n	ρ^n	ρ^{p+q}

$$Z_2' \text{ even}: \mathbf{1}_{\pm+},$$
$$Z_2' \text{ odd}: \mathbf{1}_{\pm-}, \quad \mathbf{2}_k. \tag{18.43}$$

Then, the anomaly-free conditions are written as

$$\sum_{\mathbf{1}_{-\pm}}\sum_{\boldsymbol{R}^{(f)}} T_2(\boldsymbol{R}^{(f)}) + \sum_{\mathbf{2}_{k=\text{odd}}}\sum_{\boldsymbol{R}^{(f)}} T_2(\boldsymbol{R}^{(f)}) = 0 \mod 1 , \tag{18.44}$$

for the $Z_2 - G_g - G_g$ anomaly and

$$\sum_{\mathbf{1}_{\pm-}}\sum_{\boldsymbol{R}^{(f)}} T_2(\boldsymbol{R}^{(f)}) + \sum_{\mathbf{2}_k}\sum_{\boldsymbol{R}^{(f)}} T_2(\boldsymbol{R}^{(f)}) = 0 \mod 1 , \tag{18.45}$$

for the $Z_2' - G_g - G_g$ anomaly.

- **$\Sigma(2N^2)$**

We study anomalies of $\Sigma(2N^2)$. As shown in Chap. 9, the $\Sigma(2N^2)$ group has $2N$ singlets, $\mathbf{1}_{\pm n}$, and $N(N-1)/2$ doublets, $\mathbf{2}_{p,q}$. All elements of $\Sigma(2N^2)$ can be written by products of a, a' and b. Their determinants for each representation are shown in Table 18.11, where $\rho = e^{2\pi i/N}$. Then, it is found that only the Z_2 symmetry corresponding to b and the Z_N symmetry corresponding to a can be anomalous. Another Z_N symmetry corresponding to a' is not independent of the Z_N symmetry for a. Under such Z_2 symmetry, each representation has the following behavior,

$$Z_2 \text{ even}: \mathbf{1}_{+n},$$
$$Z_2 \text{ odd}: \mathbf{1}_{-n}, \quad \mathbf{2}_{p,q}, \tag{18.46}$$

and under the Z_N symmetry corresponding to a each representation has the following Z_N charge q_N,

$$q_N = n: \mathbf{1}_{\pm n},$$
$$q_N = p + q: \mathbf{2}_{p,q}. \tag{18.47}$$

Then, the anomaly-free condition is obtained as

$$\sum_{1_{-n}}\sum_{\boldsymbol{R}^{(f)}} T_2(\boldsymbol{R}^{(f)}) + \sum_{2_{p,q}}\sum_{\boldsymbol{R}^{(f)}} T_2(\boldsymbol{R}^{(f)}) = 0 \quad \mathrm{mod}\ 1\ , \tag{18.48}$$

for the $Z_2 - G_g - G_g$ anomaly. Similarly, the anomaly-free conditions for the Z_N symmetry are obtained as

$$\sum_{1_{\pm n}}\sum_{\boldsymbol{R}^{(f)}} n T_2(\boldsymbol{R}^{(f)}) + \sum_{2_{p,q}}\sum_{\boldsymbol{R}^{(f)}} (p+q) T_2(\boldsymbol{R}^{(f)}) = 0 \quad \mathrm{mod}\ N/2\ , \tag{18.49}$$

for the $Z_N - G_g - G_g$ anomaly and

$$\sum_{1_{\pm n}}\sum_{\boldsymbol{R}^{(f)}} n \dim \boldsymbol{R}^{(f)} + \sum_{2_{p,q}}\sum_{\boldsymbol{R}^{(f)}} (p+q) \dim \boldsymbol{R}^{(f)} = 0 \quad \mathrm{mod}\ N/2\ , \tag{18.50}$$

for the Z_N-gravity-gravity anomaly.

• $\Delta(3N^2)$ (N/3 ≠ integer)
We study anomalies of $\Delta(3N^2)$ with $N/3 \neq$ integer. As shown in Chap. 10, the $\Delta(3N^2)$ group with $N/3 \neq$ integer has three singlets, $\mathbf{1}_0$, $\mathbf{1}_1$ and $\mathbf{1}_2$, and $(N^2 - 1)/3$ triplets, $\mathbf{3}_{[k][\ell]}$. All elements of $\Delta(3N^2)$ can be written by products of a, a' and b. It is found that $\det(a) = \det(a') = 1$ on all of representations. Thus, these elements are irrelevant to anomalies. On the other hand, the determinant of b is obtained as $\det(b) = 1$ for all $\mathbf{3}_{[k][\ell]}$ and $\mathbf{1}_0$, $\det(b) = \omega$ for $\mathbf{1}_1$ and $\det(b) = \omega^2$ for $\mathbf{1}_2$, with $\omega = e^{2\pi i/3}$, as shown in Table 18.12. This implies that only the Z_3 symmetry corresponding to b can be anomalous. Under such a Z_3 symmetry, each representation has the following Z_3 charge q_3,

$$\begin{aligned} q_3 &= 0 : \mathbf{1}_0, \quad \mathbf{3}_{[k][\ell]}, \\ q_3 &= 1 : \mathbf{1}_1, \\ q_3 &= 2 : \mathbf{1}_2. \end{aligned} \tag{18.51}$$

Then, the anomaly-free conditions are written as

$$\sum_{1_1}\sum_{\boldsymbol{R}^{(f)}} T_2(\boldsymbol{R}^{(f)}) + 2\sum_{1_2}\sum_{\boldsymbol{R}^{(f)}} T_2(\boldsymbol{R}^{(f)}) = 0 \quad \mathrm{mod}\ 3/2\ , \tag{18.52}$$

Table 18.12 Determinants on $\Delta(3N^2)$ representations ($N/3 \neq$ integer)

	$\mathbf{1}_k$	$\mathbf{3}_{[k][\ell]}$
$\det(b)$	ω^k	1
$\det(a)$	1	1
$\det(a')$	1	1

Table 18.13 Determinants on $\Delta(3N^2)$ representations ($N/3 =$ integer)

	$\mathbf{1}_{k,\ell}$	$\mathbf{3}_{[k][\ell]}$
$\det(b)$	ω^k	1
$\det(a)$	ω^ℓ	1
$\det(a')$	ω^ℓ	1

for the $Z_3 - G_g - G_g$ anomaly and

$$\sum_{\mathbf{1}_1} \sum_{\boldsymbol{R}^{(f)}} \dim \boldsymbol{R}^{(f)} + 2 \sum_{\mathbf{1}_2} \sum_{\boldsymbol{R}^{(f)}} \dim \boldsymbol{R}^{(f)} = 0 \quad \mod 3/2 \,, \tag{18.53}$$

for the Z_3-gravity-gravity anomaly.

- **$\Delta(3N^2)$ (N/3 = integer)**

Similarly, we can study anomalies of $\Delta(3N^2)$ with $N/3 =$ integer. As shown in Chap. 10, the $\Delta(3N^2)$ group with $N/3 =$ integer has nine singlets $\mathbf{1}_{k,\ell}$ and $(N^2 - 3)/3$ triplets, $\mathbf{3}_{[k][\ell]}$. All elements $\Delta(3N^2)$ can be written by products of a, a' and b. On all of triplet representations $\mathbf{3}_{[k][\ell]}$, their determinants are obtained as $\det(a) = \det(a') = \det(b) = 1$. On the other hand, it is found that $\det(a) = \det(a')$ on all of nine singlets. Furthermore, nine singlets are classified by values of $\det(a) = \det(a')$ and $\det(b)$. That is, the determinants of $\det(a) = \det(a')$ and $\det(b)$ are obtained as $\det(a) = \det(a') = \omega^\ell$ and $\det(b) = \omega^k$ on $\mathbf{1}_{k,\ell}$. These results are shown in Table 18.13. This implies that two independent Z_3 symmetries can be anomalous. One corresponds to b and the other corresponds to a. For the Z_3 symmetry corresponding to b, each representation has the following Z_3 charge q_3,

$$
\begin{aligned}
q_3 &= 0 : \mathbf{1}_{0,\ell}, \quad \mathbf{3}_{[k][\ell]}, \\
q_3 &= 1 : \mathbf{1}_{1,\ell}, \\
q_3 &= 2 : \mathbf{1}_{2,\ell},
\end{aligned} \tag{18.54}
$$

while for Z_3' symmetry corresponding to a, each representation has the following Z_3 charge q_3',

$$
\begin{aligned}
q_3' &= 0 : \mathbf{1}_{k,0}, \quad \mathbf{3}_{[k][\ell]}, \\
q_3' &= 1 : \mathbf{1}_{k,1}, \\
q_3' &= 2 : \mathbf{1}_{k,2}.
\end{aligned} \tag{18.55}
$$

Then, the anomaly-free conditions are written as

$$\sum_{\mathbf{1}_{1,\ell}} \sum_{\boldsymbol{R}^{(f)}} T_2(\boldsymbol{R}^{(f)}) + 2 \sum_{\mathbf{1}_{2,\ell}} \sum_{\boldsymbol{R}^{(f)}} T_2(\boldsymbol{R}^{(f)}) = 0 \quad \mod 3/2 \,, \tag{18.56}$$

Table 18.14 Determinants on T_N representations

	$\mathbf{1}_0$	$\mathbf{1}_1$	$\mathbf{1}_2$	$\mathbf{3}_m$	$\bar{\mathbf{3}}_m$
$\det(a)$	1	1	1	1	1
$\det(b)$	1	ω	ω^2	1	1

for the $Z_3 - G_g - G_g$ anomaly and

$$\sum_{\mathbf{1}_{1,\ell}} \sum_{\mathbf{R}^{(f)}} \dim \mathbf{R}^{(f)} + 2 \sum_{\mathbf{1}_{2,\ell}} \sum_{\mathbf{R}^{(f)}} \dim \mathbf{R}^{(f)} = 0 \mod 3/2 \,, \qquad (18.57)$$

for the Z_3-gravity-gravity anomaly. Similarly, for the Z_3' symmetry, the anomaly-free conditions are written as

$$\sum_{\mathbf{1}_{k,1}} \sum_{\mathbf{R}^{(f)}} T_2(\mathbf{R}^{(f)}) + 2 \sum_{\mathbf{1}_{k,2}} \sum_{\mathbf{R}^{(f)}} T_2(\mathbf{R}^{(f)}) = 0 \mod 3/2 \,, \qquad (18.58)$$

for the $Z_3' - G_g - G_g$ anomaly and

$$\sum_{\mathbf{1}_{k,1}} \sum_{\mathbf{R}^{(f)}} \dim \mathbf{R}^{(f)} + 2 \sum_{\mathbf{1}_{k,2}} \sum_{\mathbf{R}^{(f)}} \dim \mathbf{R}^{(f)} = 0 \mod 3/2 \,, \qquad (18.59)$$

for the Z_3'-gravity-gravity anomaly.

• **T_N**

We study anomalies of T_N. As shown in Chap. 11, the T_N group has three singlets, $\mathbf{1}_{0,1,2}$, and $(N - 1)/3$ triplets, $\mathbf{3}(\bar{\mathbf{3}})_m$. All elements of T_N can be written by products of a and b, where a and b correspond to the generators of Z_N and Z_3, respectively. It is found that $\det(a) = 1$ on all of representations. Thus, these elements are irrelevant to anomalies. On the other hand, the determinant of b is obtained as $\det(b) = 1$ for any triplet $\mathbf{3}(\bar{\mathbf{3}})_m$ and $\det(b) = \omega^k$ for $\mathbf{1}_k$ ($k = 0, 1, 2$), as shown in Table 18.14. These results imply that only the Z_3 symmetry corresponding to b can be anomalous. Under such a Z_3 symmetry, each representation has the following Z_3 charge q_3,

$$\begin{aligned} q_3 &= 0 : \mathbf{1}_0, \quad \mathbf{3}, \quad \bar{\mathbf{3}}, \\ q_3 &= 1 : \mathbf{1}_1, \\ q_3 &= 2 : \mathbf{1}_2. \end{aligned} \qquad (18.60)$$

Then, the anomaly-free conditions are written as

$$\sum_{\mathbf{1}_1} \sum_{\mathbf{R}^{(f)}} T_2(\mathbf{R}^{(f)}) + 2 \sum_{\mathbf{1}_2} \sum_{\mathbf{R}^{(f)}} T_2(\mathbf{R}^{(f)}) = 0 \mod 3/2 \,, \qquad (18.61)$$

Table 18.15 Determinants on $\Sigma(3N^3)$ representations

	$1_{k,\ell}$	$3_{[\ell][m][n]}$
$\det(b)$	ω^k	1
$\det(a)$	ρ^ℓ	$\rho^{\ell+m+n}$
$\det(a')$	ρ^ℓ	$\rho^{\ell+m+n}$
$\det(a'')$	ρ^ℓ	$\rho^{\ell+m+n}$

for the $Z_3 - G_g - G_g$ anomaly and

$$\sum_{1_1}\sum_{R^{(f)}} \dim R^{(f)} + 2\sum_{1_2}\sum_{R^{(f)}} \dim R^{(f)} = 0 \quad \mod 3/2 \,, \qquad (18.62)$$

for the Z_3-gravity-gravity anomaly.

- **$\Sigma(3N^3)$**

We study anomalies of $\Sigma(3N^3)$. The $\Sigma(3N^3)$ group has $3N$ singlets $1_{k,\ell}$, and $N(N^2-1)/3$ triplets, $3_{[\ell][m][n]}$. All elements of $\Sigma(3N^3)$ can be written by products of a, a', a'' and b. Their determinants for each representation are shown in Table 18.15, where $\rho = e^{2\pi i/N}$. Then, it is found that only the Z_3 symmetry corresponding to b and the Z_N symmetry corresponding to a can be anomalous. Other Z_N symmetries corresponding to a' and a'' are not independent of the Z_N symmetry for a. For the Z_3 symmetry corresponding to b, each representation has the following Z_3 charge q_3,

$$\begin{aligned} q_3 = 0 &: 1_{0,\ell}, \quad 3_{[\ell][m][n]}, \\ q_3 = 1 &: 1_{1,\ell}, \\ q_3 = 2 &: 1_{2,\ell}, \end{aligned} \qquad (18.63)$$

and under the Z_N symmetry corresponding to a each representation has the following Z_N charge q_N,

$$\begin{aligned} q_N = \ell &: 1_{k,\ell}, \\ q_N = \ell + m + n &: 3_{[\ell][m][n]}. \end{aligned} \qquad (18.64)$$

Then, the anomaly-free condition is obtained as

$$\sum_{1_{1,\ell}}\sum_{R^{(f)}} T_2(R^{(f)}) + \sum_{1_{2,\ell}}\sum_{R^{(f)}} T_2(R^{(f)}) = 0 \quad \mod 3/2 \,, \qquad (18.65)$$

for the $Z_3 - G_g - G_g$ anomaly and

$$\sum_{1_{1,\ell}}\sum_{R^{(f)}} \dim R^{(f)} + 2\sum_{1_{2,\ell}}\sum_{R^{(f)}} \dim R^{(f)} = 0 \quad \mod 3/2 \,, \qquad (18.66)$$

Table 18.16 Determinants on $\Delta(6N^2)$ with $3N \neq$ integer representations

	1_0	1_1	2	3_{1k}	3_{2k}	$6_{[[k],[\ell]]}$
$\det(a)$	1	1	1	1	1	1
$\det(a')$	1	1	1	1	1	1
$\det(b)$	1	1	1	1	1	1
$\det(c)$	1	-1	-1	-1	1	-1

for the Z_3-gravity-gravity anomaly. Similarly, the anomaly-free conditions for the Z_N symmetry are obtained as

$$\sum_{1_{k,\ell}}\sum_{R^{(f)}} \ell T_2(R^{(f)}) + \sum_{3_{[\ell][m][n]}}\sum_{R^{(f)}} (\ell + m + n)T_2(R^{(f)}) = 0 \mod N/2, \quad (18.67)$$

for the $Z_N - G_g - G_g$ anomaly and

$$\sum_{1_{k,\ell}}\sum_{R^{(f)}} \ell \dim R^{(f)} + \sum_{3_{[\ell][m][n]}}\sum_{R^{(f)}} (\ell + m + n)\dim R^{(f)} = 0 \mod N/2, \quad (18.68)$$

for the Z_N-gravity-gravity anomaly.

- $\Delta(6N^2)$ ($N/3 \neq$ **integer**)
We study anomalies of $\Delta(6N^2)$ with ($N/3 \neq$ integer). As shown in Chap. 13 for ($N/3 \neq$ integer), the group has two singlets, $1_{0,1}$, one doublet, 2, $2(N-1)$ triplets, 3_{1k} and 3_{2k}, and $N(N-3)/6$ sextets, $6_{[[k],[\ell]]}$. All elements can be written by products of a, a', b and c. Determinants of a, a' and b on any representation are obtained as $\det(a) = \det(a') = \det(b) = 1$. The determinants of c for 1_0 and 3_{2k} are obtained as $\det(c) = 1$ while the other representations lead to $\det(c) = -1$. These results are shown in Table 18.16. That implies that only the Z_2 symmetry corresponding to the generator c can be anomalous. Under such a Z_2 symmetry, each representation has the following Z_2 charge q_2,

$$q_2 = 0 : 1_0, 3_{2k},$$
$$q_2 = 1 : 1_1, 2, 3_{1k}, 6_{[[k],[\ell]]}. \quad (18.69)$$

Then, the anomaly-free conditions are written as

$$\sum_{1_1}\sum_{R^{(f)}} T_2(R^{(f)}) + \sum_{2}\sum_{R^{(f)}} T_2(R^{(f)}) + \sum_{3_{1k}}\sum_{R^{(f)}} T_2(R^{(f)}) + \sum_{6_{[[k],[\ell]]}}\sum_{R^{(f)}} T_2(R^{(f)}) = 0 \mod 1,$$
$$(18.70)$$

for the $Z_2 - G_g - G_g$ anomaly.

- $\Delta(6N^2)$ ($N/3 =$ **integer**)
In the same way, we study anomalies of $\Delta(6N^2)$ with ($N/3 =$ integer). As shown in Chap. 13 for ($N/3 =$ integer), the group has two singlets, $1_{0,1}$, four doublets, $2_{1,2,3,4}$,

Table 18.17 Determinants on $\Delta(6N^2)$ with $3N$ =integer representations

	1_0	1_1	2_1	2_2	2_3	2_4	3_{1k}	3_{2k}	$6_{[[k],[\ell]]}$
det(a)	1	1	1	1	1	1	1	1	1
det(a')	1	1	1	1	1	1	1	1	1
det(b)	1	1	1	1	1	1	1	1	1
det(c)	1	-1	-1	-1	-1	-1	-1	1	-1

$2(N-1)$ triplets, 3_{1k} and 3_{2k}, and $(N^2 - 3N + 2)/6$ sextets, $6_{[[k],[\ell]]}$. All elements of $\Delta(54)$ can be written by products of a, a', b and c. Determinants of a, a' and b on any representation are obtained as $\det(a) = \det(a') = \det(b) = 1$. The determinants of c for 1_0 and 3_{2k} are obtained as $\det(c) = 1$ while the other representations lead to $\det(c) = -1$. These results are shown in Table 18.17. That implies that only the Z_2 symmetry corresponding to the generator c can be anomalous. Under such a Z_2 symmetry, each representation has the following Z_2 charge q_2,

$$q_2 = 0 : 1_0, 3_{2k},$$
$$q_2 = 1 : 1_1, 2_{1,2,3,4}, 3_{1k}, 6_{[[k],[\ell]]}. \tag{18.71}$$

Then, the anomaly-free conditions are written as

$$\sum_{1_1}\sum_{R^{(f)}} T_2(R^{(f)}) + \sum_{2_k}\sum_{R^{(f)}} T_2(R^{(f)}) + \sum_{3_{1k}}\sum_{R^{(f)}} T_2(R^{(f)}) + \sum_{6_{[[k],[\ell]]}}\sum_{R^{(f)}} T_2(R^{(f)}) = 0 \mod 1, \tag{18.72}$$

for the $Z_2 - G_g - G_g$ anomaly.

Similarly, we can analyze on anomalies for other non-Abelian discrete symmetries.

18.3 Comments on Anomalies

Finally, we comment on the symmetry breaking effects by quantum effect. When a discrete (flavor) symmetry is anomalous, breaking terms can appear in Lagrangian, e.g. by instanton effects, such as $\frac{1}{M^n}\Lambda^m \Phi_1 \ldots \Phi_k$, where Λ is a dynamical scale and M is a typical (cut-off) scale. Within the framework of string theory discrete anomalies as well as anomalies of continuous gauge symmetries can be canceled by the Green-Schwarz (GS) mechanism [10] unless discrete symmetries are accidental. In the GS mechanism, dilaton and moduli fields, i.e. the so-called GS fields Φ_{GS}, transform non-linearly under anomalous transformation. The anomaly cancellation due to the GS mechanism imposes certain relations among anomalies. (See e.g. Ref. [3].)[1] Stringy non-perturbative effects as well as field-theoretical effects induce terms in

---

[1] See also Ref. [11].

Lagrangian such as $\frac{1}{M^n}e^{-a\Phi_{GS}}\Phi_1\ldots\Phi_k$. The GS fields Φ_{GS}, i.e. dilaton/moduli fields are expected to develop non-vanishing vacuum expectation values and the above terms correspond to breaking terms of discrete symmetries.

The above breaking terms may be small. Such approximate discrete symmetries with small breaking terms may be useful in particle physics,[2] if breaking terms are controllable. Alternatively, if exact symmetries are necessary, one has to arrange matter fields and their quantum numbers such that models are free from anomalies.

References

1. Fujikawa, K.: Phys. Rev. Lett. **42**, 1195 (1979)
2. Fujikawa, K.: Phys. Rev. D **21**, 2848 (1980)
3. Araki, T., Kobayashi, T., Kubo, J., Ramos-Sanchez, S., Ratz, M., Vaudrevange, P.K.S.: Nucl. Phys. B **805**, 124 (2008). arXiv:0805.0207 [hep-th]
4. Alvarez-Gaume, L., Witten, E.: Nucl. Phys. B **234**, 269 (1984)
5. Alvarez-Gaume, L., Ginsparg, P.H.: Ann. Phys. **161**, 423 (1985)
6. Fujikawa, K., Ojima, S., Yajima, S.: Phys. Rev. D **34**, 3223 (1986)
7. Rohlin, V.: Dokl. Akad. Nauk. **128**, 980–983 (1959)
8. Csaki, C., Murayama, H.: Nucl. Phys. B **515**, 114–162 (1998). [hep-th/9710105]
9. Araki, T.: Prog. Theor. Phys. **117**, 1119–1138 (2007). [hep-ph/0612306]
10. Green, M.B., Schwarz, J.H.: Phys. Lett. B **149**, 117–122 (1984)
11. Kobayashi, T., Nakano, H.: Nucl. Phys. B **496**, 103–131 (1997). [hep-th/9612066]
12. Fukuoka, H., Kubo, J., Suematsu, D.: Phys. Lett. B **678**, 401 (2009). arXiv:0905.2847 [hep-ph]

---

[2] See for some applications e.g. [12].

Non-Abelian Discrete Symmetry in Quark/Lepton Flavor Models

19

Non-Abelian discrete groups have been adopted for the flavor models of the quarks and leptons. In this chapter, we present some successful flavor models of these discrete symmetries. Examples will illustrate how such models are built. However, before discussing the models themselves, we briefly review the main features of the experimental data regarding neutrino flavor mixing.

19.1 Neutrino Flavor Mixing and Neutrino Mass Matrix

The experimental data of neutrino oscillations has stimulated work on the non-Abelian discrete symmetry of flavors. Both atmospheric neutrino mixing angle θ_{23} and solar neutrino mixing angle θ_{12} are quite large, especially, θ_{23} is almost maximal. These neutrino mixing angles are defined in the neutrino mixing matrix U as

$$
U = \begin{pmatrix} c_{12}c_{13} & s_{12}c_{13} & s_{13}e^{-i\delta_{CP}} \\ -s_{12}c_{23} - c_{12}s_{23}s_{13}e^{i\delta_{CP}} & c_{12}c_{23} - s_{12}s_{23}s_{13}e^{i\delta_{CP}} & s_{23}c_{13} \\ s_{12}s_{23} - c_{12}c_{23}s_{13}e^{i\delta_{CP}} & -c_{12}s_{23} - s_{12}c_{23}s_{13}e^{i\delta_{CP}} & c_{23}c_{13} \end{pmatrix}
$$
$$
\times \begin{pmatrix} 1 & 0 & 0 \\ 0 & e^{i\frac{\alpha_{21}}{2}} & 0 \\ 0 & 0 & e^{i\frac{\alpha_{31}}{2}} \end{pmatrix}, \tag{19.1}
$$

where c_{ij} and s_{ij} denote $\cos\theta_{ij}$ and $\sin\theta_{ij}$, respectively. The phases δ_{CP}, $\alpha_{i1}(i = 2, 3)$ are Dirac CP phase and Majorana phases, respectively.

The global fit of the neutrino experimental data [1] indicates roughly $\sin^2\theta_{23} \simeq 0.5$ and $\sin^2\theta_{12} \simeq 0.3$, which suggest the tri-bimaximal mixing matrix U_{tribi} of three lepton flavors [2–5], where $\sin^2\theta_{13}$ is supposed to be negligibly small. The mixing

T. Kobayashi et al., *An Introduction to Non-Abelian Discrete Symmetries for Particle Physicists*, Lecture Notes in Physics 995, https://doi.org/10.1007/978-3-662-64679-3_19

matrix is

$$U_{\text{tribi}} = \begin{pmatrix} \frac{2}{\sqrt{6}} & \frac{1}{\sqrt{3}} & 0 \\ -\frac{1}{\sqrt{6}} & \frac{1}{\sqrt{3}} & -\frac{1}{\sqrt{2}} \\ -\frac{1}{\sqrt{6}} & \frac{1}{\sqrt{3}} & \frac{1}{\sqrt{2}} \end{pmatrix}. \tag{19.2}$$

This specific mixing matrix favors the non-Abelian discrete symmetry for the lepton flavor. Indeed, various types of models leading to the tri-bimaximal mixing were proposed by assuming non-Abelian discrete flavor symmetries as seen, e.g. in Refs. [6,7].

In the tri-bimaximal mixing, θ_{13} is vanishing. Indeed, it is finite as $\sin^2 \theta_{13} \simeq$ 0.02 [1,8]. Thus the precise prediction of the neutrino mixing angles is required to test the flavor models. At first, we introduce typical models to reproduce the tri-bimaximal mixing of neutrino flavors in the flavor model with the non-Abelian discrete symmetry.

The neutrino mass matrix with the tri-bimaximal mixing of flavors is expressed by the sum of three simple mass matrices in the flavor diagonal basis of the charged lepton. In terms of neutrino mass eigenvalues m_1, m_2 and m_3, the neutrino mass matrix is given as

$$
\begin{aligned}
M_\nu &= U_{\text{tribi}}^* \begin{pmatrix} m_1 & 0 & 0 \\ 0 & m_2 & 0 \\ 0 & 0 & m_3 \end{pmatrix} U_{\text{tribi}}^\dagger \\
&= \frac{m_1 + m_3}{2} \begin{pmatrix} 1 & 0 & 0 \\ 0 & 1 & 0 \\ 0 & 0 & 1 \end{pmatrix} + \frac{m_2 - m_1}{3} \begin{pmatrix} 1 & 1 & 1 \\ 1 & 1 & 1 \\ 1 & 1 & 1 \end{pmatrix} + \frac{m_1 - m_3}{2} \begin{pmatrix} 1 & 0 & 0 \\ 0 & 0 & 1 \\ 0 & 1 & 0 \end{pmatrix}. \tag{19.3}
\end{aligned}
$$

This neutrino mass matrix can be easily realized in some non-Abelian discrete symmetry. In the following sections, we present simple realization of this neutrino mass matrix, which arises from the 5D non-renormalizable operators [9], or the see-saw mechanism [10–14].

19.2 A_4 Flavor Symmetry

Simple models realizing the tri-bimaximal mixing have been proposed based on the non-Abelian finite group A_4 [15–62]. The A_4 flavor model considered by Alterelli et al. [17,18] realizes the tri-bimaximal flavor mixing. The deviation from the tri-bimaximal mixing can also be predicted. Actually, we have investigated the deviation from the tri-bimaximal mixing including higher dimensional operators in the effective model [19,20].

Table 19.1 Assignments of $SU(2)$, A_4, and Z_3 representations, where $\omega = e^{\frac{2\pi i}{3}}$

	(l_e, l_μ, l_τ)	e^c	μ^c	τ^c	$h_{u,d}$	ϕ_l	ϕ_ν	ξ	ξ'	ξ''
$SU(2)$	2	1	1	1	2	1	1	1	1	1
A_4	3	1	$1''$	$1'$	1	3	3	1	$1'$	$1''$
Z_3	ω	ω^2	ω^2	ω^2	1	1	ω	ω	ω	ω

19.2.1 Realizing the Tri-Bimaximal Mixing of Flavors

We begin by presenting the $A_4 \times Z_3$ flavor model with the supersymmetry including the right-handed neutrinos [17,18]. In the non-Abelian finite group A_4, there are twelve group elements and four irreducible representations: 1, $1'$, $1''$ and 3. The A_4 and Z_3 charge assignments of leptons, Higgs fields and SM-singlets are listed in Table 19.1. Under the A_4 symmetry, the chiral superfields for three families of the left-handed lepton doublet $l = (l_e, l_\mu, l_\tau)$ are assumed to transform as 3, while the right-handed ones of the charged leptons e^c, μ^c and τ^c are assigned with 1, $1''$, $1'$, respectively. The third row of Table 19.1 shows how each chiral multiplet transforms under Z_3, where $\omega = e^{2\pi i/3}$. We assume flavons ϕ_l and ϕ_ν, which are A_4 triplets. In addition to these triplet flavons, we can consider singlet flavons ξ, ξ', ξ'', which are 1, $1'$, $1''$, respectively. The flavor symmetry is spontaneously broken by VEVs of two 3's, $\phi_l = (\phi_{l1}, \phi_{l2}, \phi_{l3})$, $\phi_\nu = (\phi_{\nu1}, \phi_{\nu2}, \phi_{\nu3})$, and by singlets, ξ, ξ', ξ'', which are $SU(2)_L \times U(1)_Y$ singlets.

In order to realize the tri-bimaximal mixing, we consider the case of $\langle \xi' \rangle = \langle \xi'' \rangle = 0$. The superpotential of the lepton sector which respects the gauge and the flavor symmetry is described by

$$
\begin{aligned}
w_\ell = & y^e e^c l \phi_l h_d / \Lambda + y^\mu \mu^c l \phi_l h_d / \Lambda + y^\tau \tau^c l \phi_l h_d / \Lambda \\
& + (y^\nu_{\phi_\nu} \phi_\nu + y^\nu_\xi \xi + y^\nu_{\xi'} \xi' + y^\nu_{\xi''} \xi'') l l h_u h_u / \Lambda^2 ,
\end{aligned}
\tag{19.4}
$$

where y^e, y^μ, y^τ, $y^\nu_{\phi_\nu}$, y^ν_ξ, $y^\nu_{\xi'}$, and $y^\nu_{\xi''}$ are the dimensionless coupling constants, and Λ is the cutoff scale. Hereafter, we follow the convention that the chiral superfield and its lowest component are denoted by the same letter. Decompositions into the A_4 singlet are given by using the basis in Appendix C:

$$
\begin{aligned}
e^c l \phi_l &\to e^c (l_e \phi_{l1} + l_\mu \phi_{l3} + l_\tau \phi_{l2}), \\
l l \phi_\nu &\to \begin{pmatrix} 2 l_e l_e - l_\mu l_\tau - l_\tau l_\mu \\ 2 l_\tau l_\tau - l_e l_\mu - l_\mu l_e \\ 2 l_\mu l_\mu - l_e l_\tau - l_\tau l_e \end{pmatrix} \phi_\nu \\
&\to (2 l_e l_e - l_\mu l_\tau - l_\tau l_\mu) \phi_{\nu1} + (2 l_\tau l_\tau - l_e l_\mu - l_\mu l_e) \phi_{\nu3} \\
&\quad + (2 l_\mu l_\mu - l_e l_\tau - l_\tau l_e) \phi_{\nu2}, \\
l l \xi &\to (l_e l_e + l_\mu l_\tau + l_\tau l_\mu) \xi, \\
l l \xi' &\to (l_\mu l_\mu + l_e l_\tau + l_\tau l_e) \xi', \\
l l \xi'' &\to (l_\tau l_\tau + l_e l_\mu + l_\mu l_e) \xi''.
\end{aligned}
\tag{19.5}
$$

Suppose the following vacuum alignments:

$$\langle \phi_l \rangle = \alpha_l \Lambda(1, 0, 0) , \quad \langle \phi_\nu \rangle = \alpha_\nu \Lambda(1, 1, 1) , \tag{19.6}$$

with $\langle \xi \rangle = \alpha_\xi \Lambda$. We omit the discussion of the origin of these vacuum alignments since the purpose of this section is not to present details of the model, but rather to apply the A_4 group to the neutrino mixing.

By using these vacuum alignments and Eq. (19.5), we can obtain mass matrices of the charged leptons and neutrinos. Then, the diagonal charged lepton mass matrix is given as,

$$M_l = \alpha_l v_d \begin{pmatrix} y^e & 0 & 0 \\ 0 & y^\mu & 0 \\ 0 & 0 & y^\tau \end{pmatrix} , \tag{19.7}$$

where $\langle h_{u,d} \rangle = v_{u,d}$. Also, the effective neutrino mass matrix is given as

$$
\begin{aligned}
M_\nu &= \frac{y^\nu_{\phi_\nu} \alpha_\nu v_u^2}{3\Lambda} \begin{pmatrix} 2 & -1 & -1 \\ -1 & 2 & -1 \\ -1 & -1 & 2 \end{pmatrix} + \frac{y^\nu_{\phi_\xi} \alpha_\xi v_u^2}{\Lambda} \begin{pmatrix} 1 & 0 & 0 \\ 0 & 0 & 1 \\ 0 & 1 & 0 \end{pmatrix} \\
&= a \begin{pmatrix} 1 & 0 & 0 \\ 0 & 1 & 0 \\ 0 & 0 & 1 \end{pmatrix} + b \begin{pmatrix} 1 & 1 & 1 \\ 1 & 1 & 1 \\ 1 & 1 & 1 \end{pmatrix} + c \begin{pmatrix} 1 & 0 & 0 \\ 0 & 0 & 1 \\ 0 & 1 & 0 \end{pmatrix} ,
\end{aligned} \tag{19.8}
$$

where

$$a = \frac{y^\nu_{\phi_\nu} \alpha_\nu v_u^2}{\Lambda}, \quad b = -\frac{y^\nu_{\phi_\nu} \alpha_\nu v_u^2}{3\Lambda}, \quad c = \frac{y^\nu_\xi \alpha_\xi v_u^2}{\Lambda}. \tag{19.9}$$

Thus, the tri-bimaximal mixing is easily derived in the $A_4 \times Z_3$ flavor model.

19.2.2 Breaking Tri-Bimaximal Mixing

It should be emphasized that the A_4 flavor symmetry does not necessarily give the tri-bimaximal mixing at the leading order even if the relevant alignments of VEVs of the flavons are realized. Certainly, for the neutrino mass matrix with three flavors, the A_4 symmetry can give the mass matrix with the $(2, 3)$ off diagonal matrix due to the A_4 singlet flavon, 1, in addition to the unit matrix and the democratic matrix, which leads to the tri-bimaximal mixing of flavors. However, the $(1, 3)$ off diagonal matrix and the $(1, 2)$ off diagonal matrix also appear at the leading order as far as the VEV of ξ' or ξ'' flavon does not vanish [21];

$$\begin{pmatrix} 0 & 0 & 1 \\ 0 & 1 & 0 \\ 1 & 0 & 0 \end{pmatrix} \quad \text{for } \xi', \qquad \begin{pmatrix} 0 & 1 & 0 \\ 1 & 0 & 0 \\ 0 & 0 & 1 \end{pmatrix} \quad \text{for } \xi''. \tag{19.10}$$

The tri-bimaximal mixing is modified at the leading order in such a case.

Fig. 19.1 The $\sum m_i$ dependence of $\sin\theta_{13}$ for normal mass hierarchy

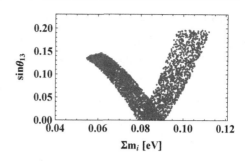

Let us consider the case of non-vanishing $\langle\xi'\rangle$, but still vanishing $\langle\xi''\rangle$. The charged lepton mass matrix is still diagonal as in Eq. (19.7). The effective neutrino mass matrix is modified as

$$
M_\nu = \frac{y^\nu_{\phi_\nu}\alpha_\nu v_u^2}{3\Lambda}\begin{pmatrix} 2 & -1 & -1 \\ -1 & 2 & -1 \\ -1 & -1 & 2 \end{pmatrix} + \frac{y^\nu_{\phi_\xi}\alpha_\xi v_u^2}{\Lambda}\begin{pmatrix} 1 & 0 & 0 \\ 0 & 0 & 1 \\ 0 & 1 & 0 \end{pmatrix} + \frac{y^\nu_{\phi_{\xi'}}\alpha_{\xi'} v_u^2}{\Lambda}\begin{pmatrix} 0 & 0 & 1 \\ 0 & 1 & 0 \\ 1 & 0 & 0 \end{pmatrix}
$$

$$
= a\begin{pmatrix} 1 & 0 & 0 \\ 0 & 1 & 0 \\ 0 & 0 & 1 \end{pmatrix} + b\begin{pmatrix} 1 & 1 & 1 \\ 1 & 1 & 1 \\ 1 & 1 & 1 \end{pmatrix} + c\begin{pmatrix} 1 & 0 & 0 \\ 0 & 0 & 1 \\ 0 & 1 & 0 \end{pmatrix} + d\begin{pmatrix} 0 & 0 & 1 \\ 0 & 1 & 0 \\ 1 & 0 & 0 \end{pmatrix}, \tag{19.11}
$$

where

$$
a = \frac{y^\nu_{\phi_\nu}\alpha_\nu v_u^2}{\Lambda}, \qquad b = -\frac{y^\nu_{\phi_\nu}\alpha_\nu v_u^2}{3\Lambda}, \qquad c = \frac{y^\nu_\xi \alpha_\xi v_u^2}{\Lambda}, \qquad d = \frac{y^\nu_{\xi'}\alpha_{\xi'} v_u^2}{\Lambda}. \tag{19.12}
$$

Therefore, the tri-bimaximal mixing is broken in the A_4 flavor model.

As seen in Eqs. (19.11) and (19.12), the non-vanishing d is generated through the coupling $ll\xi' h_u h_u$. Since the relation $a = -3b$ is given in this model, the predicted regions of the lepton mixing angles are reduced compared with the one in the previous section. In the case where the parameters a, c, d are real, they are fixed by the three neutrino masses m_1, m_2 and m_3. That is, $\sin\theta_{13}$ can be plotted as a function of the total mass $\sum m_i$.

Figure 19.1, shows the predicted $\sin\theta_{13}$ versus $\sum m_i$, where the normal hierarchy of the neutrino masses is taken. The leptonic mixing is almost tri-bimaximal, that is $\sin\theta_{13} = 0$, in the regime where $\sum m_i \simeq 0.08 - 0.09$ eV. In the case of $m_3 \gg m_2, m_1$, that is $\sum m_i \simeq 0.05$ eV, $\sin\theta_{13}$ is expected to be around 0.15.

We can also predict $\sin\theta_{13}$ versus $\sum m_i$ in the case of the inverted hierarchy of neutrino masses. We get a different prediction of $\sin\theta_{13}$ as seen in Fig. 19.2. The predicted maximal value of $\sin\theta_{13}$ is 0.2 at $\sum m_i \simeq 0.1$ eV, which corresponds to $m_3 \ll m_2, m_1$. In conclusion, the $A_4 \times Z_3$ model predicts $\sin\theta_{13} = 0.15 - 0.2$ for cases of $m_3 \gg m_2, m_1$ and $m_3 \ll m_2, m_1$, which are completely consistent with recent observed value [1,8].

Finally we comment on flavor models with other non-Abelian discrete symmetries which gives the non-vanishing d effectively. One is the flavor model based on $\Delta(27)$ group which is given by Grimus and Lavoura [63]. The trimaximal mixing is enforced

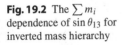

Fig. 19.2 The $\sum m_i$ dependence of $\sin\theta_{13}$ for inverted mass hierarchy

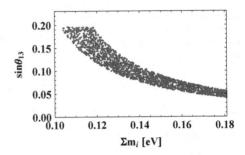

by the softly broken discrete symmetry. In this model, we find the relation $d = e^{i\pi/3}c$, where a, b, c, d are complex. As seen in Ref. [63] the large $\sin\theta_{13}$ is expected. Another example is the flavor twisting model in the 5D framework [64,65]. In this model, the flavor symmetry breaking is triggered by the boundary conditions of the bulk right-handed neutrino in the fifth spatial dimension. The parameters a, b, c, d involve the bulk neutrino masses and the volume of the extra dimension.

19.3 S_4 Flavor Model

In this section, we present a $S_4 \times Z_4$ flavor model to unify the quarks and leptons in the framework of the $SU(5)$ GUT [66]. The S_4 group has 24 distinct elements and has five irreducible representations **1**, **1′**, **2**, **3**, and **3′**. In $SU(5)$, matter fields are unified into $10(q_1, u^c, e^c)_L$ and $\bar{5}(d^c, l_e)_L$ dimensional representations. The 5-dimensional, $\bar{5}$-dimensional and 45-dimensional Higgs of the $SU(5)$, H_5, $H_{\bar{5}}$, and H_{45} are assigned **1** of S_4. Three generations of $\bar{5}$, which are denoted by F_i, are assigned **3** of S_4. On the other hand, the third generation of the 10-dimensional representation is assigned **1** of S_4, so that the top quark Yukawa coupling is allowed in the tree level. While, the first and the second generations are assigned **2** of S_4. These 10-dimensional representations are denoted by T_3 and (T_1, T_2), respectively. These assignments of S_4 for $\bar{5}$ and **10** lead to the completely different structure of quark and lepton mass matrices.

Right-handed neutrinos, which are $SU(5)$ gauge singlets, are also assigned **1′** and **2** for N_τ^c and (N_e^c, N_μ^c), respectively. These assignments are essential to realize the tri-bimaximal mixing of neutrino flavors. Assignments of $SU(5)$, S_4, Z_4 and $U(1)_{FN}$ representations are summarized in Table 19.2. Taking vacuum alignments of relevant gauge singlet scalars, we predict the quark mixing as well as the tri-bimaximal mixing of leptons. Especially, the Cabbibo angle is predicted to be around $15°$ under the relevant vacuum alignments.

We introduce new scalars $\chi_i (i = 1 - 14)$, which are supposed to be $SU(5)$ gauge singlets. Those flavons are summarized in Table 19.2. In order to obtain the mass hierarchy among quarks and leptons, the Froggatt-Nielsen mechanism [67] is introduced as an additional $U(1)_{FN}$ flavor symmetry. Θ denotes the Froggatt-Nielsen flavon. The particle assignments of $SU(5)$, S_4 and Z_4 and $U(1)_{FN}$ are presented in Table 19.2.

Table 19.2 Assignments of $SU(5)$, S_4, Z_4, and $U(1)_{FN}$ representations

	(T_1, T_2)	T_3	(F_1, F_2, F_3)	(N_e^c, N_μ^c)	N_τ^c	H_5	$H_{\bar 5}$	H_{45}	Θ
$SU(5)$	10	10	$\bar 5$	1	1	5	$\bar 5$	45	1
S_4	2	1	3	2	$1'$	1	1	1	1
Z_4	$-i$	-1	i	1	1	1	1	-1	1
$U(1)_{FN}$	1	0	0	1	0	0	0	0	-1

	(χ_1, χ_2)	(χ_3, χ_4)	(χ_5, χ_6, χ_7)	$(\chi_8, \chi_9, \chi_{10})$	$(\chi_{11}, \chi_{12}, \chi_{13})$	χ_{14}
$SU(5)$	1	1	1	1	1	1
S_4	2	2	$3'$	3	3	1
Z_4	$-i$	1	$-i$	-1	i	i
$U(1)_{FN}$	-1	-2	0	0	0	-1

We can now write down the superpotential respecting S_4, Z_4 and $U(1)_{FN}$ symmetries in terms of the S_4 cutoff scale Λ, and the $U(1)_{FN}$ cutoff scale $\bar\Lambda$. The $SU(5)$ invariant superpotential of the Yukawa sector up to the linear terms of χ_i is given as

$$
\begin{aligned}
w_{SU(5)} = [&y_1^u (T_1, T_2) \otimes T_3 \otimes (\chi_1, \chi_2) \otimes H_5/\Lambda + y_2^u T_3 \otimes T_3 \otimes H_5 \\
&+ y_1^N (N_e^c, N_\mu^c) \otimes (N_e^c, N_\mu^c) \otimes \Theta^2/\bar\Lambda \\
&+ y_2^N (N_e^c, N_\mu^c) \otimes (N_e^c, N_\mu^c) \otimes (\chi_3, \chi_4) + M N_\tau^c \otimes N_\tau^c \\
&+ y_1^D (N_e^c, N_\mu^c) \otimes (F_1, F_2, F_3) \otimes (\chi_5, \chi_6, \chi_7) \otimes H_5 \otimes \Theta/(\Lambda\bar\Lambda) \\
&+ y_2^D N_\tau^c \otimes (F_1, F_2, F_3) \otimes (\chi_5, \chi_6, \chi_7) \otimes H_5/\Lambda \\
&+ y_1 (T_1, T_2) \otimes (F_1, F_2, F_3) \otimes (\chi_8, \chi_9, \chi_{10}) \otimes H_{45} \otimes \Theta/(\Lambda\bar\Lambda) \\
&+ y_2 T_3 \otimes (F_1, F_2, F_3) \otimes (\chi_{11}, \chi_{12}, \chi_{13}) \otimes H_{\bar 5}/\Lambda]_1, \quad (19.13)
\end{aligned}
$$

where $[\ ]_1$ denotes the only trivial singlet components of S_4 extracted from tensor products. Parameters y_1^u, y_2^u, y_1^N, y_2^N, y_1^D, y_2^D, y_1, and y_2 are Yukawa couplings. We take the basis II in Appendix B(B2) for the multiplication rules to get S_4 singlet,

$$
\begin{pmatrix} T_1 \\ T_2 \end{pmatrix}_2 \otimes \begin{pmatrix} \chi_1 \\ \chi_2 \end{pmatrix}_2 \to (T_1\chi_1 + T_2\chi_2)_1
$$

$$
\begin{pmatrix} N_e^c \\ N_\mu^c \end{pmatrix}_2 \otimes \begin{pmatrix} N_e^c \\ N_\mu^c \end{pmatrix}_2 \to (N_e^c N_e^c + N_\mu^c N_\mu^c)_1
$$

$$
\begin{pmatrix} N_e^c \\ N_\mu^c \end{pmatrix}_2 \otimes \begin{pmatrix} N_e^c \\ N_\mu^c \end{pmatrix}_2 \otimes \begin{pmatrix} \chi_3 \\ \chi_4 \end{pmatrix}_2 \to \begin{pmatrix} N_e^c N_\mu^c + N_\mu^c N_e^c \\ N_e^c N_e^c - N_\mu^c N_\mu^c \end{pmatrix} \otimes \begin{pmatrix} \chi_3 \\ \chi_4 \end{pmatrix}_2
$$

$$
\to ((N_e^c N_\mu^c + N_\mu^c N_e^c)\chi_3 + (N_e^c N_e^c - N_\mu^c N_\mu^c)\chi_4)_1
$$

$$
\begin{pmatrix} N_e^c \\ N_\mu^c \end{pmatrix}_2 \otimes \begin{pmatrix} F_1 \\ F_2 \\ F_3 \end{pmatrix}_3 \otimes \begin{pmatrix} \chi_5 \\ \chi_6 \\ \chi_7 \end{pmatrix}_{3'} \to \begin{pmatrix} N_e^c \\ N_\mu^c \end{pmatrix}_2 \otimes \begin{pmatrix} \frac{1}{\sqrt{6}}(2F_1\chi_5 - F_2\chi_6 - F_3\chi_7) \\ \frac{1}{\sqrt{2}}(F_2\chi_6 - F_3\chi_7) \end{pmatrix}_2
$$

$$
\to \begin{pmatrix} \frac{N_e^c}{\sqrt{6}}(2F_1\chi_5 - F_2\chi_6 - F_3\chi_7) + \frac{N_\mu^c}{\sqrt{2}}(F_2\chi_6 - F_3\chi_7) \end{pmatrix}_1
$$

$$
\begin{pmatrix} F_1 \\ F_2 \\ F_3 \end{pmatrix}_3 \otimes \begin{pmatrix} \chi_5 \\ \chi_6 \\ \chi_7 \end{pmatrix}_{3'} \to (F_1\chi_5 + F_2\chi_6 + F_3\chi_7)_{1'}
$$

$$\begin{pmatrix} T_1 \\ T_2 \end{pmatrix}_2 \otimes \begin{pmatrix} F_1 \\ F_2 \\ F_3 \end{pmatrix}_3 \otimes \begin{pmatrix} \chi_8 \\ \chi_9 \\ \chi_{10} \end{pmatrix}_3 \rightarrow \begin{pmatrix} T_1 \\ T_2 \end{pmatrix}_2 \otimes \begin{pmatrix} \frac{1}{\sqrt{2}}(F_2\chi_9 - F_3\chi_{10}) \\ \frac{1}{\sqrt{6}}(2F_1\chi_8 - F_2\chi_9 - F_3\chi_{10}) \end{pmatrix}_2$$

$$\rightarrow \left(\frac{T_1}{\sqrt{2}}(F_2\chi_9 - F_3\chi_{10}) + \frac{T_2}{\sqrt{6}}(-2F_1\chi_8 + F_2\chi_9 + F_3\chi_{10}) \right)_1$$

$$\begin{pmatrix} F_1 \\ F_2 \\ F_3 \end{pmatrix}_3 \times \begin{pmatrix} \chi_8 \\ \chi_9 \\ \chi_{10} \end{pmatrix}_3 \rightarrow (F_1\chi_8 + F_2\chi_9 + F_3\chi_{10})_1 . \tag{19.14}$$

We discuss the quark and lepton mass matrices and flavor mixing based on this superpotential. Furthermore, we take into account the next leading superpotential in the numerical study of the flavor mixing and CP violation.

Let us start with extracting the lepton sector from the superpotential $w_{SU(5)}^{(0)}$. Denoting Higgs doublets as h_u and h_d, the superpotential of the Yukawa sector respecting the $S_4 \times Z_4 \times U(1)_{\text{FN}}$ symmetry is given for charged leptons as

$$w_l = -3y_1 \left[\frac{e^c}{\sqrt{2}}(l_\mu \chi_9 - l_\tau \chi_{10}) + \frac{\mu^c}{\sqrt{6}}(-2l_e\chi_8 + l_\mu\chi_9 + l_\tau\chi_{10}) \right] h_{45}\Theta/(\Lambda\bar{\Lambda})$$

$$+ y_2 \tau^c (l_e\chi_{11} + l_\mu\chi_{12} + l_\tau\chi_{13}) h_d/\Lambda. \tag{19.15}$$

For right-handed Majorana neutrinos, the superpotential is given as

$$w_N = y_1^N (N_e^c N_e^c + N_\mu^c N_\mu^c)\Theta^2/\bar{\Lambda}$$

$$+ y_2^N \left[(N_e^c N_\mu^c + N_\mu^c N_e^c)\chi_3 + (N_e^c N_e^c - N_\mu^c N_\mu^c)\chi_4 \right] + M N_\tau^c N_\tau^c, \tag{19.16}$$

and for Dirac neutrino Yukawa couplings, the superpotential is

$$w_D = y_1^D \left[\frac{N_e^c}{\sqrt{6}}(2l_e\chi_5 - l_\mu\chi_6 - l_\tau\chi_7) + \frac{N_\mu^c}{\sqrt{2}}(l_\mu\chi_6 - l_\tau\chi_7) \right] h_u\Theta/(\Lambda\bar{\Lambda})$$

$$+ y_2^D N_\tau^c (l_e\chi_5 + l_\mu\chi_6 + l_\tau\chi_7) h_u/\Lambda. \tag{19.17}$$

Higgs doublets h_u, h_d and gauge singlet scalars Θ and χ_i, are assumed to develop their VEVs as follows:

$$\langle h_u \rangle = v_u, \quad \langle h_d \rangle = v_d, \quad \langle h_{45} \rangle = v_{45}, \quad \langle \Theta \rangle = \theta,$$

$$\langle (\chi_3, \chi_4) \rangle = (u_3, u_4), \quad \langle (\chi_5, \chi_6, \chi_7) \rangle = (u_5, u_6, u_7),$$

$$\langle (\chi_8, \chi_9, \chi_{10}) \rangle = (u_8, u_9, u_{10}), \quad \langle (\chi_{11}, \chi_{12}, \chi_{13}) \rangle = (u_{11}, u_{12}, u_{13}), \tag{19.18}$$

which are supposed to be real. Then, we obtain the mass matrix for charged leptons as

$$M_l = -3y_1\lambda v_{45} \begin{pmatrix} 0 & \alpha_9/\sqrt{2} & -\alpha_{10}/\sqrt{2} \\ -2\alpha_8/\sqrt{6} & \alpha_9/\sqrt{6} & \alpha_{10}/\sqrt{6} \\ 0 & 0 & 0 \end{pmatrix} + y_2 v_d \begin{pmatrix} 0 & 0 & 0 \\ 0 & 0 & 0 \\ \alpha_{11} & \alpha_{12} & \alpha_{13} \end{pmatrix},$$

$$\tag{19.19}$$

while the right-handed Majorana neutrino mass matrix is given as

$$
M_N = \begin{pmatrix} y_1^N \lambda^2 \bar{\Lambda} + y_2^N \alpha_4 \Lambda & y_2^N \alpha_3 \Lambda & 0 \\ y_2^N \alpha_3 \Lambda & y_1^N \lambda^2 \bar{\Lambda} - y_2^N \alpha_4 \Lambda & 0 \\ 0 & 0 & M \end{pmatrix} . \tag{19.20}
$$

It is noticed that $(1, 3), (2, 3), (3, 1)$ and $(3, 3)$ elements of the right-handed Majorana neutrino mass matrix vanish. These are so called supersymmetric zeros. The Dirac mass matrix of neutrinos is

$$
M_D = y_1^D \lambda v_u \begin{pmatrix} 2\alpha_5/\sqrt{6} & -\alpha_6/\sqrt{6} & -\alpha_7/\sqrt{6} \\ 0 & \alpha_6/\sqrt{2} & -\alpha_7/\sqrt{2} \\ 0 & 0 & 0 \end{pmatrix} + y_2^D v_u \begin{pmatrix} 0 & 0 & 0 \\ 0 & 0 & 0 \\ \alpha_5 & \alpha_6 & \alpha_7 \end{pmatrix} , \tag{19.21}
$$

where we denote $\alpha_i \equiv u_i/\Lambda$ and $\lambda \equiv \theta/\bar{\Lambda}$.

If we can take the vacuum alignment to be

$$
(u_8, u_9, u_{10}) = (0, u_9, 0) , \qquad (u_{11}, u_{12}, u_{13}) = (0, 0, u_{13}) ,
$$

that is $\alpha_8 = \alpha_{10} = \alpha_{11} = \alpha_{12} = 0$, we obtain

$$
M_l = \begin{pmatrix} 0 & -3y_1 \lambda \alpha_9 v_{45}/\sqrt{2} & 0 \\ 0 & -3y_1 \lambda \alpha_9 v_{45}/\sqrt{6} & 0 \\ 0 & 0 & y_2 \alpha_{13} v_d \end{pmatrix} , \tag{19.22}
$$

then $M_l^\dagger M_l$ is obtained as follows:

$$
M_l^\dagger M_l = v_d^2 \begin{pmatrix} 0 & 0 & 0 \\ 0 & 6|\bar{y}_1 \lambda \alpha_9|^2 & 0 \\ 0 & 0 & |y_2|^2 \alpha_{13}^2 \end{pmatrix} , \tag{19.23}
$$

where we replace $y_1 v_{45}$ by $\bar{y}_1 v_d$. That is, the left-handed mixing angles of charged lepton mass matrix vanish. The charged lepton masses are given as

$$
m_e^2 = 0 , \quad m_\mu^2 = 6|\bar{y}_1 \lambda \alpha_9|^2 v_d^2 , \quad m_\tau^2 = |y_2|^2 \alpha_{13}^2 v_d^2 . \tag{19.24}
$$

It is remarkable that the electron mass vanishes. The electron mass is obtained in the next leading order.

Taking the vacuum alignment $(u_3, u_4) = (0, u_4)$ and $(u_5, u_6, u_7) = (u_5, u_5, u_5)$ in Eq. (19.20), the right-handed Majorana mass matrix of neutrinos turns to

$$
M_N = \begin{pmatrix} y_1^N \lambda^2 \bar{\Lambda} + y_2^N \alpha_4 \Lambda & 0 & 0 \\ 0 & y_1^N \lambda^2 \bar{\Lambda} - y_2^N \alpha_4 \Lambda & 0 \\ 0 & 0 & M \end{pmatrix} , \tag{19.25}
$$

and the Dirac mass matrix of neutrinos turns to

$$M_D = y_1^D \lambda v_u \begin{pmatrix} 2\alpha_5/\sqrt{6} & -\alpha_5/\sqrt{6} & -\alpha_5/\sqrt{6} \\ 0 & \alpha_5/\sqrt{2} & -\alpha_5/\sqrt{2} \\ 0 & 0 & 0 \end{pmatrix} + y_2^D v_u \begin{pmatrix} 0 & 0 & 0 \\ 0 & 0 & 0 \\ \alpha_5 & \alpha_5 & \alpha_5 \end{pmatrix}. \quad (19.26)$$

By using the seesaw mechanism $M_\nu = M_D^T M_N^{-1} M_D$, the left-handed Majorana neutrino mass matrix is written as

$$M_\nu = \begin{pmatrix} a + \frac{2}{3}b & a - \frac{1}{3}b & a - \frac{1}{3}b \\ a - \frac{1}{3}b & a + \frac{1}{6}b + \frac{1}{2}c & a + \frac{1}{6}b - \frac{1}{2}c \\ a - \frac{1}{3}b & a + \frac{1}{6}b - \frac{1}{2}c & a + \frac{1}{6}b + \frac{1}{2}c \end{pmatrix}, \quad (19.27)$$

where

$$a = \frac{(y_2^D \alpha_5 v_u)^2}{M}, \qquad b = \frac{(y_1^D \alpha_5 v_u \lambda)^2}{y_1^N \lambda^2 \bar{\Lambda} + y_2^N \alpha_4 \Lambda}, \qquad c = \frac{(y_1^D \alpha_5 v_u \lambda)^2}{y_1^N \lambda^2 \bar{\Lambda} - y_2^N \alpha_4 \Lambda}. \quad (19.28)$$

The neutrino mass matrix is decomposed as

$$M_\nu = \frac{b+c}{2} \begin{pmatrix} 1 & 0 & 0 \\ 0 & 1 & 0 \\ 0 & 0 & 1 \end{pmatrix} + \frac{3a-b}{3} \begin{pmatrix} 1 & 1 & 1 \\ 1 & 1 & 1 \\ 1 & 1 & 1 \end{pmatrix} + \frac{b-c}{2} \begin{pmatrix} 1 & 0 & 0 \\ 0 & 0 & 1 \\ 0 & 1 & 0 \end{pmatrix}, \quad (19.29)$$

which gives the tri-bimaximal mixing matrix U_{tribi} and mass eigenvalues as follows:

$$m_1 = b, \qquad m_2 = 3a, \qquad m_3 = c. \quad (19.30)$$

The finite mixing angle θ_{13} is obtained by including the next leading terms of the superpotential of leptons.

Let us discuss the quark sector. For down-type quarks, we can write the superpotential as follows:

$$w_d = y_1 \left[\frac{1}{\sqrt{2}}(s^c\chi_9 - b^c\chi_{10})q_1 + \frac{1}{\sqrt{6}}(-2d^c\chi_8 + s^c\chi_9 + b^c\chi_{10})q_2 \right] h_{45}\Theta/(\Lambda\bar{\Lambda})$$
$$+ y_2(d^c\chi_{11} + s^c\chi_{12} + b^c\chi_{13})q_3 h_d/\Lambda. \quad (19.31)$$

Since the vacuum alignment is fixed in the lepton sector as seen in Eq. (19.18), the down-type quark mass matrix at the leading order is given as

$$M_d = v_d \begin{pmatrix} 0 & 0 & 0 \\ \bar{y}_1\lambda\alpha_9/\sqrt{2} & \bar{y}_1\lambda\alpha_9/\sqrt{6} & 0 \\ 0 & 0 & y_2\alpha_{13} \end{pmatrix}, \quad (19.32)$$

where we denote $\bar{y}_1 v_d = y'_1 v_{45}$. Then, we have

$$
M_d^\dagger M_d = v_d^2 \begin{pmatrix} \frac{1}{2}|\bar{y}_1 \lambda \alpha_9|^2 & \frac{1}{2\sqrt{3}}|\bar{y}_1 \lambda \alpha_9|^2 & 0 \\ \frac{1}{2\sqrt{3}}|\bar{y}_1 \lambda \alpha_9|^2 & \frac{1}{6}|\bar{y}_1 \lambda \alpha_9|^2 & 0 \\ 0 & 0 & |y_2|^2 \alpha_{13}^2 \end{pmatrix}. \tag{19.33}
$$

This matrix can be diagonalized by the orthogonal matrix U_d as

$$
U_d = \begin{pmatrix} \cos 60° & \sin 60° & 0 \\ -\sin 60° & \cos 60° & 0 \\ 0 & 0 & 1 \end{pmatrix}. \tag{19.34}
$$

The down-type quark masses are given as

$$
m_d^2 = 0, \quad m_s^2 = \frac{2}{3}|\bar{y}_1 \lambda \alpha_9|^2 v_d^2, \quad m_b^2 \approx |y_2|^2 \alpha_{13}^2 v_d^2, \tag{19.35}
$$

which correspond to ones of charged lepton masses in Eq. (19.24). The down quark mass vanishes as well as the electron mass, however tiny masses appear in the next leading order.

Realistic CKM mixing matrix in the quark sector is obtained by including the next leading terms of the superpotential, in which quadratic terms of χ_i such as $d^c q_1 \chi_1 \chi_5$ are dominant. We obtain the next leading down-type quark mass matrix elements $\bar{\epsilon}_{ij}$, which are given in terms of Yukawa couplings and VEVs of flavons. Magnitudes of $\bar{\epsilon}_{ij}$'s are of $\mathcal{O}(\tilde{\alpha}^2)$, where $\tilde{\alpha}$ is a linear combination of α_i's. The down-type quark mass matrix is written in terms of $\bar{\epsilon}_{ij}$ as

$$
M_d \simeq \begin{pmatrix} \bar{\epsilon}_{11} & \bar{\epsilon}_{21} & \bar{\epsilon}_{31} \\ \frac{\sqrt{3}m_s}{2} + \bar{\epsilon}_{12} & \frac{m_s}{2} + \bar{\epsilon}_{22} & \bar{\epsilon}_{32} \\ \bar{\epsilon}_{13} & \bar{\epsilon}_{23} & m_b + \bar{\epsilon}_{33} \end{pmatrix}, \tag{19.36}
$$

where ϵ_{ij} should be the order of m_d.

By rotating $M_d^\dagger M_d$ with the mixing matrix U_d in Eq. (19.34), we have

$$
U_d^\dagger M_d^\dagger M_d U_d \simeq \begin{pmatrix} \mathcal{O}(m_d^2) & \mathcal{O}(m_d m_s) & \mathcal{O}(m_d m_b) \\ \mathcal{O}(m_d m_s) & \mathcal{O}(m_s^2) & \mathcal{O}(m_d m_b) \\ \mathcal{O}(m_d m_b) & \mathcal{O}(m_d m_b) & m_b^2 \end{pmatrix}. \tag{19.37}
$$

Then we get mixing angles θ_{12}^d, θ_{13}^d, θ_{23}^d in the mass matrix of Eq. (19.37) as

$$
\theta_{12}^d = \mathcal{O}\left(\frac{m_d}{m_s}\right), \quad \theta_{13}^d = \mathcal{O}\left(\frac{m_d}{m_b}\right), \quad \theta_{23}^d = \mathcal{O}\left(\frac{m_d}{m_b}\right), \tag{19.38}
$$

where CP violating phases are neglected.

Let us discuss the up-type quark sector. The superpotential respecting $S_4 \times Z_4 \times U(1)_{FN}$ is given as

$$w_u = y_1^u \left[(u^c \chi_1 + c^c \chi_2) q_3 + t^c (q_1 \chi_1 + q_2 \chi_2) \right] h_u / \Lambda + y_2^u t^c q_3 h_u . \quad (19.39)$$

We denote their VEVs as follows:

$$\langle (\chi_1, \chi_2) \rangle = (u_1, u_2) . \quad (19.40)$$

Then, we obtain the mass matrix for up-type quarks as

$$M_u = v_u \begin{pmatrix} 0 & 0 & y_1^u \alpha_1 \\ 0 & 0 & y_1^u \alpha_2 \\ y_1^u \alpha_1 & y_1^u \alpha_2 & y_2^u \end{pmatrix} . \quad (19.41)$$

The next order terms of the superpotential are also important to predict the CP violation in the quark sector. The relevant superpotential is given at the next leading order as

$$\Delta w_u = [y_{\Delta_a}^u (T_1, T_2) \otimes (T_1, T_2) \otimes (\chi_1, \chi_2) \otimes (\chi_1, \chi_2) \otimes H_5 / \Lambda^2$$
$$+ y_{\Delta_b}^u (T_1, T_2) \otimes (T_1, T_2) \otimes \chi_{14} \otimes \chi_{14} \otimes H_5 / \Lambda^2$$
$$+ y_{\Delta_c}^u T_3 \otimes T_3 \otimes (\chi_8, \chi_9, \chi_{10}) \otimes (\chi_8, \chi_9, \chi_{10}) \otimes H_5 / \Lambda^2]_1 . \quad (19.42)$$

The multiplication rule to get the S_4 singlet is

$$\begin{pmatrix} T_1 \\ T_2 \end{pmatrix}_2 \otimes \begin{pmatrix} T_1 \\ T_2 \end{pmatrix}_2 \otimes \begin{pmatrix} \chi_1 \\ \chi_2 \end{pmatrix}_2 \otimes \begin{pmatrix} \chi_1 \\ \chi_2 \end{pmatrix}_2 \rightarrow (T_1 T_1 + T_2 T_2)_1 \otimes (\chi_1 \chi_1 + \chi_2 \chi_2)_1$$
$$\oplus \begin{pmatrix} T_1 T_2 + T_2 T_1 \\ T_1 T_1 - T_2 T_2 \end{pmatrix}_2 \otimes \begin{pmatrix} \chi_1 \chi_2 + \chi_2 \chi_1 \\ \chi_1 \chi_1 - \chi_2 \chi_2 \end{pmatrix}_2$$
$$\rightarrow ((T_1 T_1 + T_2 T_2)(\chi_1 \chi_1 + \chi_2 \chi_2))_1$$
$$\oplus ((T_1 T_2 + T_2 T_1)(\chi_1 \chi_2 + \chi_2 \chi_1)$$
$$+ (T_1 T_1 - T_2 T_2)(\chi_1 \chi_1 - \chi_2 \chi_2))_1 . \quad (19.43)$$

We obtain the following mass matrix including the next leading terms,

$$M_u = v_u \begin{pmatrix} 2y_{\Delta_{a1}}^u \alpha_1^2 + y_{\Delta_b}^u \alpha_{14}^2 & y_{\Delta_{a2}}^u \alpha_1^2 & y_1^u \alpha_1 \\ y_{\Delta_{a2}}^u \alpha_1^2 & 2y_{\Delta_{a1}}^u \alpha_1^2 + y_{\Delta_b}^u \alpha_{14}^2 & y_1^u \alpha_1 \\ y_1^u \alpha_1 & y_1^u \alpha_1 & y_2^u + y_{\Delta_c}^u \alpha_9^2 \end{pmatrix} , \quad (19.44)$$

where we take the alignment $\alpha_1 = \alpha_2$. After rotating the mass matrix M_u by $\theta_{12} = 45°$, we get

$$\hat{M}_u \approx v_u \begin{pmatrix} (2y_{\Delta_{a1}}^u - y_{\Delta_{a2}}^u) \alpha_1^2 + y_{\Delta_b}^u \alpha_{14}^2 & 0 & 0 \\ 0 & (2y_{\Delta_{a1}}^u + y_{\Delta_{a2}}^u) \alpha_1^2 + y_{\Delta_b}^u \alpha_{14}^2 & \sqrt{2} y_1^u \alpha_1 \\ 0 & \sqrt{2} y_1^u \alpha_1 & y_2^u \end{pmatrix} .$$
$$(19.45)$$

This mass matrix is taken to be real one by removing phases. The matrix is diagonalized by the orthogonal transformation as $V_u^T \hat{M}_u V_F$, where

$$V_u \simeq \begin{pmatrix} 1 & 0 & 0 \\ 0 & r_t & r_c \\ 0 & -r_c & r_t \end{pmatrix}, \quad r_c = \sqrt{\frac{m_c}{m_c + m_t}}, \quad r_t = \sqrt{\frac{m_t}{m_c + m_t}}. \quad (19.46)$$

Now we can calculate the CKM matrix. Mixing matrices of up- and down-type quarks are summarized as

$$U_u \simeq \begin{pmatrix} \cos 45° & \sin 45° & 0 \\ -\sin 45° & \cos 45° & 0 \\ 0 & 0 & 1 \end{pmatrix} \begin{pmatrix} 1 & 0 & 0 \\ 0 & e^{-i\rho} & 0 \\ 0 & 0 & 1 \end{pmatrix} \begin{pmatrix} 1 & 0 & 0 \\ 0 & r_t & r_c \\ 0 & -r_c & r_t \end{pmatrix},$$

$$U_d \simeq \begin{pmatrix} \cos 60° & \sin 60° & 0 \\ -\sin 60° & \cos 60° & 0 \\ 0 & 0 & 1 \end{pmatrix} \begin{pmatrix} 1 & \theta_{12}^d & \theta_{13}^d \\ -\theta_{12}^d - \theta_{13}^d \theta_{23}^d & 1 & \theta_{23}^d \\ -\theta_{13}^d + \theta_{12}^d \theta_{23}^d & -\theta_{23}^d - \theta_{12}^d \theta_{13}^d & 1 \end{pmatrix}. \quad (19.47)$$

Therefore, the CKM matrix is given as $U_u^\dagger U_d$. The relevant mixing elements are

$$V_{us} \approx \theta_{12}^d \cos 15° + \sin 15°,$$

$$V_{ub} \approx \theta_{13}^d \cos 15° + \theta_{23}^d \sin 15°,$$

$$V_{cb} \approx -r_t \theta_{13}^d e^{i\rho} \sin 15° + r_t \theta_{23}^d e^{i\rho} \cos 15° - r_c,$$

$$V_{td} \approx -r_c \sin 15° e^{i\rho} - r_c(\theta_{12}^d + \theta_{13}^d \theta_{23}^d) e^{i\rho} \cos 15° + r_t(-\theta_{13}^d + \theta_{12}^d \theta_{23}^d).$$
$$(19.48)$$

We can reproduce the experimental values with a parameter set

$$\rho = 123°, \quad \theta_{12}^d = -0.0340, \quad \theta_{13}^d = 0.00626, \quad \theta_{23}^d = -0.00880, \quad (19.49)$$

by putting typical masses at the GUT scale $m_u = 1.04 \times 10^{-3}$ GeV, $m_c = 302 \times 10^{-3}$ GeV, $m_t = 129$ GeV [68].

In terms of a phase ρ, we can also estimate the magnitude of CP violation measure, Jarlskog invariant J_{CP} [69], which is given as

$$|J_{CP}| = |\text{Im}\left\{ V_{us} V_{cs}^* V_{ub} V_{cb}^* \right\}| \approx 3.06 \times 10^{-5}. \quad (19.50)$$

Our prediction is consistent with the experimental values $J_{CP} = 3.00^{+0.15}_{-0.09} \times 10^{-5}$ [8].

19.4   Alternative Flavor Mixing

In previous section, we have presented flavor models, in which the tri-bimaximal mixing of the lepton flavor is reproduced. However, there are other flavor mixing patterns as the golden ratio [70], the trimaximal, and the bimaximal [71] for the lepton flavor mixing.

Table 19.3 Assignments of $SU(2)$ and A_5 representations

	L	$\bar{e}$	H	ξ	ψ	χ
$SU(2)$	2	1	2	1	1	1
A_5	3	$3'$	1	5	5	4

At first we show the golden ratio which appears in the solar mixing angle θ_{12}. The golden ratio can be derived A_5 [72] and D_{10} [73] models. One example is proposed as $\tan\theta_{12} = 1/\phi$ where $\phi = (1 + \sqrt{5})/2 \simeq 1.62$ [70]. The rotational icosahedral group, which is isomorphic to A_5, the alternating group of five elements, provides a natural context of the golden ratio $\cos\theta_{12} = \phi/2$ [72]. In this model, the solar angle is related to the golden ratio, the atmospheric angle is maximal, and the reactor angle vanishes. The particle contents are summarized in Table 19.3.

In this model, ξ couples to the $LHLH$ operator and that ψ and χ couple to the $L\bar{e}H$ operator. This set up is derived from additional symmetries such as Z_n which forbid ξ and ψ from coupling to $L\bar{e}H$ and $LHLH$, respectively. For simplicity, it is assumed that the tree level LL term is forbidden by such additional symmetries.

With these assumptions, the mass terms are written as follows:

$$\mathcal{L}_{\text{mass}} = \frac{\alpha_{ijk}}{MM'}L_i H L_j H \xi_k + \frac{\beta_{ijk}}{M'}L_i \bar{e}_j H \psi_k + \frac{\gamma_{ijl}}{M'}L_i \bar{e}_j H \chi_l + \text{h.c.}, \quad (19.51)$$

in which M' represents the scale of flavor symmetry breaking, and α_{ijk}, β_{ijk}, and γ_{ijl} are dimensionless couplings that encode the tensor product decomposition of the icosahedral symmetry. In principle, M' has no relation to M. Taking VEV alignment as

$$\langle \xi \rangle = \frac{\sqrt{3}}{2\alpha}\left(\frac{1}{\sqrt{15}}(m_2 - m_1), \frac{2}{\sqrt{15}}(m_2 - m_1), 0, 0, -(m_1 + m_2)\right), \quad (19.52)$$

the neutrino mass matrix is written as

$$M_\nu = \frac{1}{\sqrt{5}}\begin{pmatrix} \phi m_1 + \frac{1}{\phi}m_2 & m_2 - m_1 & 0 \\ m_2 - m_1 & \frac{1}{\phi}m_1 + \phi m_2 & 0 \\ 0 & 0 & -\sqrt{5}(m_1 + m_2) \end{pmatrix}, \quad (19.53)$$

where $\phi = (1 + \sqrt{5})/2$. The neutrino mass eigenvalues are m_1, m_2, and $m_3 = -(m_1 + m_2)$. The neutrino mixing matrix U_ν, defined by $U_\nu M_\nu U_\nu^T$, satisfies $\theta_{12} = \tan^{-1}(1/\phi) = 31.72°$ as follow:

$$U_\nu = \begin{pmatrix} \sqrt{\frac{\phi}{\sqrt{5}}} & -\sqrt{\frac{1}{\sqrt{5}\phi}} & 0 \\ \sqrt{\frac{1}{\sqrt{5}\phi}} & \sqrt{\frac{\phi}{\sqrt{5}}} & 0 \\ 0 & 0 & -i \end{pmatrix}. \quad (19.54)$$

The maximal mixing between the second and third families is derived from the charged lepton sector. At leading order in flavon fields, only m_τ is finite. Taking VEV alignment as

$$\langle \psi \rangle = \frac{m_\tau}{2\sqrt{6}\beta} \left(-\sqrt{\frac{5}{3}}, 0, -\frac{2}{\sqrt{3}}\phi, 0, 1 \right) , \quad \langle \chi \rangle = \frac{m_\tau}{3\sqrt{2}\gamma} \left(0, \frac{1}{\phi}, 0, 1 \right) , \quad (19.55)$$

the charged lepton mass matrix takes the form

$$M_e = \frac{1}{\sqrt{2}} \begin{pmatrix} 0 & 0 & 0 \\ 0 & 0 & m_\tau \\ 0 & 0 & m_\tau \end{pmatrix} . \quad (19.56)$$

The left-handed states are diagonalized by the mixing matrix U_e as

$$U_e = \begin{pmatrix} 1 & 0 & 0 \\ 0 & \frac{1}{\sqrt{2}} & -\frac{1}{\sqrt{2}} \\ 0 & \frac{1}{\sqrt{2}} & \frac{1}{\sqrt{2}} \end{pmatrix} . \quad (19.57)$$

In Eqs. (19.54) and (19.57), the lepton mixing matrix U is obtained as

$$U = U_e U_\nu^\dagger = \begin{pmatrix} \sqrt{\frac{\phi}{\sqrt{5}}} & \sqrt{\frac{1}{\sqrt{5}\phi}} & 0 \\ -\frac{1}{\sqrt{2}}\sqrt{\frac{1}{\sqrt{5}\phi}} & \frac{1}{\sqrt{2}}\sqrt{\frac{\phi}{\sqrt{5}}} & -\frac{1}{\sqrt{2}} \\ -\frac{1}{\sqrt{2}}\sqrt{\frac{1}{\sqrt{5}\phi}} & \frac{1}{\sqrt{2}}\sqrt{\frac{\phi}{\sqrt{5}}} & \frac{1}{\sqrt{2}} \end{pmatrix} \mathcal{P}, \quad (19.58)$$

where $\mathcal{P} = \mathrm{Diag}(1, 1, i)$ is the Majorana phase matrix. In conclusion, the lepton mixing matrix has a vanishing reactor mixing angle, a maximal atmospheric mixing angle, and a solar angle given by $\theta_{12} = \tan^{-1}(1/\phi)$.

We now discuss another mixing pattern, namely, the trimaximal lepton mixing [63], defined by $|U_{\alpha 2}|^2 = 1/3$ for $\alpha = e, \mu$, and so the mixing matrix U_tri is given by using an arbitrary angle θ and a phase η as follows:

$$U_\mathrm{tri} = \begin{pmatrix} \frac{2}{\sqrt{6}} & \frac{1}{\sqrt{3}} & 0 \\ -\frac{1}{\sqrt{6}} & \frac{1}{\sqrt{3}} & -\frac{1}{\sqrt{2}} \\ -\frac{1}{\sqrt{6}} & \frac{1}{\sqrt{3}} & \frac{1}{\sqrt{2}} \end{pmatrix} \begin{pmatrix} \cos\theta & 0 & \sin\theta e^{-i\eta} \\ 0 & 1 & 0 \\ -\sin\theta e^{i\eta} & 0 & \cos\theta \end{pmatrix} . \quad (19.59)$$

This corresponds to a two-parameter lepton flavor mixing matrix. We have presented a model for the lepton sector in which trimaximal mixing based on the group A_4 in Sect. 19.2.2, or based on the group $\Delta(27)$. An interesting feature of the $\Delta(27)$ model is that no vacuum alignment is required. The other model is based on the S_3 [64] or S_4 [65] discrete symmetry where the symmetry breaking is triggered by the boundary conditions of the bulk right-handed neutrino in the fifth spacial dimension.

Table 19.4 Assignments of $SU(2)$, S_4, Z_4, and $U(1)_{FN}$ representations

	l	e^c	μ^c	τ^c	$h_{u,d}$	θ	φ_l	χ_l	ξ_ν	φ_ν
$SU(2)$	2	1	1	1	2	1	1	1	1	1
S_4	3	1	$1'$	1	1	1	3	$3'$	1	3
Z_4	1	-1	$-i$	$-i$	1	1	i	i	1	1
$U(1)_{FN}$	0	2	1	0	0	-1	0	0	0	0

Finally we consider the bimaximal mixing [71], which is the solar angle $\theta_{12} = \pi/4$, atmospheric angle $\theta_{23} = \pi/4$, and reactor angle $\theta_{13} = 0$. The mixing matrix U_{bi} is as follows:

$$U_{bi} = \begin{pmatrix} \frac{1}{\sqrt{2}} & \frac{1}{\sqrt{2}} & 0 \\ -\frac{1}{2} & \frac{1}{2} & -\frac{1}{\sqrt{2}} \\ -\frac{1}{2} & \frac{1}{2} & \frac{1}{\sqrt{2}} \end{pmatrix}. \tag{19.60}$$

The $S_4 \times Z_4$ flavor model is presented in Ref. [74]. The particle contents are summarized in Table 19.4.

In the charged lepton sector, the superpotential is given as follows:

$$w_l = \frac{y_e^{(1)}}{\Lambda^2} \frac{\theta^2}{\Lambda^2} e^c (l\varphi_l\varphi_l)h_d + \frac{y_e^{(2)}}{\Lambda^2} \frac{\theta^2}{\Lambda^2} e^c (l\chi_l\chi_l)h_d + \frac{y_e^{(3)}}{\Lambda^2} \frac{\theta^2}{\Lambda^2} e^c (l\varphi_l\chi_l)h_d$$
$$+ \frac{y_\mu}{\Lambda} \frac{\theta}{\Lambda} \mu^c (l\chi_l)'h_d + \frac{y_\tau}{\Lambda} \tau^c (l\varphi_l)h_d , \tag{19.61}$$

where $y_e^{(i)}$, y_μ and y_τ are yukawa couplings. In the neutrino sector, the effective superpotential is given as,

$$w_\nu^{\text{eff}} = \frac{M}{\Lambda}(lh_ulh_u) + \frac{a}{\Lambda^2}(lh_ulh_u)\xi_\nu + \frac{b}{\Lambda^2}(lh_ulh_u\varphi_l) , \tag{19.62}$$

where M, a and b are given in mass unit. Taking VEV alignment and VEVs as

$$\frac{\langle\varphi_l\rangle}{\Lambda} = A(0, 1, 0), \quad \frac{\langle\chi_l\rangle}{\Lambda} = B(0, 0, 1), \quad \frac{\langle\varphi_\nu\rangle}{\Lambda} = C(0, 1, -1),$$
$$\frac{\langle\xi_\nu\rangle}{\Lambda} = D, \quad \frac{\langle\theta\rangle}{\Lambda} = t, \quad \langle h_{u,d}\rangle = v_{u,d} , \tag{19.63}$$

the charged lepton mass matrix is given as the diagonal one;

$$m_l = \begin{pmatrix} (y_e^{(1)} B^2 - y_e^{(2)} A^2 + y_e^{(3)} AB)t^2 & 0 & 0 \\ 0 & y_\mu Bt & 0 \\ 0 & 0 & y_\tau A \end{pmatrix} v_d , \tag{19.64}$$

and the effective neutrino mass matrix is written as

$$
m_\nu^{\text{eff}} = \begin{pmatrix} 2M + 2aD & -2bC & -2bC \\ -2bC & 0 & 2M + 2aD \\ -2bC & 2M + 2aD & 0 \end{pmatrix} \frac{v_u^2}{\Lambda} .
\tag{19.65}
$$

Then, the neutrino mixing is bimaximal as seen in Eq. (19.60), and neutrino mass eigenvalues are

$$
m_1 = 2|M + aD - \sqrt{2}bC| \frac{v_u^2}{\Lambda}, \quad m_2 = 2|M + aD + \sqrt{2}bC| \frac{v_u^2}{\Lambda}, \quad m_3 = 2|M + aD| \frac{v_u^2}{\Lambda} .
\tag{19.66}
$$

It is noticed that the bimaximal neutrino mixing is also studied in the context of the quark-lepton complementarity of mixing angles [75] in the S_4 model [76].

19.5   Modular A_4 Invariance and Neutrino Mixing

The finite groups S_3, A_4, S_4 and A_5 are formed as the quotient groups of the modular group and its principal congruence subgroups [77] as discussed in the Chap. 16. Based on the finite modular group $\Gamma_3 \simeq A_4$, the new flavor model of lepton mass matrices have been presented [78]. There is a difference between the modular symmetry and the usual flavor symmetry. Coupling constants such as Yukawa couplings also transform non-trivially under the modular symmetry and are written as functions of the modulus called modular forms, which are holomorphic functions of the modulus τ. On the other hand, coupling constants are invariant under the usual flavor symmetries as seen in the A_4 and S_4 models of previous sections (Sects. 19.2, 19.3). Along with this work, $\Gamma_2 \simeq S_3$ [79], $\Gamma_4 \simeq S_4$ [80] and $\Gamma_5 \simeq A_5$ [81] have been discussed.

In this section, we present the phenomenological discussion of a simple lepton model with the $\Gamma_3 \simeq A_4$ modular symmetry, which predicts a clear correlation between the neutrino mixing angle θ_{23} and the CP violating Dirac phase [82]. The mass matrices of neutrinos and charged leptons are essentially given by fixing the expectation value of the modulus τ, which is the only source of the breaking of the modular invariance. However, there are freedoms for the assignment of irreducible representations and modular weights to leptons.

We suppose that three left-handed lepton doublets are of a triplet of the A_4 group. The three right-handed neutrinos are also of a triplet of A_4. On the other hand, the Higgs doublets are supposed to be singlets of A_4. The generic assignments of representations and modular weights to the MSSM fields as well as right-handed neutrino superfields are presented in Table 19.5.

For the charged leptons, we assign three right-handed charged leptons for three different singlets of A_4, $(1, 1', 1'')$, respectively. Therefore, there are three independent couplings in the superpotential of the charged lepton sector. Those coupling constants can be adjusted to the observed charged lepton masses.

Table 19.5 The charge assignment of SU(2), A_4, and the modular weight. The assignment of the modular forms $\mathbf{Y_3}$ (see Appendix I)

	L	e_R, μ_R, τ_R	ν_R	H_u	H_d	$\mathbf{Y_3} =$ $(Y_1, Y_2, Y_3)^T$
SU(2)	2	1	1	2	2	1
A_4	3	1, 1", 1'	3	1	1	3
$-k_I$	-1	-1	-1	0	0	$k = 2$

In terms of modular forms $\mathbf{Y_3} = (Y_1, Y_2, Y_3)^T$ in Appendix I, the modular invariant Yukawa coupling and Majorana mass terms of the leptons are given by the following superpotentials:

$$w_e = \alpha_e e_R H_d (L\mathbf{Y_3}) + \beta_e \mu_R H_d (L\mathbf{Y_3}) + \gamma_e \tau_R H_d (L\mathbf{Y_3}) , \qquad (19.67)$$

$$w_D = g_i (\nu_R H_u L\mathbf{Y_3})_1 , \qquad (19.68)$$

$$w_N = \Lambda (\nu_R \nu_R \mathbf{Y_3})_1 , \qquad (19.69)$$

where the sums of the modular weights vanish. The parameters α_e, β_e, γ_e, g_i, and Λ are constant coefficients.

VEVs of the neutral component of H_u and H_d are written as v_u and v_d, respectively. Then, the mass matrix of charged leptons is given by the superpotential Eq. (19.67) as follows:

$$M_E = v_d \, \text{diag}[\alpha_e, \beta_e, \gamma_e] \begin{pmatrix} Y_1 & Y_3 & Y_2 \\ Y_2 & Y_1 & Y_3 \\ Y_3 & Y_2 & Y_1 \end{pmatrix}_{RL} . \qquad (19.70)$$

The coefficients α_e, β_e, and γ_e are taken to be real positive by rephasing right-handed charged lepton fields without loss of generality.

Since the tensor product of $3 \otimes 3$ is decomposed into the symmetric triplet and the antisymmetric triplet as seen in Appendix C, the superpotential of the Dirac neutrino mass in Eq. (19.68) is expressed by introducing additional two parameters g_1 and g_2 as:

$$
\begin{aligned}
w_D = & v_u \begin{pmatrix} \nu_{R1} \\ \nu_{R2} \\ \nu_{R3} \end{pmatrix} \otimes \left[g_1 \begin{pmatrix} 2\nu_e Y_1 - \nu_\mu Y_3 - \nu_\tau Y_2 \\ 2\nu_\tau Y_3 - \nu_e Y_2 - \mu Y_1 \\ 2\nu_\mu Y_2 - \nu_\tau Y_1 - \nu_e Y_3 \end{pmatrix} \oplus g_2 \begin{pmatrix} \nu_\mu Y_3 - \nu_\tau Y_2 \\ \nu_e Y_2 - \nu_\mu Y_1 \\ \nu_\tau Y_1 - \nu_e Y_3 \end{pmatrix} \right] \\
= & v_u g_1 \left[\nu_{R1}(2\nu_e Y_1 - \nu_\mu Y_3 - \nu_\tau Y_2) + \nu_{R2}(2\nu_\mu Y_2 - \nu_\tau Y_1 - \nu_e Y_3) \right. \\
& \left. + \nu_{R3}(2\nu_\tau Y_3 - \nu_e Y_2 - \nu_\mu Y_1) \right] \\
& + v_u g_2 \left[\nu_{R1}(\nu_\mu Y_3 - \nu_\tau Y_2) + \nu_{R2}(\nu_\tau Y_1 - \nu_e Y_3) + \nu_{R3}(\nu_e Y_2 - \nu_\mu Y_1) \right] .
\end{aligned}
\qquad (19.71)
$$

The Dirac neutrino mass matrix is given as

$$M_D = v_u \begin{pmatrix} 2g_1 Y_1 & (-g_1 + g_2)Y_3 & (-g_1 - g_2)Y_2 \\ (-g_1 - g_2)Y_3 & 2g_1 Y_2 & (-g_1 + g_2)Y_1 \\ (-g_1 + g_2)Y_2 & (-g_1 - g_2)Y_1 & 2g_1 Y_3 \end{pmatrix}_{RL} . \tag{19.72}$$

On the other hand, since the Majorana neutrino mass terms are symmetric, the superpotential in Eq. (19.69) is expressed simply as

$$\begin{aligned} w_N &= \Lambda \begin{pmatrix} 2\nu_{R1}\nu_{R1} - \nu_{R2}\nu_{R3} - \nu_{R3}\nu_{R2} \\ 2\nu_{R3}\nu_{R3} - \nu_{R1}\nu_{R2} - \nu_{R2}\nu_{R1} \\ 2\nu_{R2}\nu_{R2} - \nu_{R3}\nu_{R1} - \nu_{R1}\nu_{R3} \end{pmatrix} \otimes \begin{pmatrix} Y_1 \\ Y_2 \\ Y_3 \end{pmatrix} \\ &= \Lambda \left[(2\nu_{R1}\nu_{R1} - \nu_{R2}\nu_{R3} - \nu_{R3}\nu_{R2})Y_1 + (2\nu_{R3}\nu_{R3} - \nu_{R1}\nu_{R2} - \nu_{R2}\nu_{R1})Y_3 \right. \\ &\quad \left. + (2\nu_{R2}\nu_{R2} - \nu_{R3}\nu_{R1} - \nu_{R1}\nu_{R3})Y_2 \right] . \end{aligned}$$

$$\tag{19.73}$$

Then, the modular invariant right-handed Majorana neutrino mass matrix is given as

$$M_N = \Lambda \begin{pmatrix} 2Y_1 & -Y_3 & -Y_2 \\ -Y_3 & 2Y_2 & -Y_1 \\ -Y_2 & -Y_1 & 2Y_3 \end{pmatrix}_{RR} . \tag{19.74}$$

Finally, the effective neutrino mass matrix is obtained by the type I seesaw as follows:

$$M_\nu = -M_D^{\mathrm{T}} M_N^{-1} M_D . \tag{19.75}$$

When τ is fixed, the modular invariance is broken, and then the lepton mass matrices give the mass eigenvalues and flavor mixing numerically. In order to determine the value of τ, we use the result of NuFIT 5.0 [1]. By inputting the data of $\Delta m_{\mathrm{atm}}^2 \equiv m_3^2 - m_1^2$, $\Delta m_{\mathrm{sol}}^2 \equiv m_2^2 - m_1^2$, and three mixing angles θ_{23}, θ_{12}, and θ_{13} with 3σ error-bar, we obtain values of the modulus τ and the other parameters, and then we can predict the CP violating Dirac phases δ_{CP}. We consider both the normal hierarchy (NH) of neutrino masses $m_1 < m_2 < m_3$ and the inverted hierarchy (IH) of neutrino masses $m_3 < m_1 < m_2$, where m_1, m_2, and m_3 denote three light neutrino masses. Since the sum of three neutrino masses $\sum m_i$ is constrained by the recent cosmological data [83–85], we exclude the predictions with $\sum m_i \geq 200\,\mathrm{meV}$ even if mixing angles are consistent with observed one.

The coefficients α_e/γ_e and β_e/γ_e in the charged lepton mass matrix are given only in terms of τ by the input of the observed values m_e/m_τ and m_μ/m_τ. As the input charged lepton masses, we take Yukawa couplings of charged leptons at the GUT scale 2×10^{16} GeV, where $\tan \beta = 5$ is taken as a bench mark [86,87]:

$$y_e = (1.97 \pm 0.024) \times 10^{-6}, \quad y_\mu = (4.16 \pm 0.050) \times 10^{-4},$$
$$y_\tau = (7.07 \pm 0.073) \times 10^{-3}. \tag{19.76}$$

Fig. 19.3 The allowed regions of Re[τ] and Im[τ] for NH. The regions (I) and (II) correspond to $\sum m_i < 160$ meV and $\sum m_i > 160$ meV, respectively. Black and grey points are in 2σ and 3σ regions, respectively

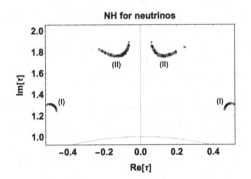

Fig. 19.4 Predictions of δ_{CP} versus $\sin^2\theta_{23}$ for NH. Vertical solid and dashed lines denote the central value, and the upper and lower bounds of the experimental data with 3σ, respectively. Notations are same as in Fig. 19.3

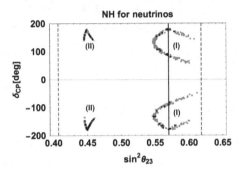

Lepton masses are given by $m_\ell = y_\ell v_H$ with $v_H = 174$ GeV.

Then, we have two free complex parameters, g_2/g_1 and the modulus τ apart from the overall factors in the neutrino sector. The value of τ is scanned in the fundamental domain of $SL(2, Z)$.

In Fig. 19.3, we show the allowed region in the Re[τ]-Im[τ] plain for NH of neutrino masses. The regions (I) and (II) denote $\sum m_i < 160$ meV and $\sum m_i > 160$ meV, respectively. The 2σ and 3σ regions are presented by black and grey points, respectively. If the cosmological observation confirms $\sum m_i < 160$ meV, the region (II) is excluded. Thus, the region of τ is severely restricted in this model.

We present the prediction of the Dirac CP violating phase δ_{CP} versus $\sin^2\theta_{23}$ for NH of neutrino masses in Fig. 19.4. The predicted regions (I) and (II) correspond to τ in Fig. 19.3. It is emphasized that the predicted $\sin^2\theta_{23}$ is larger than 0.535, and $\delta_{CP} = \pm(60°-180°)$ in the region (I). Since the correlation of $\sin^2\theta_{23}$ and δ_{CP} is characteristic, this prediction is testable in the future experiments of neutrinos. On the other hand, predicted $\sin^2\theta_{12}$ and $\sin^2\theta_{13}$ cover observed full region with 3σ error-bar, and there are no correlations with δ_{CP}.

The prediction of the effective mass m_{ee}, which is the measure of the neutrinoless double beta decay, is around 20–22 meV for (I) and 45–60 meV for (II). The sum of neutrino masses is predicted $\sum m_i = 140$–150 meV for (I), which is marginal compared with the cosmological upper bound for the sum of neutrino masses [83–85], and $\sum m_i = 170$–200 meV for (II), which may be excluded soon.

Finally, we present a sample set of parameters and outputs for NH in Table 19.6.

Table 19.6 Numerical values of parameters and output at a sample point of NH

τ	$0.4707 + 1.2916\,i$
g_2/g_1	$-0.0509 + 1.2108\,i$
α_e/γ_e	2.09×10^2
β_e/γ_e	3.44×10^3
$\sin^2 \theta_{12}$	0.320
$\sin^2 \theta_{23}$	0.571
$\sin^2 \theta_{13}$	0.0229
δ_{CP}	$278°$
$\sum m_i$	$146\,\text{meV}$
$\langle m_{ee} \rangle$	$22\,\text{meV}$
$\sum \chi_i^2$	1.86

We have also scanned the parameter space for the case of IH of neutrino masses. We have found parameter sets which fit the data of Δm_{sol}^2 and Δm_{atm}^2 reproduce the observed three mixing angles $\sin^2 \theta_{23}$, $\sin^2 \theta_{12}$, and $\sin^2 \theta_{13}$. However, the predicted $\sum m_i$ is around 200 meV. Therefore, we omit the case of IH.

It is helpful to comment on the effects of the supersymmetry breaking and the radiative corrections because we have discussed our model in the limit of exact supersymmetry. The supersymmetry breaking effect can be neglected if the separation between the supersymmetry breaking scale and the supersymmetry breaking mediator scale is sufficiently large. In our numerical results, the corrections by the renormalization are very small as far as we take the relatively small value of $\tan \beta$.

Although this model is the prototype of $\Gamma_3 \simeq A_4$ flavor symmetry, other flavor models of $\Gamma_3 \simeq A_4$ are also viable as seen in [88–90]. Furthermore, phenomenological discussions of the neutrino flavor mixing have been done based on $\Gamma_4 \simeq S_4$ [91–93] and $\Gamma_5 \simeq A_5$ [81,94]. The double covering groups T′ [95,96], S′$_4$ [97,98] and A′$_5$ [99,100] have also discussed from the modular symmetry. Furthermore, modular forms for $\Delta(96)$ and $\Delta(384)$ have been constructed [101], and the level 7 finite modular group $\Gamma_7 \simeq \text{PSL}(2, Z_7)$ has been examined for the lepton mixing [102].

On the other hand, the quark mass matrix has been also studied in the $\Gamma_3 \simeq A_4$ flavor symmetries [103,104]. Hence, the unification of quarks and leptons has been applied in the framework of the SU(5) GUT [105–109].

There are also another important physics, the baryon asymmetry in the universe, which is discussed with the modular symmetry. Indeed, the A_4 modular flavor symmetry has been examined in the leptogenesis [110–112].

The modular symmetry keeps a residual symmetry at the fixed points even if the modular symmetry is broken. It gives interesting lepton mass matrices for the flavor mixing [113,114]. It is also remarked that the hierarchical structure of quarks and leptons could be derived at the nearby fixed points of the modular symmetry [115–117].

New ideas have been proposed for the model building of flavors [118–121]. Further phenomenology has been developed in many works [122–142].

19.6   Comments on Other Applications

Supersymmetric extension is one of interesting candidates for the physics beyond the standard model. Even if the theory is supersymmetric at high energy, super-symmetry must break above the weak scale. The supersymmetry breaking induces soft supersymmetry breaking terms such as gaugino masses, sfermion masses and scalar trilinear couplings, i.e. the so-called A-terms. Flavor symmetries control not only quark/lepton mass matrices but also squark/slepton masses and their A-terms. Suppose that flavor symmetries are exact. When three families have quantum numbers different from each other under flavor symmetries, squark/slepton mass-squared matrices are diagonal. Furthermore, when three (two) of three families correspond to doublets triplets (doublets) of flavor symmetries, their diagonal squark/slepton masses are degenerate. That is, the sfermion mass squared matrix $(m^2)_{ij}$ is written as

$$(m^2)_{ij} = \begin{pmatrix} m^2 & 0 & 0 \\ 0 & m^2 & 0 \\ 0 & 0 & m^2 \end{pmatrix}, \qquad (19.77)$$

when the three families correspond to a triplet. On the other hand, the sfermion mass squared matrix $(m^2)_{ij}$ becomes

$$(m^2)_{ij} = \begin{pmatrix} m^2 & 0 & 0 \\ 0 & m^2 & 0 \\ 0 & 0 & m'^2 \end{pmatrix}, \qquad (19.78)$$

when the first two families correspond to a doublet and the third family corresponds to a singlet. These patterns would become an interesting prediction of a certain class of flavor models, which could be tested if the supersymmetry breaking scale is reachable by collider experiments.

Flavor symmetries have similar effects on A-terms. These results are very impor-tant to suppress flavor changing neutral currents, which are constrained strongly by experiments. However, the flavor symmetry must break to lead to realistic quark/lepton mass matrices. Such breaking effects deform the above predictions. How much results are changed depends on breaking patterns. If masses of super-partners are of $\mathcal{O}(100)$ GeV, some models may be ruled out e.g. by experiments on flavor changing neutral currents. See e.g. Refs. [22,23,143–146].

Soft supersymmetry breaking terms in modular flavor symmetric models have been also studied in [147,148]. Specific patterns of soft terms have been predicted.

The application to dark matter (DM) models is also one of interesting topics. It is known that DM should have a global symmetry to be stabilized or long-lived, even after the electroweak breaking. Such kind of symmetry could be a natural origin in the non-Abelian discrete flavor symmetries [149–152].

Indirect detections of DM are reported by several experiments such as PAMELA [153,154], Fermi-LAT [155], AMS-02 [156], DAMPE [157], and CALET [158,159], in which excess of positron and the total flux of electron and positron is observed in the cosmic ray. These observations can be explained by scattering/decay of TeV-scale DM particles. Since PAMELA measured negative results for anti-proton excess [160,161], leptophilic DM is preferable.

Even so, if the main final state of scattering or decay of DM is $\tau^{\pm}$, this annihilation/decay mode is disfavored because it will overproduce gamma-rays as final state radiation [162]. This may indicate that if the cosmic-ray anomalies are induced by DM scattering or decay, these processes also reflect flavor structure of the model.

In case of decaying DM models, no excess of anti-proton in the cosmic ray implies that lifetime of the DM particle should be of $\mathcal{O}$ (10^{26}) sec. This long lifetime is achieved if the TeV-scale DM; (e.g., gauge singlet fermion X) decays into leptons by dimension six operators $\bar{L}E\bar{L}X/\Lambda^2$ suppressed by GUT scale $\Lambda \sim 10^{16}$ GeV. In this case, the lifetime of the DM is estimated as $\Gamma^{-1} \sim ((\text{TeV})^5/\Lambda^4)^{-1} \sim 10^{26}$ sec. However it could be in general difficult to induce only such an operator and forbid the other undesirable operators; e.g., $\bar{L}EX$ or other four dimensional interacting terms of X. However the non-Abelian discrete symmetries can play an important role in well-selecting operator [163,164]. (Moreover it was proved in Ref. [163] that it was impossible to adapt $U(1)$ symmetries.) Bosonic DM case is also considered in a similar way [165–167].

Recently, there is a new application to stabilize DM via modular non-Abelian discrete symmetries [168,208]. Thanks to the modular weight, this quantum number can play a role in assuring the stability of DM candidate if one appropriately assigns DM to charge of the modular weight. In addition, since DM annihilates into specific flavor final states due to a non-Abelian discrete symmetry, DM would be more testable through some experiments such as Large Hadron Collider, International Linear Collider, as well as indirect detections as discussed before. It is also remarked that the SM Effective Field Theory (SMEFT) is discussed based on the modular symmetry [209,210].

19.7 Comment on Origins of Flavor Symmetries

What is the origin of non-Abelian flavor symmetries? Some of them are symmetries of geometrical solids. Thus, its origin may be geometrical symmetries of extra dimensions. For example, it is found that the two-dimensional orbifold T^2/Z_2 with proper values of moduli has discrete symmetries such as A_4 and S_4 [169,170] (See also [171]).

Superstring theory is a promising candidate for unified theory including gravity, and predicts extra six dimensions. Superstring theory on a certain type of six-dimensional compact space realizes a discrete flavor symmetry. Such a string theory leads to stringy selection rules for allowed couplings among matter fields in four-dimensional effective field theory. Such stringy selection rules and geometrical symmetries as well as broken (continuous) gauge symmetries result in discrete flavor symmetries in superstring theory. For example, discrete flavor symmetries in heterotic orbifold models are studied in Refs. [144,172,173], and D_4 and $\Delta(54)$ are realized. Magnetized/intersecting D-brane models also realize the same flavor symmetries and other types such as $\Delta(27)$ [174–176]. Different types of non-Abelian flavor symmetries may be derived in other string models. Thus, such a study is quite important.

In addition, modular symmetries of certain compact spaces such as tori and orbifolds can be origins of flavor symmetries, because quotients of modular symmetry Γ_N includes S_3, A_4, S_4, A_5, $\Delta(96)$, and $\Delta(384)$, and also their covering groups can be realized as mentioned in Chap. 16. Indeed, modular flavor symmetries were studied in heterotic orbifold models [177–180] and magnetized D-brane models [101,181–189]. Furthermore, extensions of modular symmetries, e.g. unification with CP, R-symmetry and other geometrical symmetries, and symplectic modular symmetries of multi-moduli, were studied [190–200]. See also for derivation of modular flavor symmetries from higher dimensional theory [211].

At any rate, the experimental data of quark/lepton masses and mixing angles have no symmetry. Thus, flavor symmetries must be broken to reproduce the experimentally observed masses and mixing angles. The breaking direction is important, because the forms of mass matrices are determined by along which direction the flavor symmetries break. We need a proper breaking direction to derive realistic values of quark/lepton masses and mixing angles. Chapters 14 and 15 are useful to build models.

One way to fix the breaking direction is to analyze the potential minima of scalar fields with non-trivial representations of flavor symmetries. The number of the potential minima may be finite and in one of them the realistic breaking would happen. That is rather the conventional approach.

Another scenario to fix the breaking direction could be realized in theories with extra dimensions. One can impose the boundary conditions of matter fermions [64] and/or flavon scalars [201–203] in bulk such that zero modes for some components of irreducible multiplets are projected out, that is, the symmetry breaking. If a proper component of a flavon multiplet remains, that can realize a realistic breaking direction.

In modular flavor symmetric models, the symmetry breaking is caused by fixing the modulus, i.e. the moduli stabilization. Indeed, the moduli stabilization is one of important issues in string compactification. The moduli stabilization was studied from the viewpoint of modular flavor symmetric models in [204–206], where spontaneous CP breaking was also studied. (See for possibility of spontaneous CP breaking [197,207,212].) We discuss the CP violation in the flavor symmetry in Chap. 20.

References

1. Esteban, I., Gonzalez-Garcia, M.C., Maltoni, M., Schwetz, T., Zhou, A.: JHEP **09**, 178 (2020). arXiv:2007.14792 [hep-ph]
2. Harrison, P.F., Perkins, D.H., Scott, W.G.: Phys. Lett. B **530**, 167 (2002). arXiv:hep-ph/0202074
3. Harrison, P.F., Scott, W.G.: Phys. Lett. B **535**, 163 (2002). arXiv:hep-ph/0203209
4. Harrison, P.F., Scott, W.G.: Phys. Lett. B **557**, 76 (2003). arXiv:hep-ph/0302025
5. Harrison, P.F., Scott, W.G.: arXiv:hep-ph/0402006

6. Altarelli, G., Feruglio, F.: Rev. Mod. Phys. **82**, 2701–2729 (2010). arXiv:1002.0211 [hep-ph]
7. Ishimori, H., Kobayashi, T., Ohki, H., Shimizu, Y., Okada, H., Tanimoto, M.: Prog. Theor. Phys. Suppl. **183**, 1–163 (2010). arXiv:1003.3552 [hep-th]
8. Zyla, P.A. et al.: [Particle Data Group], PTEP **2020**(8), 083C01 (2020)
9. Weinberg, S.: Phys. Rev. Lett. **43**, 1566 (1979)
10. Minkowski, P.: Phys. Lett. B **67**, 421 (1977)
11. Yanagida, T.: In: Proceedings of the Workshop on Unified Theory and Baryon Number in the Universe, KEK (1979)
12. Gell-Mann, M., Ramond, P., Slansky, R.: Supergravity. Stony Brook (1979)
13. Glashow, S.L.: In: Cargèse, Lévy, M., et al. (eds.) Quarks and Leptons, p. 707. Plenum, New York (1980)
14. Mohapatra, R.N., Senjanovic, G.: Phys. Rev. Lett. **44**, 912 (1980)
15. Ma, E., Rajasekaran, G.: Phys. Rev. D **64**, 113012 (2001). arXiv:hep-ph/0106291
16. Lin, Y.: Nucl. Phys. B **824**, 95 (2010). arXiv:0905.3534 [hep-ph]
17. Altarelli, G., Feruglio, F.: Nucl. Phys. B **720**, 64 (2005). arXiv:hep-ph/0504165
18. Altarelli, G., Feruglio, F.: Nucl. Phys. B **741**, 215 (2006). arXiv:hep-ph/0512103
19. Honda, M., Tanimoto, M.: Prog. Theor. Phys. **119**, 583 (2008). arXiv:0801.0181 [hep-ph]
20. Hayakawa, A., Ishimori, H., Shimizu, Y., Tanimoto, M.: Phys. Lett. B **680**, 334 (2009). arXiv:0904.3820 [hep-ph]
21. Brahmachari, B., Choubey, S., Mitra, M.: Phys. Rev. D **77**, 073008 (2008) [Erratum-ibid. D **77**, 119901 (2008)]. arXiv:0801.3554 [hep-ph]
22. Babu, K.S., Kobayashi, T., Kubo, J.: Phys. Rev. D **67**, 075018 (2003). arXiv:hep-ph/0212350
23. Ishimori, H., Kobayashi, T., Omura, Y., Tanimoto, M.: JHEP **0812**, 082 (2008). arXiv:0807.4625 [hep-ph]
24. Babu, K.S., Enkhbat, T., Gogoladze, I.: Phys. Lett. B **555**, 238 (2003). arXiv:hep-ph/0204246
25. Babu, K.S., Ma, E., Valle, J.W.F.: Phys. Lett. B **552**, 207 (2003). arXiv:hep-ph/0206292
26. Ma, E.: Mod. Phys. Lett. A **17**, 2361 (2002). arXiv:hep-ph/0211393
27. Hirsch, M., Romao, J.C., Skadhauge, S., Valle, J.W.F., Villanova del Moral, A.: Phys. Rev. D **69**, 093006 (2004) arXiv:hep-ph/0312265
28. Ma, E.: Phys. Rev. D **70**, 031901 (2004). arXiv:hep-ph/0404199
29. Chen, S.L., Frigerio, M., Ma, E.: Nucl. Phys. B **724**, 423 (2005). arXiv:hep-ph/0504181
30. Zee, A.: Phys. Lett. B **630**, 58 (2005). arXiv:hep-ph/0508278
31. Ma, E.: Mod. Phys. Lett. A **20**, 2601 (2005). arXiv:hep-ph/0508099
32. Ma, E.: Phys. Lett. B **632**, 352 (2006). arXiv:hep-ph/0508231
33. Ma, E.: Phys. Rev. D **73**, 057304 (2006). arXiv:hep-ph/0511133
34. He, X.G., Keum, Y.Y., Volkas, R.R.: JHEP **0604**, 039 (2006). arXiv:hep-ph/0601001
35. Adhikary, B., Brahmachari, B., Ghosal, A., Ma, E., Parida, M.K.: Phys. Lett. B **638**, 345 (2006). arXiv:hep-ph/0603059
36. Ma, E., Sawanaka, H., Tanimoto, M.: Phys. Lett. B **641**, 301 (2006) arXiv:hep-ph/0606103
37. Valle, J.W.F.: J. Phys. Conf. Ser. **53**, 473 (2006). arXiv:hep-ph/0608101
38. Adhikary, B., Ghosal, A: Phys. Rev. D **75**, 073020 (2007). arXiv:hep-ph/0609193
39. Lavoura, L., Kuhbock, H.: Mod. Phys. Lett. A **22**, 181 (2007). arXiv:hep-ph/0610050
40. King S.F., Malinsky, M.: Phys. Lett. B **645**, 351 (2007). arXiv:hep-ph/0610250
41. Hirsch, M., Joshipura, A.S., Kaneko, S., Valle, J.W.F.: Phys. Rev. Lett. **99**, 151802 (2007). arXiv:hep-ph/0703046
42. Bazzocchi, F., Kaneko, S., Morisi, S.: JHEP **0803**, 063 (2008). arXiv:0707.3032 [hep-ph]
43. Grimus, W., Kuhbock, H.: Phys. Rev. D **77**, 055008 (2008). arXiv:0710.1585 [hep-ph]
44. Adhikary, B., Ghosal, A.: Phys. Rev. D **78**, 073007 (2008). arXiv:0803.3582 [hep-ph]
45. Fukuyama, T.: arXiv:0804.2107 [hep-ph]
46. Lin, Y.: Nucl. Phys. B **813**, 91 (2009). arXiv:0804.2867 [hep-ph]
47. Frampton, P.H., Matsuzaki, S.: arXiv:0806.4592 [hep-ph]
48. Feruglio, F., Hagedorn, C., Lin, Y., Merlo, L.: Nucl. Phys. B **809**, 218 (2009). arXiv:0807.3160 [hep-ph]
49. Morisi, S.: Nuovo Cim. **123B**, 886 (2008). arXiv:0807.4013 [hep-ph]

50. Ma, E.: Phys. Lett. B **671**, 366 (2009). arXiv:0808.1729 [hep-ph]
51. Bazzocchi, F., Frigerio, M., Morisi, S.: Phys. Rev. D **78**, 116018 (2008). arXiv:0809.3573 [hep-ph]
52. Hirsch, M., Morisi, S., Valle, J.W.F.: Phys. Rev. D **79**, 016001 (2009). arXiv:0810.0121 [hep-ph]
53. Merlo, L.: arXiv:0811.3512 [hep-ph]
54. Baek, S., Oh, M.C.: arXiv:0812.2704 [hep-ph]
55. Morisi, S.: Phys. Rev. D **79**, 033008 (2009). arXiv:0901.1080 [hep-ph]
56. Ciafaloni, P., Picariello, M., Torrente-Lujan, E., Urbano, A.: Phys. Rev. D **79**, 116010 (2009). arXiv:0901.2236 [hep-ph]
57. Merlo, L.: J. Phys. Conf. Ser. **171**, 012083 (2009). arXiv:0902.3067 [hep-ph]
58. Chen, M.C., King, S.F.: JHEP **0906**, 072 (2009). arXiv:0903.0125 [hep-ph]
59. Branco, G.C., Gonzalez Felipe, R., Rebelo, M.N., Serodio, H.: Phys. Rev. D **79**, 093008 (2009). arXiv:0904.3076 [hep-ph]
60. Altarelli, G., Meloni, D.: J. Phys. G **36**, 085005 (2009). arXiv:0905.0620 [hep-ph]
61. Urbano, A.: arXiv:0905.0863 [hep-ph]
62. Hirsch, M., Morisi, S., Valle, J.W.F.: Phys. Lett. B **679**, 454 (2009). arXiv:0905.3056 [hep-ph]
63. Grimus, W., Lavoura, L.: JHEP **0809**, 106 (2008). arXiv:0809.0226 [hep-ph]
64. Haba, N., Watanabe, A., Yoshioka, K.: Phys. Rev. Lett. **97**, 041601 (2006). arXiv:hep-ph/0603116
65. Ishimori, H., Shimizu, Y., Tanimoto, M., Watanabe, A.: Phys. Rev. D **83**, 033004 (2011). arXiv:1010.3805 [hep-ph]
66. Ishimori, H., Shimizu, Y., Tanimoto, M.: Prog. Theor. Phys. **121**, 769 (2009). arXiv:0812.5031 [hep-ph]
67. Froggatt, C.D., Nielsen, H.B.: Nucl. Phys. B **147**, 277 (1979)
68. Fusaoka, H., Koide, Y.: Phys. Rev. D **57**, 3986 (1998). arXiv:hep-ph/9712201
69. Jarlskog, C.: Phys. Rev. Lett. **55**, 1039 (1985)
70. Kajiyama, Y., Raidal, M., Strumia, A.: Phys. Rev. D **76**, 117301 (2007). arXiv:0705.4559 [hep-ph]
71. Barger, V.D., Pakvasa, S., Weiler, T.J., Whisnant, K.: Phys. Lett. B **437**, 107 (1998). arXiv:hep-ph/9806387
72. Everett, L.L., Stuart, A.J.: Phys. Rev. D **79**, 085005 (2009). arXiv:0812.1057 [hep-ph]
73. Adulpravitchai, A., Blum, A., Rodejohann, W.: New J. Phys. **11**, 063026 (2009). arXiv:0903.0531 [hep-ph]
74. Altarelli, G., Feruglio, F., Merlo, L.: JHEP **0905**, 020 (2009). arXiv:0903.1940 [hep-ph]
75. Minakata, H., Smirnov, A.Y.: Phys. Rev. D **70**, 073009 (2004). arXiv:hep-ph/0405088
76. Merlo, L.: AIP Conf. Proc. **1200**(1), 948–951 (2010). arXiv:0909.2760 [hep-ph]
77. de Adelhart Toorop, R., Feruglio, F., Hagedorn, C.: Nucl. Phys. B **858**, 437 (2012) arXiv:1112.1340 [hep-ph]
78. Feruglio, F.: In: Levy, A., Stefano Forte, S., Ridolfi, G. (eds.) From My Vast Repertoire ...: Guido Altarelli's Legacy, pp. 227–266 (2019). arXiv:1706.08749 [hep-ph]
79. Kobayashi, T., Tanaka, K., Tatsuishi, T.H.: Phys. Rev. D **98**(1), 016004 (2018). arXiv:1803.10391 [hep-ph]
80. Penedo, J.T., Petcov, S.T.: Nucl. Phys. B **939**, 292 (2019). arXiv:1806.11040 [hep-ph]
81. Novichkov, P.P., Penedo, J.T., Petcov, S.T., Titov, A.V.: JHEP **1904**, 174 (2019). arXiv:1812.02158 [hep-ph]
82. Kobayashi, T., Omoto, N., Shimizu, Y., Takagi, K., Tanimoto, M., Tatsuishi, T.H.: JHEP **11**, 196 (2018). arXiv:1808.03012 [hep-ph]
83. Giusarma, E., Gerbino, M., Mena, O., Vagnozzi, S., Ho, S., Freese, K.: Phys. Rev. D **94**(8), 083522 (2016). arXiv:1605.04320 [astro-ph.CO]
84. Vagnozzi, S., Giusarma, E., Mena, O., Freese, K., Gerbino, M., Ho, S., Lattanzi, M.: Phys. Rev. D **96**(12), 123503 (2017). arXiv:1701.08172 [astro-ph.CO]
85. Aghanim, N., et al.: [Planck Collaboration]. arXiv:1807.06209 [astro-ph.CO]
86. Antusch, S., Maurer, V.: JHEP **1311**, 115 (2013). arXiv:1306.6879 [hep-ph]

87. Björkeroth, F., de Anda, F.J., de Medeiros Varzielas, I., King, S.F.: JHEP **1506**, 141 (2015). arXiv:1503.03306 [hep-ph]
88. Criado, J.C., Feruglio, F.: SciPost Phys. **5**(5), 042 (2018). arXiv:1807.01125 [hep-ph]
89. Ding, G.J., King, S.F., Liu, X.G.: JHEP **1909**, 074 (2019). arXiv:1907.11714 [hep-ph]
90. Zhang, D.: Nucl. Phys. B **952**, 114935 (2020). arXiv:1910.07869 [hep-ph]
91. Novichkov, P.P., Penedo, J.T., Petcov, S.T., Titov, A.V.: JHEP **1904**, 005 (2019). arXiv:1811.04933 [hep-ph]
92. Kobayashi, T., Shimizu, Y., Takagi, K., Tanimoto, M., Tatsuishi, T.H.: JHEP **02**, 097 (2020). arXiv:1907.09141 [hep-ph]
93. Wang, X., Zhou, S.: JHEP **05**, 017 (2020). arXiv:1910.09473 [hep-ph]
94. Ding, G.J., King, S.F., Liu, G.: Phys. Rev. D **100**(11), 115005 (2019). arXiv:1903.12588 [hep-ph]
95. Liu, X.G., Ding, G.J.: JHEP **1908**, 134 (2019). arXiv:1907.01488 [hep-ph]
96. Chen, P., Ding, G.J., Lu, J.N., Valle, J.W.F.: Phys. Rev. D **102**(9), 095014 (2020). arXiv:2003.02734 [hep-ph]
97. Novichkov, P.P., Penedo, J.T., Petcov, S.T.: Nucl. Phys. B **963**, 115301 (2021). arXiv:2006.03058 [hep-ph]
98. Liu, X.G., Yao, C.Y., Ding, G.J.: Phys. Rev. D **103**(5), 056013 (2021). arXiv:2006.10722 [hep-ph]
99. Wang, X., Yu, B., Zhou, S.: Phys. Rev. D **103**(7), 076005 (2021). arXiv:2010.10159 [hep-ph]
100. Yao, C.Y., Liu, X.G., Ding, G.J.: arXiv:2011.03501 [hep-ph]
101. Kobayashi, T., Tamba, S.: Phys. Rev. D **99**(4), 046001 (2019). arXiv:1811.11384 [hep-th]
102. Ding, G.J., King, S.F., Li, C.C., Zhou, Y.L.: JHEP **08**, 164 (2020). arXiv:2004.12662 [hep-ph]
103. Okada, H., Tanimoto, M.: Phys. Lett. B **791**, 54 (2019). arXiv:1812.09677 [hep-ph]
104. Okada, H., Tanimoto, M.: Eur. Phys. J. C **81**(1), 52 (2021). arXiv:1905.13421 [hep-ph]
105. de Anda, F.J., King, S.F., Perdomo, E.: Phys. Rev. D **101**(1), 015028 (2020). arXiv:1812.05620 [hep-ph]
106. Kobayashi, T., Shimizu, Y., Takagi, K., Tanimoto, M., Tatsuishi, T.H.: arXiv:1906.10341 [hep-ph]
107. King, S.F., Zhou, Y.L.: JHEP **04**, 291 (2021). arXiv:2103.02633 [hep-ph]
108. Chen, P., Ding, G.J., King, S.F.: JHEP **04**, 239 (2021). arXiv:2101.12724 [hep-ph]
109. Ding, G.J., King, S.F., Yao, C.Y.: arXiv:2103.16311 [hep-ph]
110. Asaka, T., Heo, Y., Tatsuishi, T.H., Yoshida, T.: JHEP **2001**, 144 (2020). arXiv:1909.06520 [hep-ph]
111. Okada, H., Shimizu, Y., Tanimoto, M., Yoshida, T.: JHEP **07**, 184 (2021). arXiv:2105.14292 [hep-ph]
112. Qu, B.Y., Liu, X.G., Chen, P.T., Ding, G.J.: arXiv:2106.11659 [hep-ph]
113. Novichkov, P.P., Petcov, S.T., Tanimoto, M.: Phys. Lett. B **793**, 247 (2019). arXiv:1812.11289 [hep-ph]
114. Ding, G.J., King, S.F., Liu, X.G., Lu, J.N.: JHEP **1912**, 030 (2019). arXiv:1910.03460 [hep-ph]
115. Okada, H., Tanimoto, M.: Phys. Rev. D **103**(1), 015005 (2021). arXiv:2009.14242 [hep-ph]
116. Feruglio, F., Gherardi, V., Romanino, A., Titov, A.: arXiv:2101.08718 [hep-ph]
117. Novichkov, P.P., Penedo, J.T., Petcov, S.T.: JHEP **04**, 206 (2021). arXiv:2102.07488 [hep-ph]
118. Kobayashi, T., Shimizu, Y., Takagi, K., Tanimoto, M., Tatsuishi, T.H., Uchida, H.: Phys. Lett. B **794**, 114 (2019). arXiv:1812.11072 [hep-ph]
119. de Medeiros Varzielas, I., King, S.F., Zhou, Y.L.: Phys. Rev. D **1015**, 055033 (2020). arXiv:1906.02208 [hep-ph]
120. King, S.J.D., King, S.F.: JHEP **09**, 043 (2020). arXiv:2002.00969 [hep-ph]
121. Kuranaga, H., Ohki, H., Uemura, S.: arXiv:2105.06237 [hep-ph]
122. Criado, J.C., Feruglio, F., King, S.J.D.: JHEP **2002**, 001 (2020). arXiv:1908.11867 [hep-ph]
123. de Medeiros Varzielas, I., Levy, M., Zhou, Y.L.: JHEP **11**, 085 (2020). arXiv:2008.05329 [hep-ph]
124. King, S.F., Zhou, Y.L.: Phys. Rev. D **101**(1), 015001 (2020). arXiv:1908.02770 [hep-ph]

125. Nomura, T., Okada, H., Patra, S.: Nucl. Phys. B **967**, 115395 (2021). arXiv:1912.00379 [hep-ph]
126. Kobayashi, T., Nomura, T., Shimomura, T.: Phys. Rev. D **102**(3), 035019 (2020). arXiv:1912.00637 [hep-ph]
127. Lu, J.N., Liu, X.G., Ding, G.J.: Phys. Rev. D **101**(11), 115020 (2020). arXiv:1912.07573 [hep-ph]
128. Wang, X.: Nucl. Phys. B **957**, 115105 (2020). arXiv:1912.13284 [hep-ph]
129. Abbas, M.: Phys. Rev. D **103**(5), 056016 (2021). arXiv:2002.01929 [hep-ph]
130. Okada, H., Shoji, Y.: Phys. Dark Univ. **31**, 100742 (2021). arXiv:2003.11396 [hep-ph]
131. Okada, H., Shoji, Y.: Nucl. Phys. B **961**, 115216 (2020). arXiv:2003.13219 [hep-ph]
132. Ding, G.J., Feruglio, F.: JHEP **06**, 134 (2020). arXiv:2003.13448 [hep-ph]
133. Nomura, T., Okada, H.: arXiv:2007.04801 [hep-ph]
134. Nomura, T., Okada, H.: arXiv:2007.15459 [hep-ph]
135. Asaka, T., Heo, Y., Yoshida, T.: Phys. Lett. B **811**, 135956 (2020). arXiv:2009.12120 [hep-ph]
136. Nomura, T., Okada, H.: Nucl. Phys. B **966**, 115372 (2021). arXiv:1906.03927 [hep-ph]
137. Okada, H., Orikasa, Y.: arXiv:1908.08409 [hep-ph]
138. Nomura, T., Okada, H., Popov, O.: Phys. Lett. B **803**, 135294 (2020). arXiv:1908.07457 [hep-ph]
139. Okada, H., Tanimoto, M.: arXiv:2005.00775 [hep-ph]
140. Nagao, K.I., Okada, H.: JCAP **05**, 063 (2021). arXiv:2008.13686 [hep-ph]
141. Nagao, K.I., Okada, H.: arXiv:2010.03348 [hep-ph]
142. Abbas, M.: Phys. Atom. Nucl. **83**(5), 764–769 (2020)
143. Kobayashi, T., Kubo, J., Terao, H.: Phys. Lett. B **568**, 83 (2003). arXiv:hep-ph/0303084
144. Ko, P., Kobayashi, T., Park, J.h., Raby, S.: Phys. Rev. D **76**, 035005 (2007) [Erratum-ibid. D **76**, 059901 (2007)]. arXiv:0704.2807 [hep-ph]
145. Ishimori, H., Kobayashi, T., Ohki, H., Omura, Y., Takahashi, R., Tanimoto, M.: Phys. Rev. D **77**, 115005 (2008). arXiv:0803.0796 [hep-ph]
146. Ishimori, H., Kobayashi, T., Okada, H., Shimizu, Y., Tanimoto, M.: JHEP **0912**, 054 (2009). arXiv:0907.2006 [hep-ph]
147. Kobayashi, T., Shimomura, T., Tanimoto, M.: Phys. Lett. B **819**, 136452 (2021). arXiv:2102.10425 [hep-ph]
148. Tanimoto, M., Yamamoto, K.: JHEP **10**, 183 (2021). arXiv:2106.10919 [hep-ph]
149. Hirsch, M., Morisi, S., Peinado, E., Valle, J.W.F.: Phys. Rev. D **82**, 116003 (2010). arXiv:1007.0871 [hep-ph]
150. Meloni, D., Morisi, S., Peinado, E.: Phys. Lett. **B697**, 339–342 (2011). arXiv:1011.1371 [hep-ph]
151. Boucenna, M.S., Hirsch, M., Morisi, S., Peinado, E., Taoso, M., Valle, J.W.F.: JHEP **1105**, 037 (2011). arXiv:1101.2874 [hep-ph]
152. Meloni, D., Morisi, S., Peinado, E.: Phys. Lett. B **703**, 281–287 (2011). arXiv:1104.0178 [hep-ph]
153. Adriani, O., et al.: [PAMELA], Nature **458**, 607–609 (2009). arXiv:0810.4995 [astro-ph]
154. Adriani, O., et al.: [PAMELA], Phys. Rev. Lett. **111**, 081102 (2013). arXiv:1308.0133 [astro-ph.HE]
155. Abdo, A.A., et al.: [Fermi-LAT], Phys. Rev. Lett. **102**, 181101 (2009). arXiv:0905.0025 [astro-ph.HE]
156. Accardo, L., et al.: AMS. Phys. Rev. Lett. **113**, 121101 (2014)
157. Ambrosi, G., et al.: [DAMPE]. Nature **552**, 63–66 (2017). arXiv:1711.10981 [astro-ph.HE]
158. Adriani, O., et al.: [CALET]. Phys. Rev. Lett. **119**(18), 181101 (2017). arXiv:1712.01711 [astro-ph.HE]
159. Adriani, O., Akaike, Y., Asano, K., Asaoka, Y., Bagliesi, M.G., Berti, E., Bigongiari, G., Binns, W.R., Bonechi, S., Bongi, M., et al.: Phys. Rev. Lett. **120**(26), 261102 (2018). arXiv:1806.09728 [astro-ph.HE]

160. Adriani, O., Barbarino, G.C., Bazilevskaya, G.A., Bellotti, R., Boezio, M., Bogomolov, E.A., Bonechi, L., Bongi, M., Bonvicini, V., Bottai, S., et al.: Phys. Rev. Lett. **102**, 051101 (2009). arXiv:0810.4994 [astro-ph]
161. Aguilar, M., et al.: [AMS]. Phys. Rev. Lett. **117**(9), 091103 (2016)
162. Papucci, M., Strumia, A.: JCAP **1003**, 014 (2010). arXiv:0912.0742 [hep-ph]
163. Haba, N., Kajiyama, Y., Matsumoto, S., Okada, H., Yoshioka, K.: Phys. Lett. B **695**, 476–481 (2011). arXiv:1008.4777 [hep-ph]
164. Kajiyama, Y., Okada, H.: Nucl. Phys. **B848**, 303–313 (2011). arXiv:1011.5753 [hep-ph]
165. Daikoku, Y., Okada, H., Toma, T.: arXiv:1106.4717 [hep-ph]
166. Kajiyama, Y., Okada, H., Toma, T.: arXiv:1109.2722 [hep-ph]
167. Kajiyama, Y., Okada, H., Toma, T.: Eur. Phys. J. C **71**, 1688 (2011). arXiv:1104.0367 [hep-ph]
168. Nomura, T., Okada, H.: Phys. Lett. B **797**, 134799 (2019). arXiv:1904.03937 [hep-ph]
169. Altarelli, G., Feruglio, F., Lin, Y.: Nucl. Phys. B **775**, 31 (2007). arXiv:hep-ph/0610165
170. Adulpravitchai, A., Blum, A., Lindner, M.: JHEP **0907**, 053 (2009). arXiv:0906.0468 [hep-ph]
171. Abe, H., Choi, K.-S., Kobayashi, T., Ohki, H., Sakai, M.: Int. J. Mod. Phys. **A26**, 4067–4082 (2011). arXiv:1009.5284 [hep-ph]
172. Kobayashi, T., Raby, S., Zhang, R.J.: Nucl. Phys. B **704**, 3 (2005). arXiv:hep-ph/0409098
173. Kobayashi, T., Nilles, H.P., Ploger, F., Raby, S., Ratz, M.: Nucl. Phys. B **768**, 135 (2007). arXiv:hep-ph/0611020
174. Abe, H., Choi, K.S., Kobayashi, T., Ohki, H.: Nucl. Phys. B **820**, 317 (2009). arXiv:0904.2631 [hep-ph]
175. Abe, H., Choi, K.S., Kobayashi, T., Ohki, H.: Phys. Rev. D **80**, 126006 (2009). arXiv:0907.5274 [hep-th]
176. Abe, H., Choi, K.S., Kobayashi, T., Ohki, H.: arXiv:1001.1788 [hep-th]
177. Lauer, J., Mas, J., Nilles, H.P.: Phys. Lett. B **226**, 251 (1989)
178. Lauer, J., Mas, J., Nilles, H.P.: Nucl. Phys. B **351**, 353 (1991)
179. Lerche, W., Lust, D., Warner, N.P.: Phys. Lett. B **231**, 417 (1989)
180. Ferrara, S., Lust, D., Theisen, S.: Phys. Lett. B **233**, 147 (1989)
181. Kikuchi, S., Kobayashi, T., Takada, S., Tatsuishi, T.H., Uchida, H.: Phys. Rev. D **102**(10), 105010 (2020). arXiv:2005.12642 [hep-th]
182. Kobayashi, T., Nagamoto, S.: Phys. Rev. D **96**(9), 096011 (2017). arXiv:1709.09784 [hep-th]
183. Kobayashi, T., Nagamoto, S., Takada, S., Tamba, S., Tatsuishi, T.H.: Phys. Rev. D **97**(11), 116002 (2018). arXiv:1804.06644 [hep-th]
184. Kariyazono, Y., Kobayashi, T., Takada, S., Tamba, S., Uchida, H.: Phys. Rev. D **100**(4), 045014 (2019). arXiv:1904.07546 [hep-th]
185. Ohki, H., Uemura, S., Watanabe, R.: Phys. Rev. D **102**(8), 085008 (2020). arXiv:2003.04174 [hep-th]
186. Kikuchi, S., Kobayashi, T., Otsuka, H., Takada, S., Uchida, H.: JHEP **11**, 101 (2020). arXiv:2007.06188 [hep-th]
187. Kikuchi, S., Kobayashi, T., Uchida, H.: arXiv:2101.00826 [hep-th]
188. Almumin, Y., Chen, M.C., Knapp-Pérez, V., Ramos-Sánchez, S., Ratz, M., Shukla, S.: JHEP **05**, 078 (2021). arXiv:2102.11286 [hep-th]
189. Tatsuta, Y.: arXiv:2104.03855 [hep-th]
190. Baur, A., Nilles, H.P., Trautner, A., Vaudrevange, P.K.S.: Phys. Lett. B **795**, 7 (2019). arXiv:1901.03251 [hep-th]
191. Baur, A., Nilles, H.P., Trautner, A., Vaudrevange, P.K.S.: arXiv:1908.00805 [hep-th]
192. Nilles, H.P., Ramos-Śanchez, S., Vaudrevange, P.K.S.: JHEP **02**, 045 (2020). arXiv:2001.01736 [hep-th]
193. Nilles, H.P., Ramos-Sánchez, S., Vaudrevange, P.K.S.: Nucl. Phys. B **957**, 115098 (2020). arXiv:2004.05200 [hep-ph]
194. Nilles, H.P., Ramos-Sánchez, S., Vaudrevange, P.K.S.: Phys. Lett. B **808**, 135615 (2020). arXiv:2006.03059 [hep-th]
195. Baur, A., Kade, M., Nilles, H.P., Ramos-Sanchez, S., Vaudrevange, P.K.S.: JHEP **02**, 018 (2021). arXiv:2008.07534 [hep-th]

196. Nilles, H.P., Ramos-Sánchez, S., Vaudrevange, P.K.S.: Nucl. Phys. B **966**, 115367 (2021). arXiv:2010.13798 [hep-th]
197. Ishiguro, K., Kobayashi, T., Otsuka, H.: arXiv:2010.10782 [hep-th]
198. Baur, A., Kade, M., Nilles, H.P., Ramos-Sanchez, S., Vaudrevange, P.K.S.: Phys. Lett. B **816**, 136176 (2021). arXiv:2012.09586 [hep-th]
199. Ding, G.J., Feruglio, F., Liu, X.G.: JHEP **01**, 037 (2021). arXiv:2010.07952 [hep-th]
200. Ishiguro, K., Kobayashi, T., Otsuka, H.: arXiv:2107.00487 [hep-th]
201. Kobayashi, T., Omura, Y., Yoshioka, K.: Phys. Rev. D **78**, 115006 (2008). arXiv:0809.3064
202. Seidl, G.: Phys. Rev. D **81**, 025004 (2010). arXiv:0811.3775 [hep-ph]
203. Adulpravitchai, A., Schmidt, M.A.: JHEP **01**, 106 (2011). arXiv:1001.3172 [hep-ph]
204. Kobayashi, T., Shimizu, Y., Takagi, K., Tanimoto, M., Tatsuishi, T.H.: Phys. Rev. D **100**(11), 115045 (2019) [erratum: Phys. Rev. D **101**(3), 039904 (2020)]. arXiv:1909.05139 [hep-ph]
205. Kobayashi, T., Shimizu, Y., Takagi, K., Tanimoto, M., Tatsuishi, T.H., Uchida, H.: Phys. Rev. D **101**(5), 055046 (2020). arXiv:1910.11553 [hep-ph]
206. Ishiguro, K., Kobayashi, T., Otsuka, H.: JHEP **03**, 161 (2021). arXiv:2011.09154 [hep-ph]
207. Kobayashi, T., Otsuka, H.: Phys. Rev. D **102**(2), 026004 (2020). arXiv:2004.04518 [hep-th]
208. Kobayashi, T., Okada, H., Orikasa, Y.: arXiv:2111.05674 [hep-ph]
209. Kobayashi, T., Otsuka, H.: arXiv:2108.02700 [hep-ph]
210. Kobayashi, T., Otsuka, H., Tanimoto, M., Yamamoto, K.: arXiv:2112.00493 [hep-ph]
211. Kikuchi, S., Kobayashi, T., Otsuka, H., Tanimoto, M., Uchida, H., Yamamoto, K.: arXiv:2201.04505 [hep-ph]
212. Novichkov, P.P., Penedo, J.T., Petcov, S.T.: arXiv:2201.02020 [hep-ph]

Generalized CP Symmetry

20

20.1 CP Transformation in Flavor Space

The CP symmetry is vital concepts in particle physics. Since the CP symmetry is the discrete one, there is a possibility of predicting the CP phase from the flavor symmetry. One can find a hint of predicting the CP phase in the $\mu - \tau$ symmetry [1]. Suppose that the elements of the PMNS matrix U satisfy $U_{\mu i} = U_{\tau i}(i = 1, 2, 3)$, which is so called $\mu - \tau$ symmetry. This condition gives us $\theta_{23} = \pi/4$ and $\sin \theta_{13} \cos \delta_{CP} = 0$ simply. Taking account of the observation of non-vanishing $\sin \theta_{13}$, $\cos \delta_{CP} = 0$ must be realized. That is $\delta_{CP} = \pm\pi/2$, which is the maximal CP violating phase. This simple example provides us a new aspect of the CP violation [2–4], where the CP is conserved in the high energy theory before the flavor symmetry is broken [5].

The progressive approach is the generalized CP symmetry, which has been discussed in the flavor space [6–10]. It is emphasized that the generalized CP symmetry can predict the CP violating phase in the framework of the flavor symmetry [11]. The modular symmetry has been also studied combining with the CP symmetries for theories of flavors [12,13]. The generalized CP symmetry provides a powerful framework to predict CP violating phases of quarks and leptons in the modular invariant flavor model.

Since the gauge interactions in a weak base do not distinguish the difference between generations of fermions, the CP transformation is non-trivial if the non-Abelian discrete flavor symmetry G is set in a Lagrangian [11]. Let us consider the Lagrangian given by chiral superfields. The CP is a discrete symmetry which involves both Hermitian conjugation of a chiral superfield $\psi(x)$ and inversion of spatial coordinates,

$$\psi(x) \rightarrow \mathbf{X_r}\overline{\psi}(x_P) \,, \tag{20.1}$$

where $x_P = (t, -\mathbf{x})$ and $\mathbf{X_r}$ is a unitary transformation of $\psi(x)$ in the irreducible representation $\mathbf{r}$ of the discrete flavor symmetry G. If $\mathbf{X_r}$ is the unit matrix, the CP

T. Kobayashi et al., *An Introduction to Non-Abelian Discrete Symmetries for Particle Physicists*, Lecture Notes in Physics 995, https://doi.org/10.1007/978-3-662-64679-3_20

transformation is the trivial one. This is the case for the continuous flavor symmetry [5]. However, in the framework of the non-Abelian discrete family symmetry, non-trivial choices of $\mathbf{X_r}$ are possible. The unbroken CP transformation $\mathbf{X_r}$s form the group H_{CP}. Then, $\mathbf{X_r}$ must be consistent with the flavor symmetry transformation,

$$\psi(x) \rightarrow \rho_\mathbf{r}(g)\psi(x) , \quad g \in G , \qquad (20.2)$$

where $\rho_\mathbf{r}(g)$ is the representation matrix for g in the irreducible representation $\mathbf{r}$.

Let us discuss the condition that $\mathbf{X_r}$ should satisfy. Perform a CP transformation $\psi(x) \rightarrow \mathbf{X_r}\overline{\psi}(x_P)$, then apply a flavor symmetry transformation, $\overline{\psi}(x_P) \rightarrow \rho_\mathbf{r}^*(g)\overline{\psi}(x_P)$, and finally an inverse CP transformation. The whole transformation is written as $\psi(x) \rightarrow \mathbf{X_r}\rho_\mathbf{r}^*(g)\mathbf{X_r}^{-1}\psi(x)$, which must be equivalent to some flavor symmetry $\psi(x) \rightarrow \rho_\mathbf{r}(g')\psi(x)$. Thus, one obtains [14]

$$\mathbf{X_r}\rho_\mathbf{r}^*(g)\mathbf{X_r}^{-1} = \rho_\mathbf{r}(g') , \qquad g,\, g' \in G . \qquad (20.3)$$

This equation defines the consistency condition, which has to be respected for consistent implementation of a generalized CP symmetry along with a flavor symmetry [15,16].

The issues of the consistent condition that warrant more attention are summarized in [12]:

- (20.3) has to be satisfied for all irreducible representation $\mathbf{r}$ simultaneously, i.e., the elements g and g' must be the same for all $\mathbf{r}$.
- For a given irreducible representation $\mathbf{r}$, the consistency condition defines $\mathbf{X_r}$ up to an overall phase and a G_f transformation.
- It follows from (20.3) that the elements g and g' must be of the same order.
- It is sufficient to impose (20.3) on the generators of a discrete group G_f.
- This chain $CP \rightarrow g \rightarrow CP^{-1}$ maps the group element g onto g' and preserves the flavor symmetry group structure. Therefore, it realizes a homomorphism $u(g) = g'$ of G. Assuming the presence of faithful representations $\mathbf{r}$, i.e., those for which ρ_r maps each element of G_f to a distinct matrix, (20.3) defines a unique mapping of G to itself. In this case, $u(g)$ is an automorphism of G as defined in (2.29).
- The automorphism $u(g) = g'$ must be class-inverting with respect to G_f, i.e. g' and g^{-1} belong to the same conjugacy class because of the group characters $\chi_\mathbf{r}$ fulfill $\chi_\mathbf{r}(u(g)) = \mathrm{tr}[\rho_\mathbf{r}(u(g))] = [\mathbf{X}_r\rho_\mathbf{r}(g)^*\mathbf{X}_r^{-1}] = \mathrm{tr}[\rho_\mathbf{r}(g)]^* = \chi_\mathbf{r}(g)^* = \chi_\mathbf{r}(g^{-1})$ for every irreducible representations [16]. The group of generalized CP transformations is given by the outer automorphism group as defined in (2.30), meaning no $h \in G_f$ exists such that $g' = h^{-1}gh$.

The group structure of automorphism for the relevant groups are summarized in Table 17.1 of Chap. 17.

It has been also shown that the full symmetry group is isomorphic to a semi-direct product of G and H_{CP}, that is $G \rtimes H_{CP}$, where $H_{CP} \simeq Z_2^{CP}$, is the group generated by the generalized CP transformation under the assumption of $\mathbf{X_r}$ being a symmetric matrix [15].

20.2 CP Violating Phase and Its Group Theoretical Origin

Suppose the full symmetry including the CP symmetry and the flavor symmetry are broken to the subgroups in the neutrino sector and the charged lepton sector, respectively. The CP symmetry gives us the relations as to the neutrino mass matrix $m_{\nu LL}$ and the charged lepton mass matrix m_ℓ as follows:

$$\mathbf{X}_{\mathbf{r_i}}^{\nu T} m_{\nu LL}\mathbf{X}_{\mathbf{r_i}}^{\nu} = m_{\nu LL} , \qquad \mathbf{X}_{\mathbf{r_i}}^{\ell\dagger} m_\ell^\dagger m_\ell \mathbf{X}_{\mathbf{r_i}}^{\ell} = m_\ell^\dagger m_\ell . \tag{20.4}$$

Once the subgroups of G and H_{CP} are chosen to satisfy the conditions of (20.3) and (20.4) for the neutrino sector and the charged lepton sector, respectively, one can predict the CP phase, δ_{CP}. Some examples have been given in these years [17–25].

We present an example of a neutrino mass matrix in the S_4 symmetry [20,21], which satisfies the consistency condition (20.3). The neutrino mass matrix is given in general as

$$m_{\nu LL} = \alpha \begin{pmatrix} 2 & -1 & -1 \\ -1 & 2 & -1 \\ -1 & -1 & 2 \end{pmatrix} + \beta \begin{pmatrix} 1 & 0 & 0 \\ 0 & 0 & 1 \\ 0 & 1 & 0 \end{pmatrix} + \gamma \begin{pmatrix} 0 & 1 & 1 \\ 1 & 1 & 0 \\ 1 & 0 & 1 \end{pmatrix} + \epsilon \begin{pmatrix} 0 & 1 & -1 \\ 1 & -1 & 0 \\ -1 & 0 & 1 \end{pmatrix} , \tag{20.5}$$

where α, β, γ and ϵ are arbitrary complex parameters. Imposing the CP symmetry for the neutrino mass matrix, one finds α, β and γ to be real, and ϵ to be pure imaginary. Then, it is diagonalized by the unitary matrix:

$$V_\nu = \begin{pmatrix} 2c/\sqrt{6} & 1/\sqrt{3} & 2s/\sqrt{6} \\ -c/\sqrt{6}+is/\sqrt{2} & 1/\sqrt{3} & -s/\sqrt{6}-ic/\sqrt{2} \\ -c/\sqrt{6}+is/\sqrt{2} & 1/\sqrt{3} & -s/\sqrt{6}+ic/\sqrt{2} \end{pmatrix} , \tag{20.6}$$

where c and s are denoted by $c = \cos\phi$ and $s = \sin\phi$, respectively, in terms of an arbitrary mixing angle ϕ. Since the charged lepton mass matrix is diagonal, we obtain the neutrino mixing matrix elements in (19.1) as follows:

$$\sin^2\theta_{13} = \frac{2}{3}\sin^2\phi, \qquad \sin^2\theta_{12} = \frac{1}{2+\cos 2\phi},$$

$$\sin^2\theta_{23} = \frac{1}{2}, \qquad |\sin\delta_{CP}| = 1, \tag{20.7}$$

which correspond to the maximal CP violation, $\delta = \pm\pi/2$.

The prediction of the CP phase depends on the respected "Generators" of the flavor symmetry and the CP symmetry. Typically, it is simple values 0, $\pm\pi/2$ or π [20–23,26] although other predictions are possible [24,25].

It is useful to summarize the comprehensive work by Chen et al. [16] in the relation between the discrete symmetries and the physical CP invariance guaranteed by generalized CP transformations. They have studied the CP violation by the automorphisms of G carefully.

The origin of the CP violation with a discrete flavor symmetry is categorized into three types: (i) Groups explicitly violate CP, which can be related to the complexity of some Clebsch-Gordan (CG) coefficients. An example is the $\Delta(27)$ group. (ii) Groups for which one can find a CP basis in which all the CG coefficients are real. For such groups, imposing CP invariance restricts the phases of coupling coefficients. The examples are A_4, T' and S_4. (iii) Groups that do not admit real CG coefficients, but can define the generalized CP transformation. For such groups, imposing the CP invariance can lead to an additional symmetry [27]. An example is $\Sigma(72)$.

20.3   Modular Symmetry with Generalized CP Symmetry

20.3.1   CP Transformation of the Modulus τ

The CP transformation in the modular symmetry was given by using the generalized CP symmetry [12]. Consider the CP and modular transformation γ of the chiral superfield $\psi(x)$ assigned to an irreducible unitary representation $\mathbf{r}$ of Γ_N. Under the modular transformation, the chiral superfields ψ_i (i denotes flavors) with weight $-k$ transform as [28],

$$\psi_i \longrightarrow (c\tau + d)^{-k} \rho(\gamma)_{ij}\psi_j \, . \tag{20.8}$$

By using the notation of the CP transformation in Eq. (20.1), the chain $CP \to \gamma \to CP^{-1} = \gamma' \in \bar{\Gamma}$ is expressed as:

$$\psi(x) \xrightarrow{CP} \mathbf{X_r}\overline{\psi}(x_P) \xrightarrow{\gamma} (c\tau^* + d)^{-k}\mathbf{X_r}\,\rho_{\mathbf{r}}^*(\gamma)\overline{\psi}(x_P)$$
$$\xrightarrow{CP^{-1}} (c\tau_{CP^{-1}}^* + d)^{-k}\mathbf{X_r}\,\rho_{\mathbf{r}}^*(\gamma)\mathbf{X_r}^{-1}\psi(x) \, , \tag{20.9}$$

where $\tau_{CP^{-1}}$ is the operation of CP^{-1} on τ. The result of this chain transformation should be equivalent to a modular transformation γ' which maps $\psi(x)$ to $(c'\tau + d')^{-k}\rho_{\mathbf{r}}(\gamma')\psi(x)$. Therefore, we obtain

$$\mathbf{X_r}\rho_{\mathbf{r}}^*(\gamma)\mathbf{X_r}^{-1} = \left(\frac{c'\tau + d'}{c\tau_{CP^{-1}}^* + d}\right)^{-k} \rho_{\mathbf{r}}(\gamma') \, . \tag{20.10}$$

Since $\mathbf{X_r}$, $\rho_{\mathbf{r}}$ and $\rho_{\mathbf{r}'}$ are independent of τ, the overall coefficient on the right-hand side of (20.10) has to be a constant for non-zero weight k:

$$\frac{c'\tau + d'}{c\tau_{CP^{-1}}^* + d} = \frac{1}{\lambda^*} \, , \tag{20.11}$$

where λ is a complex and $|\lambda| = 1$ due to the unitarity of $\rho_{\mathbf{r}}$ and $\rho_{\mathbf{r}'}$. The values of λ, c' and d' depend on γ.

Taking $\gamma = S$ ($c = 1, d = 0$), and denoting $c'(S) = C$, $d'(S) = D$ while keeping $\lambda(S) = \lambda$, we find $\tau = (\lambda \tau^*_{CP^{-1}} - D)/C$ from (20.11), and consequently,

$$\tau \xrightarrow{CP^{-1}} \tau_{CP^{-1}} = \lambda(C\tau^* + D), \qquad \tau \xrightarrow{CP} \tau_{CP} = \frac{1}{C}(\lambda \tau^* - D). \quad (20.12)$$

Let us act with chain $CP \to T \to CP^{-1}$ on the modular τ itself:

$$\tau \xrightarrow{CP} \tau_{CP} = \frac{1}{C}(\lambda \tau^* - D) \xrightarrow{T} \frac{1}{C}(\lambda(\tau^* + 1) - D) \xrightarrow{CP^{-1}} \tau + \frac{\lambda}{C}. \quad (20.13)$$

The resulting transformation has to be a modular transformation, therefore λ/C is an integer. Since $|\lambda| = 1$, we find $|C| = 1$ and $\lambda = \pm 1$. After choosing the sign of C as $C = \mp 1$ so that $\text{Im}\tau_{CP} > 0$, the CP transformation rule Eq. (20.12) simplifies to

$$\tau \xrightarrow{CP} n - \tau^*, \quad (20.14)$$

where n is an integer. The chain $CP \to S \to CP^{-1} = \gamma'(S)$ imposes no further restrictions on τ_{CP}. Since S and T generate the entire modular group, (20.14) is the most general CP transformation of the modulus τ compatible with the modular symmetry. It is always possible to redefine the CP transformation in such a way that $n = 0$ by using the freedom of T transformation. Therefore, we define that the modulus τ transforms under CP as

$$\tau \xrightarrow{CP} -\tau^*, \quad (20.15)$$

without loss of generality.

The same transformation of τ was also derived from the superstring theory [13,29,30]. Higher dimensional theories such as higher dimensional super Yang–Mills theory and superstring theory conserve CP. The four-dimensional CP symmetry can be embedded into $(4 + d)$ dimensions as higher dimensional proper Lorentz symmetry with positive determinant [31–36]. That is, one can combine the four-dimensional CP transformation and d-dimensional transformation with negative determinant so as to obtain $(4 + d)$ dimensional proper Lorentz transformation. For example in six-dimensional theory, we denote the two extra coordinates by a complex coordinate z. The four-dimensional CP symmetry with $z \to z^*$ or $z \to -z^*$ is a six-dimensional proper Lorentz symmetry. Note that $z = x + \tau y$, where x and y are real coordinates. The latter transformation $z \to -z^*$ maps the upper half plane $\text{Im}[\tau] > 0$ to the same half plane. Hence, we consider the transformation $z \to -z^*$ as the CP symmetry. That means that the CP transforms $\tau \to -\tau^*$.

20.3.2 CP Transformation of Modular Multiplets

Under a modular transformation, modular forms transform as in (I.1) of Appendix I. The CP transformation of modular forms were given in [12]. Define a modular multiplet of the irreducible representation $\mathbf{r}$ of Γ_N with weight k as $\mathbf{Y}_{\mathbf{r}}^{(k)}(\tau)$. Under the action of CP, we have

$$\mathbf{Y}_{\mathbf{r}}^{(k)}(\tau) \xrightarrow{CP} Y_{\mathbf{r}}^{(k)}(-\tau^*) \,. \tag{20.16}$$

The complex conjugated CP transformed modular forms $\mathbf{Y}_{\mathbf{r}}^{(k)*}(-\tau^*)$ transform almost like the original multiplets $\mathbf{Y}_{\mathbf{r}}^{(k)}(\tau)$ under a modular transformation, namely:

$$\mathbf{Y}_{\mathbf{r}}^{(k)*}(-\tau^*) \xrightarrow{\gamma} \mathbf{Y}_{\mathbf{r}}^{(k)*}(-(\gamma\tau)^*) = (c\tau+d)^k \rho_{\mathbf{r}}^*(u(\gamma)) \mathbf{Y}_{\mathbf{r}}^{(k)*}(-\tau^*) \,, \tag{20.17}$$

where

$$u(\gamma) \equiv CP\gamma CP^{-1} = \begin{pmatrix} a & -b \\ -c & d \end{pmatrix} \,, \tag{20.18}$$

and u acts on the generator as $u(S) = S$ and $u(T) = T^{-1}$ [12]. Using the consistency condition of Eq. (20.3), which gives $\mathbf{X}_{\mathbf{r}}^{\mathbf{T}} \rho_{\mathbf{r}}^*(u(\gamma)) = \rho_{\mathbf{r}}(\gamma) \mathbf{X}_{\mathbf{r}}^{\mathbf{T}}$, we obtain

$$\mathbf{X}_r^T Y_{\mathbf{r}}^{(k)*}(-\tau^*) \xrightarrow{\gamma} (c\tau+d)^k \rho_{\mathbf{r}}(\gamma) \mathbf{X}_{\mathbf{r}}^{\mathbf{T}} Y_{\mathbf{r}}^{(k)*}(-\tau^*) \,. \tag{20.19}$$

Therefore, if there exists a unique modular multiplet at a level N, weight k and representation $\mathbf{r}$, which is satisfied for $N = 2$–5 with weight 2, we can express the modular form $\mathbf{Y}_{\mathbf{r}}^{(k)}(\tau)$ as:

$$\mathbf{Y}_{\mathbf{r}}^{(k)}(\tau) = \kappa \mathbf{X}_{\mathbf{r}}^{\mathbf{T}} \mathbf{Y}_{\mathbf{r}}^{(k)*}(-\tau^*) \,, \tag{20.20}$$

where κ is a proportional coefficient. Make $\mathbf{Y}_{\mathbf{r}}^{(k)*}(-\tau^*)$ by using Eq. (20.20) and substitute it for $\mathbf{Y}_{\mathbf{r}}^{(k)*}(-\tau^*)$ in the right hand side of Eq. (20.20). Then, one obtains $\mathbf{X}_r^* \mathbf{X}_r = |\kappa|^2 \mathbf{1}_r$ since $\mathbf{Y}_{\mathbf{r}}^{(k)}(-(-\tau^*)^*) = \mathbf{Y}_{\mathbf{r}}^{(k)}(\tau)$. Therefore, the unitary matrix $\mathbf{X}_{\mathbf{r}}$ is symmetric one, and $\kappa = e^{i\phi}$ is a phase, which can be absorbed in the normalization of modular forms. Thus, the modular symmetry restricts $\mathbf{X}_{\mathbf{r}}$ being symmetric. In conclusion, the CP transformation of modular forms is given as:

$$\mathbf{Y}_{\mathbf{r}}^{(k)}(\tau) \xrightarrow{CP} \mathbf{Y}_{\mathbf{r}}^{(k)}(-\tau^*) = \mathbf{X}_{\mathbf{r}} \mathbf{Y}_{\mathbf{r}}^{(k)*}(\tau) \,. \tag{20.21}$$

It is also emphasized that $\mathbf{X}_{\mathbf{r}} = \mathbf{1}_{\mathbf{r}}$ satisfies the consistency condition of (20.3) in a basis that generators of S and T of Γ_N are represented by symmetric matrices because of $\rho_{\mathbf{r}}^*(S) = \rho_{\mathbf{r}}^{\dagger}(S) = \rho_{\mathbf{r}}(S^{-1}) = \rho_{\mathbf{r}}(S)$ and $\rho_{\mathbf{r}}^*(T) = \rho_{\mathbf{r}}^{\dagger}(T) = \rho_{\mathbf{r}}(T^{-1})$.

We summarize the CP transformations of chiral superfields and modular multiplets as follows:

$$\tau \xrightarrow{CP} -\tau^*, \quad \psi(x) \xrightarrow{CP} X_r \overline{\psi}(x_P), \quad \mathbf{Y}_\mathbf{r}^{(k)}(\tau) \xrightarrow{CP} \mathbf{Y}_\mathbf{r}^{(k)}(-\tau^*) = X_r \mathbf{Y}_\mathbf{r}^{(k)*}(\tau),$$
(20.22)

where $\mathbf{X_r} = \mathbf{1_r}$ in a basis of symmetric generators of S and T. We can use this CP transformation of modular forms to construct the CP invariant mass matrices in the modular symmetry.

20.4 CP Violation by Modulus τ in A_4 Model of Leptons

There are the modular A_4 invariant models with the generalized CP symmetry [37, 38]. Both CP and modular symmetries are broken spontaneously by VEV of the modulus τ. We discuss the phenomenological implication of a simple model [37], that is the neutrino mixing angles and the CP violating Dirac phase of leptons, which is expected to be observed in the future.

We assign the A_4 representation and weight for superfields of leptons in Table 20.1, where the three left-handed lepton doublets compose a A_4 triplet L, and the right-handed charged leptons e^c, μ^c and τ^c are A_4 singlets. The weights of the superfields of left-handed leptons and right-handed charged leptons are -2 and 0, respectively. Then, the simple lepton mass matrices for charged leptons and neutrinos are obtained [37].

The superpotential of the charged lepton mass term is given in terms of modular forms of weight 2, $\mathbf{Y}_3^{(2)}$. It is given as:

$$w_E = \alpha_e e^c H_d \mathbf{Y}_3^{(2)} L + \beta_e \mu^c H_d \mathbf{Y}_3^{(2)} L + \gamma_e \tau^c H_d \mathbf{Y}_3^{(2)} L ,$$
(20.23)

where L is the left-handed A_4 triplet leptons. We can take real for α_e, β_e and γ_e. Under CP, the superfields transform as:

$$e^c \xrightarrow{CP} \mathbf{X}_1^* \overline{e}^c , \quad \mu^c \xrightarrow{CP} \mathbf{X}_{1''}^* \overline{\mu}^c , \quad \tau^c \xrightarrow{CP} \mathbf{X}_{1'}^* \overline{\tau}^c , \quad L \xrightarrow{CP} \mathbf{X}_3 \overline{L} ,$$
(20.24)

and $H_d \xrightarrow{CP} \overline{H}_d$. Taking the representations of symmetric S and T as seen in Eq. (C.2) of Appendix C, we can choose $\mathbf{X_3} = \mathbf{1_3}$ and $\mathbf{X_1} = \mathbf{X_{1'}} = \mathbf{X_{1''}} = 1$.

Table 20.1 Representations and weights k for MSSM fields and modular forms of weight 2 and 4

	L	(e^c, μ^c, τ^c)	H_u	H_d	$\mathbf{Y}_\mathbf{r}^{(2)}$, $\mathbf{Y}_\mathbf{r}^{(4)}$
$SU(2)$	2	1	2	2	1
A_4	3	$(1, 1'', 1')$	1	1	$3, \{3, 1, 1'\}$
k	-2	$(0, 0, 0)$	0	0	2, 4

Taking (e_L, μ_L, τ_L) in the flavor base, the charged lepton mass matrix M_E is simply written as:

$$M_E(\tau) = v_d \begin{pmatrix} \alpha_e & 0 & 0 \\ 0 & \beta_e & 0 \\ 0 & 0 & \gamma_e \end{pmatrix} \begin{pmatrix} Y_1(\tau) & Y_3(\tau) & Y_2(\tau) \\ Y_2(\tau) & Y_1(\tau) & Y_3(\tau) \\ Y_3(\tau) & Y_2(\tau) & Y_1(\tau) \end{pmatrix} , \qquad (20.25)$$

where v_d is VEV of the neutral component of H_d, and coefficients α_e, β_e and γ_e are taken to be real without loss of generality. Under CP transformation, the mass matrix M_E is transformed following from (20.22) as:

$$M_E(\tau) \xrightarrow{CP} M_E(-\tau^*) = M_E^*(\tau) = v_d \begin{pmatrix} \alpha_e & 0 & 0 \\ 0 & \beta_e & 0 \\ 0 & 0 & \gamma_e \end{pmatrix} \begin{pmatrix} Y_1(\tau)^* & Y_3(\tau)^* & Y_2(\tau)^* \\ Y_2(\tau)^* & Y_1(\tau)^* & Y_3(\tau)^* \\ Y_3(\tau)^* & Y_2(\tau)^* & Y_1(\tau)^* \end{pmatrix} .$$
$$(20.26)$$

Let us discuss the neutrino mass matrix. Suppose neutrinos to be Majorana particles. By using the Weinberg operator, the superpotential of the neutrino mass term, w_ν is given as:

$$w_\nu = -\frac{1}{\Lambda} (H_u H_u L L \mathbf{Y}_{\mathbf{r}}^{(4)})_{\mathbf{1}} , \qquad (20.27)$$

where Λ is a relevant cutoff scale. Since the left-handed lepton doublet has weight -2, the superpotential is given in terms of modular forms of weight 4, $\mathbf{Y}_{\mathbf{1}}^{(4)}(\tau)$, $\mathbf{Y}_{\mathbf{1}'}^{(4)}(\tau)$, $\mathbf{Y}_{\mathbf{3}}^{(4)}(\tau)$:

$$\mathbf{Y}_{\mathbf{1}}^{(4)}(\tau) = Y_1(\tau)^2 + 2Y_2(\tau)Y_3(\tau) , \qquad \mathbf{Y}_{\mathbf{1}'}^{(4)}(\tau) = Y_3(\tau)^2 + 2Y_1(\tau)Y_2(\tau) ,$$

$$\mathbf{Y}_{\mathbf{3}}^{(4)}(\tau) = \begin{pmatrix} Y_1^{(4)}(\tau) \\ Y_2^{(4)}(\tau) \\ Y_3^{(4)}(\tau) \end{pmatrix} = \begin{pmatrix} Y_1(\tau)^2 - Y_2(\tau)Y_3(\tau) \\ Y_3(\tau)^2 - Y_1(\tau)Y_2(\tau) \\ Y_2(\tau)^2 - Y_1(\tau)Y_3(\tau) \end{pmatrix} . \qquad (20.28)$$

By putting v_u for VEV of the neutral component of H_u and using the tensor products of A_4 in Appendix A, we have

$$\begin{aligned} w_\nu = \frac{v_u^2}{\Lambda} & \left[\begin{pmatrix} 2\nu_e\nu_e - \nu_\mu\nu_\tau - \nu_\tau\nu_\mu \\ 2\nu_\tau\nu_\tau - \nu_e\nu_\mu - \nu_\mu\nu_\tau \\ 2\nu_\mu\nu_\mu - \nu_\tau\nu_e - \nu_e\nu_\tau \end{pmatrix} \otimes \mathbf{Y}_{\mathbf{3}}^{(4)} \right. \\ & + (\nu_e\nu_e + \nu_\mu\nu_\tau + \nu_\tau\nu_\mu) \otimes g_1^\nu \mathbf{Y}_{\mathbf{1}}^{(4)} + (\nu_e\nu_\tau + \nu_\mu\nu_\mu + \nu_\tau\nu_e) \otimes g_2^\nu \mathbf{Y}_{\mathbf{1}'}^{(4)} \right] \\ = \frac{v_u^2}{\Lambda} & \left[(2\nu_e\nu_e - \nu_\mu\nu_\tau - \nu_\tau\nu_\mu) Y_1^{(4)} + (2\nu_\tau\nu_\tau - \nu_e\nu_\mu - \nu_\mu\nu_e) Y_3^{(4)} \right. \\ & + (2\nu_\mu\nu_\mu - \nu_\tau\nu_e - \nu_e\nu_\tau) Y_2^{(4)} \qquad\qquad\qquad\qquad (20.29) \\ & + (\nu_e\nu_e + \nu_\mu\nu_\tau + \nu_\tau\nu_\mu) g_1^\nu \mathbf{Y}_{\mathbf{1}}^{(4)} + (\nu_e\nu_\tau + \nu_\mu\nu_\mu + \nu_\tau\nu_e) g_2^\nu \mathbf{Y}_{\mathbf{1}'}^{(4)} \right] , \end{aligned}$$

where $\mathbf{Y}_3^{(4)}$, $\mathbf{Y}_1^{(4)}$ and $\mathbf{Y}_{1'}^{(4)}$ in Eq. (20.28) are given by $Y_1(\tau)$, $Y_2(\tau)$, $Y_3(\tau)$ in Appendix I, and g_1^ν, g_2^ν are complex parameters in general. The neutrino mass matrix is written as follows:

$$
M_\nu(\tau) = \frac{v_u^2}{\Lambda}\left[\begin{pmatrix} 2Y_1^{(4)}(\tau) & -Y_3^{(4)}(\tau) & -Y_2^{(4)}(\tau) \\ -Y_3^{(4)}(\tau) & 2Y_2^{(4)}(\tau) & -Y_1^{(4)}(\tau) \\ -Y_2^{(4)}(\tau) & -Y_1^{(4)}(\tau) & 2Y_3^{(4)}(\tau) \end{pmatrix} + g_1^\nu \mathbf{Y}_1^{(4)}(\tau) \begin{pmatrix} 1 & 0 & 0 \\ 0 & 0 & 1 \\ 0 & 1 & 0 \end{pmatrix} \right.
$$
$$
\left. + g_2^\nu \mathbf{Y}_{1'}^{(4)}(\tau) \begin{pmatrix} 0 & 0 & 1 \\ 0 & 1 & 0 \\ 1 & 0 & 0 \end{pmatrix} \right]. \tag{20.30}
$$

Under CP transformation, the mass matrix M_ν is transformed following from (20.22) as:

$$
M_\nu(\tau) \xrightarrow{CP} M_\nu(-\tau^*) = M_\nu^*(\tau)
$$
$$
= \frac{v_u^2}{\Lambda}\left[\begin{pmatrix} 2Y_1^{(4)*}(\tau) & -Y_3^{(4)*}(\tau) & -Y_2^{(4)*}(\tau) \\ -Y_3^{(4)*}(\tau) & 2Y_2^{(4)*}(\tau) & -Y_1^{(4)*}(\tau) \\ -Y_2^{(4)*}(\tau) & -Y_1^{(4)*}(\tau) & 2Y_3^{(4)*}(\tau) \end{pmatrix} + g_1^{\nu*} \mathbf{Y}_1^{(4)*}(\tau) \begin{pmatrix} 1 & 0 & 0 \\ 0 & 0 & 1 \\ 0 & 1 & 0 \end{pmatrix} \right.
$$
$$
\left. + g_2^{\nu*} \mathbf{Y}_{1'}^{(4)*}(\tau) \begin{pmatrix} 0 & 0 & 1 \\ 0 & 1 & 0 \\ 1 & 0 & 0 \end{pmatrix} \right]. \tag{20.31}
$$

In a CP conserving modular invariant theory, both CP and modular symmetries are broken spontaneously by VEV of the modulus τ. However, there exist certain values of τ which conserve CP while breaking the modular symmetry. Obviously, this is the case if τ is left invariant by CP, i.e.

$$
\tau \xrightarrow{CP} -\tau^* = \tau , \tag{20.32}
$$

which indicates τ lies on the imaginary axis, $\text{Re}[\tau] = 0$. In addition to $\text{Re}[\tau] = 0$, CP is conserved at the boundary of the fundamental domain. Then, one has

$$
M_E(\tau) = M_E^*(\tau) , \qquad M_\nu(\tau) = M_\nu^*(\tau) , \tag{20.33}
$$

which leads to g_1^ν and g_2^ν being real. Since parameters α_e, β_e, γ_e are also real, the source of the CP violation is only non-trivial $\text{Re}[\tau]$ after breaking the modular symmetry.

If the CP violation will be confirmed at the experiments of neutrino oscillations, the CP symmetry should be broken spontaneously by VEV of the modulus τ. Thus, VEV of τ breaks the CP symmetry as well as the modular invariance. The source of the CP violation is only the real part of τ. It is interesting to ask whether the spontaneous CP violation is realized due to the value of τ, which is consistent with observed lepton mixing angles and neutrino masses. If this is the case, the CP violating Dirac phase and Majorana phases are predicted clearly under the fixed value of τ.

The framework of our calculations is given as follows. Parameter ratios α_e/γ_e and β_e/γ_e are given in terms of charged lepton masses and τ. Therefore, the lepton mixing angles, the Dirac phase and Majorana phases are given by our model parameters g_1^ν and g_2^ν in addition to the value of τ.

The input charged lepton masses are given in Eq. (19.76) at the GUT scale 2×10^{16} GeV, where $\tan \beta = 5$ is taken as a bench mark [39,40]. We also input the lepton mixing angles and neutrino mass parameters which are given by NuFit 5.0 [41]. In our analysis, the Dirac CP phase δ is output because its observed range is too wide at $3\,\sigma$ confidence level. We investigate two possible cases of neutrino masses m_i, which are the normal hierarchy (NH), $m_3 > m_2 > m_1$, and the inverted hierarchy (IH), $m_2 > m_1 > m_3$. Neutrino masses and mixings are obtained by diagonalizing $M_E^\dagger M_E$ and $M_\nu^\dagger M_\nu$. We also investigate the effective mass for the $0\nu\beta\beta$ decay, $\langle m_{ee} \rangle$ and the sum of three neutrino masses $\sum m_i$ since it is constrained by the recent cosmological data, which is the upper-bound $\sum m_i \leq 120\,\mathrm{meV}$ obtained at the 95% confidence level [42,43].

Let us discuss numerical results for NH of neutrino masses. Since the spontaneous CP violation in Type IIB string theory is possibly realized at nearby fixed points, where the moduli stabilization is performed in a controlled way [44,45]. There are two fixed points in the fundamental domain of $PSL(2, Z)$, $\tau = i$ and $\tau = \omega$. Indeed, the viable τ of our lepton mass matrices is found around $\tau = i$. We scan τ around i while neutrino couplings g_1^ν and g_2^ν are scanned in the real space of $[-10, 10]$. As a measure of good-fit, we adopt the sum of one-dimensional χ^2 function for four accurately known dimensionless observables $\Delta m_{\mathrm{atm}}^2/\Delta m_{\mathrm{sol}}^2$, $\sin^2 \theta_{12}$, $\sin^2 \theta_{23}$ and $\sin^2 \theta_{13}$ in NuFit 5.0 [41]. In addition, we employ Gaussian approximations for fitting m_e/m_τ and m_μ/m_τ by using the data of PDG [46].

In Fig. 20.1 we show the allowed region on the Re $[\tau]$–Im $[\tau]$ plane, where three mixing angles and $\Delta m_{\mathrm{atm}}^2/\Delta m_{\mathrm{sol}}^2$ are consistent with observed ones. The Dark, middle and light region correspond to 2σ, 3σ and 5σ confidence levels, respectively.

The predicted range of τ is in Re $[\tau] = \pm[0.073, 0.083]$ and Im $[\tau] = [1.006, 1.014]$ at $3\,\sigma$ confidence level (middle), which are close to the fixed point $\tau = i$.

Due to restricted Re $[\tau]$, the CP violating Dirac phase δ_{CP}, which is defined in PDG, is predicted clearly. In Fig. 20.2, we show prediction of δ_{CP} versus the sum of neutrino masses $\sum m_i$. It is remarked that δ_{CP} is almost independent of $\sum m_i$. The predicted ranges of δ_{CP} are narrow such as $[98°, 110°]$ and $[250°, 262°]$ at $3\,\sigma$

Fig. 20.1 Allowed regions of τ for NH. The dark, middle and light regions correspond to $2\sigma, 3\sigma, 5\sigma$ confidence levels, respectively. The solid curve is the boundary of the fundamental domain, $|\tau| = 1$

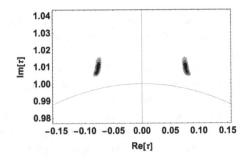

Fig. 20.2 The prediction of δ_{CP} versus $\sum m_i$ for NH. The dark, middle and light regions denote same ones in Fig. 20.1

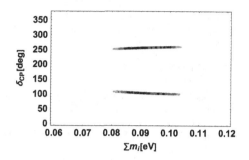

confidence level. The predicted ranges $[98°, 110°]$ and $[250°, 262°]$ correspond to $\mathrm{Re}\,[\tau] = (0.073–0.083)$ and $\mathrm{Re}\,[\tau] = -(0.073–0.083)$, respectively. The predicted $\sum m_i$ is in $[82, 102]\,\mathrm{meV}$ for $3\,\sigma$ confidence level (middle). The minimal cosmological model, CDM $+ \sum m_i$, provides the upper-bound $\sum m_i < 120\,\mathrm{meV}$ [42,43]. Thus, our predicted sum of neutrino masses is consistent with the cosmological bound 120 meV.

We can calculate the effective mass $\langle m_{ee} \rangle$ for the $0\nu\beta\beta$ decay. The predicted $\langle m_{ee} \rangle$ is in $[12.5, 20.5]\,\mathrm{meV}$ for $3\,\sigma$ confidence level. The prediction of $\langle m_{ee} \rangle \simeq 20\,\mathrm{meV}$ will be testable in the future experiments of the neutrinoless double beta decay.

For the case of inverted hierarchy (IH) of neutrino masses, there are no allowed regions of $2\,\sigma$ and $3\,\sigma$ confidence levels, but only $5\,\sigma$ confidence level. Moreover, the sum of neutrino masses are in the range of $[134, 180]\,\mathrm{meV}$. Therefore, the IH of neutrino masses is unfavored in this model.

In our numerical calculations, we have not included the effects of the renormalization group equation in the lepton mixing angles and neutrino mass ratio $\Delta m_{\mathrm{sol}}^2/\Delta m_{\mathrm{atm}}^2$. Those corrections are very small between the electroweak and GUT scales in the case of $\tan\beta \leq 5$ unless neutrino masses are almost degenerate.

In this model. both CP and modular symmetries are broken spontaneously by VEV of the modulus τ. The source of the CP violation is a non-trivial value of $\mathrm{Re}[\tau]$ while parameters of neutrinos g_1^ν and g_2^ν are real. The value of τ close to the fixed point $\tau = i$ is consistent with the observed lepton mixing angles and lepton masses for NH. The CP violating Dirac phase δ_{CP} is predicted to be around $\pm 100°$ and $\sum m_i$ is in $[82, 102]\,\mathrm{meV}$.

Although we have constrained τ by observables of leptons phenomenologically, one also should pay attention to the recent theoretical work of the moduli stabilization from the viewpoint of modular flavor symmetries [13,45,47,48]. The study of modulus τ is interesting to reveal the flavor theory in both theoretical and phenomenological aspects.

References

1. Ferreira, P.M., Grimus, W., Lavoura, L., Ludl, P.O.: JHEP **1209**, 128 (2012) arXiv:1206.7072 [hep-ph]
2. de Medeiros Varzielas, I., Emmanuel-Costa, D.: Phys. Rev. D **84**, 117901 (2011). arXiv:1106.5477 [hep-ph]

3. de Medeiros Varzielas, I.: JHEP **1208**, 055 (2012). arXiv:1205.3780 [hep-ph]
4. de Medeiros Varzielas, I., Emmanuel-Costa, D., Leser, P.: Phys. Lett. B **716**, 193 (2012). arXiv:1204.3633 [hep-ph]
5. Branco, G.C., Felipe, R.G., Joaquim, F.R.: Rev. Mod. Phys. **84**, 515 (2012). arXiv:1111.5332 [hep-ph]
6. Ecker, G., Grimus, W., Konetschny, W.: Nucl. Phys. B **191**, 465–492 (1981)
7. Ecker, G., Grimus, W., Neufeld, H.: Nucl. Phys. B **247**, 70–82 (1984)
8. Ecker, G., Grimus, W., Neufeld, H.: J. Phys. A **20**, L807 (1987)
9. Neufeld, H., Grimus, W., Ecker, G.: Int. J. Mod. Phys. A **3**, 603–616 (1988)
10. Grimus, W., Rebelo, M.N.: Phys. Rept. **281**, 239–308 (1997). arXiv:hep-ph/9506272 [hep-ph]
11. Grimus, W., Lavoura, L.: Phys. Lett. B **579**, 113 (2004). [hep-ph/0305309]
12. Novichkov, P.P., Penedo, J.T., Petcov, S.T., Titov, A.V.: JHEP **1907**, 165 (2019). arXiv:1905.11970 [hep-ph]
13. Kobayashi, T., Shimizu, Y., Takagi, K., Tanimoto, M., Tatsuishi, T.H., Uchida, H.: Phys. Rev. D **101**(5), 055046 (2020). arXiv:1910.11553 [hep-ph]
14. Holthausen, M., Lindner, M., Schmidt, M.A.: JHEP **1304**, 122 (2013). arXiv:1211.6953 [hep-ph]
15. Feruglio, F., Hagedorn, C., Ziegler, R.: JHEP **07**, 027 (2013). arXiv:1211.5560 [hep-ph]
16. Chen, M.C., Fallbacher, M., Mahanthappa, K.T., Ratz, M., Trautner, A.: Nucl. Phys. B **883**, 267 (2014). arXiv:1402.0507 [hep-ph]
17. Mohapatra, R.N., Nishi, C.C.: Phys. Rev. D **86**, 073007 (2012). arXiv:1208.2875 [hep-ph]
18. Ma, E.: Phys. Lett. B **723**, 161 (2013). arXiv:1304.1603 [hep-ph]
19. Girardi, I., Meroni, A., Petcov, S.T., Spinrath, M.: JHEP **1402**, 050 (2014). arXiv:1312.1966 [hep-ph]
20. Feruglio, F., Hagedorn, C., Ziegler, R.: Eur. Phys. J. C **74**, 2753 (2014). arXiv:1303.7178 [hep-ph]
21. Ding, G.J., King, S.F., Luhn, C., Stuart, A.J.: JHEP **1305**, 084 (2013). arXiv:1303.6180 [hep-ph]
22. Meroni, A., Petcov, S.T., Spinrath, M.: Phys. Rev. D **86**, 113003 (2012). arXiv:1205.5241 [hep-ph]
23. Li, C.C., Ding, G.J.: Nucl. Phys. B **881**, 206 (2014). arXiv:1312.4401 [hep-ph]
24. Luhn, C.: Nucl. Phys. B **875**, 80 (2013). arXiv:1306.2358 [hep-ph]
25. Ding, G.J., Zhou, Y.L.: Chin. Phys. C **39**(2), 021001 (2015). arXiv:1312.5222 [hep-ph]
26. Ding, G.J., King, S.F., Stuart, A.J.: JHEP **1312**, 006 (2013). arXiv:1307.4212 [hep-ph]
27. Chen, M.C., Mahanthappa, K.T.: Phys. Lett. B **681**, 444 (2009). arXiv:0904.1721 [hep-ph]
28. Ferrara, S., Lust, D., Shapere, A.D., Theisen, S.: Phys. Lett. B **225**, 363 (1989)
29. Baur, A., Nilles, H.P., Trautner, A., Vaudrevange, P.K.S.: Phys. Lett. B **795**, 7 (2019). arXiv:1901.03251 [hep-th]
30. Baur, A., Nilles, H.P., Trautner, A., Vaudrevange, P.K.S.: arXiv:1908.00805 [hep-th]
31. Green, M.B., Schwarz, J.H., Witten, E.: Cambridge Monographs On Mathematical Physics. Cambridge, Uk: Univ. Pr. (1987) 596 P
32. Strominger, A., Witten, E.: Commun. Math. Phys. **101**, 341 (1985)
33. Dine, M., Leigh, R.G., MacIntire, D.A.: Phys. Rev. Lett. **69**, 2030 (1992). [hep-th/9205011]
34. Choi, K.w., Kaplan, D.B., Nelson, A.E.: Nucl. Phys. B **391**, 515 (1993) [bhep-ph/9205202]
35. Lim, C.S.: Phys. Lett. B **256**, 233 (1991)
36. Kobayashi, T., Lim, C.S.: Phys. Lett. B **343**, 122 (1995). [hep-th/9410023]
37. Okada, H., Tanimoto, M.: JHEP **03**, 010 (2021). arXiv:2012.01688 [hep-ph]
38. Yao, C.Y., Lu, J.N., Ding, G.J.: JHEP **05**, 102 (2021). arXiv:2012.13390 [hep-ph]
39. Antusch, S., Maurer, V.: JHEP **1311**, 115 (2013). arXiv:1306.6879 [hep-ph]
40. Björkeroth, F., de Anda, F.J., de Medeiros Varzielas, I., King, S.F.: JHEP **1506**, 141 (2015). arXiv:1503.03306 [hep-ph]
41. Esteban, I., Gonzalez-Garcia, M.C., Maltoni, M., Schwetz, T., Zhou, A.: JHEP **09**, 178 (2020). arXiv:2007.14792 [hep-ph]
42. Vagnozzi, S., Giusarma, E., Mena, O., Freese, K., Gerbino, M., Ho, S., Lattanzi, M.: Phys. Rev. D **96**(12), 123503 (2017). arXiv:1701.08172 [astro-ph.CO]

43. Aghanim, N., et al.: [Planck Collaboration], Astron. Astrophys. **641**, A6 (2020) [erratum: Astron. Astrophys. **652**, C4 (2021)]. [arXiv:1807.06209 [astro-ph.CO]]
44. Kobayashi, T., Otsuka, H.: Phys. Rev. D **102**(2), 026004 (2020). arXiv:2004.04518 [hep-th]
45. Abe, H., Kobayashi, T., Uemura, S., Yamamoto, J.: Phys. Rev. D **102**(4), 045005 (2020). arXiv:2003.03512 [hep-th]
46. Zyla, P.A., et al.: [Particle Data Group], PTEP **2020**(8), 083C01 (2020)
47. Kobayashi, T., Shimizu, Y., Takagi, K., Tanimoto, M., Tatsuishi, T.H.: Phys. Rev. D **100**(11), 115045 (2019) [erratum: Phys. Rev. D **101**(3), 039904 (2020)]. arXiv:1909.05139 [hep-ph]
48. Ishiguro, K., Kobayashi, T., Otsuka, H.: JHEP **03**, 161 (2021). arXiv:2011.09154 [hep-ph]

Useful Theorems

A

In this appendix, we give simple proofs of useful theorems. (See also e.g. Refs. [1–4].)

• Lagrange's theorem

The order N_H of a subgroup of a finite group G is a divisor of the order N_G of G.

Proof)

If $H = G$, the claim is trivial, $N_H = H_G$. Thus, we consider $H \neq G$. Let a_1 be an element of G, but be not contained in H. Here, we denote all of elements in H by $\{e = h_0, h_1, \ldots, h_{N_H-1}\}$. Then, we consider the products of a_1 and elements of H,

$$a_1 H = \{a_1, a_1 h_1, \ldots, a_1 h_{N_h-1}\}. \tag{A.1}$$

All of $a_1 h_i$ are different from each other. None of $a_1 h_i$ are contained in H. If $a_1 h_i = h_j$, we could find $a_1 = h_j h_i^{-1}$, that is, a_1 would be an element in H. Thus, the set $a_1 H$ includes the N_H elements. Next, let a_2 be an element of G, but be contained in neither H nor $a_1 H$. If $a_2 h_i = a_1 h_j$, the element a_2 would be written as $a_2 = a_1 h_j h_i^{-1}$, that is, an element of $a_1 H$. Thus, when $a_2 \notin H$ and $a_2 \notin a_1 H$, the set $a_2 H$ yields N_H new elements. We repeat this process. Then, we can decompose

$$G = H + a_1 H + \cdots + a_{m-1} H. \tag{A.2}$$

That implies $N_G = m N_H$. ∎

• Theorem

For a finite group, every representation is equivalent to a unitary representation.

Proof)

Every group element a is represented by a matrix $D(a)$, which acts on the vector space. We denote the basis of the representation vector space by $\{e_1, \ldots, e_d\}$. We consider two vectors, v and w,

$$v = \sum_{i=1}^{d} v_i e_i, \qquad w = \sum_{i=1}^{d} w_i e_i. \tag{A.3}$$

© Springer-Verlag GmbH Germany, part of Springer Nature 2022
T. Kobayashi et al., *An Introduction to Non-Abelian Discrete Symmetries for Particle Physicists,* Lecture Notes in Physics 995,
https://doi.org/10.1007/978-3-662-64679-3

We define the scalar product between $\mathbf{v}$ and $\mathbf{w}$ as

$$(\boldsymbol{v}, \boldsymbol{w}) = \sum_{i=1}^{d} v_i^* w_i. \tag{A.4}$$

Here, we define another scalar product by

$$\langle \boldsymbol{v}, \boldsymbol{w} \rangle = \frac{1}{N_G} \sum_{a \in G} (D(a)\boldsymbol{v}, D(a)\boldsymbol{w}). \tag{A.5}$$

Then, we find

$$\begin{aligned}
\langle D(b)\boldsymbol{v}, D(b)\boldsymbol{w} \rangle &= \frac{1}{N_G} \sum_{a \in G} (D(b)D(a)\boldsymbol{v}, D(b)D(a)\boldsymbol{w}) \\
&= \frac{1}{N_G} \sum_{a \in G} (D(ba)\boldsymbol{v}, D(ba)\boldsymbol{w}) \\
&= \frac{1}{N_G} \sum_{c \in G} (D(c)\boldsymbol{v}, D(c)\boldsymbol{w}) \\
&= \langle \boldsymbol{v}, \boldsymbol{w} \rangle.
\end{aligned} \tag{A.6}$$

That implies that $D(b)$ is unitary with respect to the scalar product $\langle \boldsymbol{v}, \boldsymbol{w} \rangle$. The orthogonal bases $\{\boldsymbol{e}_i\}$ and $\{\boldsymbol{e}_i'\}$ for the two scalar products $(\boldsymbol{v}, \boldsymbol{w})$ and $\langle \boldsymbol{v}, \boldsymbol{w} \rangle$ can be related by the linear transformation T as $\boldsymbol{e}_i' = T\boldsymbol{e}_i$, i.e. $(\boldsymbol{v}, \boldsymbol{w}) = \langle T\boldsymbol{v}, T\boldsymbol{w} \rangle$. We define $D'(g) = T^{-1}D(g)T$. Then, it is found that

$$\begin{aligned}
(T^{-1}D(a)T\boldsymbol{v}, T^{-1}D(a)T\boldsymbol{w}) &= \langle D(a)T\boldsymbol{v}, D(a)T\boldsymbol{w} \rangle \\
&= \langle T\boldsymbol{v}, T\boldsymbol{w} \rangle \\
&= (\boldsymbol{v}, \boldsymbol{w}).
\end{aligned} \tag{A.7}$$

That is, the matrix $D'(g)$ is unitary and is equivalent to $D(g)$. ∎

- **Schur's lemma**
(I) Let $D_1(g)$ and $D_2(g)$ be irreducible representations of G, which are inequivalent to each other. If

$$A D_1(g) = D_2(g) A, \qquad \forall g \in G, \tag{A.8}$$

the matrix A should vanish, $A = 0$.
    (II) If

$$D(g)A = AD(g), \qquad \forall g \in G, \tag{A.9}$$

the matrix A should be proportional to the identity matrix I, i.e. $A = \lambda I$.

Proof) (I) We denote the representation vector spaces for $D_1(g)$ and $D_2(g)$ by V and W, respectively. Let the map A be a map $A : V \to W$ such that it satisfies (A.8). We consider the kernel of A,

$$Ker(A) = \{v \in V | Av = 0\}. \tag{A.10}$$

Let $v \in Ker(A)$. Then, we have

$$AD_1(g)v = D_2(g)Av = 0. \tag{A.11}$$

It is found that $D_1(g)Ker(A) \subset Ker(A)$, that is, $Ker(A)$ is invariant under $D_1(g)$. Because $D_1(g)$ is irreducible, that implies that

$$Ker(A) = \{0\}, \quad \text{or} \quad Ker(A) = V. \tag{A.12}$$

The later, $Ker(A) = V$, can not be realized unless $A = 0$. Next, we consider the image

$$Im(A) = \{Av | v \in V\}. \tag{A.13}$$

We find

$$D_2(g)Av = AD_1(g)v \in Im(A). \tag{A.14}$$

That is, $Im(A)$ is invariant under $D_2(g)$. Because $D_2(g)$ is irreducible, that implies that

$$Im(A) = \{0\}, \quad \text{or} \quad Im(A) = W. \tag{A.15}$$

The former, $Im(A) = \{0\}$, can not be realized unless $A = 0$. As a result, it is found that A should satisfy

$$A = 0, \quad \text{or} \quad AD_1(g)A^{-1} = D_2(g). \tag{A.16}$$

The later means that the representations, $D_1(g)$ and $D_2(g)$, are equivalent to each other. Therefore, A should vanish, $A = 0$, if $D_1(g)$ and $D_2(g)$ are not equivalent. ∎

Proof)(II) Now, we consider the case with $D(g) = D_1(g) = D_2(g)$ and $V = W$. Here, A is a linear operators on V. The finite dimensional matrix A has at least one eigenvalue, because the characteristic equation $det(A - \lambda I) = 0$ has at lease one root, where λ is an eigenvalue. Then, Eq. (A.9) leads to

$$D(g)(A - \lambda I) = (A - \lambda I)D(g), \quad \forall g \in G. \tag{A.17}$$

Using the above proof of Schur's lemma (I) and $Ker(A - \lambda I) \neq \{0\}$, we find $Ker(A - \lambda I) = V$, that is, $A - \lambda I = 0$. ∎

● **Theorem**

Let $D_\alpha(g)$ and $D_\beta(g)$ be irreducible representations of a group G on the d_α and d_β dimensional vector spaces. Then, they satisfy the following orthogonality relation,

$$\sum_{a \in G} D_\alpha(a)_{i\ell} D_\beta(a^{-1})_{mj} = \frac{N_G}{d_\alpha} \delta_{\alpha\beta} \delta_{ij} \delta_{\ell m}. \tag{A.18}$$

Proof)

We define

$$A = \sum_{a \in G} D_\alpha(a) B D_\alpha(a^{-1}), \tag{A.19}$$

where B is a $(d_\alpha \times d_\alpha)$ arbitrary matrix. We find $D(b)A = AD(b)$, since

$$
\begin{aligned}
D_\alpha(b)A &= \sum_{a \in G} D_\alpha(b) D_\alpha(a) B D_\alpha(a^{-1}) \\
&= \sum_{a \in G} D_\alpha(ba) B D_\alpha((ba)^{-1}) D_\alpha(b) \\
&= \sum_{c \in G} D_\alpha(c) B D_\alpha(c^{-1}) D_\alpha(b). \tag{A.20}
\end{aligned}
$$

That is, by use of Schur's lemma (II) it is found that the matrix A should be proportional to the $(d_\alpha \times d_\alpha)$ identity matrix. We choose $B_{ij} = \delta_{i\ell} \delta_{jm}$. Then, we obtain

$$A_{ij} = \sum_{a \in G} D_\alpha(a)_{i\ell} D_\alpha(a^{-1})_{mj}, \tag{A.21}$$

and right hand side (RHS) should be written by $\lambda(\ell, m)\delta_{ij}$, that is,

$$\sum_{a \in G} D_\alpha(a)_{i\ell} D_\alpha(a^{-1})_{mj} = \lambda(\ell, m)\delta_{ij}. \tag{A.22}$$

Furthermore, we compute the trace of both sides. The trace of RHS is computed as

$$\lambda(\ell, m)\mathrm{tr}\delta_{ij} = d_\alpha \lambda(\ell, m), \tag{A.23}$$

while the trace of left hand side (LHS) is obtained as

$$
\begin{aligned}
\sum_{i=1}^{d} \sum_{a \in G} D_\alpha(a)_{i\ell} D_\alpha(a^{-1})_{mi} &= \sum_{a \in G} D_\alpha(aa^{-1})_{\ell m} \\
&= N_G \delta_{\ell m}. \tag{A.24}
\end{aligned}
$$

By comparing these results, we obtain $\lambda(\ell, m) = \frac{N_G}{d_\alpha}\delta_{\ell m}$. Then, we find

$$\sum_{a \in G} D_\alpha(a)_{i\ell} D_\alpha(a^{-1})_{mj} = \frac{N_G}{d_\alpha}\delta_{ij}\delta_{\ell m}. \qquad (A.25)$$

Similarly, we define

$$A^{(\alpha\beta)} = \sum_{a \in G} D_\alpha(a) B D_\beta(a^{-1}), \qquad (A.26)$$

where $D_\alpha(a)$ and $D_\beta(a)$ are inequivalent to each other. Then, we find $D_\alpha(a)A = A D_\beta(a)$. Similarly to the previous analysis, using Schur's lemma (I), we can obtain

$$\sum_{a \in G} D_\alpha(a)_{i\ell} D_\beta(a^{-1})_{mj} = 0. \qquad (A.27)$$

Thus, we can obtain Eq. (A.18). Furthermore, if the representation is unitary, Eq. (A.18) is written as

$$\sum_{a \in G} D_\alpha(a)_{i\ell} D_\beta^*(a)_{jm} = \frac{N_G}{d_\alpha}\delta_{\alpha\beta}\delta_{ij}\delta_{\ell m}. \qquad \blacksquare \qquad (A.28)$$

Because of this orthogonality, we can expand an arbitrary function of a, $F(a)$, in terms of the matrix elements of irreducible representations

$$F(a) = \sum_{\alpha,j,k} c_{j,k}^\alpha D_\alpha(a)_{jk}. \qquad (A.29)$$

• **Theorem**
The characters for $D_\alpha(g)$ and $D_\beta(g)$ representations, $\chi_\alpha(g)$ and $\chi_\beta(g)$, satisfy the following orthogonality relation,

$$\sum_{g \in G} \chi_{D_\alpha}(g)^* \chi_{D_\beta}(g) = N_G \delta_{\alpha\beta}. \qquad (A.30)$$

Proof)
From Eq. (A.28) we obtain

$$\sum_{g \in G} D_\alpha(g)_{ii} D_\beta^*(g)_{jj} = \frac{N_G}{d_\alpha}\delta_{\alpha\beta}\delta_{ij}. \qquad (A.31)$$

Thus, by summing over all i and j, we obtain Eq. (A.30). $\blacksquare$

The **class function** is defined as a function of a, $F(a)$, which satisfies

$$F(g^{-1}ag) = F(a), \qquad \forall g \in G. \tag{A.32}$$

• **Theorem**
The number of irreducible representations is equal to the number of conjugacy classes.

Proof) The class function can also be expanded in terms of the matrix elements of the irreducible representations as (A.29). Then, it is found that

$$\begin{aligned}
F(a) &= \frac{1}{N_G} \sum_{g \in G} F(g^{-1}ag) \\
&= \frac{1}{N_G} \sum_{g \in G} \sum_{\alpha, j, k} c^{\alpha}_{j,k} D_{\alpha}(g^{-1}ag)_{jk} \\
&= \frac{1}{N_G} \sum_{g \in G} \sum_{\alpha, j, k} c^{\alpha}_{j,k} \left(D_{\alpha}(g^{-1}) D_{\alpha}(a) D_{\alpha}(g) \right)_{jk}.
\end{aligned} \tag{A.33}$$

By using the orthogonality relation (A.28), we obtain

$$\begin{aligned}
F(a) &= \sum_{\alpha, j, \ell} \frac{1}{d_{\alpha}} c^{\alpha}_{j,j} D_{\alpha}(a)_{\ell\ell} \\
&= \sum_{\alpha, j} \frac{1}{d_{\alpha}} c^{\alpha}_{j,j} \chi_{\alpha}(a).
\end{aligned} \tag{A.34}$$

That is, any class function, $F(a)$, which is constant on conjugacy classes, can be expanded by the characters $\chi_{\alpha}(a)$. That implies that the number of irreducible representations is equal to the number of conjugacy classes. ∎

• **Theorem**
The characters satisfy the following orthogonality relation,

$$\sum_{\alpha} \chi_{D_{\alpha}}(C_i)^* \chi_{D_{\alpha}}(C_j) = \frac{N_G}{n_i} \delta_{C_i C_j}, \tag{A.35}$$

where C_i and C_j denote the conjugacy classes and n_i is the number of elements in the conjugacy class C_i.

Proof)
We define the following matrix $V_{i\alpha}$,

$$V_{i\alpha} = \sqrt{\frac{n_i}{N_G}} \chi_{\alpha}(C_i), \tag{A.36}$$

where n_i is the number of elements in the conjugacy class C_i. Note that i and α label the conjugacy class C_i and the irreducible representation, respectively. The matrix $V_{i\alpha}$ is a square matrix because the number of irreducible representations is equal to the number of conjugacy classes. By use of $V_{i\alpha}$, the orthogonality relation (A.30) can be rewritten as $V^\dagger V = 1$, that is, V is unitary. Thus, we also obtain $VV^\dagger = 1$. That means Eq. (A.35). ■

Representations of S_4 in Several Bases

B

For the S_4 group, several bases of representations have been used in the literature. Most of group-theoretical aspects such as conjugacy classes and characters are independent of the basis of representations. Tensor products are also independent of the basis. For example, we always have

$$2 \otimes 2 = \mathbf{1}_1 \oplus \mathbf{1}_2 \oplus \mathbf{2}, \tag{B.1}$$

in any basis. However, it depends on the basis of representation how this equation is written by components. For example, the singlets $\mathbf{1}_1$ and $\mathbf{1}_2$ in RHS are represented by components of $\mathbf{2}$ in LHS, but their forms depend on the basis of representations as we will see below. For applications, it is useful to show explicitly the transformation of bases and tensor products for several bases. That is shown below.

B.1   The Basis I

First, we show the basis in Sect. 3.2. All of the S_4 elements are written by products of the generators b_1 and d_4, which satisfy

$$(b_1)^3 = (d_4)^4 = e, \quad d_4(b_1)^2 d_4 = b_1, \quad d_4 b_1 d_4 = b_1 (d_4)^2 b_1 . \tag{B.2}$$

These generators are represented on $\mathbf{2}$, $\mathbf{3}$ and $\mathbf{3}'$ as follows,

$$b_1 = \begin{pmatrix} \omega & 0 \\ 0 & \omega^2 \end{pmatrix}, \quad d_4 = \begin{pmatrix} 0 & 1 \\ 1 & 0 \end{pmatrix}, \qquad \text{on } \mathbf{2}, \tag{B.3}$$

$$b_1 = \begin{pmatrix} 0 & 0 & 1 \\ 1 & 0 & 0 \\ 0 & 1 & 0 \end{pmatrix}, \quad d_4 = \begin{pmatrix} -1 & 0 & 0 \\ 0 & 0 & -1 \\ 0 & 1 & 0 \end{pmatrix}, \qquad \text{on } \mathbf{3}, \tag{B.4}$$

© Springer-Verlag GmbH Germany, part of Springer Nature 2022
T. Kobayashi et al., *An Introduction to Non-Abelian Discrete Symmetries for Particle Physicists,* Lecture Notes in Physics 995,
https://doi.org/10.1007/978-3-662-64679-3

$$b_1 = \begin{pmatrix} 0 & 0 & 1 \\ 1 & 0 & 0 \\ 0 & 1 & 0 \end{pmatrix}, \quad d_4 = \begin{pmatrix} 1 & 0 & 0 \\ 0 & 0 & 1 \\ 0 & -1 & 0 \end{pmatrix}, \quad \text{on } \mathbf{3}'. \tag{B.5}$$

Therefore, the multiplication rules are obtained as follows:

$$\begin{pmatrix} a_1 \\ a_2 \end{pmatrix}_2 \otimes \begin{pmatrix} b_1 \\ b_2 \end{pmatrix}_2 = (a_1 b_2 + a_2 b_1)_1 \oplus (a_1 b_2 - a_2 b_1)_{1'} \oplus \begin{pmatrix} a_2 b_2 \\ a_1 b_1 \end{pmatrix}_2, \tag{B.6}$$

$$\begin{pmatrix} a_1 \\ a_2 \end{pmatrix}_2 \otimes \begin{pmatrix} b_1 \\ b_2 \\ b_3 \end{pmatrix}_3 = \begin{pmatrix} a_1 b_1 + a_2 b_1 \\ \omega^2 a_1 b_2 + \omega a_2 b_2 \\ \omega a_1 b_3 + \omega^2 a_2 b_3 \end{pmatrix}_3 \oplus \begin{pmatrix} a_1 b_1 - a_2 b_1 \\ \omega^2 a_1 b_2 - \omega a_2 b_2 \\ \omega a_1 b_3 - \omega^2 a_2 b_3 \end{pmatrix}_{3'}, \tag{B.7}$$

$$\begin{pmatrix} a_1 \\ a_2 \end{pmatrix}_2 \otimes \begin{pmatrix} b_1 \\ b_2 \\ b_3 \end{pmatrix}_{3'} = \begin{pmatrix} a_1 b_1 - a_2 b_1 \\ \omega^2 a_1 b_2 - \omega a_2 b_2 \\ \omega a_1 b_3 - \omega^2 a_2 b_3 \end{pmatrix}_3 \oplus \begin{pmatrix} a_1 b_1 + a_2 b_1 \\ \omega^2 a_1 b_2 + \omega a_2 b_2 \\ \omega a_1 b_3 + \omega^2 a_2 b_3 \end{pmatrix}_{3'}, \tag{B.8}$$

$$\begin{pmatrix} a_1 \\ a_2 \\ a_3 \end{pmatrix}_3 \otimes \begin{pmatrix} b_1 \\ b_2 \\ b_3 \end{pmatrix}_3 = (a_1 b_1 + a_2 b_2 + a_3 b_3)_1 \oplus \begin{pmatrix} a_1 b_1 + \omega a_2 b_2 + \omega^2 a_3 b_3 \\ a_1 b_1 + \omega^2 a_2 b_2 + \omega a_3 b_3 \end{pmatrix}_2$$
$$\oplus \begin{pmatrix} a_2 b_3 + a_3 b_2 \\ a_3 b_1 + a_1 b_3 \\ a_1 b_2 + a_2 b_1 \end{pmatrix}_3 \oplus \begin{pmatrix} a_2 b_3 - a_3 b_2 \\ a_3 b_1 - a_1 b_3 \\ a_1 b_2 - a_2 b_1 \end{pmatrix}_{3'}, \tag{B.9}$$

$$\begin{pmatrix} a_1 \\ a_2 \\ a_3 \end{pmatrix}_{3'} \otimes \begin{pmatrix} b_1 \\ b_2 \\ b_3 \end{pmatrix}_{3'} = (a_1 b_1 + a_2 b_2 + a_3 b_3)_1 \oplus \begin{pmatrix} a_1 b_1 + \omega a_2 b_2 + \omega^2 a_3 b_3 \\ a_1 b_1 + \omega^2 a_2 b_2 + \omega a_3 b_3 \end{pmatrix}_2$$
$$\oplus \begin{pmatrix} a_2 b_3 + a_3 b_2 \\ a_3 b_1 + a_1 b_3 \\ a_1 b_2 + a_2 b_1 \end{pmatrix}_3 \oplus \begin{pmatrix} a_2 b_3 - a_3 b_2 \\ a_3 b_1 - a_1 b_3 \\ a_1 b_2 - a_2 b_1 \end{pmatrix}_{3'}, \tag{B.10}$$

$$\begin{pmatrix} a_1 \\ a_2 \\ a_3 \end{pmatrix}_3 \otimes \begin{pmatrix} b_1 \\ b_2 \\ b_3 \end{pmatrix}_{3'} = (a_1 b_1 + a_2 b_2 + a_3 b_3)_{1'} \oplus \begin{pmatrix} a_1 b_1 + \omega a_2 b_2 + \omega^2 a_3 b_3 \\ -a_1 b_1 - \omega^2 a_2 b_2 - \omega a_3 b_3 \end{pmatrix}_2$$
$$\oplus \begin{pmatrix} a_2 b_3 - a_3 b_2 \\ a_3 b_1 - a_1 b_3 \\ a_1 b_2 - a_2 b_1 \end{pmatrix}_3 \oplus \begin{pmatrix} a_2 b_3 + a_3 b_2 \\ a_3 b_1 + a_1 b_3 \\ a_1 b_2 + a_2 b_1 \end{pmatrix}_{3'}, \tag{B.11}$$

$$a_1 \otimes \begin{pmatrix} b_1 \\ b_2 \\ b_3 \end{pmatrix}_{3,3'} = \begin{pmatrix} ab_1 \\ ab_2 \\ ab_3 \end{pmatrix}_{3,3'}, \quad a_{1'} \otimes \begin{pmatrix} b_1 \\ b_2 \\ b_3 \end{pmatrix}_{3,3'} = \begin{pmatrix} ab_1 \\ ab_2 \\ ab_3 \end{pmatrix}_{3',3}, \tag{B.12}$$

$$a_1 \otimes \begin{pmatrix} b_1 \\ b_2 \end{pmatrix}_2 = \begin{pmatrix} ab_1 \\ ab_2 \end{pmatrix}_2, \quad a_{1'} \otimes \begin{pmatrix} b_1 \\ b_2 \end{pmatrix}_2 = \begin{pmatrix} ab_1 \\ -ab_2 \end{pmatrix}_2. \tag{B.13}$$

B.2   The Basis II

Next, we consider another basis, which is used e.g. in Ref. [5]. Following Ref. [5], we denote the generators b_1 and d_4 by $b = b_1$ and $a = d_4$. In this basis, the generators, a and b, are represented as

$$a = \begin{pmatrix} -1 & 0 \\ 0 & 1 \end{pmatrix}, \quad b = -\frac{1}{2}\begin{pmatrix} 1 & \sqrt{3} \\ -\sqrt{3} & 1 \end{pmatrix}, \qquad \text{on } \mathbf{2}, \qquad (B.14)$$

$$a = \begin{pmatrix} -1 & 0 & 0 \\ 0 & 0 & -1 \\ 0 & 1 & 0 \end{pmatrix}, \quad b = \begin{pmatrix} 0 & 0 & 1 \\ 1 & 0 & 0 \\ 0 & 1 & 0 \end{pmatrix}, \qquad \text{on } \mathbf{3}_1, \qquad (B.15)$$

$$a = \begin{pmatrix} 1 & 0 & 0 \\ 0 & 0 & 1 \\ 0 & -1 & 0 \end{pmatrix}, \quad b = \begin{pmatrix} 0 & 0 & 1 \\ 1 & 0 & 0 \\ 0 & 1 & 0 \end{pmatrix}, \qquad \text{on } \mathbf{3}_2, \qquad (B.16)$$

where we define as $\mathbf{3}_1 \equiv \mathbf{3}$ and $\mathbf{3}_2 \equiv \mathbf{3}'$ hereafter. These generators, a and b, are represented in the real basis. On the other hand, the above generators, b_1 and d_4, are represented in the complex basis. These bases for $\mathbf{2}$ are transformed by the unitary transformation, $U^\dagger g U$, where

$$U = \frac{1}{\sqrt{2}}\begin{pmatrix} 1 & i \\ -1 & i \end{pmatrix}. \qquad (B.17)$$

That is, the elements a and b are written by b_1 and d_4 as

$$b = U^\dagger b_1 U = -\frac{1}{2}\begin{pmatrix} 1 & \sqrt{3} \\ -\sqrt{3} & 1 \end{pmatrix}, \quad a = U^\dagger d_4 U = \begin{pmatrix} -1 & 0 \\ 0 & 1 \end{pmatrix}, \qquad (B.18)$$

in the real basis. For the triplets, the (b_1, d_4) basis is the same as the (b, a) basis. Therefore, the multiplication rules are obtained as follows:

$$\begin{pmatrix} a_1 \\ a_2 \end{pmatrix}_2 \otimes \begin{pmatrix} b_1 \\ b_2 \end{pmatrix}_2 = (a_1 b_1 + a_2 b_2)_{1_1} \oplus (-a_1 b_2 + a_2 b_1)_{1_2} \oplus \begin{pmatrix} a_1 b_2 + a_2 b_1 \\ a_1 b_1 - a_2 b_2 \end{pmatrix}_2,$$

$$(B.19)$$

$$\begin{pmatrix} a_1 \\ a_2 \end{pmatrix}_2 \otimes \begin{pmatrix} b_1 \\ b_2 \\ b_3 \end{pmatrix}_{3_1} = \begin{pmatrix} a_2 b_1 \\ -\frac{1}{2}(\sqrt{3}a_1 b_2 + a_2 b_2) \\ \frac{1}{2}(\sqrt{3}a_1 b_3 - a_2 b_3) \end{pmatrix}_{3_1} \oplus \begin{pmatrix} a_1 b_1 \\ \frac{1}{2}(\sqrt{3}a_2 b_2 - a_1 b_2) \\ -\frac{1}{2}(\sqrt{3}a_2 b_3 + a_1 b_3) \end{pmatrix}_{3_2},$$

$$(B.20)$$

$$\begin{pmatrix} a_1 \\ a_2 \end{pmatrix}_2 \otimes \begin{pmatrix} b_1 \\ b_2 \\ b_3 \end{pmatrix}_{3_2} = \begin{pmatrix} a_1 b_1 \\ \frac{1}{2}(\sqrt{3}a_2 b_2 - a_1 b_2) \\ -\frac{1}{2}(\sqrt{3}a_2 b_3 + a_1 b_3) \end{pmatrix}_{3_1} \oplus \begin{pmatrix} a_2 b_1 \\ -\frac{1}{2}(\sqrt{3}a_1 b_2 + a_2 b_2) \\ \frac{1}{2}(\sqrt{3}a_1 b_3 - a_2 b_3) \end{pmatrix}_{3_2},$$

$$(B.21)$$

$$\begin{pmatrix} a_1 \\ a_2 \\ a_3 \end{pmatrix}_{3_1} \otimes \begin{pmatrix} b_1 \\ b_2 \\ b_3 \end{pmatrix}_{3_1} = (a_1 b_1 + a_2 b_2 + a_3 b_3)_{1_1} \oplus \begin{pmatrix} \frac{1}{\sqrt{2}}(a_2 b_2 - a_3 b_3) \\ \frac{1}{\sqrt{6}}(-2a_1 b_1 + a_2 b_2 + a_3 b_3) \end{pmatrix}_2$$

$$\oplus \begin{pmatrix} a_2 b_3 + a_3 b_2 \\ a_1 b_3 + a_3 b_1 \\ a_1 b_2 + a_2 b_1 \end{pmatrix}_{3_1} \oplus \begin{pmatrix} a_3 b_2 - a_2 b_3 \\ a_1 b_3 - a_3 b_1 \\ a_2 b_1 - a_1 b_2 \end{pmatrix}_{3_2} , \tag{B.22}$$

$$\begin{pmatrix} a_1 \\ a_2 \\ a_3 \end{pmatrix}_{3_2} \otimes \begin{pmatrix} b_1 \\ b_2 \\ b_3 \end{pmatrix}_{3_2} = (a_1 b_1 + a_2 b_2 + a_3 b_3)_{1_1} \oplus \begin{pmatrix} \frac{1}{\sqrt{2}}(a_2 b_2 - a_3 b_3) \\ \frac{1}{\sqrt{6}}(-2a_1 b_1 + a_2 b_2 + a_3 b_3) \end{pmatrix}_2$$

$$\oplus \begin{pmatrix} a_2 b_3 + a_3 b_2 \\ a_1 b_3 + a_3 b_1 \\ a_1 b_2 + a_2 b_1 \end{pmatrix}_{3_1} \oplus \begin{pmatrix} a_3 b_2 - a_2 b_3 \\ a_1 b_3 - a_3 b_1 \\ a_2 b_1 - a_1 b_2 \end{pmatrix}_{3_2} , \tag{B.23}$$

$$\begin{pmatrix} a_1 \\ a_2 \\ a_3 \end{pmatrix}_{3_1} \otimes \begin{pmatrix} b_1 \\ b_2 \\ b_3 \end{pmatrix}_{3_2} = (a_1 b_1 + a_2 b_2 + a_3 b_3)_{1_2} \oplus \begin{pmatrix} \frac{1}{\sqrt{6}}(2a_1 b_1 - a_2 b_2 - a_3 b_3) \\ \frac{1}{\sqrt{2}}(a_2 b_2 - a_3 b_3) \end{pmatrix}_2$$

$$\oplus \begin{pmatrix} a_3 b_2 - a_2 b_3 \\ a_1 b_3 - a_3 b_1 \\ a_2 b_1 - a_1 b_2 \end{pmatrix}_{3_1} \oplus \begin{pmatrix} a_2 b_3 + a_3 b_2 \\ a_1 b_3 + a_3 b_1 \\ a_1 b_2 + a_2 b_1 \end{pmatrix}_{3_2} , \tag{B.24}$$

$$a_1 \otimes \begin{pmatrix} b_1 \\ b_2 \\ b_3 \end{pmatrix}_{3,3'} = \begin{pmatrix} ab_1 \\ ab_2 \\ ab_3 \end{pmatrix}_{3,3'} , \quad a_{1'} \otimes \begin{pmatrix} b_1 \\ b_2 \\ b_3 \end{pmatrix}_{3,3'} = \begin{pmatrix} ab_1 \\ ab_2 \\ ab_3 \end{pmatrix}_{3',3} , \tag{B.25}$$

$$a_1 \otimes \begin{pmatrix} b_1 \\ b_2 \end{pmatrix}_2 = \begin{pmatrix} ab_1 \\ ab_2 \end{pmatrix}_2 , \quad a_{1'} \otimes \begin{pmatrix} b_1 \\ b_2 \end{pmatrix}_2 = \begin{pmatrix} -ab_2 \\ ab_1 \end{pmatrix}_2 . \tag{B.26}$$

B.3   The Basis III

Next, we consider a different basis, which is used, e.g. in Ref. [6], with the generator s and t corresponding to d_4 and b_1, respectively. These generators are represented

$$s = \begin{pmatrix} 0 & 1 \\ 1 & 0 \end{pmatrix}, \quad t = \begin{pmatrix} \omega & 0 \\ 0 & \omega^2 \end{pmatrix}, \qquad \text{on } \mathbf{2}, \tag{B.27}$$

$$s = \frac{1}{3} \begin{pmatrix} -1 & 2\omega & 2\omega^2 \\ 2\omega & 2\omega^2 & -1 \\ 2\omega^2 & -1 & 2\omega \end{pmatrix}, \quad t = \begin{pmatrix} 1 & 0 & 0 \\ 0 & \omega^2 & 0 \\ 0 & 0 & \omega \end{pmatrix}, \qquad \text{on } \mathbf{3_1}, \tag{B.28}$$

$$s = \frac{1}{3} \begin{pmatrix} 1 & -2\omega & -2\omega^2 \\ -2\omega & -2\omega^2 & 1 \\ -2\omega^2 & 1 & -2\omega \end{pmatrix}, \quad t = \begin{pmatrix} 1 & 0 & 0 \\ 0 & \omega^2 & 0 \\ 0 & 0 & \omega \end{pmatrix}, \qquad \text{on } \mathbf{3_2}. \tag{B.29}$$

The doublet of this basis [6] is the same as the (d_4, b_1) basis. In the representations $\mathbf{3}_1$ and $\mathbf{3}_2$, the (s, t) basis and (d_4, b_1) basis are transformed by the following unitary matrix U_ω:

$$
U_\omega = \frac{1}{\sqrt{3}} \begin{pmatrix} 1 & 1 & 1 \\ 1 & \omega & \omega^2 \\ 1 & \omega^2 & \omega \end{pmatrix}, \tag{B.30}
$$

which is the so-called magic matrix. That is, the elements s and t are written by d_4 and b_1 as

$$
s = U_\omega^\dagger d_4 U_\omega = \frac{1}{3} \begin{pmatrix} -1 & 2\omega & 2\omega^2 \\ 2\omega & 2\omega^2 & -1 \\ 2\omega^2 & -1 & 2\omega \end{pmatrix}, \quad t = U_\omega^\dagger b_1 U_\omega = \begin{pmatrix} 1 & 0 & 0 \\ 0 & \omega^2 & 0 \\ 0 & 0 & \omega \end{pmatrix}. \tag{B.31}
$$

For $\mathbf{3}_2$, we also find s and t in the same way.

Therefore, the multiplication rules are obtained as follows:

$$
\begin{pmatrix} a_1 \\ a_2 \end{pmatrix}_2 \otimes \begin{pmatrix} b_1 \\ b_2 \end{pmatrix}_2 = (a_1 b_2 + a_2 b_1)_{1_1} \oplus (a_1 b_2 - a_2 b_1)_{1_2} \oplus \begin{pmatrix} a_2 b_2 \\ a_1 b_1 \end{pmatrix}_2, \tag{B.32}
$$

$$
\begin{pmatrix} a_1 \\ a_2 \end{pmatrix}_2 \otimes \begin{pmatrix} b_1 \\ b_2 \\ b_3 \end{pmatrix}_{3_1} = \begin{pmatrix} a_1 b_2 + a_2 b_3 \\ a_1 b_3 + a_2 b_1 \\ a_1 b_1 + a_2 b_2 \end{pmatrix}_{3_1} \oplus \begin{pmatrix} a_1 b_2 - a_2 b_3 \\ a_1 b_3 - a_2 b_1 \\ a_1 b_1 - a_2 b_2 \end{pmatrix}_{3_2}, \tag{B.33}
$$

$$
\begin{pmatrix} a_1 \\ a_2 \end{pmatrix}_2 \otimes \begin{pmatrix} b_1 \\ b_2 \\ b_3 \end{pmatrix}_{3_2} = \begin{pmatrix} a_1 b_2 - a_2 b_3 \\ a_1 b_3 - a_2 b_1 \\ a_1 b_1 - a_2 b_2 \end{pmatrix}_{3_1} \oplus \begin{pmatrix} a_1 b_2 + a_2 b_3 \\ a_1 b_3 + a_2 b_1 \\ a_1 b_1 + a_2 b_2 \end{pmatrix}_{3_2}, \tag{B.34}
$$

$$
\begin{pmatrix} a_1 \\ a_2 \\ a_3 \end{pmatrix}_{3_1} \otimes \begin{pmatrix} b_1 \\ b_2 \\ b_3 \end{pmatrix}_{3_1} = (a_1 b_1 + a_2 b_3 + a_3 b_2)_{1_1} \oplus \begin{pmatrix} a_2 b_2 + a_1 b_3 + a_3 b_1 \\ a_3 b_3 + a_1 b_2 + a_2 b_1 \end{pmatrix}_2
$$
$$
\oplus \begin{pmatrix} 2 a_1 b_1 - a_2 b_3 - a_3 b_2 \\ 2 a_3 b_3 - a_1 b_2 - a_2 b_1 \\ 2 a_2 b_2 - a_1 b_3 - a_3 b_1 \end{pmatrix}_{3_1} \oplus \begin{pmatrix} a_2 b_3 - a_3 b_2 \\ a_1 b_2 - a_2 b_1 \\ a_3 b_1 - a_1 b_3 \end{pmatrix}_{3_2}, \tag{B.35}
$$

$$
\begin{pmatrix} a_1 \\ a_2 \\ a_3 \end{pmatrix}_{3_2} \otimes \begin{pmatrix} b_1 \\ b_2 \\ b_3 \end{pmatrix}_{3_2} = (a_1 b_1 + a_2 b_3 + a_3 b_2)_{1_1} \oplus \begin{pmatrix} a_2 b_2 + a_1 b_3 + a_3 b_1 \\ a_3 b_3 + a_1 b_2 + a_2 b_1 \end{pmatrix}_2
$$
$$
\oplus \begin{pmatrix} 2 a_1 b_1 - a_2 b_3 - a_3 b_2 \\ 2 a_3 b_3 - a_1 b_2 - a_2 b_1 \\ 2 a_2 b_2 - a_1 b_3 - a_3 b_1 \end{pmatrix}_{3_1} \oplus \begin{pmatrix} a_2 b_3 - a_3 b_2 \\ a_1 b_2 - a_2 b_1 \\ a_3 b_1 - a_1 b_3 \end{pmatrix}_{3_2}, \tag{B.36}
$$

$$\begin{pmatrix} a_1 \\ a_2 \\ a_3 \end{pmatrix}_{3_1} \otimes \begin{pmatrix} b_1 \\ b_2 \\ b_3 \end{pmatrix}_{3_2} = (a_1b_1 + a_2b_3 + a_3b_2)_{1_2} \oplus \begin{pmatrix} a_2b_2 + a_1b_3 + a_3b_1 \\ -a_3b_3 - a_1b_2 - a_2b_1 \end{pmatrix}_2$$

$$\oplus \begin{pmatrix} a_2b_3 - a_3b_2 \\ a_1b_2 - a_2b_1 \\ a_3b_1 - a_1b_3 \end{pmatrix}_{3_1} \oplus \begin{pmatrix} 2a_1b_1 - a_2b_3 - a_3b_2 \\ 2a_3b_3 - a_1b_2 - a_2b_1 \\ 2a_2b_2 - a_1b_3 - a_3b_1 \end{pmatrix}_{3_2} \ , \quad (B.37)$$

$$a_1 \otimes \begin{pmatrix} b_1 \\ b_2 \\ b_3 \end{pmatrix}_{3,3'} = \begin{pmatrix} ab_1 \\ ab_2 \\ ab_3 \end{pmatrix}_{3,3'} \ , \quad a_{1'} \otimes \begin{pmatrix} b_1 \\ b_2 \\ b_3 \end{pmatrix}_{3,3'} = \begin{pmatrix} ab_1 \\ ab_2 \\ ab_3 \end{pmatrix}_{3',3} \ , \quad (B.38)$$

$$a_1 \otimes \begin{pmatrix} b_1 \\ b_2 \end{pmatrix}_2 = \begin{pmatrix} ab_1 \\ ab_2 \end{pmatrix}_2 \ , \quad a_{1'} \otimes \begin{pmatrix} b_1 \\ b_2 \end{pmatrix}_2 = \begin{pmatrix} ab_1 \\ -ab_2 \end{pmatrix}_2 \ . \quad (B.39)$$

B.4 The Basis IV

Here, we consider another basis, which is used, e.g. in Ref. [7], with the generator $\tilde{t}$ and $\tilde{s}$ satisfying

$$\tilde{t}^4 = \tilde{s}^2 = (\tilde{s}\tilde{t})^3 = (\tilde{t}\tilde{s})^3 = e. \quad (B.40)$$

These generators are represented as

$$\tilde{t} = \begin{pmatrix} 1 & 0 \\ 0 & -1 \end{pmatrix}, \quad \tilde{s} = \frac{1}{2}\begin{pmatrix} -1 & \sqrt{3} \\ \sqrt{3} & 1 \end{pmatrix}, \quad \tilde{s}\tilde{t} = \frac{1}{2}\begin{pmatrix} -1 & -\sqrt{3} \\ \sqrt{3} & -1 \end{pmatrix}, \quad \text{on } \mathbf{2}, \quad (B.41)$$

$$\tilde{t} = \begin{pmatrix} -1 & 0 & 0 \\ 0 & -i & 0 \\ 0 & 0 & i \end{pmatrix}, \quad \tilde{s} = \begin{pmatrix} 0 & -\frac{1}{\sqrt{2}} & -\frac{1}{\sqrt{2}} \\ -\frac{1}{\sqrt{2}} & \frac{1}{2} & -\frac{1}{2} \\ -\frac{1}{\sqrt{2}} & -\frac{1}{2} & \frac{1}{2} \end{pmatrix},$$

$$\tilde{s}\tilde{t} = \begin{pmatrix} 0 & \frac{i}{\sqrt{2}} & -\frac{i}{\sqrt{2}} \\ \frac{1}{\sqrt{2}} & -\frac{i}{2} & -\frac{i}{2} \\ \frac{1}{\sqrt{2}} & \frac{i}{2} & \frac{i}{2} \end{pmatrix}, \quad \text{on } \mathbf{3}_1, \quad (B.42)$$

$$\tilde{t} = \begin{pmatrix} 1 & 0 & 0 \\ 0 & i & 0 \\ 0 & 0 & -i \end{pmatrix}, \quad \tilde{s} = \begin{pmatrix} 0 & \frac{1}{\sqrt{2}} & \frac{1}{\sqrt{2}} \\ \frac{1}{\sqrt{2}} & -\frac{1}{2} & \frac{1}{2} \\ \frac{1}{\sqrt{2}} & \frac{1}{2} & -\frac{1}{2} \end{pmatrix},$$

$$\tilde{s}\tilde{t} = \begin{pmatrix} 0 & \frac{i}{\sqrt{2}} & -\frac{i}{\sqrt{2}} \\ \frac{1}{\sqrt{2}} & -\frac{i}{2} & -\frac{i}{2} \\ \frac{1}{\sqrt{2}} & \frac{i}{2} & \frac{i}{2} \end{pmatrix}, \qquad \text{on } \mathbf{3}_2. \tag{B.43}$$

For the representation $\mathbf{2}$, the following unitary transformation matrix U_{doublet}:

$$U_{\text{doublet}} = \frac{1}{\sqrt{2}} \begin{pmatrix} 1 & i \\ 1 & -i \end{pmatrix}, \tag{B.44}$$

is used and the elements $\tilde{t}$ and $\tilde{s}\tilde{t}$ are written by d_1 and b_1 as

$$\tilde{t} = U_{\text{doublet}}^\dagger d_4 U_{\text{doublet}} = \begin{pmatrix} 1 & 0 \\ 0 & -1 \end{pmatrix}, \quad \tilde{s}\tilde{t} = U_{\text{doublet}}^\dagger b_1 U_{\text{doublet}} = \frac{1}{2} \begin{pmatrix} -1 & -\sqrt{3} \\ \sqrt{3} & -1 \end{pmatrix}. \tag{B.45}$$

On the other hand, for the representation $\mathbf{3}_1$ and $\mathbf{3}_2$, the following unitary transformation matrix U_{triplet}:

$$U_{\text{triplet}} = \begin{pmatrix} 1 & 0 & 0 \\ 0 & \frac{1}{\sqrt{2}} & \frac{1}{\sqrt{2}} \\ 0 & \frac{i}{\sqrt{2}} & -\frac{i}{\sqrt{2}} \end{pmatrix}, \tag{B.46}$$

is used. For $\mathbf{3}_1$, the elements $\tilde{t}$ and $\tilde{s}\tilde{t}$ are written by d_4 and b_1 as

$$\tilde{t} = U_{\text{triplet}}^\dagger d_4 U_{\text{triplet}} = \begin{pmatrix} -1 & 0 & 0 \\ 0 & -i & 0 \\ 0 & 0 & i \end{pmatrix},$$

$$\tilde{s}\tilde{t} = U_{\text{triplet}}^\dagger b_1 U_{\text{triplet}} = \begin{pmatrix} 0 & \frac{i}{\sqrt{2}} & -\frac{i}{\sqrt{2}} \\ \frac{1}{\sqrt{2}} & -\frac{i}{2} & -\frac{i}{2} \\ \frac{1}{\sqrt{2}} & \frac{i}{2} & \frac{i}{2} \end{pmatrix}. \tag{B.47}$$

For $\mathbf{3}_2$, we also find the same transformations.

Therefore, the multiplication rules are as follows:

$$\begin{pmatrix} a_1 \\ a_2 \end{pmatrix}_2 \otimes \begin{pmatrix} b_1 \\ b_2 \end{pmatrix}_2 = (a_1 b_1 + a_2 b_2)_{\mathbf{1}_1} \oplus (a_1 b_2 - a_2 b_1)_{\mathbf{1}_2} \oplus \begin{pmatrix} a_2 b_2 - a_1 b_1 \\ a_1 b_2 + a_2 b_1 \end{pmatrix}_2, \tag{B.48}$$

$$\begin{pmatrix} a_1 \\ a_2 \end{pmatrix}_2 \otimes \begin{pmatrix} b_1 \\ b_2 \\ b_3 \end{pmatrix}_{\mathbf{3}_1} = \begin{pmatrix} a_1 b_1 \\ \frac{\sqrt{3}}{2} a_2 b_3 - \frac{1}{2} a_1 b_2 \\ \frac{\sqrt{3}}{2} a_2 b_2 - a_1 b_3 \end{pmatrix}_{\mathbf{3}_1} \oplus \begin{pmatrix} -a_2 b_1 \\ \frac{\sqrt{3}}{2} a_1 b_3 + \frac{1}{2} a_2 b_2 \\ \frac{\sqrt{3}}{2} a_1 b_2 + \frac{1}{2} a_2 b_3 \end{pmatrix}_{\mathbf{3}_2}, \tag{B.49}$$

$$\begin{pmatrix} a_1 \\ a_2 \end{pmatrix}_2 \otimes \begin{pmatrix} b_1 \\ b_2 \\ b_3 \end{pmatrix}_{\mathbf{3}_2} = \begin{pmatrix} -a_2 b_1 \\ \frac{\sqrt{3}}{2} a_1 b_3 + \frac{1}{2} a_2 b_2 \\ \frac{\sqrt{3}}{2} a_1 b_2 + \frac{1}{2} a_2 b_3 \end{pmatrix}_{\mathbf{3}_1} \oplus \begin{pmatrix} a_1 b_1 \\ \frac{\sqrt{3}}{2} a_2 b_3 - \frac{1}{2} a_1 b_2 \\ \frac{\sqrt{3}}{2} a_2 b_2 - a_1 b_3 \end{pmatrix}_{\mathbf{3}_2}. \tag{B.50}$$

$$\begin{pmatrix} a_1 \\ a_2 \\ a_3 \end{pmatrix}_{3_1} \otimes \begin{pmatrix} b_1 \\ b_2 \\ b_3 \end{pmatrix}_{3_1} = (a_1b_1 + a_2b_3 + a_3b_2)_{1_1} \oplus \begin{pmatrix} a_1b_1 - \frac{1}{2}(a_2b_3 + a_3b_2) \\ \frac{\sqrt{3}}{2}(a_2b_2 + a_3b_3) \end{pmatrix}_2$$

$$\oplus \begin{pmatrix} a_3b_3 - a_2b_2 \\ a_1b_3 + a_3b_1 \\ -a_1b_2 - a_2b_1 \end{pmatrix}_{3_1} \oplus \begin{pmatrix} a_3b_2 - a_2b_3 \\ a_2b_1 - a_1b_2 \\ a_1b_3 - a_3b_1 \end{pmatrix}_{3_2}, \qquad (B.51)$$

$$\begin{pmatrix} a_1 \\ a_2 \\ a_3 \end{pmatrix}_{3_2} \otimes \begin{pmatrix} b_1 \\ b_2 \\ b_3 \end{pmatrix}_{3_2} = (a_1b_1 + a_2b_3 + a_3b_2)_{1_1} \oplus \begin{pmatrix} a_1b_1 - \frac{1}{2}(a_2b_3 + a_3b_2) \\ \frac{\sqrt{3}}{2}(a_2b_2 + a_3b_3) \end{pmatrix}_2$$

$$\oplus \begin{pmatrix} a_3b_3 - a_2b_2 \\ a_1b_3 + a_3b_1 \\ -a_1b_2 - a_2b_1 \end{pmatrix}_{3_1} \oplus \begin{pmatrix} a_3b_2 - a_2b_3 \\ a_2b_1 - a_1b_2 \\ a_1b_3 - a_3b_1 \end{pmatrix}_{3_2}, \qquad (B.52)$$

$$\begin{pmatrix} a_1 \\ a_2 \\ a_3 \end{pmatrix}_{3_1} \otimes \begin{pmatrix} b_1 \\ b_2 \\ b_3 \end{pmatrix}_{3_2} = (a_1b_1 + a_2b_3 + a_3b_2)_{1_2}$$

$$\oplus \left(\frac{\sqrt{3}}{2}(a_2b_2 + a_3b_3) - a_1b_1 + \frac{1}{2}(a_2b_3 + a_3b_2) \right)_2$$

$$\oplus \begin{pmatrix} a_3b_2 - a_2b_3 \\ a_2b_1 - a_1b_2 \\ a_1b_3 - a_3b_1 \end{pmatrix}_{3_1} \oplus \begin{pmatrix} a_3b_3 - a_2b_2 \\ a_1b_3 + a_3b_1 \\ -a_1b_2 - a_2b_1 \end{pmatrix}_{3_2}, \qquad (B.53)$$

$$a_1 \otimes \begin{pmatrix} b_1 \\ b_2 \\ b_3 \end{pmatrix}_{3,3'} = \begin{pmatrix} ab_1 \\ ab_2 \\ ab_3 \end{pmatrix}_{3,3'}, \quad a_{1'} \otimes \begin{pmatrix} b_1 \\ b_2 \\ b_3 \end{pmatrix}_{3,3'} = \begin{pmatrix} ab_1 \\ ab_2 \\ ab_3 \end{pmatrix}_{3',3}, \qquad (B.54)$$

$$a_1 \otimes \begin{pmatrix} b_1 \\ b_2 \end{pmatrix}_2 = \begin{pmatrix} ab_1 \\ ab_2 \end{pmatrix}_2, \quad a_{1'} \otimes \begin{pmatrix} b_1 \\ b_2 \end{pmatrix}_2 = \begin{pmatrix} -ab_2 \\ ab_1 \end{pmatrix}_2. \qquad (B.55)$$

Representations of A_4 in Different Basis

C

We show another basis for representations of the A_4 group, which is used, e.g. in [8]. In this basis, we denote the generators S and T, which correspond to s and t in Sect. 4.1, respectively. These bases are transformed by the following unitary transformation matrix U_ω as

$$U_\omega = \frac{1}{\sqrt{3}} \begin{pmatrix} 1 & 1 & 1 \\ 1 & \omega & \omega^2 \\ 1 & \omega^2 & \omega \end{pmatrix}, \tag{C.1}$$

and the elements S and T are explicitly written as

$$S = U_\omega^\dagger s\, U_\omega = \frac{1}{3} \begin{pmatrix} -1 & 2 & 2 \\ 2 & -1 & 2 \\ 2 & 2 & -1 \end{pmatrix}, \quad T = U_\omega^\dagger t\, U_\omega = \begin{pmatrix} 1 & 0 & 0 \\ 0 & \omega^2 & 0 \\ 0 & 0 & \omega \end{pmatrix}. \tag{C.2}$$

Therefore, the multiplication rule of the triplet is obtained as follows,

$$\begin{pmatrix} a_1 \\ a_2 \\ a_3 \end{pmatrix}_3 \otimes \begin{pmatrix} b_1 \\ b_2 \\ b_3 \end{pmatrix}_3 = (a_1 b_1 + a_2 b_3 + a_3 b_2)_1 \oplus (a_3 b_3 + a_1 b_2 + a_2 b_1)_{1'}$$

$$\oplus\, (a_2 b_2 + a_1 b_3 + a_3 b_1)_{1''}$$

$$\oplus \frac{1}{3} \begin{pmatrix} 2a_1 b_1 - a_2 b_3 - a_3 b_2 \\ 2a_3 b_3 - a_1 b_2 - a_2 b_1 \\ 2a_2 b_2 - a_1 b_3 - a_3 b_1 \end{pmatrix}_3 \oplus \frac{1}{2} \begin{pmatrix} a_2 b_3 - a_3 b_2 \\ a_1 b_2 - a_2 b_1 \\ a_3 b_1 - a_1 b_3 \end{pmatrix}_3, \tag{C.3}$$

© Springer-Verlag GmbH Germany, part of Springer Nature 2022
T. Kobayashi et al., *An Introduction to Non-Abelian Discrete Symmetries for Particle Physicists,* Lecture Notes in Physics 995,
https://doi.org/10.1007/978-3-662-64679-3

$$a_1 \otimes \begin{pmatrix} b_1 \\ b_2 \\ b_3 \end{pmatrix}_3 = \begin{pmatrix} ab_1 \\ ab_2 \\ ab_3 \end{pmatrix}_3 , \quad a_{1'} \otimes \begin{pmatrix} b_1 \\ b_2 \\ b_3 \end{pmatrix}_3 = \begin{pmatrix} ab_2 \\ ab_3 \\ ab_1 \end{pmatrix}_3 , \quad a_{1''} \otimes \begin{pmatrix} b_1 \\ b_2 \\ b_3 \end{pmatrix}_3 = \begin{pmatrix} ab_3 \\ ab_1 \\ ab_2 \end{pmatrix}_3 .$$
$$\text{(C.4)}$$

The products of singlets are given as

$$\mathbf{1} \otimes \mathbf{1} = \mathbf{1} , \quad \mathbf{1}' \otimes \mathbf{1}' = \mathbf{1}'' , \quad \mathbf{1}'' \otimes \mathbf{1}'' = \mathbf{1}' , \quad \mathbf{1}' \otimes \mathbf{1}'' = \mathbf{1} , \quad \text{(C.5)}$$

where

$$T(\mathbf{1}') = \omega^2 , \quad T(\mathbf{1}'') = \omega . \quad \text{(C.6)}$$

Representations of A_5 in Different Basis

D

D.1 The Basis I

Here, we show another basis for representations of the A_5 group. First, we show the basis in Sect. 4.2. All of the A_5 elements are written by products of the generators, s and t, which satisfy

$$s^2 = t^5 = (t^2st^3st^{-1}stst^{-1})^3 = e. \tag{D.1}$$

The generators, s and t, are represented as [9],

$$s = \frac{1}{2}\begin{pmatrix} -1 & \phi & \frac{1}{\phi} \\ \phi & \frac{1}{\phi} & 1 \\ \frac{1}{\phi} & 1 & -\phi \end{pmatrix}, \quad t = \frac{1}{2}\begin{pmatrix} 1 & \phi & \frac{1}{\phi} \\ -\phi & \frac{1}{\phi} & 1 \\ \frac{1}{\phi} & -1 & \phi \end{pmatrix}, \quad \text{on } \mathbf{3}, \tag{D.2}$$

$$s = \frac{1}{2}\begin{pmatrix} -\phi & \frac{1}{\phi} & 1 \\ \frac{1}{\phi} & -1 & \phi \\ 1 & \phi & \frac{1}{\phi} \end{pmatrix}, \quad t = \frac{1}{2}\begin{pmatrix} -\phi & -\frac{1}{\phi} & 1 \\ \frac{1}{\phi} & 1 & \phi \\ -1 & \phi & -\frac{1}{\phi} \end{pmatrix}, \quad \text{on } \mathbf{3'}, \tag{D.3}$$

$$s = \frac{1}{4}\begin{pmatrix} -1 & -1 & -3 & -\sqrt{5} \\ -1 & 3 & 1 & -\sqrt{5} \\ -3 & 1 & -1 & \sqrt{5} \\ -\sqrt{5} & -\sqrt{5} & \sqrt{5} & -1 \end{pmatrix},$$

$$t = \frac{1}{4}\begin{pmatrix} -1 & 1 & -3 & \sqrt{5} \\ -1 & -3 & 1 & \sqrt{5} \\ 3 & 1 & 1 & \sqrt{5} \\ \sqrt{5} & -\sqrt{5} & -\sqrt{5} & -1 \end{pmatrix}, \quad \text{on } \mathbf{4}, \tag{D.4}$$

© Springer-Verlag GmbH Germany, part of Springer Nature 2022
T. Kobayashi et al., *An Introduction to Non-Abelian Discrete Symmetries for Particle Physicists*, Lecture Notes in Physics 995,
https://doi.org/10.1007/978-3-662-64679-3

$$s = \frac{1}{2} \begin{pmatrix} \frac{1-3\phi}{4} & \frac{\phi^2}{2} & -\frac{1}{2\phi^2} & \frac{\sqrt{5}}{2} & \frac{\sqrt{3}}{4\phi} \\ \frac{\phi^2}{2} & 1 & 1 & 0 & \frac{\sqrt{3}}{2\phi} \\ -\frac{1}{2\phi^2} & 1 & 0 & -1 & -\frac{\sqrt{3}\phi}{2} \\ \frac{\sqrt{5}}{2} & 0 & -1 & 1 & -\frac{\sqrt{3}}{2} \\ \frac{\sqrt{3}}{4\phi} & \frac{\sqrt{3}}{2\phi} & -\frac{\sqrt{3}\phi}{2} & -\frac{\sqrt{3}}{2} & \frac{3\phi-1}{4} \end{pmatrix},$$

$$t = \frac{1}{2} \begin{pmatrix} \frac{1-3\phi}{4} & -\frac{\phi^2}{2} & -\frac{1}{2\phi^2} & -\frac{\sqrt{5}}{2} & \frac{\sqrt{3}}{4\phi} \\ \frac{\phi^2}{2} & -1 & 1 & 0 & \frac{\sqrt{3}}{2\phi} \\ \frac{1}{2\phi^2} & 1 & 0 & -1 & \frac{\sqrt{3}\phi}{2} \\ -\frac{\sqrt{5}}{2} & 0 & 1 & 1 & \frac{\sqrt{3}}{2} \\ \frac{\sqrt{3}}{4\phi} & -\frac{\sqrt{3}}{2\phi} & -\frac{\sqrt{3}\phi}{2} & \frac{\sqrt{3}}{2} & \frac{3\phi-1}{4} \end{pmatrix}, \quad \text{on } \mathbf{5}, \quad \text{(D.5)}$$

where $\phi = \frac{1+\sqrt{5}}{2}$. Here we show the tensor products as

$$\begin{pmatrix} x_1 \\ x_2 \\ x_3 \end{pmatrix}_3 \otimes \begin{pmatrix} y_1 \\ y_2 \\ y_3 \end{pmatrix}_3 = (x_1 y_1 + x_2 y_2 + x_3 y_3)_1 \oplus \begin{pmatrix} x_3 y_2 - x_2 y_3 \\ x_1 y_3 - x_3 y_1 \\ x_2 y_1 - x_1 y_2 \end{pmatrix}_3$$

$$\oplus \begin{pmatrix} x_2 y_2 - x_1 y_1 \\ x_2 y_1 + x_1 y_2 \\ x_3 y_2 + x_2 y_3 \\ x_1 y_3 + x_3 y_1 \\ -\frac{1}{\sqrt{3}}(x_1 y_1 + x_2 y_2 - 2 x_3 y_3) \end{pmatrix}_5, \quad \text{(D.6)}$$

$$\begin{pmatrix} x_1 \\ x_2 \\ x_3 \end{pmatrix}_{3'} \otimes \begin{pmatrix} y_1 \\ y_2 \\ y_3 \end{pmatrix}_{3'} = (x_1 y_1 + x_2 y_2 + x_3 y_3)_1 \oplus \begin{pmatrix} x_3 y_2 - x_2 y_3 \\ x_1 y_3 - x_3 y_1 \\ x_2 y_1 - x_1 y_2 \end{pmatrix}_{3'}$$

$$\oplus \begin{pmatrix} \frac{1}{2}(-\frac{1}{\phi}x_1 y_1 - \phi x_2 y_2 + \sqrt{5} x_3 y_3) \\ x_2 y_1 + x_1 y_2 \\ -(x_3 y_1 + x_1 y_3) \\ x_2 y_3 + x_3 y_2 \\ \frac{1}{2\sqrt{3}}((1 - 3\phi)x_1 y_1 + (3\phi - 2)x_2 y_2 + x_3 y_3) \end{pmatrix}_5 \quad \text{(D.7)}$$

$$\begin{pmatrix} x_1 \\ x_2 \\ x_3 \end{pmatrix}_3 \otimes \begin{pmatrix} y_1 \\ y_2 \\ y_3 \end{pmatrix}_{3'} = \begin{pmatrix} \frac{1}{\phi}x_3 y_2 - \phi x_1 y_3 \\ \phi x_3 y_1 + \frac{1}{\phi}x_2 y_3 \\ -\frac{1}{\phi}x_1 y_1 + \phi x_2 y_2 \\ x_2 y_1 - x_1 y_2 + x_3 y_3 \end{pmatrix}_4 \oplus \begin{pmatrix} \frac{1}{2}(\phi^2 x_2 y_1 + \frac{1}{\phi^2}x_1 y_2 - \sqrt{5} x_3 y_3) \\ -(\phi x_1 y_1 + \frac{1}{\phi}x_2 y_2) \\ \frac{1}{\phi}x_3 y_1 - \phi x_2 y_3 \\ \phi x_3 y_2 + \frac{1}{\phi}x_1 y_3 \\ \frac{\sqrt{3}}{2}(\frac{1}{\phi}x_2 y_1 + \phi x_1 y_2 + x_3 y_3) \end{pmatrix}_5,$$

$$\text{(D.8)}$$

$$
\begin{pmatrix} x_1 \\ x_2 \\ x_3 \end{pmatrix}_3 \otimes \begin{pmatrix} y_1 \\ y_2 \\ y_3 \\ y_4 \end{pmatrix}_4 = \begin{pmatrix} -\frac{1}{\phi^2}x_1y_3 + \frac{1}{\phi}x_2y_4 + x_3y_2 \\ -\frac{1}{\phi}x_1y_4 + x_2y_3 + \frac{1}{\phi^2}x_3y_1 \\ -x_1y_1 + \frac{1}{\phi^2}x_2y_2 + \frac{1}{\phi}x_3y_4 \end{pmatrix}_{3'} \oplus \begin{pmatrix} -x_1y_3 + x_2y_4 - x_3y_2 \\ -x_1y_4 - x_2y_3 + x_3y_1 \\ x_1y_1 + x_2y_2 + x_3y_4 \\ x_1y_2 - x_2y_1 - x_3y_3 \end{pmatrix}_4
$$

$$
\oplus \begin{pmatrix} \frac{1}{2}\left((6\phi+5)x_1y_2 + (3\phi+4)x_2y_1 + (3\phi+1)x_3y_3\right) \\ -x_1y_1 + (3\phi+2)x_2y_2 - (3\phi+1)x_3y_4 \\ -(3\phi+1)x_1y_4 - x_2y_3 - (3\phi+2)x_3y_1 \\ -(3\phi+2)x_1y_3 - (3\phi+1)x_2y_4 + x_3y_2 \\ \frac{\sqrt{3}}{2}\left(x_1y_2 - (3\phi+2)x_2y_1 + 3(\phi+1)x_3y_3\right) \end{pmatrix}_5, \quad (D.9)
$$

$$
\begin{pmatrix} x_1 \\ x_2 \\ x_3 \end{pmatrix}_{3'} \otimes \begin{pmatrix} y_1 \\ y_2 \\ y_3 \\ y_4 \end{pmatrix}_4 = \begin{pmatrix} x_1y_3 + \phi x_2y_4 + \phi^2 x_3y_1 \\ -\phi x_1y_4 - \phi^2 x_2y_3 - x_3y_2 \\ -\phi^2 x_1y_2 - x_2y_1 - \phi x_3y_4 \end{pmatrix}_3 \oplus \begin{pmatrix} x_1y_4 - x_2y_3 + x_3y_2 \\ x_1y_3 + x_2y_4 - x_3y_1 \\ -x_1y_2 + x_2y_1 + x_3y_4 \\ -(x_1y_1 + x_2y_2 + x_3y_3) \end{pmatrix}_4
$$

$$
\oplus \begin{pmatrix} x_1y_1 - \phi^4 x_2y_2 + \phi^2(2\phi-1)x_3y_3 \\ x_1y_2 - \phi^4 x_2y_1 + \phi^2(2\phi-1)x_3y_4 \\ \phi^4 x_1y_3 - \phi^2(2\phi-1)x_2y_4 + x_3y_1 \\ \phi^2(2\phi-1)x_1y_4 - x_2y_3 - \phi^4 x_3y_2 \\ -\sqrt{3}\phi\left(\phi^2 x_1y_1 - x_2y_2 - \phi x_3y_3\right) \end{pmatrix}_5, \quad (D.10)
$$

$$
\begin{pmatrix} x_1 \\ x_2 \\ x_3 \end{pmatrix}_3 \otimes \begin{pmatrix} y_1 \\ y_2 \\ y_3 \\ y_4 \\ y_5 \end{pmatrix}_5 = \begin{pmatrix} x_1(y_1 + \frac{1}{\sqrt{3}}y_5) - x_2y_2 - x_3y_4 \\ -x_1y_2 - x_2(y_1 - \frac{1}{\sqrt{3}}y_5) - x_3y_3 \\ -x_1y_4 - x_2y_3 - \frac{2}{\sqrt{3}}x_3y_5 \end{pmatrix}_3
$$

$$
\oplus \begin{pmatrix} x_1y_2 - \frac{\phi}{2}x_2y_1 - \frac{\sqrt{3}}{2\phi^2}x_2y_5 - \frac{1}{\phi^2}x_3y_3 \\ -\frac{\sqrt{3}}{2}x_1y_5 - \frac{1}{2\phi^3}x_1y_1 + \frac{1}{\phi^2}x_2y_2 - x_3y_4 \\ -\frac{1}{\phi^2}x_1y_4 + x_2y_3 + \frac{\sqrt{5}}{2\phi}x_3y_1 - \frac{\sqrt{3}}{2\phi}x_3y_5 \end{pmatrix}_{3'}
$$

$$
\oplus \begin{pmatrix} \frac{1}{\phi^2}x_1y_2 + \frac{\phi^2-6}{2}x_2y_1 + \frac{\sqrt{3}}{2}\phi^2 x_2y_5 + \phi^2 x_3y_3 \\ -\frac{\phi+4}{2}x_1y_1 - \frac{\sqrt{3}}{2\phi^2}x_1y_5 - \phi^2 x_2y_2 - \frac{1}{\phi^2}x_3y_4 \\ \phi^2 x_1y_4 + \frac{1}{\phi^2}x_2y_3 - \frac{\sqrt{5}}{2}x_3y_1 - \frac{3\sqrt{3}}{2}x_3y_5 \\ \sqrt{5}(x_1y_3 + x_2y_4 + x_3y_2) \end{pmatrix}_4
$$

$$
\oplus \begin{pmatrix} x_1y_3 + x_2y_4 - 2x_3y_2 \\ x_1y_4 - x_2y_3 + 2x_3y_1 \\ -x_1y_1 + x_2y_2 - x_3y_4 + \sqrt{3}x_1y_5 \\ -x_1y_2 - x_2y_1 + x_3y_3 - \sqrt{3}x_2y_5 \\ -\sqrt{3}(x_1y_3 - x_2y_4) \end{pmatrix}_5, \quad (D.11)
$$

$$\begin{pmatrix} x_1 \\ x_2 \\ x_3 \end{pmatrix}_{3'} \otimes \begin{pmatrix} y_1 \\ y_2 \\ y_3 \\ y_4 \\ y_5 \end{pmatrix}_5 = \begin{pmatrix} -\phi^2 x_1 y_2 + \frac{1}{2\phi} x_2 y_1 + \frac{\sqrt{3}}{2}\phi^2 x_2 y_5 + x_3 y_4 \\ \frac{2\phi+1}{2} x_1 y_1 + \frac{\sqrt{3}}{2} x_1 y_5 - x_2 y_2 - \phi^2 x_3 y_3 \\ x_1 y_3 + \phi^2 x_2 y_4 - \frac{\sqrt{5}}{2}\phi x_3 y_1 + \frac{\sqrt{3}}{2}\phi x_3 y_5 \end{pmatrix}_3$$

$$\oplus \begin{pmatrix} \frac{1}{2\phi} x_1 y_1 - x_2 y_2 + x_3 y_3 + \frac{3\phi-1}{2\sqrt{3}} x_1 y_5 \\ -x_1 y_2 + \frac{\phi}{2} x_2 y_1 - x_3 y_4 - \frac{3\phi-2}{2\sqrt{3}} x_2 y_5 \\ x_1 y_3 - x_2 y_4 - \frac{\sqrt{5}}{2} x_3 y_1 - \frac{1}{2\sqrt{3}} x_3 y_5 \end{pmatrix}_{3'}$$

$$\oplus \begin{pmatrix} \frac{1}{\sqrt{5}}\left(\frac{1}{\phi^2} x_1 y_1 + \phi^2 x_2 y_2 + \frac{1}{\phi^2} x_3 y_3 - \sqrt{3}\phi x_1 y_5\right) \\ \frac{1}{\sqrt{5}}\left(-\frac{1}{\phi^2} x_1 y_2 - \phi^2 x_2 y_1 - \frac{\sqrt{3}(\phi-3)}{\sqrt{5}} x_2 y_5 - \phi^2 x_3 y_4\right) \\ \frac{1}{\sqrt{5}}\left(\phi^2 x_1 y_3 - \frac{1}{\phi^2} x_2 y_4 + \sqrt{5} x_3 y_1 + \sqrt{3} x_3 y_5\right) \\ x_1 y_4 - x_2 y_3 + x_3 y_2 \end{pmatrix}_4$$

$$\oplus \begin{pmatrix} -(3\phi-1) x_1 y_4 + (2-3\phi) x_2 y_3 + x_3 y_2 \\ -2 x_1 y_3 - 2 x_2 y_4 - x_3 y_1 + \sqrt{15} x_3 y_5 \\ 2 x_1 y_2 - (2-3\phi) x_2 y_1 - 2 x_3 y_4 + \sqrt{3}\phi x_2 y_5 \\ (3\phi-1) x_1 y_1 + 2 x_2 y_2 + 2 x_3 y_3 - \frac{\sqrt{3}}{\phi} x_1 y_5 \\ \frac{\sqrt{3}}{\phi} x_1 y_4 - \phi\sqrt{3} x_2 y_3 - \sqrt{15} x_3 y_2 \end{pmatrix}_5 , \qquad \text{(D.12)}$$

$$\begin{pmatrix} x_1 \\ x_2 \\ x_3 \\ x_4 \end{pmatrix}_4 \otimes \begin{pmatrix} y_1 \\ y_2 \\ y_3 \\ y_4 \end{pmatrix}_4 = (x_1 y_1 + x_2 y_2 + x_3 y_3 + x_4 y_4) \mathbf{1} \oplus \begin{pmatrix} x_1 y_3 + x_2 y_4 - x_3 y_1 - x_4 y_2 \\ -x_1 y_4 + x_2 y_3 - x_3 y_2 + x_4 y_1 \\ x_1 y_2 - x_2 y_1 - x_3 y_4 + x_4 y_3 \end{pmatrix}_3$$

$$\oplus \begin{pmatrix} x_1 y_4 + x_2 y_3 - x_3 y_2 - x_4 y_1 \\ -x_1 y_3 + x_2 y_4 + x_3 y_1 - x_4 y_2 \\ x_1 y_2 - x_2 y_1 + x_3 y_4 - x_4 y_3 \end{pmatrix}_{3'}$$

$$\oplus \begin{pmatrix} x_1 y_4 - \sqrt{5} x_2 y_3 - \sqrt{5} x_3 y_2 + x_4 y_1 \\ -\sqrt{5} x_1 y_3 + x_2 y_4 - \sqrt{5} x_3 y_1 + x_4 y_2 \\ -\sqrt{5} x_1 y_2 - \sqrt{5} x_2 y_1 + x_3 y_4 + x_4 y_3 \\ x_1 y_1 + x_2 y_2 + x_3 y_3 - 3 x_4 y_4 \end{pmatrix}_4$$

$$\oplus \begin{pmatrix} -\frac{\phi^2}{\sqrt{5}} x_1 y_1 + \frac{1}{\sqrt{5}\phi^2} x_2 y_2 + x_3 y_3 \\ -\frac{1}{\sqrt{5}} x_1 y_2 - \frac{1}{\sqrt{5}} x_2 y_1 - x_3 y_4 - x_4 y_3 \\ \frac{1}{\sqrt{5}} x_1 y_3 + x_2 y_4 + \frac{1}{\sqrt{5}} x_3 y_1 + x_4 y_2 \\ -x_1 y_4 - \frac{1}{\sqrt{5}} x_2 y_3 - \frac{1}{\sqrt{5}} x_3 y_2 - x_4 y_1 \\ -\sqrt{\frac{3}{5}}\left(\frac{1}{\phi} x_1 y_1 - \phi x_2 y_2 + x_3 y_3\right) \end{pmatrix}_5 , \qquad \text{(D.13)}$$

$$\begin{pmatrix} x_1 \\ x_2 \\ x_3 \\ x_4 \end{pmatrix}_4 \otimes \begin{pmatrix} y_1 \\ y_2 \\ y_3 \\ y_4 \\ y_5 \end{pmatrix}_5 = \begin{pmatrix} \frac{2}{\phi^2}x_1y_2 - (\phi+4)x_2y_1 + 2\phi^2 x_3y_4 + 2\sqrt{5}x_4y_3 - \frac{\sqrt{3}}{\phi^2}x_2y_5 \\ (\phi-5)x_1y_1 + \sqrt{3}\phi^2 x_1y_5 - 2\phi^2 x_2y_2 + \frac{2}{\phi^2}x_3y_3 + 2\sqrt{5}x_4y_4 \\ 2\phi^2 x_1y_3 - \frac{2}{\phi^2}x_2y_4 - \sqrt{5}x_3y_1 - 3\sqrt{3}x_3y_5 + 2\sqrt{5}x_4y_2 \end{pmatrix}_3$$

$$\oplus \begin{pmatrix} \frac{1}{\phi^2}x_1y_1 - \sqrt{3}\phi x_1y_5 - \frac{1}{\phi^2}x_2y_2 + \phi^2 x_3y_3 + \sqrt{5}x_4y_4 \\ -\phi^2 x_1y_2 - \phi^2 x_2y_1 + \frac{\sqrt{3}}{\phi}x_2y_5 - \frac{1}{\phi^2}x_3y_4 - \sqrt{5}x_4y_3 \\ \frac{1}{\phi^2}x_1y_3 - \phi^2 x_2y_4 + \sqrt{5}x_3y_1 + \sqrt{3}x_3y_5 + \sqrt{5}x_4y_2 \end{pmatrix}_{3'}$$

$$\oplus \begin{pmatrix} -\frac{\phi^2}{\sqrt{5}}x_1y_1 - \frac{1}{\sqrt{5}}x_2y_2 + \frac{1}{\sqrt{5}}x_3y_3 - x_4y_4 - \frac{\sqrt{3}}{\sqrt{5}\phi}x_1y_5 \\ -\frac{1}{\sqrt{5}}x_1y_2 + \frac{1}{\sqrt{5}\phi^2}x_2y_1 + \sqrt{\frac{3}{5}}\phi x_2y_5 - \frac{1}{\sqrt{5}}x_3y_4 + x_4y_3 \\ \frac{1}{\sqrt{5}}x_1y_3 - \frac{1}{\sqrt{5}}x_2y_4 + x_3y_1 - \sqrt{\frac{3}{5}}x_3y_5 - x_4y_2 \\ -x_1y_4 + x_2y_3 - x_3y_2 \end{pmatrix}_4$$

$$\oplus \begin{pmatrix} \frac{1}{2}\left(\phi^2 x_1y_4 - \frac{1}{\phi^2}x_2y_3 - 3x_3y_2 + 3x_4y_1 + \sqrt{\frac{5}{3}}x_4y_5\right) \\ \phi x_1y_3 - \frac{1}{\phi}x_2y_4 - x_3y_1 - x_4y_2 + \sqrt{\frac{5}{3}}x_3y_5 \\ \frac{1}{\phi}x_1y_2 + \frac{1}{\phi}x_2y_1 + \frac{1}{\sqrt{3}}\phi^2 x_2y_5 + \phi x_3y_4 - x_4y_3 \\ \phi x_1y_1 + \frac{1}{\sqrt{3}\phi^2}x_1y_5 - \phi x_2y_2 + \frac{1}{2}x_3y_3 - x_4y_4 \\ \frac{1}{2\sqrt{3}}\left(-(\phi-5)x_1y_4 + (\phi+4)x_2y_3 + \sqrt{5}x_3y_2 - \sqrt{5}x_4y_1\right) + \frac{3}{2}x_4y_5 \end{pmatrix}_5,$$

$$\oplus \begin{pmatrix} x_1y_4 - x_2y_3 - 2x_3y_2 + \frac{3}{2}x_4y_1 + \frac{\sqrt{15}}{2}x_4y_5 \\ \phi^2 x_1y_3 + \frac{1}{\phi^2}x_2y_4 - \frac{1}{2}x_3y_1 + \frac{\sqrt{15}}{2}x_3y_5 - x_4y_2 \\ -\frac{1}{\phi^2}x_1y_2 + \frac{3\phi-2}{2}x_2y_1 + \frac{\sqrt{3}}{2}\phi x_2y_5 + \phi^2 x_3y_4 - x_4y_3 \\ \frac{3\phi-1}{2}x_1y_1 - \phi^2 x_2y_2 - \frac{1}{\phi^2}x_3y_3 - x_4y_4 - \frac{\sqrt{3}}{2\phi}x_1y_5 \\ \sqrt{3}x_1y_4 + \sqrt{3}x_2y_3 + \frac{3}{2}x_4y_5 - \frac{\sqrt{15}}{2}x_4y_1 \end{pmatrix}_5, \qquad \text{(D.14)}$$

$$\begin{pmatrix} x_1 \\ x_2 \\ x_3 \\ x_4 \\ x_5 \end{pmatrix}_5 \otimes \begin{pmatrix} y_1 \\ y_2 \\ y_3 \\ y_4 \\ y_5 \end{pmatrix}_5$$

$$= (x_1y_1 + x_2y_2 + x_3y_3 + x_4y_4 + x_5y_5)_1$$

$$\oplus \begin{pmatrix} x_1y_3 - x_3y_1 + x_2y_4 - x_4y_2 + \sqrt{3}(x_3y_5 - x_5y_3) \\ x_1y_4 - x_4y_1 - (x_2y_3 - x_3y_2) - \sqrt{3}(x_4y_5 - x_5y_4) \\ -2(x_1y_2 - x_2y_1) - (x_3y_4 - x_4y_3) \end{pmatrix}_3$$

$$\oplus \begin{pmatrix} (2\phi+3)(x_1y_4 - x_4y_1) + 2\phi(x_2y_3 - x_3y_2) + \sqrt{3}(x_4y_5 - x_5y_4) \\ (\phi+3)(x_1y_3 - x_3y_1) + 2\phi(x_2y_4 - x_4y_2) - \sqrt{3}\phi^2(x_3y_5 - x_5y_3) \\ -\phi(x_1y_2 - x_2y_1) - \sqrt{15}\phi(x_2y_5 - x_5y_2) + 2\phi(x_3y_4 - x_4y_3) \end{pmatrix}_{3'}$$

$$\oplus \begin{pmatrix} \sqrt{5}\left(\frac{\phi^2}{2}(x_1y_4 + x_4y_1) + x_2y_3 + x_3y_2 + \frac{\sqrt{3}}{2\phi}(x_4y_5 + x_5y_4)\right) \\ \sqrt{5}\left(\frac{1}{2\phi^2}(x_1y_3 + x_3y_1) - (x_2y_4 + x_4y_2) + \frac{\sqrt{3}}{2}\phi(x_3y_5 + x_5y_3)\right) \\ \frac{\sqrt{5}}{2}\left(-\sqrt{5}(x_1y_2 + x_2y_1) + \sqrt{3}(x_2y_5 + x_5y_2) + 2(x_3y_4 + x_4y_3)\right) \\ 3x_1y_1 - 2(x_2y_2 + x_3y_3 + x_4y_4) + 3x_5y_5 \end{pmatrix}_4$$

$$\oplus \begin{pmatrix} \frac{1}{\sqrt{5}}\left(\frac{1}{2\phi}(x_1y_4 - x_4y_1) - (x_2y_3 - x_3y_2) + \frac{\phi^2}{2\sqrt{3}}(x_4y_5 - x_5y_4)\right) \\ \frac{1}{\sqrt{5}}\left(\frac{\phi}{2}(x_1y_3 - x_3y_1) - (x_2y_4 - x_4y_2) + \frac{1}{2\sqrt{3}\phi^2}(x_3y_5 - x_5y_3)\right) \\ \frac{1}{2\sqrt{5}}(x_1y_2 - x_2y_1) - \frac{1}{2\sqrt{3}}(x_2y_5 - x_5y_2) - \frac{1}{\sqrt{5}}(x_3y_4 - x_4y_3) \\ -\frac{1}{\sqrt{3}}(x_1y_5 - x_5y_1) \end{pmatrix}_4$$

$$\oplus \begin{pmatrix} -x_1y_1 - \frac{11}{3\sqrt{15}}(x_1y_5 + x_5y_1) + \frac{4}{3}x_2y_2 - \frac{4\sqrt{5}}{15}(\phi x_3y_3 + \frac{1}{\phi}x_4y_4) + x_5y_5 \\ \frac{4}{3}\left(x_1y_2 + x_2y_1 + \frac{1}{\sqrt{15}}(x_2y_5 + x_5y_2) + \frac{2}{\sqrt{5}}(x_3y_4 + x_4y_3)\right) \\ \frac{4}{3\sqrt{5}}\left(-\phi(x_1y_3 + x_3y_1) + 2(x_2y_4 + x_4y_2) - \frac{2-3\phi}{\sqrt{3}}(x_3y_5 + x_5y_3)\right) \\ \frac{4\sqrt{5}}{15}\left(-\frac{1}{\phi}(x_1y_4 + x_4y_1) + 2(x_2y_3 + x_3y_2) - \frac{\sqrt{3}}{3}(3\phi - 1)(x_4y_5 + x_5y_4)\right) \\ x_1y_5 + x_5y_1 + \frac{1}{3\sqrt{15}}(-11x_1y_1 + 4x_2y_2 + 11x_5y_5) - \frac{4\sqrt{15}}{45}((2 - 3\phi)x_3y_3 + (3\phi - 1)x_4y_4) \end{pmatrix}_5$$

$$\oplus \begin{pmatrix} -\frac{3\sqrt{5}}{4}(x_1y_1 - x_5y_5) - \frac{\sqrt{3}}{4}(x_1y_5 + x_5y_1) + \sqrt{5}x_2y_2 - \phi^2x_3y_3 + \frac{1}{\phi}x_4y_4 \\ \sqrt{5}(x_1y_2 + x_2y_1) + \sqrt{3}(x_2y_5 + x_5y_2) + x_3y_4 + x_4y_3 \\ -\phi^2(x_1y_3 + x_3y_1) + x_2y_4 + x_4y_2 + \frac{\sqrt{3}}{\phi}(x_3y_5 + x_5y_3) \\ \frac{1}{\phi^2}(x_1y_4 + x_4y_1) + (x_2y_3 + x_3y_2) - \sqrt{3}\phi(x_4y_5 + x_5y_4) \\ -\frac{\sqrt{3}}{4}(x_1y_1 - 4x_2y_2 - x_5y_5) + \frac{3\sqrt{5}}{4}(x_1y_5 + x_5y_1) + \frac{\sqrt{3}}{\phi}x_3y_3 - \sqrt{3}\phi x_4y_4 \end{pmatrix}_5 \qquad (D.15)$$

D.2   The Basis II

Next, we consider another basis, which is used, e.g. in Ref. [10], with the generator a and b satisfying

$$a^2 = b^3 = (ab)^5 = e. \qquad (D.16)$$

The a and b are represented by s and t in Sect. D.1 as

$$a = st^3st^2s, \quad ab = t^4. \qquad (D.17)$$

In above transformation, ab is not diagonal but a is diagonal. In order to change the diagonal ab basis, we diagonalize ab with unitary transformation U_ϕ as

$$a = U_\phi^\dagger st^3st^2sU_\phi, \quad ab = U_\phi^\dagger t^4 U_\phi. \qquad (D.18)$$

These generators are represented as follows:

$$a = \frac{1}{\sqrt{5}}\begin{pmatrix} 1 & -\sqrt{2} & -\sqrt{2} \\ -\sqrt{2} & -\phi & \frac{1}{\phi} \\ -\sqrt{2} & \frac{1}{\phi} & -\phi \end{pmatrix}, \quad ab = \begin{pmatrix} 1 & 0 & 0 \\ 0 & \rho & 0 \\ 0 & 0 & \rho^4 \end{pmatrix}, \quad \text{on } \mathbf{3}, \qquad (D.19)$$

with

$$U_\phi = \frac{1}{\sqrt{2}5^{1/4}}\begin{pmatrix} -\sqrt{\frac{2}{\phi}} & -\sqrt{\phi} & -\sqrt{\phi} \\ 0 & i5^{1/4} & -i5^{1/4} \\ -\sqrt{2\phi} & \frac{1}{\sqrt{\phi}} & \frac{1}{\sqrt{\phi}} \end{pmatrix}, \qquad (D.20)$$

$$a = \frac{1}{\sqrt{5}} \begin{pmatrix} -1 & \sqrt{2} & \sqrt{2} \\ \sqrt{2} & -\frac{1}{\phi} & \phi \\ \sqrt{2} & \phi & -\frac{1}{\phi} \end{pmatrix}, \quad ab = \begin{pmatrix} 1 & 0 & 0 \\ 0 & \rho^2 & 0 \\ 0 & 0 & \rho^3 \end{pmatrix}, \quad \text{on } \mathbf{3'}, \qquad (D.21)$$

with

$$U_\phi = \frac{1}{\sqrt{2}5^{1/4}} \begin{pmatrix} 0 & i5^{1/4} & -i5^{1/4} \\ \sqrt{2\phi} & -\frac{1}{\sqrt{\phi}} & -\frac{1}{\sqrt{\phi}} \\ \sqrt{\frac{2}{\phi}} & \sqrt{\phi} & \sqrt{\phi} \end{pmatrix}, \qquad (D.22)$$

$$a = \frac{1}{\sqrt{5}} \begin{pmatrix} 1 & \frac{1}{\phi} & \phi & -1 \\ \frac{1}{\phi} & -1 & 1 & \phi \\ \phi & 1 & -1 & \frac{1}{\phi} \\ -1 & \phi & \frac{1}{\phi} & 1 \end{pmatrix}, \quad ab = \begin{pmatrix} \rho & 0 & 0 & 0 \\ 0 & \rho^2 & 0 & 0 \\ 0 & 0 & \rho^3 & 0 \\ 0 & 0 & 0 & \rho^4 \end{pmatrix}, \quad \text{on } \mathbf{4}, \qquad (D.23)$$

with

$$U_\phi = \frac{1}{2} \begin{pmatrix} 1 & -1 & -1 & 1 \\ \frac{i}{5^{1/4}\phi^{3/2}} & \frac{i\phi^{3/2}}{5^{1/4}} & -\frac{i\phi^{3/2}}{5^{1/4}} & -\frac{i}{5^{1/4}\phi^{3/2}} \\ \frac{i\phi^{3/2}}{5^{1/4}} & -\frac{i}{5^{1/4}\phi^{3/2}} & \frac{i}{5^{1/4}\phi^{3/2}} & -\frac{i\phi^{3/2}}{5^{1/4}} \\ 1 & 1 & 1 & 1 \end{pmatrix}, \qquad (D.24)$$

and

$$a = \frac{1}{5} \begin{pmatrix} -1 & \sqrt{6} & \sqrt{6} & \sqrt{6} & \sqrt{6} \\ \sqrt{6} & \frac{1}{\phi^2} & -2\phi & \frac{2}{\phi} & \phi^2 \\ \sqrt{6} & -2\phi & \phi^2 & \frac{1}{\phi^2} & \frac{2}{\phi} \\ \sqrt{6} & \frac{2}{\phi} & \frac{1}{\phi^2} & \phi^2 & -2\phi \\ \sqrt{6} & \phi^2 & \frac{2}{\phi} & -2\phi & \frac{1}{\phi^2} \end{pmatrix}, \quad ab = \begin{pmatrix} 1 & 0 & 0 & 0 & 0 \\ 0 & \rho & 0 & 0 & 0 \\ 0 & 0 & \rho^2 & 0 & 0 \\ 0 & 0 & 0 & \rho^3 & 0 \\ 0 & 0 & 0 & 0 & \rho^4 \end{pmatrix}, \quad \text{on } \mathbf{5}, \quad (D.25)$$

with

$$U_\phi = \frac{1}{\sqrt{10}} \begin{pmatrix} -\frac{1}{\phi}\sqrt{\frac{3}{2}} & 1 & -\frac{1}{2}\sqrt{\phi^4+8} & -\frac{1}{2}\sqrt{\phi^4+8} & 1 \\ 0 & \frac{i5^{1/4}}{\sqrt{\phi}} & -i5^{1/4}\sqrt{\phi} & i5^{1/4}\sqrt{\phi} & -\frac{i5^{1/4}}{\sqrt{\phi}} \\ 0 & i5^{1/4}\sqrt{\phi} & \frac{i5^{1/4}}{\sqrt{\phi}} & -\frac{i5^{1/4}}{\sqrt{\phi}} & -i5^{1/4}\sqrt{\phi} \\ \sqrt{6} & -1 & -1 & -1 & -1 \\ \frac{\phi^2}{\sqrt{2}} & \sqrt{3} & \frac{\sqrt{3}}{2\phi} & \frac{\sqrt{3}}{2\phi} & \sqrt{3} \end{pmatrix},$$

$$(D.26)$$

where $\phi = \frac{1+\sqrt{5}}{2}$ and $\rho = e^{2i\pi/5}$. Here we show the tensor products as

$$\begin{pmatrix} x_1 \\ x_2 \\ x_3 \end{pmatrix}_3 \otimes \begin{pmatrix} y_1 \\ y_2 \\ y_3 \end{pmatrix}_3 = (x_1 y_1 + x_2 y_3 + x_3 y_2)_1 \oplus \begin{pmatrix} x_2 y_3 - x_3 y_2 \\ x_1 y_2 - x_2 y_1 \\ x_3 y_1 - x_1 y_3 \end{pmatrix}_3$$

$$\oplus \begin{pmatrix} 2x_1 y_1 - x_2 y_3 - x_3 y_2 \\ -\sqrt{3}x_1 y_2 - \sqrt{3}x_2 y_1 \\ \sqrt{6}x_2 y_2 \\ \sqrt{6}x_3 y_3 \\ -\sqrt{3}x_1 y_3 - \sqrt{3}x_3 y_1 \end{pmatrix}_5, \qquad (D.27)$$

$$\begin{pmatrix} x_1 \\ x_2 \\ x_3 \end{pmatrix}_{3'} \otimes \begin{pmatrix} y_1 \\ y_2 \\ y_3 \end{pmatrix}_{3'} = (x_1 y_1 + x_2 y_3 + x_3 y_2)_1 \oplus \begin{pmatrix} x_2 y_3 - x_3 y_2 \\ x_1 y_2 - x_2 y_1 \\ x_3 y_1 - x_1 y_3 \end{pmatrix}_{3'}$$

$$\oplus \begin{pmatrix} 2x_1 y_1 - x_2 y_3 - x_3 y_2 \\ \sqrt{6}x_3 y_3 \\ -\sqrt{3}x_1 y_2 - \sqrt{3}x_2 y_1 \\ -\sqrt{3}x_1 y_3 - \sqrt{3}x_3 y_1 \\ \sqrt{6}x_2 y_2 \end{pmatrix}_5, \qquad (D.28)$$

$$\begin{pmatrix} x_1 \\ x_2 \\ x_3 \end{pmatrix}_3 \otimes \begin{pmatrix} y_1 \\ y_2 \\ y_3 \end{pmatrix}_{3'} = \begin{pmatrix} \sqrt{2}x_2 y_1 + x_3 y_2 \\ -\sqrt{2}x_1 y_2 - x_3 y_3 \\ -\sqrt{2}x_1 y_3 - x_2 y_2 \\ \sqrt{2}x_3 y_1 + x_2 y_3 \end{pmatrix}_4 \oplus \begin{pmatrix} \sqrt{3}x_1 y_1 \\ x_2 y_1 - \sqrt{2}x_3 y_2 \\ x_1 y_2 - \sqrt{2}x_3 y_3 \\ x_1 y_3 - \sqrt{2}x_2 y_2 \\ x_3 y_1 - \sqrt{2}x_2 y_3 \end{pmatrix}_5, \quad (D.29)$$

$$\begin{pmatrix} x_1 \\ x_2 \\ x_3 \end{pmatrix}_3 \otimes \begin{pmatrix} y_1 \\ y_2 \\ y_3 \\ y_4 \end{pmatrix}_4 = \begin{pmatrix} -\sqrt{2}x_2 y_4 - \sqrt{2}x_3 y_1 \\ \sqrt{2}x_1 y_2 - x_2 y_1 + x_3 y_3 \\ \sqrt{2}x_1 y_3 + x_2 y_2 - x_3 y_4 \end{pmatrix}_{3'} \oplus \begin{pmatrix} x_1 y_1 - \sqrt{2}x_3 y_2 \\ -x_1 y_2 - \sqrt{2}x_2 y_1 \\ x_1 y_3 + \sqrt{2}x_3 y_4 \\ -x_1 y_4 + \sqrt{2}x_2 y_3 \end{pmatrix}_4$$

$$\oplus \begin{pmatrix} \sqrt{6}x_2 y_4 - \sqrt{6}x_3 y_1 \\ 2\sqrt{2}x_1 y_1 + 2x_3 y_2 \\ -\sqrt{2}x_1 y_2 + x_2 y_1 + 3x_3 y_3 \\ \sqrt{2}x_1 y_3 - 3x_2 y_2 - x_3 y_4 \\ -2\sqrt{2}x_1 y_4 - 2x_2 y_3 \end{pmatrix}_5, \qquad (D.30)$$

$$\begin{pmatrix} x_1 \\ x_2 \\ x_3 \end{pmatrix}_{3'} \otimes \begin{pmatrix} y_1 \\ y_2 \\ y_3 \\ y_4 \end{pmatrix}_4 = \begin{pmatrix} -\sqrt{2}x_2 y_3 - \sqrt{2}x_3 y_2 \\ \sqrt{2}x_1 y_1 + x_2 y_4 - x_3 y_3 \\ \sqrt{2}x_1 y_4 - x_2 y_2 + x_3 y_1 \end{pmatrix}_3 \oplus \begin{pmatrix} x_1 y_1 + \sqrt{2}x_3 y_3 \\ x_1 y_2 - \sqrt{2}x_3 y_4 \\ -x_1 y_3 + \sqrt{2}x_2 y_1 \\ -x_1 y_4 - \sqrt{2}x_2 y_2 \end{pmatrix}_4$$

$$\oplus \begin{pmatrix} \sqrt{6}x_2y_3 - \sqrt{6}x_3y_2 \\ \sqrt{2}x_1y_1 - 3x_2y_4 - x_3y_3 \\ 2\sqrt{2}x_1y_2 + 2x_3y_4 \\ -2\sqrt{2}x_1y_3 - 2x_2y_1 \\ -\sqrt{2}x_1y_4 + x_2y_2 + 3x_3y_1 \end{pmatrix}_5 , \qquad \text{(D.31)}$$

$$\begin{pmatrix} x_1 \\ x_2 \\ x_3 \end{pmatrix}_3 \otimes \begin{pmatrix} y_1 \\ y_2 \\ y_3 \\ y_4 \\ y_5 \end{pmatrix}_5 = \begin{pmatrix} -2x_1y_1 + \sqrt{3}x_2y_5 + \sqrt{3}x_3y_2 \\ \sqrt{3}x_1y_2 + x_2y_1 - \sqrt{6}x_3y_3 \\ \sqrt{3}x_1y_5 - \sqrt{6}x_2y_4 + x_3y_1 \end{pmatrix}_3$$

$$\oplus \begin{pmatrix} \sqrt{3}x_1y_1 + x_2y_5 + x_3y_2 \\ x_1y_3 - \sqrt{2}x_2y_2 - \sqrt{2}x_3y_4 \\ x_1y_4 - \sqrt{2}x_2y_3 - \sqrt{2}x_3y_5 \end{pmatrix}_{3'}$$

$$\oplus \begin{pmatrix} 2\sqrt{2}x_1y_2 - \sqrt{6}x_2y_1 + x_3y_3 \\ -\sqrt{2}x_1y_3 + 2x_2y_2 - 3x_3y_4 \\ \sqrt{2}x_1y_4 + 3x_2y_3 - 2x_3y_5 \\ -2\sqrt{2}x_1y_5 - x_2y_4 + \sqrt{6}x_3y_1 \end{pmatrix}_4$$

$$\oplus \begin{pmatrix} \sqrt{3}x_2y_5 - \sqrt{3}x_3y_2 \\ -x_1y_2 - \sqrt{3}x_2y_1 - \sqrt{2}x_3y_3 \\ -2x_1y_3 - \sqrt{2}x_2y_2 \\ 2x_1y_4 + \sqrt{2}x_3y_5 \\ x_1y_5 + \sqrt{2}x_2y_4 + \sqrt{3}x_3y_1 \end{pmatrix}_5 , \qquad \text{(D.32)}$$

$$\begin{pmatrix} x_1 \\ x_2 \\ x_3 \end{pmatrix}_{3'} \otimes \begin{pmatrix} y_1 \\ y_2 \\ y_3 \\ y_4 \\ y_5 \end{pmatrix}_5 = \begin{pmatrix} \sqrt{3}x_1y_1 + x_2y_4 + x_3y_3 \\ x_1y_2 - \sqrt{2}x_2y_5 - \sqrt{2}x_3y_4 \\ x_1y_5 - \sqrt{2}x_2y_3 - \sqrt{2}x_3y_2 \end{pmatrix}_3$$

$$\oplus \begin{pmatrix} -2x_1y_1 + \sqrt{3}x_2y_4 + \sqrt{3}x_3y_3 \\ \sqrt{3}x_1y_3 + x_2y_1 - \sqrt{6}x_3y_5 \\ \sqrt{3}x_1y_4 - \sqrt{6}x_2y_2 + x_3y_1 \end{pmatrix}_{3'}$$

$$\oplus \begin{pmatrix} \sqrt{2}x_1y_2 + 3x_2y_5 - 2x_3y_4 \\ 2\sqrt{2}x_1y_3 - \sqrt{6}x_2y_1 + x_3y_5 \\ -2\sqrt{2}x_1y_4 - x_2y_2 + \sqrt{6}x_3y_1 \\ -\sqrt{2}x_1y_5 + 2x_2y_3 - 3x_3y_2 \end{pmatrix}_4$$

$$\oplus \begin{pmatrix} \sqrt{3}x_2y_4 - \sqrt{3}x_3y_3 \\ 2x_1y_2 + \sqrt{2}x_3y_4 \\ -x_1y_3 - \sqrt{3}x_2y_1 - \sqrt{2}x_3y_5 \\ x_1y_4 + \sqrt{2}x_2y_2 + \sqrt{3}x_3y_1 \\ -2x_1y_5 - \sqrt{2}x_2y_3 \end{pmatrix}_5 , \tag{D.33}$$

$$\begin{pmatrix} x_1 \\ x_2 \\ x_3 \\ x_4 \end{pmatrix}_4 \otimes \begin{pmatrix} y_1 \\ y_2 \\ y_3 \\ y_4 \end{pmatrix}_4 = (x_1y_4 + x_2y_3 + x_3y_2 + x_4y_1)_1 \oplus \begin{pmatrix} -x_1y_4 + x_2y_3 - x_3y_2 + x_4y_1 \\ \sqrt{2}x_2y_4 - \sqrt{2}x_4y_2 \\ \sqrt{2}x_1y_3 - \sqrt{2}x_3y_1 \end{pmatrix}_3$$

$$\oplus \begin{pmatrix} x_1y_4 + x_2y_3 - x_3y_2 - x_4y_1 \\ \sqrt{2}x_3y_4 - \sqrt{2}x_4y_3 \\ \sqrt{2}x_1y_2 - \sqrt{2}x_2y_1 \end{pmatrix}_{3'}$$

$$\oplus \begin{pmatrix} x_2y_4 + x_3y_3 + x_4y_2 \\ x_1y_1 + x_3y_4 + x_4y_3 \\ x_1y_2 + x_2y_1 + x_4y_4 \\ x_1y_3 + x_2y_2 + x_3y_1 \end{pmatrix}_4$$

$$\oplus \begin{pmatrix} \sqrt{3}x_1y_4 - \sqrt{3}x_2y_3 - \sqrt{3}x_3y_2 + \sqrt{3}x_4y_1 \\ -\sqrt{2}x_2y_4 + 2\sqrt{2}x_3y_3 - \sqrt{2}x_4y_2 \\ -2\sqrt{2}x_1y_1 + \sqrt{2}x_3y_4 + \sqrt{2}x_4y_3 \\ \sqrt{2}x_1y_2 + \sqrt{2}x_2y_1 - 2\sqrt{2}x_4y_4 \\ -\sqrt{2}x_1y_3 + 2\sqrt{2}x_2y_2 - \sqrt{2}x_3y_1 \end{pmatrix}_5 , \tag{D.34}$$

$$\begin{pmatrix} x_1 \\ x_2 \\ x_3 \\ x_4 \end{pmatrix}_4 \otimes \begin{pmatrix} y_1 \\ y_2 \\ y_3 \\ y_4 \\ y_5 \end{pmatrix}_5 = \begin{pmatrix} 2\sqrt{2}x_1y_5 - \sqrt{2}x_2y_4 + \sqrt{2}x_3y_3 - 2\sqrt{2}x_4y_2 \\ -\sqrt{6}x_1y_1 + 2x_2y_5 + 3x_3y_4 - x_4y_3 \\ x_1y_4 - 3x_2y_3 - 2x_3y_2 + \sqrt{6}x_4y_1 \end{pmatrix}_3$$

$$\oplus \begin{pmatrix} \sqrt{2}x_1y_5 + 2\sqrt{2}x_2y_4 - 2\sqrt{2}x_3y_3 - \sqrt{2}x_4y_2 \\ 3x_1y_2 - \sqrt{6}x_2y_1 - x_3y_5 + 2x_4y_4 \\ -2x_1y_3 + x_2y_2 + \sqrt{6}x_3y_1 - 3x_4y_5 \end{pmatrix}_{3'}$$

$$\oplus \begin{pmatrix} \sqrt{3}x_1y_1 - \sqrt{2}x_2y_5 + \sqrt{2}x_3y_4 - 2\sqrt{2}x_4y_3 \\ -\sqrt{2}x_1y_2 - \sqrt{3}x_2y_1 + 2\sqrt{2}x_3y_5 + \sqrt{2}x_4y_4 \\ \sqrt{2}x_1y_3 + 2\sqrt{2}x_2y_2 - \sqrt{3}x_3y_1 - \sqrt{2}x_4y_5 \\ -2\sqrt{2}x_1y_4 + \sqrt{2}x_2y_3 - \sqrt{2}x_3y_2 + \sqrt{3}x_4y_1 \end{pmatrix}_4$$

$$\oplus \begin{pmatrix} \sqrt{2}x_1y_5 - \sqrt{2}x_2y_4 - \sqrt{2}x_3y_3 + \sqrt{2}x_4y_2 \\ -\sqrt{2}x_1y_1 - \sqrt{3}x_3y_4 - \sqrt{3}x_4y_3 \\ \sqrt{3}x_1y_2 + \sqrt{2}x_2y_1 + \sqrt{3}x_3y_5 \\ \sqrt{3}x_2y_2 + \sqrt{2}x_3y_1 + \sqrt{3}x_4y_5 \\ -\sqrt{3}x_1y_4 - \sqrt{3}x_2y_3 - \sqrt{2}x_4y_1 \end{pmatrix}_5$$

$$\oplus \begin{pmatrix} 2x_1y_5 + 4x_2y_4 + 4x_3y_3 + 2x_4y_2 \\ 4x_1y_1 + 2\sqrt{6}x_2y_5 \\ -\sqrt{6}x_1y_2 + 2x_2y_1 - \sqrt{6}x_3y_5 + 2\sqrt{6}x_4y_4 \\ 2\sqrt{6}x_1y_3 - \sqrt{6}x_2y_2 + 2x_3y_1 - \sqrt{6}x_4y_5 \\ 2\sqrt{6}x_3y_2 + 4x_4y_1 \end{pmatrix}_5 , \quad (D.35)$$

$$\begin{pmatrix} x_1 \\ x_2 \\ x_3 \\ x_4 \\ x_5 \end{pmatrix}_5 \otimes \begin{pmatrix} y_1 \\ y_2 \\ y_3 \\ y_4 \\ y_5 \end{pmatrix}_5 = (x_1y_1 + x_2y_5 + x_3y_4 + x_4y_3 + x_5y_2)_1$$

$$\oplus \begin{pmatrix} x_2y_5 + 2x_3y_4 - 2x_4y_3 - x_5y_2 \\ -\sqrt{3}x_1y_2 + \sqrt{3}x_2y_1 + \sqrt{2}x_3y_5 - \sqrt{2}x_5y_3 \\ \sqrt{3}x_1y_5 + \sqrt{2}x_2y_4 - \sqrt{2}x_4y_2 - \sqrt{3}x_5y_1 \end{pmatrix}_3$$

$$\oplus \begin{pmatrix} 2x_2y_5 - x_3y_4 + x_4y_3 - 2x_5y_2 \\ \sqrt{3}x_1y_3 - \sqrt{3}x_3y_1 + \sqrt{2}x_2y_5 - \sqrt{2}x_5y_4 \\ -\sqrt{3}x_1y_4 + \sqrt{2}x_2y_3 - \sqrt{2}x_3y_2 + \sqrt{3}x_4y_1 \end{pmatrix}_{3'}$$

$$\oplus \begin{pmatrix} 3\sqrt{2}x_1y_2 + 3\sqrt{2}x_2y_1 - \sqrt{3}x_3y_5 + 4\sqrt{3}x_4y_4 - \sqrt{3}x_5y_3 \\ 3\sqrt{2}x_1y_3 + 4\sqrt{3}x_2y_2 + 3\sqrt{2}x_3y_1 - \sqrt{3}x_4y_5 - \sqrt{3}x_5y_4 \\ 3\sqrt{2}x_1y_4 - \sqrt{3}x_2y_3 - \sqrt{3}x_3y_2 + 3\sqrt{2}x_4y_1 + 4\sqrt{3}x_5y_5 \\ 3\sqrt{2}x_1y_5 - \sqrt{3}x_2y_4 + 4\sqrt{3}x_3y_3 - \sqrt{3}x_4y_2 + 3\sqrt{2}x_5y_1 \end{pmatrix}_4$$

$$\oplus \begin{pmatrix} \sqrt{2}x_1y_2 - \sqrt{2}x_2y_1 + \sqrt{3}x_3y_5 - \sqrt{3}x_5y_3 \\ -\sqrt{2}x_1y_3 + \sqrt{2}x_3y_1 + \sqrt{3}x_4y_5 - \sqrt{3}x_5y_4 \\ -\sqrt{2}x_1y_4 - \sqrt{3}x_2y_3 + \sqrt{3}x_3y_2 + \sqrt{2}x_4y_1 \\ \sqrt{2}x_1y_5 - \sqrt{3}x_2y_4 + \sqrt{3}x_4y_2 - \sqrt{2}x_5y_1 \end{pmatrix}_4$$

$$\oplus \begin{pmatrix} 2x_1y_1 + x_2y_5 - 2x_3y_4 - 2x_4y_3 + x_5y_2 \\ x_1y_2 + x_2y_1 + \sqrt{6}x_3y_5 + \sqrt{6}x_5y_3 \\ -2x_1y_3 + \sqrt{6}x_2y_2 - 2x_3y_1 \\ -2x_1y_4 - 2x_4y_1 + \sqrt{6}x_5y_5 \\ x_1y_5 + \sqrt{6}x_2y_4 + \sqrt{6}x_4y_2 + x_5y_1 \end{pmatrix}_5$$

$$\oplus \begin{pmatrix} 2x_1y_1 - 2x_2y_5 + x_3y_4 + x_4y_3 - 2x_5y_2 \\ -2x_1y_2 - 2x_2y_1 + \sqrt{6}x_4y_4 \\ x_1y_3 + x_3y_1 + \sqrt{6}x_4y_5 + \sqrt{6}x_5y_4 \\ x_1y_4 + \sqrt{6}x_2y_3 + \sqrt{6}x_3y_2 + x_4y_1 \\ -2x_1y_5 + \sqrt{6}x_3y_3 - 2x_5y_1 \end{pmatrix}_5 . \quad (D.36)$$

Representations of T' in Different Basis

Here, we show another basis for representations of the T' group. All of the T' elements are written by products of the generators, s and t, which satisfy

$$s^2 = r, \quad r^2 = t^3 = (st)^3 = e, \quad rt = tr. \tag{E.1}$$

In Chap. 5, the doublet and triplet representations are as follow:

$$t = \begin{pmatrix} \omega^2 & 0 \\ 0 & \omega \end{pmatrix}, \quad r = \begin{pmatrix} -1 & 0 \\ 0 & -1 \end{pmatrix}, \quad s = -\frac{1}{\sqrt{3}} \begin{pmatrix} i & \sqrt{2}p \\ -\sqrt{2}\bar{p} & -i \end{pmatrix} \quad \text{on } \mathbf{2}, \text{(E.2)}$$

$$t = \begin{pmatrix} 1 & 0 \\ 0 & \omega^2 \end{pmatrix}, \quad r = \begin{pmatrix} -1 & 0 \\ 0 & -1 \end{pmatrix}, \quad s = -\frac{1}{\sqrt{3}} \begin{pmatrix} i & \sqrt{2}p \\ -\sqrt{2}\bar{p} & -i \end{pmatrix} \quad \text{on } \mathbf{2'}, \text{(E.3)}$$

$$t = \begin{pmatrix} \omega & 0 \\ 0 & 1 \end{pmatrix}, \quad r = \begin{pmatrix} -1 & 0 \\ 0 & -1 \end{pmatrix}, \quad s = -\frac{1}{\sqrt{3}} \begin{pmatrix} i & \sqrt{2}p \\ -\sqrt{2}\bar{p} & -i \end{pmatrix} \quad \text{on } \mathbf{2''}, \text{(E.4)}$$

$$t = \begin{pmatrix} 1 & 0 & 0 \\ 0 & \omega & 0 \\ 0 & 0 & \omega^2 \end{pmatrix}, \quad r = \begin{pmatrix} 1 & 0 & 0 \\ 0 & 1 & 0 \\ 0 & 0 & 1 \end{pmatrix}, \quad s = \frac{1}{3} \begin{pmatrix} -1 & 2p_1 & 2p_1 p_2 \\ 2\bar{p}_1 & -1 & 2p_2 \\ 2\bar{p}_1 \bar{p}_2 & 2\bar{p}_2 & -1 \end{pmatrix} \quad \text{on } \mathbf{3}, \text{(E.5)}$$

where $p_1 = e^{i\phi_1}$ and $p_2 = e^{i\phi_2}$.

© Springer-Verlag GmbH Germany, part of Springer Nature 2022
T. Kobayashi et al., *An Introduction to Non-Abelian Discrete Symmetries for Particle Physicists,* Lecture Notes in Physics 995,
https://doi.org/10.1007/978-3-662-64679-3

E.1   The Basis I

First, we show the basis in Chap. 5. We take the parameters $p = i$ and $p_1 = p_2 = 1$, then the generator s is written as

$$s = -\frac{i}{\sqrt{3}} \begin{pmatrix} 1 & \sqrt{2} \\ \sqrt{2} & -1 \end{pmatrix}, \quad \text{on } \mathbf{2}, \mathbf{2'}, \mathbf{2''}, \tag{E.6}$$

$$s = \begin{pmatrix} -1 & 2 & 2 \\ 2 & -1 & 2 \\ 2 & 2 & -1 \end{pmatrix}, \quad \text{on } \mathbf{3}. \tag{E.7}$$

Here we show the tensor products as

$$\begin{pmatrix} x_1 \\ x_2 \end{pmatrix}_{2(2')} \otimes \begin{pmatrix} y_1 \\ y_2 \end{pmatrix}_{2(2'')} = \left(\frac{x_1 y_2 - x_2 y_1}{\sqrt{2}} \right)_1 \oplus \begin{pmatrix} \frac{x_1 y_2 + x_2 y_1}{\sqrt{2}} \\ -x_1 y_1 \\ x_2 y_2 \end{pmatrix}_3, \tag{E.8}$$

$$\begin{pmatrix} x_1 \\ x_2 \end{pmatrix}_{2'(2)} \otimes \begin{pmatrix} y_1 \\ y_2 \end{pmatrix}_{2'(2'')} = \left(\frac{x_1 y_2 - x_2 y_1}{\sqrt{2}} \right)_{1''} \oplus \begin{pmatrix} -x_1 y_1 \\ x_2 y_2 \\ \frac{x_1 y_2 + x_2 y_1}{\sqrt{2}} \end{pmatrix}_3, \tag{E.9}$$

$$\begin{pmatrix} x_1 \\ x_2 \end{pmatrix}_{2''(2)} \otimes \begin{pmatrix} y_1 \\ y_2 \end{pmatrix}_{2''(2')} = \left(\frac{x_1 y_2 - x_2 y_1}{\sqrt{2}} \right)_{1'} \oplus \begin{pmatrix} x_2 y_2 \\ \frac{x_1 y_2 + x_2 y_1}{\sqrt{2}} \\ -x_1 y_1 \end{pmatrix}_3, \tag{E.10}$$

$$\begin{pmatrix} x_1 \\ x_2 \\ x_3 \end{pmatrix}_3 \otimes \begin{pmatrix} y_1 \\ y_2 \\ y_3 \end{pmatrix}_3 = [x_1 y_1 + x_2 y_3 + x_3 y_2]_1$$
$$\oplus [x_3 y_3 + x_1 y_2 + x_2 y_1]_{1'} \oplus (x_2 y_2 + x_1 y_3 + x_3 y_1)_{1''}$$
$$\oplus \begin{pmatrix} 2x_1 y_1 - x_2 y_3 - x_3 y_3 \\ 2x_3 y_3 - x_1 y_2 - x_2 y_1 \\ 2x_2 y_2 - x_1 y_3 - x_3 y_1 \end{pmatrix}_3$$
$$\oplus \begin{pmatrix} x_2 y_3 - x_3 y_2 \\ x_1 y_2 - x_2 y_1 \\ x_3 y_1 - x_1 y_3 \end{pmatrix}_3, \tag{E.11}$$

$$\begin{pmatrix} x_1 \\ x_2 \end{pmatrix}_{2,2',2''} \otimes \begin{pmatrix} y_1 \\ y_2 \\ y_3 \end{pmatrix}_3 = \begin{pmatrix} \sqrt{2}x_2 y_2 + x_1 y_1 \\ \sqrt{2}x_1 y_3 - x_2 y_1 \end{pmatrix}_{2,2',2''} \oplus \begin{pmatrix} \sqrt{2}x_2 y_3 + x_1 y_2 \\ \sqrt{2}x_1 y_1 - x_2 y_2 \end{pmatrix}_{2',2'',2}$$
$$\oplus \begin{pmatrix} \sqrt{2}x_2 y_1 + x_1 y_3 \\ \sqrt{2}x_1 y_2 - x_2 y_3 \end{pmatrix}_{2'',2,2'}, \tag{E.12}$$

$$(x)_{1'(1'')} \otimes \begin{pmatrix} y_1 \\ y_2 \end{pmatrix}_{2,2',2''} = \begin{pmatrix} xy_1 \\ xy_2 \end{pmatrix}_{2'(2''),2''(2),2(2')}, \tag{E.13}$$

$$(x)_{1'} \otimes \begin{pmatrix} y_1 \\ y_2 \\ y_3 \end{pmatrix}_3 = \begin{pmatrix} xy_3 \\ xy_1 \\ xy_2 \end{pmatrix}_3, \quad (x)_{1''} \otimes \begin{pmatrix} y_1 \\ y_2 \\ y_3 \end{pmatrix}_3 = \begin{pmatrix} xy_2 \\ xy_3 \\ xy_1 \end{pmatrix}_3. \tag{E.14}$$

E.2 The Basis II

Next, we consider another basis, which is used, e.g. in Ref. [11], We take the parameters $p = e^{i\pi/12}$ and $p_1 = p_2 = \omega$, then the generator s is written as

$$s = -\frac{1}{\sqrt{3}} \begin{pmatrix} i & \sqrt{2}e^{i\pi/12} \\ -\sqrt{2}e^{-i\pi/12} & -i \end{pmatrix}, \quad \text{on } 2, 2', 2'', \tag{E.15}$$

$$s = \begin{pmatrix} -1 & 2\omega & 2\omega^2 \\ 2\omega^2 & -1 & 2\omega \\ 2\omega & 2\omega^2 & -1 \end{pmatrix}, \quad \text{on } 3. \tag{E.16}$$

Here we show the tensor products as

$$\begin{pmatrix} x_1 \\ x_2 \end{pmatrix}_{2(2')} \otimes \begin{pmatrix} y_1 \\ y_2 \end{pmatrix}_{2(2'')} = [x_1 y_2 - x_2 y_1]_1 \oplus \begin{pmatrix} \frac{1-i}{2}(x_1 y_2 + x_2 y_1) \\ i x_1 y_1 \\ x_2 y_2 \end{pmatrix}_3, \tag{E.17}$$

$$\begin{pmatrix} x_1 \\ x_2 \end{pmatrix}_{2'(2)} \otimes \begin{pmatrix} y_1 \\ y_2 \end{pmatrix}_{2'(2'')} = [x_1 y_2 - x_2 y_1]_{1''} \oplus \begin{pmatrix} i x_1 y_1 \\ x_2 y_2 \\ \frac{1-i}{2}(x_1 y_2 + x_2 y_1) \end{pmatrix}_3, \tag{E.18}$$

$$\begin{pmatrix} x_1 \\ x_2 \end{pmatrix}_{2''(2)} \otimes \begin{pmatrix} y_1 \\ y_2 \end{pmatrix}_{2''(2')} = [x_1 y_2 - x_2 y_1]_{1'} \oplus \begin{pmatrix} x_2 y_2 \\ \frac{1-i}{2}(x_1 y_2 + x_2 y_1) \\ i x_1 y_1 \end{pmatrix}_3, \tag{E.19}$$

$$\begin{pmatrix} x_1 \\ x_2 \\ x_3 \end{pmatrix}_3 \otimes \begin{pmatrix} y_1 \\ y_2 \\ y_3 \end{pmatrix}_3 = [x_1 y_1 + x_2 y_3 + x_3 y_2]_1$$
$$\oplus [x_3 y_3 + x_1 y_2 + x_2 y_1]_{1'} \oplus (x_2 y_2 + x_1 y_3 + x_3 y_1)_{1''}$$
$$\oplus \frac{1}{3} \begin{pmatrix} 2x_1 y_1 - x_2 y_3 - x_3 y_2 \\ 2x_3 y_3 - x_1 y_2 - x_2 y_1 \\ 2x_2 y_2 - x_1 y_3 - x_3 y_1 \end{pmatrix}_3$$
$$\oplus \frac{1}{2} \begin{pmatrix} x_2 y_3 - x_3 y_2 \\ x_1 y_2 - x_2 y_1 \\ x_3 y_1 - x_1 y_3 \end{pmatrix}_3, \tag{E.20}$$

$$\begin{pmatrix} x_1 \\ x_2 \end{pmatrix}_{2,2',2''} \otimes \begin{pmatrix} y_1 \\ y_2 \\ y_3 \end{pmatrix}_3 = \begin{pmatrix} (1+i)x_2 y_2 + x_1 y_1 \\ (1-i)x_1 y_3 - x_2 y_1 \end{pmatrix}_{2,2',2''}$$

$$\oplus \begin{pmatrix} (1+i)x_2 y_3 + x_1 y_2 \\ (1-i)x_1 y_1 - x_2 y_2 \end{pmatrix}_{2',2'',2}$$

$$\oplus \begin{pmatrix} (1+i)x_2 y_1 + x_1 y_3 \\ (1-i)x_1 y_2 - x_2 y_3 \end{pmatrix}_{2'',2,2'}, \qquad (E.21)$$

$$(x)_{\mathbf{1}'(\mathbf{1}'')} \otimes \begin{pmatrix} y_1 \\ y_2 \end{pmatrix}_{\mathbf{2},\mathbf{2}',\mathbf{2}''} = \begin{pmatrix} xy_1 \\ xy_2 \end{pmatrix}_{\mathbf{2}'(\mathbf{2}''),\mathbf{2}''(\mathbf{2}),\mathbf{2}(\mathbf{2}')}, \qquad (E.22)$$

$$(x)_{\mathbf{1}'} \otimes \begin{pmatrix} y_1 \\ y_2 \\ y_3 \end{pmatrix}_{\mathbf{3}} = \begin{pmatrix} xy_3 \\ xy_1 \\ xy_2 \end{pmatrix}_{\mathbf{3}}, \quad (x)_{\mathbf{1}''} \otimes \begin{pmatrix} y_1 \\ y_2 \\ y_3 \end{pmatrix}_{\mathbf{3}} = \begin{pmatrix} xy_2 \\ xy_3 \\ xy_1 \end{pmatrix}_{\mathbf{3}}. \qquad (E.23)$$

Representations of $\Delta(96)$ in Different Basis

We show another basis for representations of the $\Delta(96)$ group. First, we show the basis in Sect. 13.4. All of the $\Delta(96)$ elements are written by products of the generators, S, T, and U that satisfy [12]

$$U^2 = T^3 = (UT)^8 = (UT^{-1}UT)^3 = e, \quad S = U(UT)^4 U(UT)^4. \tag{F.1}$$

In Appendix of Ref. [12], all the multiplication rules in the above bases are found. S, T, U are constructed in terms of our basis a, a', b, c as follows:

$$S = V^\dagger(a'^2)V, \quad T = V^\dagger(ba)V, \quad U = V^\dagger(bbca')V, \tag{F.2}$$

where the unitary matrix V depends on a kind of representation. Hereafter, we thus denote $V \equiv V_{\text{Rep}}$. Then, all the representations $\mathbf{R}'$ in [12] are uniquely transformed to our representations $\mathbf{R}$ except the overall coefficients by V_{Rep} as follows:

$$\mathbf{R} \propto V_{\text{Rep}} \cdot \mathbf{R}'. \tag{F.3}$$

Below, we show V_{Rep} for each of the representations. On the representation $\mathbf{2}$, V_2 is represented as

$$V_2 = \begin{pmatrix} 0 & \omega \\ \omega^2 & 0 \end{pmatrix} \begin{pmatrix} 0 & 1 \\ 1 & 0 \end{pmatrix}. \tag{F.4}$$

On the representations $\mathbf{3}$, there are six unitary matrices, and they are represented as

$$V_{3_{21}} = V_{3_{11}} = \frac{1}{\sqrt{3}} \begin{pmatrix} -i\omega^2 & -i & -\omega \\ \omega & 1 & \omega^2 \\ 1 & 1 & 1 \end{pmatrix} \begin{pmatrix} 1 & 0 & 0 \\ 0 & \omega^2 & 0 \\ 0 & 0 & \omega \end{pmatrix}, \tag{F.5}$$

© Springer-Verlag GmbH Germany, part of Springer Nature 2022
T. Kobayashi et al., *An Introduction to Non-Abelian Discrete Symmetries
for Particle Physicists*, Lecture Notes in Physics 995,
https://doi.org/10.1007/978-3-662-64679-3

$$V_{322} = V_{312} = \frac{1}{\sqrt{3}} \begin{pmatrix} -\omega^2 & -1 & -\omega \\ \omega & 1 & \omega^2 \\ 1 & 1 & 1 \end{pmatrix} \begin{pmatrix} 1 & 0 & 0 \\ 0 & \omega^2 & 0 \\ 0 & 0 & \omega \end{pmatrix}, \tag{F.6}$$

$$V_{323} = V_{313} = \frac{1}{\sqrt{3}} \begin{pmatrix} i\omega & i & i\omega^2 \\ \omega^2 & 1 & \omega \\ 1 & 1 & 1 \end{pmatrix} \begin{pmatrix} 1 & 0 & 0 \\ 0 & \omega^2 & 0 \\ 0 & 0 & \omega \end{pmatrix}. \tag{F.7}$$

On the representation $\mathbf{6}_{[[k],[\ell]]}$, there are six unitary matrices. However since all of them give the equivalent representations, it is enough to show V_6 in the case of $(k, \ell) = (1, 1)$. Then, it is represented by

$$V_6 = \frac{1}{\sqrt{3}} \begin{pmatrix} 0 & 0 & 0 & -i\omega^2 & -i & -i\omega \\ 0 & 0 & 0 & -\omega & -1 & -\omega^2 \\ 0 & 0 & 0 & 1 & 1 & 1 \\ -i\omega & -i & -i\omega^2 & 0 & 0 & 0 \\ i\omega^2 & i & i\omega & 0 & 0 & 0 \\ 1 & 1 & 1 & 0 & 0 & 0 \end{pmatrix} \begin{pmatrix} 1 & 0 & 0 & 0 & 0 & 0 \\ 0 & 1 & 0 & 0 & 0 & 0 \\ 0 & 0 & 1 & 0 & 0 & 0 \\ 0 & 0 & 0 & -i\omega & 0 & 0 \\ 0 & 0 & 0 & 0 & -i & 0 \\ 0 & 0 & 0 & 0 & 0 & -i\omega^2 \end{pmatrix}. \tag{F.8}$$

Notice here that there is another basis in [13], even though we do not show the basis transformations.

Other Smaller Groups

G

In this appendix, we study finite groups, whose orders are less than 31 [14,15]. Such groups are summarized in Table G.1, where g denotes the order. Here, we have omitted Abelian groups Z_N as well as their direct products. Most of the finite groups in Table G.1 are non-Abelian groups mentioned in the text and their extenstions by direct products with Abelian groups such as $Z_2 \times D_4$, $Z_2 \times Q_4$, etc. However, the table includes non-Abelian groups, which are not mentioned in the text, i.e. $Z_4 \rtimes Z_4$, $Z_8 \rtimes Z_2$, $(Z_4 \times Z_2) \rtimes Z_2(I)$, $(Z_4 \times Z_2) \rtimes Z_2(II)$, $Z_3 \rtimes Z_8$, $(Z_6 \times Z_2) \rtimes Z_2$, and $Z_9 \rtimes Z_3$. In this appendix, we explain these groups.

G.1 $Z_4 \rtimes Z_4$

Here, we study $Z_4 \rtimes Z_4$. We denote the first and second Z_4 generators by a and b, respectively. That is, they satisfy $a^4 = b^4 = e$. Then, we consider the relation $ab = ba^m$. That leads to $ab^2 = b^2 a^{m^2}$. We require $m \neq 0$ and $m^2 \neq 0$ mod 4. If $m = 1$ mod 4, the generators, a and b, are commutable each other and the group becomes the direct product, $Z_4 \times Z_4$. Thus, we require $m \neq 1$ mod 4. Then, these requirements are satisfied for $m = 3$ mod 4, i.e.

$$ab = ba^3. \tag{G.1}$$

Using them, all of $Z_4 \rtimes Z_4$ elements are written as

$$g = b^m a^n, \tag{G.2}$$

with $n, m = 0, 1, 2, 3$.

© Springer-Verlag GmbH Germany, part of Springer Nature 2022
T. Kobayashi et al., *An Introduction to Non-Abelian Discrete Symmetries for Particle Physicists,* Lecture Notes in Physics 995,
https://doi.org/10.1007/978-3-662-64679-3

Table G.1 Classification of the Non-Abelian groups with $g \leq 30$. Note that there are two finite groups isomorphic to $(Z_4 \times Z_2) \rtimes Z_2$ except D_8

g	Groups
6	$S_3 \equiv D_3$
8	$D_4, \ Q_4$
10	D_5
12	$A_4, \ D_6, \ Q_6$
14	D_7
16	$D_8, \ Q_8, \ QD_{16}, \ Z_2 \times D_4, \ Z_2 \times Q_4, \ Z_4 \rtimes Z_4,$ $Z_8 \rtimes Z_2, \ (Z_4 \times Z_2) \rtimes Z_2(I), \ (Z_4 \times Z_2) \rtimes Z_2(II)$
18	$D_9, \ Z_3 \times D_3, \ \Sigma(18) \equiv (Z_3 \times Z_3) \rtimes Z_2$
20	$D_{10}, \ Q_{10},$
21	$T_7 \equiv Z_7 \rtimes Z_3$
22	D_{11}
24	$D_{12}, \ S_4, \ Q_{12}, \ T' \simeq SL(2,3), \ Z_2 \times Z_2 \times S_3, \ Z_4 \times S_3,$ $Z_2 \times Q_6, \ Z_3 \times D_4, \ Z_3 \times Q_4, \ Z_2 \times A_4, \ Z_3 \rtimes Z_8, \ (Z_6 \times Z_2) \rtimes Z_2,$
26	D_{13}
27	$\Delta(27) \equiv (Z_3 \times Z_3) \rtimes Z_3, \ Z_9 \rtimes Z_3$
28	Q_{14}, D_{14}
30	$Z_5 \times S_3, Z_3 \times D_5, D_{15}$

All of the elements, $b^m a^n$, are classified into ten conjugacy classes,

$$\begin{aligned}
C_1 : & \quad \{e\}, & h = 1, \\
C_1^{(1)} : & \quad \{a^2\}, & h = 2, \\
C_1^{(2)} : & \quad \{b^2\}, & h = 2, \\
C_1^{(3)} : & \quad \{b^2 a^2\}, & h = 2, \\
C_2^{(1)} : & \quad \{b \ , \ ba^2\}, & h = 4, \\
C_2^{(2)} : & \quad \{ba \ , \ ba^3\}, & h = 4, \\
C_2^{(3)} : & \quad \{b^3 \ , \ b^3 a^2\}, & h = 4, \\
C_2^{(4)} : & \quad \{b^3 a \ , \ b^3 a^3\}, & h = 4, \\
C_2^{\prime(1)} : & \quad \{a \ , \ a^3\}, & h = 4, \\
C_2^{\prime(2)} : & \quad \{b^2 a \ , \ b^2 a^3\}, & h = 4.
\end{aligned} \tag{G.3}$$

The $Z_4 \rtimes Z_4$ group has eight singlets $\mathbf{1}_{\pm,k}$ with $k = 0, 1, 2, 3$ and two doublets $\mathbf{2}_1$ and $\mathbf{2}_2$. The generators, b and a, are represented as

$$b = i^k, \quad a = \pm 1, \quad \text{on } \mathbf{1}_{\pm,k}, \tag{G.4}$$

Table G.2 Characters of $Z_4 \rtimes Z_4$

	n	h	$\chi_{1+,0}$	$\chi_{1+,1}$	$\chi_{1+,2}$	$\chi_{1+,3}$	$\chi_{1-,0}$	$\chi_{1-,1}$	$\chi_{1-,2}$	$\chi_{1-,3}$	χ_{2_1}	χ_{2_2}
C_1	1	1	1	1	1	1	1	1	1	1	2	2
$C_1^{(1)}$	1	2	1	1	1	1	1	1	1	1	-2	-2
$C_1^{(2)}$	1	2	1	1	1	1	-1	-1	-1	-1	2	-2
$C_1^{(3)}$	1	2	1	1	1	1	-1	-1	-1	-1	-2	2
$C_2^{(1)}$	2	4	1	-1	-1	1	i	$-i$	i	$-i$	0	0
$C_2^{(2)}$	2	4	1	1	-1	-1	$-i$	i	i	$-i$	0	0
$C_2^{(3)}$	2	4	1	-1	-1	1	$-i$	i	$-i$	i	0	0
$C_2^{(4)}$	2	4	1	1	-1	-1	i	$-i$	$-i$	i	0	0
$C_2'^{(1)}$	2	4	1	-1	1	-1	-1	-1	1	1	0	0
$C_2'^{(2)}$	2	4	1	-1	1	-1	1	1	-1	-1	0	0

and

$$b = \begin{pmatrix} 0 & 1 \\ 1 & 0 \end{pmatrix}, \quad a = \begin{pmatrix} i & 0 \\ 0 & -i \end{pmatrix}, \quad \text{on } \mathbf{2}_1, \tag{G.5}$$

$$b = \begin{pmatrix} 0 & 1 \\ -1 & 0 \end{pmatrix}, \quad a = \begin{pmatrix} -i & 0 \\ 0 & i \end{pmatrix}, \quad \text{on } \mathbf{2}_2. \tag{G.6}$$

The characters are shown in Table G.2.

The tensor products between doublets are obtained as

$$\begin{pmatrix} x_1 \\ x_3 \end{pmatrix}_{\mathbf{2}_1} \otimes \begin{pmatrix} y_1 \\ y_3 \end{pmatrix}_{\mathbf{2}_1} = \begin{pmatrix} x_3 \\ x_1 \end{pmatrix}_{\mathbf{2}_2} \otimes \begin{pmatrix} y_3 \\ y_1 \end{pmatrix}_{\mathbf{2}_2}$$
$$= (x_1 y_3 + x_3 y_1)_{\mathbf{1}_{+,0}} \oplus (x_1 y_3 - x_3 y_1)_{\mathbf{1}_{+,2}}$$
$$\oplus (x_1 y_1 + x_3 y_3)_{\mathbf{1}_{+,3}} \oplus (x_1 y_1 - x_3 y_3)_{\mathbf{1}_{+,1}}, \tag{G.7}$$

$$\begin{pmatrix} x_1 \\ x_3 \end{pmatrix}_{\mathbf{2}_1} \otimes \begin{pmatrix} y_3 \\ y_1 \end{pmatrix}_{\mathbf{2}_2} = (x_1 y_3 + x_3 y_1)_{\mathbf{1}_{-,2}} \oplus (x_1 y_3 - x_3 y_1)_{\mathbf{1}_{-,0}}$$
$$\oplus (x_1 y_1 + x_3 y_3)_{\mathbf{1}_{-,3}} \oplus (x_1 y_1 - x_3 y_3)_{\mathbf{1}_{-,1}}. \tag{G.8}$$

The tensor products between singlets and doublets are obtained as

$$(x)_{1_{\pm,0}} \otimes \begin{pmatrix} y_{1(3)} \\ y_{3(1)} \end{pmatrix}_{2_1(2_2)} = \begin{pmatrix} xy_{1(3)} \\ xy_{3(1)} \end{pmatrix}_{2_1(2_2)}, \tag{G.9}$$

$$(x)_{1_{\pm,2}} \otimes \begin{pmatrix} y_1 \\ y_3 \end{pmatrix}_{2_1} = \begin{pmatrix} xy_1 \\ xy_3 \end{pmatrix}_{2_2}, \tag{G.10}$$

$$(x)_{1_{\pm,2}} \otimes \begin{pmatrix} y_3 \\ y_1 \end{pmatrix}_{2_2} = \begin{pmatrix} xy_3 \\ xy_1 \end{pmatrix}_{2_1}, \tag{G.11}$$

$$(x)_{1_{\pm,1}} \otimes \begin{pmatrix} y_{1(3)} \\ y_{3(1)} \end{pmatrix}_{2_1(2_2)} = (xy_1)_{1_{\pm,2}} \oplus (xy_3)_{1_{\pm,0}}, \tag{G.12}$$

$$(x)_{1_{\pm,3}} \otimes \begin{pmatrix} y_{1(3)} \\ y_{3(1)} \end{pmatrix}_{2_1(2_2)} = (xy_1)_{1_{\pm,0}} \oplus (xy_3)_{1_{\pm,2}}. \tag{G.13}$$

The tensor products between singlets are found as

$$\mathbf{1}_{\pm,i} \otimes \mathbf{1}_{\pm,j} = \mathbf{1}_{\pm,i+j(\mathrm{mod}\ 4)}, \quad \mathbf{1}_{\pm,i} \otimes \mathbf{1}_{\mp,j} = \mathbf{1}_{\mp,i+j(\mathrm{mod}\ 4)}, \tag{G.14}$$

where $i,\ j = 0, 1, 2, 3$.

G.2 $Z_8 \rtimes Z_2$

Here, we study the group $Z_8 \rtimes Z_2$ other than D_8 and QD_{16}. We denote the generators of Z_8 and Z_2 by a and b, respectively. That is, they satisfy

$$a^8 = e, \quad b^2 = e. \tag{G.15}$$

In addition, we require

$$bab = a^m, \tag{G.16}$$

where $m \neq 0$. That leads to $b^2 a b^2 = a^{m^2}$. Because $b^2 = e$, the consistency requires $m^2 = 1 \bmod 8$. Then, the possible values are obtained as $m = 1, 3, 5, 7$. However, the groups with $m = 7$ and 3 correspond to D_8 and QD_{16}, respectively, while the group with $m = 1$ is just a direct product, $Z_8 \times Z_2$. Thus, here we study the group with $m = 5$.

Table G.3 Characters of $Z_8 \rtimes Z_2$

	n	h	χ_{1+0}	χ_{1-0}	χ_{1+1}	χ_{1-1}	χ_{1+2}	χ_{1-2}	χ_{1+3}	χ_{1-3}	χ_{2_1}	χ_{2_2}
C_1	1	1	1	1	1	1	1	1	1	1	2	2
$C_2^{(1)}$	2	8	1	1	i	i	-1	-1	$-i$	$-i$	0	0
$C_2^{(2)}$	2	8	1	1	$-i$	$-i$	-1	-1	i	i	0	0
$C_1^{(1)}$	1	4	1	1	-1	-1	1	1	-1	-1	$2i$	$-2i$
$C_1^{(2)}$	1	2	1	1	1	1	1	1	1	1	-2	-2
$C_1^{(3)}$	1	4	1	1	-1	-1	1	1	i	i	$-2i$	$2i$
$C_2^{\prime(1)}$	2	2	1	-1	1	-1	1	-1	1	-1	0	0
$C_2^{\prime(2)}$	2	2	1	-1	i	$-i$	-1	1	$-i$	i	0	0
$C_2^{\prime(3)}$	2	2	1	-1	-1	1	1	1	-1	-1	1	0	0
$C_2^{\prime(4)}$	2	2	1	-1	$-i$	i	-1	1	i	$-i$	0	0

All of the group elements are written as $b^k a^\ell$ with $k = 0, 1$ and $\ell = 0, \ldots, 7$. These elements are classified into ten conjugacy classes,

$$
\begin{aligned}
C_1 : &\quad \{e\}, &\quad h = 1, \\
C_2^{(1)} : &\quad \{a, a^5\}, &\quad h = 8, \\
C_2^{(2)} : &\quad \{a^3, a^7\}, &\quad h = 8, \\
C_1^{(1)} : &\quad \{a^2\}, &\quad h = 4, \\
C_1^{(2)} : &\quad \{a^4\}, &\quad h = 2, \\
C_1^{(3)} : &\quad \{a^6\}, &\quad h = 4, \\
C_2^{\prime(1)} : &\quad \{b, ba^4\}, &\quad h = 2, \\
C_2^{\prime(2)} : &\quad \{ba, ba^5\}, &\quad h = 2, \\
C_2^{\prime(3)} : &\quad \{ba^2, ba^6\}, &\quad h = 2, \\
C_2^{\prime(4)} : &\quad \{ba^3, ba^7\}, &\quad h = 2.
\end{aligned} \tag{G.17}
$$

The $Z_8 \rtimes Z_2$ group has eight singlets $\mathbf{1}_{\pm,k}$ with $k = 0, 1, 2, 3$ and two doublets $\mathbf{2}_1$ and $\mathbf{2}_2$. The characters are shown in Table G.3.

The generators a and b can be represented as

$$
a = (i)^k, \quad b = \pm 1, \quad \text{on} \quad \mathbf{1}_{\pm,k}. \tag{G.18}
$$

The generators a and b can be represented by the following matrices as

$$
a = \begin{pmatrix} \rho & 0 \\ 0 & \rho^5 \end{pmatrix}, \quad b = \begin{pmatrix} 0 & 1 \\ 1 & 0 \end{pmatrix}, \quad \text{on} \quad \mathbf{2}_1, \tag{G.19}
$$

$$
a = \begin{pmatrix} \rho^7 & 0 \\ 0 & \rho^3 \end{pmatrix}, \quad b = \begin{pmatrix} 0 & 1 \\ 1 & 0 \end{pmatrix}, \quad \text{on} \quad \mathbf{2}_2. \tag{G.20}
$$

The tensor products between doublets are obtained as

$$\begin{pmatrix} x_1 \\ x_2 \end{pmatrix}_{2_1} \otimes \begin{pmatrix} y_1 \\ y_2 \end{pmatrix}_{2_1} = (x_1 y_1 + x_2 y_2)_{1_{+,1}} \oplus (x_1 y_1 - x_2 y_2)_{1_{-,1}}$$
$$\oplus (x_1 y_2 + x_2 y_1)_{1_{+,3}} \oplus (x_1 y_2 - x_2 y_1)_{1_{-,3}}, \quad \text{(G.21)}$$

$$\begin{pmatrix} x_1 \\ x_2 \end{pmatrix}_{2_1} \otimes \begin{pmatrix} y_1 \\ y_2 \end{pmatrix}_{2_2} = (x_1 y_1 + x_2 y_2)_{1_{+,0}} \oplus (x_1 y_1 - x_2 y_2)_{1_{-,0}}$$
$$\oplus (x_1 y_2 + x_2 y_1)_{1_{+,2}} \oplus (x_1 y_2 - x_2 y_1)_{1_{-,2}}, \quad \text{(G.22)}$$

$$\begin{pmatrix} x_1 \\ x_2 \end{pmatrix}_{2_2} \otimes \begin{pmatrix} y_1 \\ y_2 \end{pmatrix}_{2_2} = (x_1 y_1 + x_2 y_2)_{1_{+,3}} \oplus (x_1 y_1 - x_2 y_2)_{1_{-,3}}$$
$$\oplus (x_1 y_2 + x_2 y_1)_{1_{+,1}} \oplus (x_1 y_2 - x_2 y_1)_{1_{-,1}}. \quad \text{(G.23)}$$

The tensor products between doublets and singlets are found as

$$(x)_{1_{\pm,0}} \otimes \begin{pmatrix} y_1 \\ y_2 \end{pmatrix}_{2_i} = \begin{pmatrix} x y_1 \\ \pm x y_2 \end{pmatrix}_{2_i}, \quad \text{(G.24)}$$

$$(x)_{1_{\pm,1}} \otimes \begin{pmatrix} y_1 \\ y_2 \end{pmatrix}_{2_1} = \begin{pmatrix} x y_2 \\ \pm x y_1 \end{pmatrix}_{2_2}, \ (x)_{1_{\pm,1}} \otimes \begin{pmatrix} y_1 \\ y_2 \end{pmatrix}_{2_2} = \begin{pmatrix} x y_1 \\ \pm x y_2 \end{pmatrix}_{2_1}, \ \text{(G.25)}$$

$$(x)_{1_{\pm,2}} \otimes \begin{pmatrix} y_1 \\ y_2 \end{pmatrix}_{2_1} = \begin{pmatrix} x y_2 \\ \pm x y_1 \end{pmatrix}_{2_1}, \ (x)_{1_{\pm,2}} \otimes \begin{pmatrix} y_1 \\ y_2 \end{pmatrix}_{2_2} = \begin{pmatrix} x y_2 \\ \pm x y_1 \end{pmatrix}_{2_2}, \ \text{(G.26)}$$

$$(x)_{1_{\pm,3}} \otimes \begin{pmatrix} y_1 \\ y_2 \end{pmatrix}_{2_1} = \begin{pmatrix} x y_1 \\ \pm x y_2 \end{pmatrix}_{2_2}, \ (x)_{1_{\pm,3}} \otimes \begin{pmatrix} y_1 \\ y_2 \end{pmatrix}_{2_2} = \begin{pmatrix} x y_2 \\ \pm x y_1 \end{pmatrix}_{2_1}. \ \text{(G.27)}$$

The tensor products between singlets are found as

$$\mathbf{1}_{s,k} \otimes \mathbf{1}_{s',k'} = \mathbf{1}_{s'',k+k'}, \quad \text{(G.28)}$$

where $s'' = ss'$.

G.3 $(Z_2 \times Z_4) \rtimes Z_2$ (I)

Here, we discuss the group $(Z_2 \times Z_4) \rtimes Z_2(\text{I})$. We denote the first and second Z_2 generators by a and b, respectively, and the generator of Z_4 is written by $\tilde{a}$. The generators a, $\tilde{a}$ and b satisfy following conditions

$$a^2 = e, \ \tilde{a}^4 = e, \ b^2 = e, \ bab = a\tilde{a}^2, \ a\tilde{a} = \tilde{a}a, \ \tilde{a}b = b\tilde{a}. \quad \text{(G.29)}$$

Table G.4 Characters of $(Z_2 \times Z_4) \rtimes Z_2(\mathrm{I})$

	n	h	$\chi_{1_{+++}}$	$\chi_{1_{++-}}$	$\chi_{1_{+-+}}$	$\chi_{1_{+--}}$	$\chi_{1_{-++}}$	$\chi_{1_{-+-}}$	$\chi_{1_{--+}}$	$\chi_{1_{---}}$	χ_{2_1}	χ_{2_2}
C_1	1	1	1	1	1	1	1	1	1	1	2	2
$C_1^{(1)}$	1	4	1	1	-1	-1	1	1	-1	-1	$2i$	$-2i$
$C_1^{(2)}$	1	2	1	1	1	1	1	1	1	1	-2	-2
$C_1^{(3)}$	1	4	1	1	-1	-1	1	1	-1	-1	$-2i$	$2i$
$C_2^{(1)}$	2	2	1	1	1	1	-1	-1	-1	-1	0	0
$C_2^{(2)}$	2	4	1	1	-1	-1	-1	-1	1	1	0	0
$C_2^{(3)}$	2	2	1	-1	1	-1	1	-1	1	-1	0	0
$C_2^{(4)}$	2	4	1	-1	1	-1	-1	1	-1	1	0	0
$C_2^{(5)}$	2	4	1	-1	-1	1	1	-1	-1	1	0	0
$C_2^{(6)}$	2	2	1	-1	-1	1	-1	1	1	-1	0	0

All of elements are written as $b^k a^\ell \tilde{a}^m$ with $k = 0, 1$, $\ell = 0, 1$ and $m = 0, 1, 2, 3$. These elements are classified into ten conjugacy classes,

$$
\begin{aligned}
C_1 : & \quad \{e\}, & h &= 1, \\
C_1^{(1)} : & \quad \{\tilde{a}\}, & h &= 4, \\
C_1^{(2)} : & \quad \{\tilde{a}^2\}, & h &= 2, \\
C_1^{(3)} : & \quad \{\tilde{a}^3\}, & h &= 4, \\
C_2^{(1)} : & \quad \{a, a\tilde{a}^2\}, & h &= 2, \\
C_2^{(2)} : & \quad \{a\tilde{a}, a\tilde{a}^3\}, & h &= 4, \\
C_2^{(3)} : & \quad \{b, \tilde{a}^2 b\}, & h &= 2, \\
C_2^{(4)} : & \quad \{ab, a\tilde{a}^2 b\}, & h &= 4, \\
C_2^{(5)} : & \quad \{\tilde{a}b, \tilde{a}^3 b\}, & h &= 4, \\
C_2^{(6)} : & \quad \{a\tilde{a}b, a\tilde{a}^3 b\}, & h &= 2.
\end{aligned}
\tag{G.30}
$$

The $(Z_2 \rtimes Z_4) \ltimes Z_2(\mathrm{I})$ group has eight singlets $1_{\pm\pm\pm}$ and two doublets 2_1 and 2_2. The characters are shown in Table G.4.

Concerned about singelts, the generators a, $\tilde{a}$ and b can be represented as

$$
a = \pm 1, \quad \text{on } 1_{\pm ss'}, \tag{G.31}
$$

for any s and s',

$$
\tilde{a} = \pm 1, \quad \text{on } 1_{s \pm s'}, \tag{G.32}
$$

for any s and s', and

$$
b = \pm 1, \quad \text{on } 1_{ss' \pm}, \tag{G.33}
$$

for any s and s'. For the doublets, the generator $\tilde{a}$ can be represented as

$$
\tilde{a} = \begin{pmatrix} i & 0 \\ 0 & i \end{pmatrix}, \quad \text{on } 2_1, \tag{G.34}
$$

$$\tilde{a} = \begin{pmatrix} -i & 0 \\ 0 & -i \end{pmatrix}, \quad \text{on } \mathbf{2}_2. \tag{G.35}$$

The generators a and b can be represented as

$$a = \begin{pmatrix} 1 & 0 \\ 0 & -1 \end{pmatrix}, \quad b = \begin{pmatrix} 0 & 1 \\ 1 & 0 \end{pmatrix}, \tag{G.36}$$

on both doublets.

The tensor products between doublets are obtained as

$$\begin{pmatrix} x_1 \\ x_2 \end{pmatrix}_{\mathbf{2}_1} \otimes \begin{pmatrix} y_1 \\ y_2 \end{pmatrix}_{\mathbf{2}_1} = (x_1 y_1 + x_2 y_2) \mathbf{1}_{+-+} \oplus (x_1 y_1 - x_2 y_2) \mathbf{1}_{+--}$$
$$\oplus (x_1 y_2 + x_2 y_1) \mathbf{1}_{--+} \oplus (x_1 y_2 - x_2 y_1) \mathbf{1}_{---}, \tag{G.37}$$

$$\begin{pmatrix} x_1 \\ x_2 \end{pmatrix}_{\mathbf{2}_1} \otimes \begin{pmatrix} y_1 \\ y_2 \end{pmatrix}_{\mathbf{2}_2} = (x_1 y_1 + x_2 y_2) \mathbf{1}_{+++} \oplus (x_1 y_1 - x_2 y_2) \mathbf{1}_{++-}$$
$$\oplus (x_1 y_2 + x_2 y_1) \mathbf{1}_{-++} \oplus (x_1 y_2 - x_2 y_1) \mathbf{1}_{-+-}, \tag{G.38}$$

$$\begin{pmatrix} x_1 \\ x_2 \end{pmatrix}_{\mathbf{2}_2} \otimes \begin{pmatrix} y_1 \\ y_2 \end{pmatrix}_{\mathbf{2}_2} = (x_1 y_1 + x_2 y_2) \mathbf{1}_{+-+} \oplus (x_1 y_1 - x_2 y_2) \mathbf{1}_{+--}$$
$$\oplus (x_1 y_2 + x_2 y_1) \mathbf{1}_{--+} \oplus (x_1 y_2 - x_2 y_1) \mathbf{1}_{---}. \tag{G.39}$$

The tensor products between doublets and singlets are found as

$$(x)_{\mathbf{1}_{++\pm}} \otimes \begin{pmatrix} y_1 \\ y_2 \end{pmatrix}_{\mathbf{2}_i} = \begin{pmatrix} x y_1 \\ \pm x y_2 \end{pmatrix}_{\mathbf{2}_i}, \tag{G.40}$$

$$(x)_{\mathbf{1}_{+-\pm}} \otimes \begin{pmatrix} y_1 \\ y_2 \end{pmatrix}_{\mathbf{2}_1} = \begin{pmatrix} x y_1 \\ \pm x y_2 \end{pmatrix}_{\mathbf{2}_2}, \; (x)_{\mathbf{1}_{+-\pm}} \otimes \begin{pmatrix} y_1 \\ y_2 \end{pmatrix}_{\mathbf{2}_2} = \begin{pmatrix} x y_1 \\ \pm x y_2 \end{pmatrix}_{\mathbf{2}_1}, \tag{G.41}$$

$$(x)_{\mathbf{1}_{-+\pm}} \otimes \begin{pmatrix} y_1 \\ y_2 \end{pmatrix}_{\mathbf{2}_i} = \begin{pmatrix} x y_2 \\ \pm x y_1 \end{pmatrix}_{\mathbf{2}_i}, \tag{G.42}$$

$$(x)_{\mathbf{1}_{--\pm}} \otimes \begin{pmatrix} y_1 \\ y_2 \end{pmatrix}_{\mathbf{2}_1} = \begin{pmatrix} x y_2 \\ \pm x y_1 \end{pmatrix}_{\mathbf{2}_2}, \; (x)_{\mathbf{1}_{--\pm}} \otimes \begin{pmatrix} y_1 \\ y_2 \end{pmatrix}_{\mathbf{2}_2} = \begin{pmatrix} x y_2 \\ \pm x y_1 \end{pmatrix}_{\mathbf{2}_1}. \tag{G.43}$$

The tensor products between singlets are found as

$$\mathbf{1}_{s_1 s_2 s_3} \otimes \mathbf{1}_{s_1' s_2' s_3'} = \mathbf{1}_{s_1'' s_2'' s_3''}, \tag{G.44}$$

where $s_1'' = s_1 s_1''$, $s_2'' = s_2 s_2'$ and $s_3'' = s_3 s_3'$.

Table G.5 Characters of $(Z_2 \times Z_4) \rtimes Z_2(\mathrm{II})$

	n	h	$\chi_{1_{+0}}$	$\chi_{1_{+1}}$	$\chi_{1_{+2}}$	$\chi_{1_{+3}}$	$\chi_{1_{-0}}$	$\chi_{1_{-1}}$	$\chi_{1_{-2}}$	$\chi_{1_{-3}}$	χ_{2_1}	χ_{2_2}
C_1	1	1	1	1	1	1	1	1	1	1	2	2
$C_1^{(1)}$	1	2	1	1	1	1	1	1	1	1	-2	-2
$C_1^{(2)}$	1	2	1	-1	1	-1	1	-1	1	-1	2	-2
$C_1^{(3)}$	1	2	1	-1	1	-1	1	-1	1	-1	-2	2
$C_2^{(1)}$	2	4	1	i	-1	-1	1	i	-1	$-i$	0	0
$C_2^{(2)}$	2	4	1	$-i$	-1	i	1	$-i$	-1	i	0	0
$C_2^{(3)}$	2	2	1	1	1	1	-1	-1	-1	-1	0	0
$C_2^{(4)}$	2	4	1	i	-1	$-i$	-1	$-i$	1	i	0	0
$C_2^{(5)}$	2	4	1	-1	1	-1	-1	1	-1	1	0	0
$C_2^{(6)}$	2	4	1	$-i$	-1	i	-1	i	1	$-i$	0	0

G.4 $(Z_2 \times Z_4) \rtimes Z_2$ (II)

Here, we discuss $(Z_2 \times Z_4) \rtimes Z_2(\mathrm{II})$. We denote the first and second Z_2 generators by a and b and the generator of Z_4 is written by $\tilde{a}$. The generators a, $\tilde{a}$ and b satisfy following conditions

$$a^2 = e, \quad \tilde{a}^4 = e, \quad b^2 = e, \quad b\tilde{a}b = a\tilde{a}, \quad a\tilde{a} = \tilde{a}a, \quad ab = ba. \tag{G.45}$$

All of elements are written as $b^k a^\ell \tilde{a}^m$ with $k = 0, 1$, $\ell = 0, 1$ and $m = 0, 1, 2, 3$. These elements are classified into ten conjugacy classes,

$$
\begin{aligned}
C_1 : & \quad \{e\}, & h &= 1, \\
C_1^{(1)} : & \quad \{a\}, & h &= 2, \\
C_1^{(2)} : & \quad \{\tilde{a}^2\}, & h &= 2, \\
C_1^{(3)} : & \quad \{a\tilde{a}^2\}, & h &= 2, \\
C_2^{(1)} : & \quad \{\tilde{a}, a\tilde{a}\}, & h &= 4, \\
C_2^{(2)} : & \quad \{\tilde{a}^3, a\tilde{a}^3\}, & h &= 4, \\
C_2^{(3)} : & \quad \{b, ab\}, & h &= 2, \\
C_2^{(4)} : & \quad \{\tilde{a}b, a\tilde{a}b\}, & h &= 4, \\
C_2^{(5)} : & \{\tilde{a}^2 b, a\tilde{a}^2 b\}, & h &= 2, \\
C_2^{(6)} : & \{\tilde{a}^3 b, a\tilde{a}^3 b\}, & h &= 2.
\end{aligned}
\tag{G.46}
$$

The $(Z_2 \rtimes Z_4) \ltimes Z_2(\mathrm{II})$ group has eight singlets $\mathbf{1}_{\pm,k}$ with $k = 0, 1, 2, 3$ and two doublets $\mathbf{2}_1$ and $\mathbf{2}_2$. The characters are shown in Table G.5.

The generators, a, $\tilde{a}$ and b, can be represented as

$$a = 1, \quad \tilde{a} = (i)^k, \quad b = \pm 1, \quad \text{on } \mathbf{1}_{\pm,k}. \tag{G.47}$$

In addition, the generator $\tilde{a}$ can be represented as

$$\tilde{a} = \begin{pmatrix} 1 & 0 \\ 0 & -1 \end{pmatrix}, \quad \text{on } \mathbf{2}_1, \tag{G.48}$$

$$\tilde{a} = \begin{pmatrix} i & 0 \\ 0 & -i \end{pmatrix}, \quad \text{on } \mathbf{2}_2, \tag{G.49}$$

and for both doublets the generators, a and b, can be represented as

$$a = \begin{pmatrix} -1 & 0 \\ 0 & -1 \end{pmatrix}, \quad b = \begin{pmatrix} 0 & 1 \\ 1 & 0 \end{pmatrix}. \tag{G.50}$$

The tensor products between doublets are obtained as

$$\begin{pmatrix} x_1 \\ x_2 \end{pmatrix}_{\mathbf{2}_1} \otimes \begin{pmatrix} y_1 \\ y_2 \end{pmatrix}_{\mathbf{2}_1} = (x_1 y_1 + x_2 y_2)\mathbf{1}_{+,0} \oplus (x_1 y_1 - x_2 y_2)\mathbf{1}_{-,0}$$
$$\oplus (x_1 y_2 + x_2 y_1)\mathbf{1}_{+,2} \oplus (x_1 y_2 - x_2 y_1)\mathbf{1}_{-,2}, \tag{G.51}$$

$$\begin{pmatrix} x_1 \\ x_2 \end{pmatrix}_{\mathbf{2}_1} \otimes \begin{pmatrix} y_1 \\ y_2 \end{pmatrix}_{\mathbf{2}_2} = (x_1 y_1 + x_2 y_2)\mathbf{1}_{+,1} \oplus (x_1 y_1 - x_2 y_2)\mathbf{1}_{-,1}$$
$$\oplus (x_1 y_2 + x_2 y_1)\mathbf{1}_{+,3} \oplus (x_1 y_2 - x_2 y_1)\mathbf{1}_{-,3}, \tag{G.52}$$

$$\begin{pmatrix} x_1 \\ x_2 \end{pmatrix}_{\mathbf{2}_2} \otimes \begin{pmatrix} y_1 \\ y_2 \end{pmatrix}_{\mathbf{2}_2} = (x_1 y_1 + x_2 y_2)\mathbf{1}_{+,2} \oplus (x_1 y_1 - x_2 y_2)\mathbf{1}_{-,2}$$
$$\oplus (x_1 y_2 + x_2 y_1)\mathbf{1}_{+,0} \oplus (x_1 y_2 - x_2 y_1)\mathbf{1}_{-,0}. \tag{G.53}$$

The tensor products between doublets and singlets are found as

$$(x)\mathbf{1}_{\pm,0} \otimes \begin{pmatrix} y_1 \\ y_2 \end{pmatrix}_{\mathbf{2}_i} = \begin{pmatrix} x y_1 \\ \pm x y_2 \end{pmatrix}_{\mathbf{2}_i}, \tag{G.54}$$

$$(x)\mathbf{1}_{\pm,1} \otimes \begin{pmatrix} y_1 \\ y_2 \end{pmatrix}_{\mathbf{2}_1} = \begin{pmatrix} x y_1 \\ \pm x y_2 \end{pmatrix}_{\mathbf{2}_2}, \quad (x)\mathbf{1}_{\pm,1} \otimes \begin{pmatrix} y_1 \\ y_2 \end{pmatrix}_{\mathbf{2}_2} = \begin{pmatrix} x y_2 \\ \pm x y_1 \end{pmatrix}_{\mathbf{2}_1}, \tag{G.55}$$

$$(x)\mathbf{1}_{\pm,2} \otimes \begin{pmatrix} y_1 \\ y_2 \end{pmatrix}_{\mathbf{2}_i} = \begin{pmatrix} x y_2 \\ \pm x y_1 \end{pmatrix}_{\mathbf{2}_i}, \tag{G.56}$$

$$(x)\mathbf{1}_{\pm,3} \otimes \begin{pmatrix} y_1 \\ y_2 \end{pmatrix}_{\mathbf{2}_1} = \begin{pmatrix} x y_2 \\ \pm x y_1 \end{pmatrix}_{\mathbf{2}_2}, \quad (x)\mathbf{1}_{\pm,3} \otimes \begin{pmatrix} y_1 \\ y_2 \end{pmatrix}_{\mathbf{2}_2} = \begin{pmatrix} x y_1 \\ \pm x y_2 \end{pmatrix}_{\mathbf{2}_1}, \tag{G.57}$$

The tensor products between singlets are found as

$$\mathbf{1}_{s,k} \otimes \mathbf{1}_{s',k'} = \mathbf{1}_{s'',k+k'}, \tag{G.58}$$

where $s'' = ss'$.

Table G.6 Characters of $Z_3 \rtimes Z_8$

	n	h	χ_{1_0}	χ_{1_1}	χ_{1_2}	χ_{1_3}	χ_{1_4}	χ_{1_5}	χ_{1_6}	χ_{1_7}	χ_{2_1}	χ_{2_2}	χ_{2_3}	χ_{2_4}
C_1	1	1	1	1	1	1	1	1	1	1	2	2	2	2
$C_1^{(1)}$	1	8	1	i	-1	$-i$	1	i	-1	$-i$	2	$2i$	-2	$-2i$
$C_1^{(2)}$	1	8	1	-1	1	-1	1	-1	1	-1	2	-2	2	-2
$C_1^{(3)}$	1	8	1	$-i$	-1	i	1	$-i$	-1	i	2	$-2i$	-2	$2i$
$C_2^{(1)}$	2	3	1	1	1	1	1	1	1	1	-1	-1	-1	-1
$C_2^{(2)}$	2	12	1	i	-1	$-i$	1	i	-1	$-i$	-2	$-2i$	2	$2i$
$C_2^{(3)}$	2	3	1	-1	1	-1	1	-1	1	-1	-2	2	-2	2
$C_2^{(4)}$	2	3	1	$-i$	-1	i	1	$-i$	-1	i	-2	$2i$	2	$-2i$
$C_3^{(1)}$	3	8	1	ρ	i	$i\rho$	-1	$-\rho$	$-i$	$-i\rho$	0	0	0	0
$C_3^{(2)}$	3	12	1	$i\rho$	$-i$	ρ	-1	$-i\rho$	i	$-\rho$	0	0	0	0
$C_3^{(3)}$	3	8	1	$-\rho$	i	$-i\rho$	-1	ρ	$-i$	$i\rho$	0	0	0	0
$C_3^{(4)}$	3	8	1	$-i\rho$	$-i$	$-\rho$	-1	$i\rho$	i	ρ	0	0	0	0

G.5 $\quad Z_3 \rtimes Z_8$

Here, we discuss the group $Z_3 \rtimes Z_8$. We denote the Z_3 and Z_8 generators by a and b, respectively. The generators a and b satisfy following conditions

$$a^3 = e, \quad b^8 = e, \quad b^{-1}ab = a^2. \tag{G.59}$$

All of elements are written by $b^k a^\ell$ with $k = 0, \ldots, 7$ and $\ell = 0, 1, 2$. These elements are classified into twelve conjugacy classes,

$$
\begin{aligned}
C_1 : & \quad \{e\}, & h &= 1, \\
C_1^{(1)} : & \quad \{b^2\}, & h &= 4, \\
C_1^{(2)} : & \quad \{b^4\}, & h &= 2, \\
C_1^{(3)} : & \quad \{b^6\}, & h &= 4, \\
C_2^{(1)} : & \quad \{a, a^2\}, & h &= 3, \\
C_2^{(2)} : & \quad \{b^2a, b^2a^2\}, & h &= 12, \\
C_2^{(3)} : & \quad \{b^4a, b^4a^2\}, & h &= 3, \\
C_2^{(4)} : & \quad \{b^6a, b^6a^2\}, & h &= 3, \\
C_3^{(1)} : & \quad \{b, ba, ba^2\}, & h &= 8, \\
C_3^{(2)} : & \{b^3, b^3a, b^3a^2\}, & h &= 8, \\
C_3^{(3)} : & \{b^5, b^5a, b^5a^2\}, & h &= 8, \\
C_3^{(4)} : & \{b^7, b^7a, b^7a^2\}, & h &= 8.
\end{aligned}
\tag{G.60}
$$

The $Z_3 \rtimes Z_8$ group has eight singlets $\mathbf{1}_r$ with $r = 0, \ldots, 7$, and four doublets $\mathbf{2}_k$ with $k = 1, 2, 3, 4$. The characters are shown in Table G.6.

The generators a and b can be represented as

$$a = 1, \quad b = \rho^k, \quad \text{on } \mathbf{1}_k, \tag{G.61}$$

where $\rho = e^{\pi i/4}$. The generators a and b can be represented by the following matrices as

$$a = \begin{pmatrix} \omega & 0 \\ 0 & \omega^2 \end{pmatrix}, \qquad b = \begin{pmatrix} 0 & 1 \\ 1 & 0 \end{pmatrix}, \quad \text{on } \mathbf{2}_1, \tag{G.62}$$

$$a = \begin{pmatrix} \omega & 0 \\ 0 & \omega^2 \end{pmatrix}, \qquad b = \begin{pmatrix} 0 & 1 \\ i & 0 \end{pmatrix}, \quad \text{on } \mathbf{2}_2, \tag{G.63}$$

$$a = \begin{pmatrix} \omega & 0 \\ 0 & \omega^2 \end{pmatrix}, \qquad b = \begin{pmatrix} 0 & 1 \\ -1 & 0 \end{pmatrix}, \quad \text{on } \mathbf{2}_3, \tag{G.64}$$

$$a = \begin{pmatrix} \omega & 0 \\ 0 & \omega^2 \end{pmatrix}, \qquad b = \begin{pmatrix} 0 & 1 \\ -i & 0 \end{pmatrix}, \quad \text{on } \mathbf{2}_4, \tag{G.65}$$

where $\omega = e^{2\pi i/3}$.

The tensor products between doublets are obtained as

$$\begin{pmatrix} x_1 \\ x_2 \end{pmatrix}_{\mathbf{2}_1} \otimes \begin{pmatrix} y_1 \\ y_2 \end{pmatrix}_{\mathbf{2}_1} = \begin{pmatrix} x_2 y_2 \\ x_1 y_1 \end{pmatrix}_{\mathbf{2}_1} \oplus (x_1 y_2 + x_2 y_1)_{\mathbf{1}_0} \oplus (x_1 y_2 - x_2 y_1)_{\mathbf{1}_4}, \tag{G.66}$$

$$\begin{pmatrix} x_1 \\ x_2 \end{pmatrix}_{\mathbf{2}_1} \otimes \begin{pmatrix} y_1 \\ y_2 \end{pmatrix}_{\mathbf{2}_2} = \begin{pmatrix} x_2 y_1 \\ x_1 y_2 \end{pmatrix}_{\mathbf{2}_2} \oplus (x_1 y_1 + i\rho x_2 y_2)_{\mathbf{1}_1} \oplus (x_1 y_1 - i\rho x_2 y_2)_{\mathbf{1}_5}, \tag{G.67}$$

$$\begin{pmatrix} x_1 \\ x_2 \end{pmatrix}_{\mathbf{2}_1} \otimes \begin{pmatrix} y_1 \\ y_2 \end{pmatrix}_{\mathbf{2}_3} = \begin{pmatrix} x_2 y_2 \\ -x_1 y_1 \end{pmatrix}_{\mathbf{2}_3} \oplus (x_1 y_2 + i x_2 y_1)_{\mathbf{1}_2} \oplus (x_1 y_2 - i x_2 y_1)_{\mathbf{1}_6}, \tag{G.68}$$

$$\begin{pmatrix} x_1 \\ x_2 \end{pmatrix}_{\mathbf{2}_1} \otimes \begin{pmatrix} y_1 \\ y_2 \end{pmatrix}_{\mathbf{2}_4} = \begin{pmatrix} x_2 y_1 \\ x_1 y_2 \end{pmatrix}_{\mathbf{2}_1} \oplus (x_1 y_2 + i\rho x_2 y_1)_{\mathbf{1}_3} \oplus (x_1 y_2 - i\rho x_2 y_1)_{\mathbf{1}_7}, \tag{G.69}$$

$$\begin{pmatrix} x_1 \\ x_2 \end{pmatrix}_{\mathbf{2}_2} \otimes \begin{pmatrix} y_1 \\ y_2 \end{pmatrix}_{\mathbf{2}_2} = \begin{pmatrix} x_2 y_2 \\ -x_1 y_1 \end{pmatrix}_{\mathbf{2}_3} \oplus (x_1 y_2 + x_2 y_1)_{\mathbf{1}_2} \oplus (x_1 y_2 - x_2 y_1)_{\mathbf{1}_6}, \tag{G.70}$$

$$\begin{pmatrix} x_1 \\ x_2 \end{pmatrix}_{\mathbf{2}_2} \otimes \begin{pmatrix} y_1 \\ y_2 \end{pmatrix}_{\mathbf{2}_3} = \begin{pmatrix} x_2 y_2 \\ -i x_1 y_1 \end{pmatrix}_{\mathbf{2}_4} \oplus (x_1 y_2 + i\rho x_2 y_1)_{\mathbf{1}_3} \oplus (x_1 y_2 - i\rho x_2 y_1)_{\mathbf{1}_7}, \tag{G.71}$$

$$\begin{pmatrix} x_1 \\ x_2 \end{pmatrix}_{\mathbf{2}_2} \otimes \begin{pmatrix} y_1 \\ y_2 \end{pmatrix}_{\mathbf{2}_4} = \begin{pmatrix} x_2 y_2 \\ x_1 y_1 \end{pmatrix}_{\mathbf{2}_1} \oplus (x_1 y_2 + i x_2 y_1)_{\mathbf{1}_0} \oplus (x_1 y_2 - i x_2 y_1)_{\mathbf{1}_4}, \tag{G.72}$$

$$\begin{pmatrix} x_1 \\ x_2 \end{pmatrix}_{2_3} \otimes \begin{pmatrix} y_1 \\ y_2 \end{pmatrix}_{2_3} = \begin{pmatrix} x_2 y_2 \\ x_1 y_1 \end{pmatrix}_{2_1} \oplus (x_1 y_2 + x_2 y_1)_{1_0} \oplus (x_1 y_2 - x_2 y_1)_{1_4}, \quad \text{(G.73)}$$

$$\begin{pmatrix} x_1 \\ x_2 \end{pmatrix}_{2_3} \otimes \begin{pmatrix} y_1 \\ y_2 \end{pmatrix}_{2_4} = \begin{pmatrix} x_2 y_2 \\ i x_1 y_1 \end{pmatrix}_{2_2} \oplus (x_1 y_2 + \rho x_2 y_1)_{1_1} \oplus (x_1 y_2 - \rho x_2 y_1)_{1_5},$$

$$\text{(G.74)}$$

$$\begin{pmatrix} x_1 \\ x_2 \end{pmatrix}_{2_4} \otimes \begin{pmatrix} y_1 \\ y_2 \end{pmatrix}_{2_4} = \begin{pmatrix} x_2 y_2 \\ -x_1 y_1 \end{pmatrix}_{2_3} \oplus (x_1 y_2 + x_2 y_1)_{1_2} \oplus (x_1 y_2 - x_2 y_1)_{1_6}. \quad \text{(G.75)}$$

The tensor products between doublets and singlets are found as

$$(x)_{1_i} \otimes \begin{pmatrix} y_1 \\ y_2 \end{pmatrix}_{2_j} = \begin{pmatrix} x y_1 \\ \rho^i x y_2 \end{pmatrix}_{2_{i+j} \mod 4}. \quad \text{(G.76)}$$

The tensor products between singlets are found as

$$\mathbf{1}_k \otimes \mathbf{1}_{k'} = \mathbf{1}_{k+k'} \mod 8. \quad \text{(G.77)}$$

G.6 $(Z_6 \times Z_2) \rtimes Z_2$

Here, we discuss the group $(Z_6 \times Z_2) \rtimes Z_2$. We denote the generator of Z_6 by a, and the first and second Z_2 generators are written by b and c, respectively. The generators a, b and c satisfy the following conditions

$$a^6 = e, \quad b^2 = e, \quad c^2 = e, \quad c^{-1}ab = a^5, \quad c^{-1}bc = a^3 b, \quad ab = ba. \quad \text{(G.78)}$$

All of elements are written as $a^k b^\ell c^m$ with $k = 0, \ldots, 5$, $\ell, m = 0, 1$. These elements are classified into nine conjugacy classes,

$$\begin{array}{llll}
C_1 : & \{e\}, & h = 1, & \\
C_1^{(1)} : & \{a^3\}, & h = 2, & \\
C_2^{(1)} : & \{a, a^5\}, & h = 2, & \\
C_2^{(2)} : & \{a^2, a^4\}, & h = 2, & \\
C_2^{(3)} : & \{b, a^3 b\}, & h = 2, & \text{(G.79)} \\
C_2^{(4)} : & \{ab, a^2 b\}, & h = 6, & \\
C_2^{(5)} : & \{a^4 b, a^5 b\}, & h = 6, & \\
C_6^{(1)} : & \{c, ac, a^2 c, a^3 c, a^4 c, a^5 c\}, & h = 2, & \\
C_6^{(2)} : & \{bc, abc, a^2 bc, a^3 bc, a^4 bc, a^5 bc\}, & h = 6. &
\end{array}$$

The $(Z_6 \times Z2) \rtimes Z_2$ group has four singlets $\mathbf{1}_{\pm\pm}$ and five doublets $\mathbf{2}_k$ with $k = 1, \ldots, 5$. The characters are shown in Table G.7.

Table G.7 Characters of $(Z_6 \times Z_2) \rtimes Z_2$

	n	h	$\chi_{1_{++}}$	$\chi_{1_{+-}}$	$\chi_{1_{-+}}$	$\chi_{1_{--}}$	χ_{2_1}	χ_{2_2}	χ_{2_3}	χ_{2_4}	χ_{2_5}
C_1	1	1	1	1	1	1	2	2	2	2	2
$C_1^{(1)}$	1	2	1	1	1	1	2	2	-2	-2	-2
$C_2^{(1)}$	2	6	1	1	1	1	-1	-1	1	1	-2
$C_2^{(2)}$	2	3	1	1	1	1	-1	-1	-1	-1	2
$C_2^{(3)}$	2	2	1	1	-1	-1	2	-2	0	0	0
$C_2^{(4)}$	2	6	1	1	-1	-1	-1	1	$\sqrt{3}i$	$-\sqrt{3}i$	0
$C_2^{(5)}$	2	6	1	1	-1	-1	-1	1	$-\sqrt{3}i$	$\sqrt{3}i$	0
$C_6^{(1)}$	6	2	1	-1	1	-1	0	0	0	0	0
$C_6^{(2)}$	6	6	1	-1	-1	1	0	0	0	0	0

For singlets, the generators, b and c, can be represented as

$$b = \pm 1 \quad \text{on} \quad \mathbf{1}_{\pm s}, \tag{G.80}$$

for both $s = \pm$, and

$$c = \pm 1 \quad \text{on} \quad \mathbf{1}_{s \pm}, \tag{G.81}$$

for both $s = \pm$, while the generator a is represented as $a = 1$ for all of the singlets. For doublets, we use the following representations for the generators a and b

$$a = \begin{pmatrix} \omega & 0 \\ 0 & \omega^2 \end{pmatrix}, \quad b = \begin{pmatrix} 1 & 0 \\ 0 & 1 \end{pmatrix}, \quad \text{on} \quad \mathbf{2}_1, \tag{G.82}$$

$$a = \begin{pmatrix} \omega & 0 \\ 0 & \omega^2 \end{pmatrix}, \quad b = \begin{pmatrix} -1 & 0 \\ 0 & -1 \end{pmatrix}, \quad \text{on} \quad \mathbf{2}_2, \tag{G.83}$$

$$a = \begin{pmatrix} -\omega^2 & 0 \\ 0 & -\omega \end{pmatrix}, \quad b = \begin{pmatrix} 1 & 0 \\ 0 & -1 \end{pmatrix}, \quad \text{on} \quad \mathbf{2}_3, \tag{G.84}$$

$$a = \begin{pmatrix} -\omega & 0 \\ 0 & -\omega^2 \end{pmatrix}, \quad b = \begin{pmatrix} 1 & 0 \\ 0 & -1 \end{pmatrix}, \quad \text{on} \quad \mathbf{2}_4, \tag{G.85}$$

$$a = \begin{pmatrix} -1 & 0 \\ 0 & -1 \end{pmatrix}, \quad b = \begin{pmatrix} 1 & 0 \\ 0 & -1 \end{pmatrix}, \quad \text{on} \quad \mathbf{2}_5, \tag{G.86}$$

where $\omega = e^{2\pi i/3}$. while c is represented as

$$c = \begin{pmatrix} 0 & 1 \\ 1 & 0 \end{pmatrix}, \tag{G.87}$$

on all of the doublets.

The tensor products between doublets are obtained as

$$\begin{pmatrix} x_1 \\ x_2 \end{pmatrix}_{2_1} \otimes \begin{pmatrix} y_1 \\ y_2 \end{pmatrix}_{2_1} = \begin{pmatrix} x_2 y_2 \\ x_1 y_1 \end{pmatrix}_{2_1} \oplus (x_1 y_2 + x_2 y_1)_{1_{++}} \oplus (x_1 y_2 - x_2 y_1)_{1_{+-}},$$
(G.88)

$$\begin{pmatrix} x_1 \\ x_2 \end{pmatrix}_{2_1} \otimes \begin{pmatrix} y_1 \\ y_2 \end{pmatrix}_{2_2} = \begin{pmatrix} x_2 y_2 \\ x_1 y_1 \end{pmatrix}_{2_2} \oplus (x_1 y_2 + x_2 y_1)_{1_{-+}} \oplus (x_1 y_2 - x_2 y_1)_{1_{--}},$$
(G.89)

$$\begin{pmatrix} x_1 \\ x_2 \end{pmatrix}_{2_1} \otimes \begin{pmatrix} y_1 \\ y_2 \end{pmatrix}_{2_3} = \begin{pmatrix} x_2 y_1 \\ x_1 y_2 \end{pmatrix}_{2_4} \oplus \begin{pmatrix} x_1 y_1 \\ x_2 y_2 \end{pmatrix}_{2_5},$$
(G.90)

$$\begin{pmatrix} x_1 \\ x_2 \end{pmatrix}_{2_1} \otimes \begin{pmatrix} y_1 \\ y_2 \end{pmatrix}_{2_4} = \begin{pmatrix} x_1 y_1 \\ x_2 y_2 \end{pmatrix}_{2_3} \oplus \begin{pmatrix} x_2 y_1 \\ x_1 y_2 \end{pmatrix}_{2_5},$$
(G.91)

$$\begin{pmatrix} x_1 \\ x_2 \end{pmatrix}_{2_1} \otimes \begin{pmatrix} y_1 \\ y_2 \end{pmatrix}_{2_5} = \begin{pmatrix} x_2 y_1 \\ x_1 y_2 \end{pmatrix}_{2_3} \oplus \begin{pmatrix} x_1 y_1 \\ x_2 y_2 \end{pmatrix}_{2_4},$$
(G.92)

$$\begin{pmatrix} x_1 \\ x_2 \end{pmatrix}_{2_2} \otimes \begin{pmatrix} y_1 \\ y_2 \end{pmatrix}_{2_2} = \begin{pmatrix} x_2 y_2 \\ x_1 y_1 \end{pmatrix}_{2_1} \oplus (x_1 y_2 + x_2 y_1)_{1_{++}} \oplus (x_1 y_2 - x_2 y_1)_{1_{+-}},$$
(G.93)

$$\begin{pmatrix} x_1 \\ x_2 \end{pmatrix}_{2_2} \otimes \begin{pmatrix} y_1 \\ y_2 \end{pmatrix}_{2_3} = \begin{pmatrix} x_1 y_2 \\ x_2 y_1 \end{pmatrix}_{2_3} \oplus \begin{pmatrix} x_2 y_2 \\ x_1 y_1 \end{pmatrix}_{2_5},$$
(G.94)

$$\begin{pmatrix} x_1 \\ x_2 \end{pmatrix}_{2_2} \otimes \begin{pmatrix} y_1 \\ y_2 \end{pmatrix}_{2_4} = \begin{pmatrix} x_2 y_2 \\ x_1 y_1 \end{pmatrix}_{2_4} \oplus \begin{pmatrix} x_1 y_2 \\ x_2 y_1 \end{pmatrix}_{2_5},$$
(G.95)

$$\begin{pmatrix} x_1 \\ x_2 \end{pmatrix}_{2_2} \otimes \begin{pmatrix} y_1 \\ y_2 \end{pmatrix}_{2_5} = \begin{pmatrix} x_2 y_2 \\ x_1 y_1 \end{pmatrix}_{2_3} \oplus \begin{pmatrix} x_1 y_2 \\ x_2 y_1 \end{pmatrix}_{2_4},$$
(G.96)

$$\begin{pmatrix} x_1 \\ x_2 \end{pmatrix}_{2_3} \otimes \begin{pmatrix} y_1 \\ y_2 \end{pmatrix}_{2_3} = \begin{pmatrix} x_1 y_1 \\ x_2 y_2 \end{pmatrix}_{2_1} \oplus (x_1 y_2 + x_2 y_1)_{1_{-+}} \oplus (x_1 y_2 - x_2 y_1)_{1_{--}},$$
(G.97)

$$\begin{pmatrix} x_1 \\ x_2 \end{pmatrix}_{2_3} \otimes \begin{pmatrix} y_1 \\ y_2 \end{pmatrix}_{2_4} = \begin{pmatrix} x_1 y_2 \\ x_2 y_1 \end{pmatrix}_{2_2} \oplus (x_1 y_1 + x_2 y_2)_{1_{++}} \oplus (x_1 y_1 - x_2 y_2)_{1_{+-}},$$
(G.98)

$$\begin{pmatrix} x_1 \\ x_2 \end{pmatrix}_{2_3} \otimes \begin{pmatrix} y_1 \\ y_2 \end{pmatrix}_{2_5} = \begin{pmatrix} x_2 y_2 \\ x_1 y_1 \end{pmatrix}_{2_1} \oplus \begin{pmatrix} x_2 y_1 \\ x_1 y_2 \end{pmatrix}_{2_2},$$
(G.99)

$$\begin{pmatrix} x_1 \\ x_2 \end{pmatrix}_{2_4} \otimes \begin{pmatrix} y_1 \\ y_2 \end{pmatrix}_{2_4} = \begin{pmatrix} x_2 y_1 \\ x_1 y_1 \end{pmatrix}_{2_1} \oplus (x_1 y_2 + x_2 y_1)_{1_{-+}} \oplus (x_1 y_2 - x_2 y_1)_{1_{--}},$$
(G.100)

$$\begin{pmatrix} x_1 \\ x_2 \end{pmatrix}_{2_4} \otimes \begin{pmatrix} y_1 \\ y_2 \end{pmatrix}_{2_5} = \begin{pmatrix} x_1 y_1 \\ x_2 y_2 \end{pmatrix}_{2_1} \oplus \begin{pmatrix} x_1 y_2 \\ x_2 y_1 \end{pmatrix}_{2_2}, \tag{G.101}$$

$$\begin{pmatrix} x_1 \\ x_2 \end{pmatrix}_{2_5} \otimes \begin{pmatrix} y_1 \\ y_2 \end{pmatrix}_{2_5} = (x_1 y_1 + x_2 y_2)_{1_{++}} \oplus (x_1 y_1 - x_2 y_2)_{1_{+-}}$$
$$\oplus (x_1 y_2 + x_2 y_1)_{1_{-+}} \oplus (x_1 y_2 - x_2 y_1)_{1_{--}}. \tag{G.102}$$

The tensor products between doublets and singlets are found as

$$(x)_{1_{+\pm}} \otimes \begin{pmatrix} y_1 \\ y_2 \end{pmatrix}_{2_i} = \begin{pmatrix} x y_1 \\ \pm x y_2 \end{pmatrix}_{2_i}, \tag{G.103}$$

$$(x)_{1_{-\pm}} \otimes \begin{pmatrix} y_1 \\ y_2 \end{pmatrix}_{2_1} = \begin{pmatrix} x y_1 \\ \pm x y_2 \end{pmatrix}_{2_2}, \quad (x)_{1_{-\pm}} \otimes \begin{pmatrix} y_1 \\ y_2 \end{pmatrix}_{2_2} = \begin{pmatrix} x y_1 \\ \pm x y_2 \end{pmatrix}_{2_1} \tag{G.104}$$

$$(x)_{1_{-\pm}} \otimes \begin{pmatrix} y_1 \\ y_2 \end{pmatrix}_{2_3} = \begin{pmatrix} x y_2 \\ \pm x y_1 \end{pmatrix}_{2_4}, \quad (x)_{1_{-\pm}} \otimes \begin{pmatrix} y_1 \\ y_2 \end{pmatrix}_{2_4} = \begin{pmatrix} x y_2 \\ \pm x y_1 \end{pmatrix}_{2_3} \tag{G.105}$$

$$(x)_{1_{-\pm}} \otimes \begin{pmatrix} y_1 \\ y_2 \end{pmatrix}_{2_5} = \begin{pmatrix} x y_2 \\ \pm x y_1 \end{pmatrix}_{2_5}. \tag{G.106}$$

The tensor products between singlets are found as

$$\mathbf{1}_{s_1 s_2} \otimes \mathbf{1}_{s_1' s_2'} = \mathbf{1}_{s_1'' s_2''}, \tag{G.107}$$

where $s_1'' = s_1 s_1'$ and $s_2'' = s_2 s_2'$.

G.7 $Z_9 \rtimes Z_3$

We denote the Z_9 and Z_3 generators by a and b, respectively. They satisfy

$$a^9 = 1, \quad ab = ba^7. \tag{G.108}$$

Using them, all of $Z_9 \rtimes Z_3$ elements are written as

$$g = b^m a^n, \tag{G.109}$$

with $m = 0, 1, 2$ and $n = 0, \ldots, 8$.

Table G.8 Characters of T_9

	n	h	$\chi_{1_{00}}$	$\chi_{1_{01}}$	$\chi_{1_{02}}$	$\chi_{1_{10}}$	$\chi_{1_{11}}$	$\chi_{1_{12}}$	$\chi_{1_{20}}$	$\chi_{1_{21}}$	$\chi_{1_{22}}$	χ_{3_1}	χ_{3_2}
$C_1^{(0)}$	1	1	1	1	1	1	1	1	1	1	1	3	3
$C_1^{(2)}$	1	3	1	1	1	1	1	1	1	1	1	3ω	$3\omega^2$
$C_1^{(2)}$	1	3	1	1	1	1	1	1	1	1	1	$3\omega^2$	3ω
$C_3^{(1)}$	3	9	1	ω	ω^2	1	ω	ω^2	1	ω	ω^2	0	0
$C_3^{(2)}$	3	9	1	ω	ω^2	ω	ω^2	1	ω^2	1	ω	0	0
$C_3^{(3)}$	3	3	1	ω	ω^2	ω^2	1	ω^2	ω	ω^2	1	0	0
$C_3^{(4)}$	3	9	1	ω^2	ω	1	ω^2	ω	1	ω^2	ω	0	0
$C_3^{(5)}$	3	9	1	ω^2	ω	ω	1	ω^2	ω^2	ω	1	0	0
$C_3^{(6)}$	3	3	1	ω^2	ω	ω^2	ω	1	ω	1	ω^2	0	0
$C_3^{(7)}$	3	3	1	1	1	ω	ω	ω	ω^2	ω^2	ω^2	0	0
$C_3^{(8)}$	3	3	1	1	1	ω^2	ω^2	ω^2	ω	ω	ω	0	0

These elements are classified into eleven conjugacy classes,

$$
\begin{aligned}
C_1 : &\quad \{e\}, &\quad h &= 1, \\
C_1^{(2)} : &\quad \{a^3\}, &\quad h &= 3, \\
C_1^{(3)} : &\quad \{a^6\}, &\quad h &= 3, \\
C_3^{(1)} : &\quad \{b, ba^3, ba^6\}, &\quad h &= 3, \\
C_3^{(2)} : &\quad \{ba, ba^4, ba^7\}, &\quad h &= 9, \\
C_3^{(3)} : &\quad \{ba^2, ba^5, ba^8\}, &\quad h &= 9, \\
C_3^{(4)} : &\quad \{b^2, b^2a^3, b^2a^6\}, &\quad h &= 3, \\
C_3^{(5)} : &\quad \{b^2a, b^2a^4, b^2a^7\}, &\quad h &= 9, \\
C_3^{(6)} : &\quad \{b^2a^2, b^2a^5, b^2a^8\}, &\quad h &= 9, \\
C_3^{(7)} : &\quad \{a, a^4, a^7\}, &\quad h &= 3, \\
C_3^{(8)} : &\quad \{a^2, a^5, a^8\}, &\quad h &= 3.
\end{aligned}
\tag{G.110}
$$

This group has nine singlets $\mathbf{1}_{n,k}$ with n, $k = 0$, 1, 2 and two triplets $\mathbf{3}_1$ and $\mathbf{3}_2$. The characters are shown in Table G.8, where $\omega = e^{2\pi i/3}$.

The generators, a and b, are represented as

$$
a = \omega^k, \qquad b = \omega^\ell, \qquad \text{on } \mathbf{1}_{k,\ell}. \tag{G.111}
$$

We use the following representations for the generator a

$$
a = \begin{pmatrix} \rho & 0 & 0 \\ 0 & \rho^4 & 0 \\ 0 & 0 & \rho^7 \end{pmatrix}, \quad \text{on } \mathbf{3}_1, \qquad a = \begin{pmatrix} \rho^2 & 0 & 0 \\ 0 & \rho^8 & 0 \\ 0 & 0 & \rho^5 \end{pmatrix}, \quad \text{on } \mathbf{3}_2, \tag{G.112}
$$

where $\rho = e^{2\pi i/9}$. and the following representation of b

$$b = \begin{pmatrix} 0 & 1 & 0 \\ 0 & 0 & 1 \\ 1 & 0 & 0 \end{pmatrix}, \tag{G.113}$$

for both triplets.

The tensor products between triplets are obtained as

$$\begin{pmatrix} x_1 \\ x_2 \\ x_3 \end{pmatrix}_{3_1} \otimes \begin{pmatrix} y_1 \\ y_2 \\ y_3 \end{pmatrix}_{3_1} = \begin{pmatrix} x_1 y_1 \\ x_2 y_2 \\ x_3 y_3 \end{pmatrix}_{3_2} \oplus \begin{pmatrix} x_2 y_3 \\ x_3 y_1 \\ x_1 y_2 \end{pmatrix}_{3_2} \oplus \begin{pmatrix} x_3 y_2 \\ x_1 y_3 \\ x_2 y_1 \end{pmatrix}_{3_2} , \tag{G.114}$$

$$\begin{pmatrix} x_1 \\ x_2 \\ x_3 \end{pmatrix}_{3_2} \otimes \begin{pmatrix} y_1 \\ y_2 \\ y_3 \end{pmatrix}_{3_2} = \begin{pmatrix} x_3 y_3 \\ x_1 y_1 \\ x_2 y_2 \end{pmatrix}_{3_1} \oplus \begin{pmatrix} x_1 y_2 \\ x_2 y_3 \\ x_3 y_1 \end{pmatrix}_{3_1} \oplus \begin{pmatrix} x_2 y_1 \\ x_3 y_2 \\ x_1 y_3 \end{pmatrix}_{3_1} , \tag{G.115}$$

$$\begin{pmatrix} x_1 \\ x_2 \\ x_3 \end{pmatrix}_{3_1} \otimes \begin{pmatrix} y_1 \\ y_2 \\ y_3 \end{pmatrix}_{3_2} = \sum_{k=0,1,2} (x_1 y_2 + \omega^{2k} x_2 y_3 + \omega^k x_3 y_1) 1_{0k} \tag{G.116}$$

$$\oplus \sum_{k=0,1,2} (x_1 y_1 + \omega^{2k} x_2 y_2 + \omega^k x_3 y_3) 1_{1k}$$

$$\oplus \sum_{k=0,1,2} (x_1 y_3 + \omega^{2k} x_2 y_1 + \omega^k x_3 y_2) 1_{2k}. \tag{G.117}$$

The tensor products between triplets and singlets are found as

$$(x)_{1_{0k}} \otimes \begin{pmatrix} y_1 \\ y_2 \\ y_3 \end{pmatrix}_{3_1} = \begin{pmatrix} x y_1 \\ \omega^k x y_2 \\ \omega^{2k} x y_3 \end{pmatrix}_{3_1} , \quad (x)_{1_{0k}} \otimes \begin{pmatrix} y_1 \\ y_2 \\ y_3 \end{pmatrix}_{3_2} = \begin{pmatrix} x y_1 \\ \omega^k x y_2 \\ \omega^{2k} x y_3 \end{pmatrix}_{3_2} , \tag{G.118}$$

$$(x)_{1_{1k}} \otimes \begin{pmatrix} y_1 \\ y_2 \\ y_3 \end{pmatrix}_{3_1} = \begin{pmatrix} x y_3 \\ \omega^k x y_1 \\ \omega^{2k} x y_2 \end{pmatrix}_{3_1} , \quad (x)_{1_{1k}} \otimes \begin{pmatrix} y_1 \\ y_2 \\ y_3 \end{pmatrix}_{3_2} = \begin{pmatrix} x y_2 \\ \omega^k x y_3 \\ \omega^{2k} x y_1 \end{pmatrix}_{3_2} , \tag{G.119}$$

$$(x)_{1_{2k}} \otimes \begin{pmatrix} y_1 \\ y_2 \\ y_3 \end{pmatrix}_{3_1} = \begin{pmatrix} x y_2 \\ \omega^k x y_3 \\ \omega^{2k} x y_1 \end{pmatrix}_{3_1} , \quad (x)_{1_{2k}} \otimes \begin{pmatrix} y_1 \\ y_2 \\ y_3 \end{pmatrix}_{3_2} = \begin{pmatrix} x y_3 \\ \omega^k x y_1 \\ \omega^{2k} x y_2 \end{pmatrix}_{3_2} . \tag{G.120}$$

The tensor products between singlets are found as

$$1_{n,k} \otimes 1_{n',k'} = 1_{n+n',k+k'}. \tag{G.121}$$

Generators of the Modular Group

H

Here, we show that the modular group is generated by the two elements, S and T,

$$S = \begin{pmatrix} 0 & 1 \\ -1 & 0 \end{pmatrix}, \qquad T = \begin{pmatrix} 1 & 1 \\ 0 & 1 \end{pmatrix}. \tag{H.1}$$

First of all, among the $SL(2, Z)$ elements,

$$\gamma = \begin{pmatrix} a & b \\ c & d \end{pmatrix}, \tag{H.2}$$

we consider the elements with $c = 0$. They can be classified to two classes,

$$\begin{pmatrix} 1 & b \\ 0 & 1 \end{pmatrix}, \qquad \begin{pmatrix} -1 & b \\ 0 & -1 \end{pmatrix}. \tag{H.3}$$

Obviously, the first classes can be written by T^b, i.e.

$$\begin{pmatrix} 1 & b \\ 0 & 1 \end{pmatrix} = \begin{pmatrix} 1 & 1 \\ 0 & 1 \end{pmatrix}^b = T^b. \tag{H.4}$$

Similarly, we can write the second class as

$$\begin{pmatrix} -1 & b \\ 0 & -1 \end{pmatrix} = \begin{pmatrix} -1 & 0 \\ 0 & -1 \end{pmatrix} \begin{pmatrix} 1 & b \\ 0 & 1 \end{pmatrix} = S^2 T^{-b}. \tag{H.5}$$

Next, we consider the elements γ with $c = 1$. Such elements can be decomposed as

$$\begin{pmatrix} a & ad-1 \\ 1 & d \end{pmatrix} = \begin{pmatrix} 1 & a \\ 0 & 1 \end{pmatrix} \begin{pmatrix} 0 & -1 \\ 1 & 0 \end{pmatrix} \begin{pmatrix} 1 & d \\ 0 & 1 \end{pmatrix} = T^a S^3 T^d. \tag{H.6}$$

© Springer-Verlag GmbH Germany, part of Springer Nature 2022
T. Kobayashi et al., *An Introduction to Non-Abelian Discrete Symmetries for Particle Physicists,* Lecture Notes in Physics 995,
https://doi.org/10.1007/978-3-662-64679-3

Now, let us study the elements γ with $c > 1$. The entries c and d are coprime, because $ad - bc = 1$. Then, we can always find integers, q, r, which satisfy $d = cq + r$, where $0 < r < c$. We multiply the γ and the unit matrix $e = T^{-q}S^3ST^q$,

$$\begin{pmatrix} a & b \\ c & d \end{pmatrix} = \begin{pmatrix} a & b \\ c & d \end{pmatrix} T^{-q}S^3ST^q = \gamma' ST^q, \tag{H.7}$$

where

$$\gamma' = \begin{pmatrix} -aq + b & -a \\ r & -c \end{pmatrix}. \tag{H.8}$$

Note that $\det\gamma' = 1$, because $\det\gamma = (\det\gamma')(\det S)(\det T^q)$ and $\det\gamma = \det S = \det T^q = 1$. Then, γ' is obviously one of the $SL(2, Z)$ element with $0 < r < c$. We repeat this procedure. Then, finally we obtain the $SL(2, Z)$ element with $r = 1$, which can be written by S and T as shown in Eq. (H.6).

Similarly, we can decompose the $SL(2, Z)$ elements with $c \leq 1$. Thus, we can write all of the $SL(2, Z)$ elements by S and T.

Modular Forms

Here, we define the modular forms [16–19]. The modular forms of the level N are holomorphic functions $f_i(\tau)$ of the modulus τ, which transform under $\Gamma(N)$ as

$$f_i(\gamma\tau) = (c\tau + d)^k f_i(\tau), \tag{I.1}$$

where $\gamma \in \Gamma(N)$. Here, k is the modular weight and it must be even integer because the element $S^2 = e$ on $\bar{\Gamma}$ corresponds to $(a, b, c, d) = (-1, 0, -1, 0)$ in the $SL(2, Z)$ matrix. The modular forms of the weight k and the level N transforms under $\bar{\Gamma}$,

$$f_i(\gamma\tau) = (c\tau + d)^k \rho_{ij}(\gamma) f_j(\tau), \tag{I.2}$$

where $\rho_{ij}(\gamma)$ is unitary matrix representing Γ_N. The linear space of the modular forms of the weight k and the level N is denoted by $\mathcal{M}_k(\Gamma(N))$. Its dimension $\dim\mathcal{M}_k(\Gamma(N))$ is written by

$$\dim\mathcal{M}_k(\Gamma(N)) = \frac{(k-1)N + 6}{24} N^2 \prod_p \left(1 - \frac{1}{p^2}\right) \tag{I.3}$$

for $N > 2$ and $k \geq 2$, where p is the prime divisors of N, while for $N = 2$ we have

$$\dim\mathcal{M}_k(\Gamma(N)) = k + 1. \tag{I.4}$$

The second column of Table I.1 shows $\dim\mathcal{M}_k(\Gamma(N))$ for $N < 6$. The third and fourth columns show groups isomorphic to Γ_N and their orders, respectively.

© Springer-Verlag GmbH Germany, part of Springer Nature 2022
T. Kobayashi et al., *An Introduction to Non-Abelian Discrete Symmetries for Particle Physicists*, Lecture Notes in Physics 995,
https://doi.org/10.1007/978-3-662-64679-3

Table I.1 The dimension $\dim \mathcal{M}_k(\Gamma(N))$ of the modular form space of the weight k and the level N

N	$\dim \mathcal{M}_k(\Gamma(N))$	Γ_N	$|\Gamma_N|$
2	$k/2+1$	S_3	6
3	$k+1$	A_4	12
4	$2k+1$	S_4	24
5	$5k+1$	A_5	60

For example, the A_4 triplet modular forms of the weight $k=2$, $\mathbf{Y}_3^{(2)} = (Y_1(\tau),$ $Y_2(\tau), Y_3(\tau))^T$, are constructed as [20]

$$
Y_1(\tau) = \frac{i}{2\pi} \left(\frac{\eta'(\tau/3)}{\eta(\tau/3)} + \frac{\eta'((\tau+1)/3)}{\eta((\tau+1)/3)} + \frac{\eta'((\tau+2)/3)}{\eta((\tau+2)/3)} - \frac{27\eta'(3\tau)}{\eta(3\tau)} \right),
$$

$$
Y_2(\tau) = \frac{-i}{\pi} \left(\frac{\eta'(\tau/3)}{\eta(\tau/3)} + \omega^2 \frac{\eta'((\tau+1)/3)}{\eta((\tau+1)/3)} + \omega \frac{\eta'((\tau+2)/3)}{\eta((\tau+2)/3)} \right), \tag{I.5}
$$

$$
Y_3(\tau) = \frac{-i}{\pi} \left(\frac{\eta'(\tau/3)}{\eta(\tau/3)} + \omega \frac{\eta'((\tau+1)/3)}{\eta((\tau+1)/3)} + \omega^2 \frac{\eta'((\tau+2)/3)}{\eta((\tau+2)/3)} \right),
$$

where $\omega = e^{2\pi i/3}$, and $\eta(\tau)$ is the Dedekind eta function defined by

$$
\eta(\tau) = q^{1/24} \prod_{n=1}^{\infty} (1 - q^n), \tag{I.6}
$$

where $q = e^{2\pi i \tau}$. The $Y_i(\tau)$ satisfy the constraint [20]:

$$
Y_2(\tau)^2 + 2Y_1(\tau)Y_3(\tau) = 0 . \tag{I.7}
$$

The modular forms of higher modular weights can be constructed by products of Y_1, Y_2, and Y_3.

Similarly, modular forms for S_3[21], S_4[22], A_5[23], $\Delta(96)$ and $\Delta(384)$[24] have been studied. Also, modular forms of their covering groups have also been studied [25–28]. For double covering groups, the modular weights k are not always even, but odd integers are possible. For other covering groups, fractional modular weights are also possible.

References

1. Hamermesh, M.: Group Theory and its Application to Physical Problems. Addison-Wesley, Reading, Mass (1962)
2. Georgi, H.: Lie algebras in particle physics. from isospin to unified theories. Front. Phys. **54**, 1 (1982)

3. Ludl, P.O.: arXiv:0907.5587 [hep-ph]
4. Ramond, P.: Group Theory: A Physicist's Survey. Cambridge University Press, Cambridge (2010)
5. Hagedorn, C., Lindner, M., Mohapatra, R.N.: JHEP **0606**, 042 (2006). arXiv:hep-ph/0602244
6. Bazzocchi, F., Merlo, L., Morisi, S.: Nucl. Phys. B **816**, 204 (2009). arXiv:0901.2086 [hep-ph]
7. Altarelli, G., Feruglio, F., Merlo, L.: JHEP **0905**, 020 (2009). arXiv:0903.1940 [hep-ph]
8. Altarelli, G., Feruglio, F.: Nucl. Phys. B **741**, 215 (2006). arXiv:hep-ph/0512103
9. Shirai, K.: J. Phys. Soc. Jpn. **61**, 2735 (1992)
10. Ding, G.-J., Everett, L.L., Stuart, A.J.: arXiv:1110.1688 [hep-ph]
11. Feruglio, F., Hagedorn, C., Lin, Y., Merlo, L.: Nucl. Phys. B **775**, 120 (2007). arXiv:hep-ph/0702194
12. King, S.F., Luhn, C., Stuart, A.J.: Nucl. Phys. B **867**, 203–235 (2013). arXiv:1207.5741 [hep-ph]
13. Ding, G.J., King, S.F.: Phys. Rev. D **89**(9), 093020 (2014). arXiv:1403.5846 [hep-ph]
14. Frampton, P.H., Kephart, T.W.: Int. J. Mod. Phys. A **10**, 4689 (1995). arXiv:hep-ph/9409330
15. Frampton, P.H., Kephart, T.W., Rohm, R.M.: Phys. Lett. B **679**, 478 (2009). arXiv:0904.0420 [hep-ph]
16. Gunning, R.C.: Lectures on Modular Forms. Princeton University Press, Princeton, NJ (1962)
17. Schoeneberg, B.: Elliptic Modular Functions. Springer (1974)
18. Koblitz, N.: Introduction to Elliptic Curves and Modular Forms. Springer (1984)
19. Bruinier, J.H., Geer, G.V.D., Harder, G., Zagier, D.: The 1-2-3 of Modular Forms. Springer (2008)
20. Feruglio, F.: In: Levy, A., Stefano Forte, S., Ridolfi, G. (eds.) From My Vast Repertoire ...: Guido Altarelli's Legacy, pp. 227–266 (2019). arXiv:1706.08749 [hep-ph]
21. Kobayashi, T., Tanaka, K., Tatsuishi, T.H.: Phys. Rev. D **98**(1), 016004 (2018). arXiv:1803.10391 [hep-ph]
22. Penedo, J.T., Petcov, S.T.: Nucl. Phys. B **939**, 292 (2019). arXiv:1806.11040 [hep-ph]
23. Novichkov, P.P., Penedo, J.T., Petcov, S.T., Titov, A.V.: JHEP **1904**, 174 (2019). arXiv:1812.02158 [hep-ph]
24. Kobayashi, T., Tamba, S.: Phys. Rev. D **99**(4), 046001 (2019). arXiv:1811.11384 [hep-th]
25. Liu, X.G., Ding, G.J.: JHEP **1908**, 134 (2019). arXiv:1907.01488 [hep-ph]
26. Kikuchi, S., Kobayashi, T., Takada, S., Tatsuishi, T.H., Uchida, H.: Phys. Rev. D **102**(10), 105010 (2020). arXiv:2005.12642 [hep-th]
27. Liu, X.G., Yao, C.Y., Qu, B.Y., Ding, G.J.: Phys. Rev. D **102**(11), 115035 (2020). arXiv:2007.13706 [hep-ph]
28. Kikuchi, S., Kobayashi, T., Uchida, H.: Phys. Rev. D **104**(6), 065008 (2021). [arXiv:2101.00826 [hep-th]]

Printed in the United States
by Baker & Taylor Publisher Services